Modelling and Control of Mechatronic and Robotic Systems

Modelling and Control of Mechatronic and Robotic Systems

Editors

Alessandro Gasparetto
Stefano Seriani
Lorenzo Scalera

MDPI • Basel • Beijing • Wuhan • Barcelona • Belgrade • Manchester • Tokyo • Cluj • Tianjin

Editors
Alessandro Gasparetto
University of Udine
Italy

Stefano Seriani
University of Trieste
Italy

Lorenzo Scalera
University of Udine
Italy

Editorial Office
MDPI
St. Alban-Anlage 66
4052 Basel, Switzerland

This is a reprint of articles from the Special Issue published online in the open access journal *Applied Sciences* (ISSN 2076-3417) (available at: https://www.mdpi.com/journal/applsci/special_issues/Mechatronic_Robotic_Systems).

For citation purposes, cite each article independently as indicated on the article page online and as indicated below:

LastName, A.A.; LastName, B.B.; LastName, C.C. Article Title. *Journal Name* **Year**, *Volume Number*, Page Range.

ISBN 978-3-0365-1122-1 (Hbk)
ISBN 978-3-0365-1123-8 (PDF)

Contents

About the Editors

Alessandro Gasparetto (MSc in Electronic Engineering, University of Padova, Italy, 1992; MSc in Mathematics, University of Padova, Italy, 2003; PhD in Mechanics of Machines, University of Brescia, Italy, 1996) is Full Professor of Mechanics of Machines at the Polytechnic Department of Engineering and Architecture, University of Udine (Udine, Italy), where he is the head of the research group in Mechatronics and Robotics. He has been included in the ranking of the top 2% most quoted and authoritative scientists in the world, published by researchers at Stanford University. He is Deputy Rector for Research at the University of Udine and Chair of IFToMM Italy, the Italian branch of IFToMM (the International Federation for the Promotion of Mechanism and Machine Science). His research interests are in the fields of modelling and control of mechatronic systems, robotics, mechanical design, industrial automation, and mechanical vibrations. He is author of more than 200 international publications.

Stefano Seriani received his BSc (2010) and his MSc in Mechanical Engineering (2012) from the University of Trieste, Italy. He received his PhD in 2016 at the University of Trieste, Italy. In 2016–2017, he was Research Fellow at the Institute of Robotics and Mechatronics of the German Space Agency (DLR), working mainly on compliant mechanisms for rovers and modular robotics for space exploration. He is now Assistant Professor at University of Trieste. His research interests include space robotics, applied mechanics, and mobile robotics.

Lorenzo Scalera achieved the BSc in Industrial Engineering and the MSc in Mechanical Engineering (both cum laude) at University of Trieste (Trieste, Italy) in 2012 and 2015, respectively, and the PhD in Industrial and Information Engineering at University of Udine (Udine, Italy) in 2019. In 2018, he was a visiting PhD student at the Stevens Institute of Technology in Hoboken (NJ, USA). In 2019, he was a Post Doc Research Fellow at the Free University of Bozen-Bolzano (Bolzano, Italy). In 2019, he was a visiting Research Fellow for one month at the Chiang Mai University (Chiang Mai, Thailand). Since 2020, he has been Assistant Professor of Mechanics Applied to Machines at the Polytechnic Department of Engineering and Architecture of the University of Udine. His research interests include modelling and control of mechatronic and robotic systems, trajectory planning, collaborative robotics, and mechanical vibrations.

Editorial

Modelling and Control of Mechatronic and Robotic Systems

Alessandro Gasparetto [1], Stefano Seriani [2] and Lorenzo Scalera [1,*]

1   Polytechnic Department of Engineering and Architecture, University of Udine, 33100 Udine, Italy; alessandro.gasparetto@uniud.it
2   Department of Engineering and Architecture, University of Trieste, 34127 Trieste, Italy; sseriani@units.it
*   Correspondence: lorenzo.scalera@uniud.it

Citation: Gasparetto, A.; Seriani, S.; Scalera, L. Modelling and Control of Mechatronic and Robotic Systems. *Appl. Sci.* **2021**, *11*, 3242. https://doi.org/10.3390/app11073242

Received: 30 March 2021
Accepted: 2 April 2021
Published: 4 April 2021

Publisher's Note: MDPI stays neutral with regard to jurisdictional claims in published maps and institutional affiliations.

1. Introduction

Nowadays, the modelling and control of mechatronic and robotic systems is an open and challenging field of investigation in both industry and academia [1]. The mathematical modelling of a mechanical system is indeed fundamental for the development of experimental prototypes. In particular, the kinematic model of a mechatronic or robotic system is essential for the proper definition of the path and motion law that the system has to follow during its operation. On the other hand, dynamic models are required to predict both the forces and torques acting on the system and those required by the actuators. These are ultimately useful to simulate scenarios and working conditions of interest. Furthermore, the dynamic modelling of a mechatronic or robotic system allows us to evaluate its time-dependent evolution and response under different input conditions. Indeed, a proper model can be implemented to improve the design and performance with different objectives, for instance, energy efficiency [2], and vibration reduction [3]. Furthermore, the modelling of a robotic system is essential in path and trajectory planning [4] to enhance control capabilities [5], and to increase kinematic and dynamic performance [6]. Within this framework, proper control of an automatic system is essential for successfully completing a predefined task. This is especially the case when external disturbances are present, or when dealing with flexible systems in which mechanical vibrations have to be considered.

2. Modelling and Control of Mechatronic and Robotic Systems

This Special Issue of the journal *Applied Sciences* encompasses the kinematic and dynamic modelling, analysis, design, and control of mechatronic and robotic systems, with the scope of improving their performance, as well as simulating and testing novel devices and control architectures. A broad range of disciplines are included, such as robotic manipulation, mobile systems, cable-driven robots, wearable and rehabilitation devices, variable stiffness safety-oriented mechanisms, optimization of robot performance, and energy-saving systems.

Several papers deal with control problems, applied to both robotic systems and mechanical vibrations. In paper [7], a novel sliding mode control algorithm is presented and applied for the trajectory tracking on a robotic manipulator with 3 degrees of freedom (DOF). The proposed controller runs without a precise dynamic model in the presence of uncertainties and enhances the response, fast convergence time, and accuracy of the tracking position.

The main goal of paper [8] is the design, simulation, and experimental verification of an adaptive feed-forward motion control for a hydraulic differential cylinder. The proposed solution is implemented on a hydraulic loader crane. Experimental results show the advantage of the proposed controller in reducing the cylinder position error and the adaptation to model uncertainties.

The authors in [9] present the modelling and control of a cable-suspended sling-like parallel robot for throwing a mass at a suitable time. The mass is carried at the end-effector by a gripper, which releases the mass so that it can reach a given target point. Mathematical

Appl. Sci. **2021**, *11*, 3242. https://doi.org/10.3390/app11073242   https://www.mdpi.com/journal/applsci

models, which also account for the effect of cable elasticity, provide guidelines for planning the trajectory.

In [10], a 8-DOF integrated model of a semi-active seat suspension with a human model over a quarter is presented. A fuzzy logic-based self-tuning PID controller allows to regulate the controlled force on the basis of sprung mass velocity error and its derivative as input. Simulation results show that the semi-active seat suspension improves the ride comfort significantly by reducing the head acceleration compared to passive seat suspensions. Moreover, in [11], the authors investigate the control of the over-critical vibration of a transmission shaft system with a device named Smart Spring, an active vibration control mechanism based on piezoelectric material. Simulation results based on a model implemented in ADAMS and MATLAB show the feasibility of the approach in reducing the vibration of the shaft system.

Papers [12,13] deal respectively with trajectory optimization and with a novel kinematic directional index for industrial manipulators. In particular, in [12] a "whip-lashing" method is proposed to optimize the trajectory in order to increase the velocity of individual robot parts, thereby minimizing motion cycle times and exploiting the torque of the joints more effectively. The efficiency of the method is confirmed by simulation results on a 5-DOF RV-2AJ manipulator. In [13], a novel method is proposed for evaluating the maximum speed that a serial robot can reach with respect to both the position of the robot and its direction of motion. This approach, called Kinematic Directional Index, is experimentally applied to a SCARA robot and to an articulated robot with 6-DOF to outline their performances.

Paper [14] presents an energy-saving system based on a prefill system and a buffer system to improve the energy efficiency and the processing performance of hydraulic presses. Experimental results demonstrate that the proposed system could reduce the installed power and pressure shock, increase energy efficiency, and provide the same processing characteristics as the traditional hydraulic press.

Several papers in this Special Issue are focused on the design, modelling, and control of mobile robots and systems. In [15], a hierarchical driving force distribution and control strategy for a six-wheel unmanned ground vehicle with independent drive motors is presented to improve vehicle maneuverability and stability. In [16], authors investigate the effects of wheel slip compensation in the trajectory planning for mobile tractor-trailer robot applications. Experimental results on the prototype Epi.q show that with the proposed approach it is possible to kinematically compensate trajectories that otherwise would be subject to large lateral slip.

Papers [17,18] are focused on unmanned aerial vehicles (UAV). In [17], a high-precision attitude control is proposed for a quadrotor drone equipped with a 2-DOF robotic arm and subject to varied external disturbances. The research presented in [18] aims to develop an automatic UAV-based indoor environmental monitoring system for detecting rapid changes in the features of plants growing in greenhouses. Another research work on a mobile system included in this Special Issue is given by [19], which describes a motion planning and robust coordinated control scheme for the trajectory tracking of an underwater vehicle-manipulator system. Model uncertainties, time-varying external disturbances, payload, and sensory noises are considered and evaluated numerically.

Other works are focused on legged mobile robots. In [20], a Central Pattern Generator network controller is applied to the trot gait of a quadruped robot. The work of [21] outlines a method for the identification of optimal trajectories of quadruped robots through genetic algorithms. Furthermore, the authors in [22] illustrate an optimization of energy consumption and cost of transport using heuristic algorithms applied to a hexapod robot.

The Special Issue also includes interesting papers dealing with mobile robotic systems for planetary exploration. Paper [23] presents a general approach to endow a robot with the ability to sense the terrain being traversed. In [24], the authors propose a novel design of a robotic legged lander based on variable radius drums actuated by cables. Thanks to this device, the robotic system can effectively decelerate with constant force during impact with ground.

Other papers are focused on robotic systems and devices that are designed to physically interact with humans for rehabilitation and safety. Paper [25] presents a novel exoskeleton mechanism for finger motion assistance. The exoskeleton is designed as a serial 2-DOF wearable mechanism that is able to guide human finger motion. The authors of paper [26] discuss the mechanical redesign of a finger rehabilitation device based on a slider-crank mechanism, starting from an existing prototype. The purpose is to obtain a portable device that can recreate the motion trajectories of a finger. Finally, paper [27] addresses the design and experimental validation of a variable stiffness safety-oriented mechanism for physical human–robot interaction.

3. Final Remarks

This Special Issue contains valuable research works focused on the modelling and control of mechatronic and robotic systems, covering a wide area of applications. This collection shows the significant research interest in these topics with a high impact and potential for future developments. We sincerely hope that this Special Issue on *"Modelling and Control of Mechatronic and Robotic Systems"* can be a valuable source of information for researchers working on these topics. We hope that you enjoy reading this Special Issue!

Author Contributions: Conceptualization, A.G., S.S., L.S.; writing—original draft preparation, L.S.; writing—review and editing, A.G., S.S., L.S.; supervision, A.G. All authors have read and agreed to the published version of the manuscript.

Funding: This research received no external funding.

Acknowledgments: This Special Issue would not be possible without the contributions of various talented authors, hardworking and professional reviewers, and the dedicated editorial team of *Applied Sciences*. We thank the authors, who have contributed interesting papers on several subjects, covering many fields of *"Modelling and Control of Mechatronic and Robotic Systems"*. We would like to take this opportunity to record our sincere gratefulness to all reviewers, whose feedback and suggestions helped the authors to improve their papers. Finally, we place on record our gratitude to the publisher and editorial team of *Applied Sciences*, and special thanks to Wing Wang, SI Managing Editor from the MDPI Branch Office, Wuhan, for her precious support and help with the publication of this Special Issue.

Conflicts of Interest: The authors declare no conflict of interest.

References

1. Siciliano, B.; Sciavicco, L.; Villani, L.; Oriolo, G. *Robotics: Modelling, Planning and Control*; Springer Science & Business Media: Berlin, Germany, 2010.
2. Scalera, L.; Carabin, G.; Vidoni, R.; Wongratanaphisan, T. Energy efficiency in a 4-DOF parallel robot featuring compliant elements. *Int. J. Mech. Control* **2019**, *20*, 49–57.
3. Vidoni, R.; Gasparetto, A.; Giovagnoni, M. A method for modelling three-dimensional flexible mechanisms based on an equivalent rigid-link system. *J. Vib. Control* **2014**, *20*, 483–500. [CrossRef]
4. Trigatti, G.; Boscariol, P.; Scalera, L.; Pillan, D.; Gasparetto, A. A look-ahead trajectory planning algorithm for spray painting robots with non-spherical wrists. In *Mechanisms and Machine Science*; Springer: Berlin/Heidelberg, Germany, 2019; Volume 66, pp. 235–242.
5. Boscariol, P.; Gasparetto, A.; Zanotto, V. Model predictive control of a flexible links mechanism. *J. Intell. Robot. Syst.* **2010**, *58*, 125–147. [CrossRef]
6. Vidussi, F.; Boscariol, P.; Scalera, L.; Gasparetto, A. Local and trajectory-based indexes for task-related energetic performance optimization of robotic manipulators. *J. Mech. Robot.* **2021**, *13*, 021018. [CrossRef]
7. Doan, Q.V.; Vo, A.T.; Le, T.D.; Kang, H.J.; Nguyen, N.H.A. A novel fast terminal sliding mode tracking control methodology for robot manipulators. *Appl. Sci.* **2020**, *10*, 3010. [CrossRef]
8. Jensen, K.J.; Ebbesen, M.K.; Hansen, M.R. Adaptive Feedforward Control of a Pressure Compensated Differential Cylinder. *Appl. Sci.* **2020**, *10*, 7847. [CrossRef]
9. Lin, D.; Mottola, G.; Carricato, M.; Jiang, X. modelling and Control of a Cable-Suspended Sling-Like Parallel Robot for Throwing Operations. *Appl. Sci.* **2020**, *10*, 9067. [CrossRef]

10. Jain, S.; Saboo, S.; Pruncu, C.I.; Unune, D.R. Performance investigation of integrated model of quarter car semi-active seat suspension with human model. *Appl. Sci.* **2020**, *10*, 3185. [CrossRef]

11. Li, M.M.; Ma, L.L.; Wu, C.G.; Zhu, R.P. Study on the vibration active control of three-support shafting with smart spring while accelerating over the critical speed. *Appl. Sci.* **2020**, *10*, 6100. [CrossRef]

12. Benotsmane, R.; Dudás, L.; Kovács, G. Trajectory Optimization of Industrial Robot Arms Using a Newly Elaborated "Whip-Lashing" Method. *Appl. Sci.* **2020**, *10*, 8666. [CrossRef]

13. Boschetti, G. A Novel Kinematic Directional Index for Industrial Serial Manipulators. *Appl. Sci.* **2020**, *10*, 5953. [CrossRef]

14. Yan, X.; Chen, B. Energy-Efficiency Improvement and Processing Performance Optimization of Forging Hydraulic Presses Based on an Energy-Saving Buffer System. *Appl. Sci.* **2020**, *10*, 6020. [CrossRef]

15. Zhang, H.; Liang, H.; Tao, X.; Ding, Y.; Yu, B.; Bai, R. Driving Force Distribution and Control for Maneuverability and Stability of a 6WD Skid-Steering EUGV with Independent Drive Motors. *Appl. Sci.* **2021**, *11*, 961. [CrossRef]

16. Botta, A.; Cavallone, P.; Tagliavini, L.; Carbonari, L.; Visconte, C.; Quaglia, G. An Estimator for the Kinematic Behaviour of a Mobile Robot Subject to Large Lateral Slip. *Appl. Sci.* **2021**, *11*, 1594. [CrossRef]

17. Jiao, R.; Chou, W.; Rong, Y.; Dong, M. Anti-disturbance control for quadrotor UAV manipulator attitude system based on fuzzy adaptive saturation super-twisting sliding mode observer. *Appl. Sci.* **2020**, *10*, 3719. [CrossRef]

18. Solis, J.; Karlsson, C.; Johansson, S.; Richardsson, K. Towards the Development of an Automatic UAV-Based Indoor Environmental Monitoring System: Distributed Off-Board Control System for a Micro Aerial Vehicle. *Appl. Sci.* **2021**, *11*, 2347. [CrossRef]

19. Han, H.; Wei, Y.; Ye, X.; Liu, W. Motion planning and coordinated control of underwater vehicle-manipulator systems with inertial delay control and fuzzy compensator. *Appl. Sci.* **2020**, *10*, 3944. [CrossRef]

20. Liu, M.; Li, M.; Zha, F.; Wang, P.; Guo, W.; Sun, L. Local CPG Self Growing Network Model with Multiple Physical Properties. *Appl. Sci.* **2020**, *10*, 5497. [CrossRef]

21. Pepe, G.; Laurenza, M.; Belfiore, N.P.; Carcaterra, A. Quadrupedal Robots' Gaits Identification via Contact Forces Optimization. *Appl. Sci.* **2021**, *11*, 2102. [CrossRef]

22. Luneckas, M.; Luneckas, T.; Kriaučiūnas, J.; Udris, D.; Plonis, D.; Damaševičius, R.; Maskeliūnas, R. Hexapod Robot Gait Switching for Energy Consumption and Cost of Transport Management Using Heuristic Algorithms. *Appl. Sci.* **2021**, *11*, 1339. [CrossRef]

23. Dimastrogiovanni, M.; Cordes, F.; Reina, G. Terrain Estimation for Planetary Exploration Robots. *Appl. Sci.* **2020**, *10*, 6044. [CrossRef]

24. Caruso, M.; Scalera, L.; Gallina, P.; Seriani, S. Dynamic modelling and Simulation of a Robotic Lander Based on Variable Radius Drums. *Appl. Sci.* **2020**, *10*, 8862. [CrossRef]

25. Carbone, G.; Gerding, E.C.; Corves, B.; Cafolla, D.; Russo, M.; Ceccarelli, M. Design of a Two-DOFs driving mechanism for a motion-assisted finger exoskeleton. *Appl. Sci.* **2020**, *10*, 2619. [CrossRef]

26. Zapatero-Gutiérrez, A.; Castillo-Castañeda, E.; Laribi, M.A. On the Optimal Synthesis of a Finger Rehabilitation Slider-Crank-Based Device with a Prescribed Real Trajectory: Motion Specifications and Design Process. *Appl. Sci.* **2021**, *11*, 708. [CrossRef]

27. Ayoubi, Y.; Laribi, M.A.; Arsicault, M.; Zeghloul, S. Safe pHRI via the Variable Stiffness Safety-Oriented Mechanism (V2SOM): Simulation and Experimental Validations. *Appl. Sci.* **2020**, *10*, 3810. [CrossRef]

Article

A Novel Fast Terminal Sliding Mode Tracking Control Methodology for Robot Manipulators

Quang Vinh Doan [1], Anh Tuan Vo [2], Tien Dung Le [1,*], Hee-Jun Kang [3] and Ngoc Hoai An Nguyen [2]

[1] The University of Danang—University of Science and Technology, 54 Nguyen Luong Bang street, Danang 550000, Vietnam; dqvinh@dut.udn.vn
[2] Electrical and Electronic Engineering Department, The University of Danang—University of Technology and Education, Danang 550000, Vietnam; voanhtuan2204@gmail.com (A.T.V.); hoaian2206@gmail.com (N.H.A.N.)
[3] School of Electrical Engineering, University of Ulsan, Ulsan 44610, Korea; hjkang@ulsan.ac.kr
* Correspondence: ltdung@dut.udn.vn; Tel.: +84-912-483-535

Received: 23 March 2020; Accepted: 21 April 2020; Published: 26 April 2020

Featured Application: Featured Application: This paper proposed the control synthesis, which can be performed in the trajectory tracking control for various robot manipulators as well as in other mechanical systems, the control of the higher-order system, several uncertain nonlinear systems, or chaotic systems.

Abstract: This paper comes up with a novel Fast Terminal Sliding Mode Control (FTSMC) for robot manipulators. First, to enhance the response, fast convergence time, against uncertainties, and accuracy of the tracking position, the novel Fast Terminal Sliding Mode Manifold (FTSMM) is developed. Then, a Supper-Twisting Control Law (STCL) is applied to combat the unknown nonlinear functions in the control system. By using this technique, the exterior disturbances and uncertain dynamics are compensated more rapidly and more correctly with the smooth control torque. Finally, the proposed controller is launched from the proposed sliding mode manifold and the STCL to provide the desired performance. Consequently, the stabilization and robustness criteria are guaranteed in the designed system with high-performance and limited chattering. The proposed controller runs without a precise dynamic model, even in the presence of uncertain components. The numerical examples are simulated to evaluate the effectiveness of the proposed control method for trajectory tracking control of a 3-Degrees of Freedom (DOF) robotic manipulator.

Keywords: super-twisting control law; robot manipulators; fast terminal sliding mode control

1. Introduction

Robots are gradually replacing people in the fields of social life, manufacturing, exploring, and performing complex tasks. In order to improve productivity, product quality, system reliability, electronics, measurement, and mechanical systems of robotic systems, more advanced designs are required. Therefore, this leads to an increase in the complexity of the structural and mathematical model when there is an additional occurrence of uncertain components.

Sliding Mode Control (SMC) [1–14] is capable of handling high non-linearity and external noise when it possesses outstanding features such as fast response, and robustness towards the existing uncertainties. However, the chattering problem in the SMC causes oscillations in the control input system leading to vibrations in the mechanical system, heat, and even causing instability. Furthermore, the SMC does not yield convergence in a defined period and provides a slow convergence time when it uses a linear sliding manifold. As a result of the linear sliding manifold, the SMC only ensures asymptotic convergence [15]. Therefore, in case that the pressure of a large control force is unavailable, the asymptotic stability status is less likely to converge fast with high-precision control. Terminal Sliding Mode Control (TSMC) [16,17] is proposed to solve convergence problems in finite time, enhancing the transient performance. However, in several situations, TSMC does not offer the desired performance with initial state variables far from the equilibrium point. Additionally, it has not solved the chattering and slow convergence, as well as creating a new problem that is the singularity phenomenon.

To resolve the issues of SMC and TSMC in a synchronized manner, Nonsingular Fast Terminal Sliding Mode Control (NFTSMC) has been evolved successfully in robotic systems [10,12,18–24]. It thoroughly solves the problems, including singularity, convergence in finite-time, and slow convergence. Unfortunately, the chattering has not yet been addressed as these controllers still use a robust reaching control law to deal with uncertain components, and also require the upper limit value of unknown components. For chattering, many solutions have been proposed that can be mentioned, such as the Boundary Layer Method (BLM) [6,25,26], the Adaptive Super-Twisting Method (ASTM) [27], the Second-Order SMC (SOSMC) or the Third-Order SMC (TOSMC) [28–31], the Full-Order SMC (FOSMC) [32–35], and the Fuzzy-SMC (F-SMC) [5,36–38]. For the requirement of the upper limit value of unknown components, many control methods were based on a combination of Neural Networks (NNs), Fuzzy Logic Systems (FLSs), or Adaptive Control Laws (ALCs) with NFTSMC. Although the behavior of unknown components can be well learned by NNs or FLSs. However, it increases the complexity of the control method design because there are more parameters to be adjusted. ACLs are more applicable than the other two methods because they use simple and effective updating rules.

Based on the mentioned analysis, our paper attempts to propose an advanced Fast Terminal Sliding Mode Control (FTSMC) with contributions for robot manipulators: (1) inherits the benefits of the FTSMC and Supper-Twisting Control Law (STCL) in the characteristics of robustness towards the existing uncertainties, finite-time convergence, singularity elimination, estimation capability, and good transient performance; (2) proposes a new Fast Terminal Sliding Mode Manifold (FTSMM) and provides sufficient evidence of finite-time convergence; (3) further improves the precision in the trajectory tracking control; (4) the control torque commands are smooth with less oscillation.

The rest of our paper has the following arrangement. The issue statements are outlined in Section 2. Section 3 presents a synthesis of the designed controller. Continued after Section 3, simulation examples are conducted to assess the influence of the designed controller for a 3-Degrees of Freedom (DOF) robot manipulator in Section 4. Its control performance was then evaluated along with the performance of different control algorithms, including SMC and NFTSMC. Section 5 presents some noteworthy conclusions.

2. Issue Description

Consider the dynamic equation of robot manipulators without the loss of generality:

$$M(\theta)\ddot{\theta} + Q(\theta,\dot{\theta})\dot{\theta} + G(\theta) + f_r(\dot{\theta}) + \delta_d(t) = \tau \tag{1}$$

where $\theta = \begin{bmatrix} \theta_1, & \dots, & \theta_n \end{bmatrix}^T \in R^{n\times 1}$, $\dot{\theta} = \begin{bmatrix} \dot{\theta}_1, & \dots, & \dot{\theta}_n \end{bmatrix}^T \in R^{n\times 1}$, and $\ddot{\theta} = \begin{bmatrix} \ddot{\theta}_1, & \dots, & \ddot{\theta}_n \end{bmatrix}^T \in R^{n\times 1}$ declare the position angle vector, the velocity angle vector, and the acceleration angle vector, respectively. $M(\theta) = \hat{M}(\theta) + \Delta M(\theta) \in R^{n\times n}$ declares the real inertia matrix, $Q(\theta,\dot{\theta}) = \hat{Q}(\theta,\dot{\theta}) + \Delta Q(\theta,\dot{\theta}) \in R^{n\times n}$ declares the real Coriolis and centrifugal force matrix, and $G(\theta) = \hat{G}(\theta) + \Delta G(\theta) \in R^{n\times n}$ represents the real gravitational matrix. $\hat{M}(\theta) \in R^{n\times n}$ declares the estimated inertia matrix, $\hat{Q}(\theta,\dot{\theta}) \in R^{n\times n}$ represents the estimated Coriolis and centrifugal force matrix, and $\hat{G}(\theta) \in R^{n\times n}$ declares the estimated gravitational matrix. $f_r(\dot{\theta}) \in R^{n\times 1}$ and $\delta_d(t) \in R^{n\times 1}$ declare the friction vector and disturbance vector, respectively. $\Delta M(\theta) \in R^{n \times n}$ and $\Delta Q(\theta, \dot{\theta}) \in R^{n \times n}$ are the errors of the real dynamic model.

Consequently, the real dynamic equation of robot manipulators is achieved as:

$$\hat{M}(\theta)\ddot{\theta} + \hat{Q}(\theta, \dot{\theta})\dot{\theta} + \hat{G}(\theta) + \Delta U = \tau \tag{2}$$

The lumped uncertain component Δu in Equation (2) is given as:

$$\Delta u = \Delta M(\theta)\ddot{\theta} + \Delta Q(\theta, \dot{\theta})\dot{\theta} + \Delta G(\theta) + f_r + \delta_d(t) \tag{3}$$

Accordingly, the robotic dynamic in Equation (1) is reorganized as:

$$\ddot{\theta} = \hat{M}^{-1}(\theta)\tau - \hat{M}^{-1}(\theta)\left[\hat{Q}(\theta, \dot{\theta})\dot{\theta} + \hat{G}(\theta)\right] - \hat{M}^{-1}(\theta)\Delta u \tag{4}$$

Then, the dynamic Equation (4) can be transferred into the following state-space form as:

$$\begin{cases} \dot{x}_1 = x_2 \\ \dot{x}_2 = \Pi(x, t) + \Phi(x, t)u + D(\theta, \Delta u) \end{cases} \tag{5}$$

where, we set $u = \tau$ as the control input and $x = [x_1, x_2]^T$ as the state vector in which x_1, x_2 correspond to θ, $\dot{\theta} \in R^{n \times 1}$. $\Pi(x, t) = -\hat{M}^{-1}(\theta)\left[\hat{Q}(\theta, \dot{\theta})\dot{\theta} + \hat{G}(\theta)\right]$, $\Phi(x, t) = \hat{M}^{-1}(\theta)$, and $D(\theta, \Delta U) = -\hat{M}^{-1}(\theta)\Delta u$.

The control target of the system is to further increase the response speed and accuracy of the trajectory tracking control for robot manipulators, even if the effects of uncertain dynamics and external perturbations are valid. First, to enhance the response, fast convergence time against uncertainties, and accuracy of the tracking position, the novel FTSMM is developed. Then, STCL is applied to combat the unknown nonlinear functions in the control system. By using this combined technique, the exterior disturbances and uncertain dynamics will be compensated more rapidly and more correctly with the smooth control torque. Finally, the designed controller is launched from the proposed sliding mode manifold and the STCL to obtain the control efficiency.

3. Main Results

The position control error and the velocity control error on each joint are, respectively, defined as follows:

$$x_{ei} = x_{1i} - x_{di} \tag{6}$$

$$x_{dei} = x_{2i} - \dot{x}_{di}; i = 1, \ldots n \tag{7}$$

where $x_d \in R^{n \times 1}$ represents the angle of the expected position.

3.1. The Designed FTSMM

To enhance the response, fast convergence time, and accuracy of the tracking position, the novel FTSMM is developed as follows:

$$s_i = x_{dei} + \frac{2\gamma_1}{1 + e^{-\mu_1(|x_{ei}|-\phi)}}x_{ei} + \frac{2\gamma_2}{1 + e^{\mu_2(|x_{ei}|-\phi)}}|x_{ei}|^{\alpha}\text{sgn}(x_{ei}) \tag{8}$$

where s_i is the proposed sliding mode manifold, $\gamma_1, \gamma_2, \mu_1, \mu_2$ are the positive constants, $0 < \alpha < 1$, and $\phi = \left(\frac{\gamma_2}{\gamma_1}\right)^{1/(1-\alpha)}$.

Based on the SMC, when the control errors operate in the sliding mode, the following constrain is satisfied [1]:

$$\begin{aligned} s_i &= 0; \\ \dot{s}_i &= 0 \end{aligned} \tag{9}$$

From condition in Equation (9), it is pointed out that:

$$x_{dei} = -\frac{2\gamma_1}{1+e^{-\mu_1(|x_{ei}|-\phi)}}x_{ei} - \frac{2\gamma_2}{1+e^{\mu_2(|x_{ei}|-\phi)}}|x_{ei}|^{\alpha}\mathrm{sgn}(x_{ei}) \tag{10}$$

Remark 1. *When the control error of $|x_{ei}|$ is much greater than ϕ, the first component of Equation (10) offers the role of providing a quick convergence rate and the second component has a smaller role. Contrariwise, when the control error of $|x_{ei}|$ is much smaller than ϕ, the second component of Equation (10) offers a greater role than the first one.*

The following theorem is launched to guarantee that convergence takes place within the defined time.

Theorem 1. *Let us consider dynamic of Equation (10). $x_{ei} = 0$ is defined as the equilibrium point and the state variables of the dynamic of Equation (10), including x_{ei} and x_{dei} stabilize to zero in finite-time.*

Proof. To validate the correctness of Theorem 1, the Lyapunov function candidate is proposed as follows:

$$L_1 = 0.5x_{ei}^2, \tag{11}$$

and its time derivative is

$$\begin{aligned}\dot{L}_1 &= x_{ei}x_{dei}\\ &= -\frac{2\gamma_1}{1+e^{-\mu_1(|x_{ei}|-\phi)}}x_{ei}^2 - \frac{2\gamma_2}{1+e^{\mu_2(|x_{ei}|-\phi)}}|x_{ei}|^{\alpha+1}\mathrm{sgn}(x_{ei})\\ &< 0 \end{aligned} \tag{12}$$

It is shown that $\dot{L}_1 < 0$, hence, x_{ei} and x_{dei} concentrate on the equilibrium state in finite time. When $|x_{ei}(0)| > \phi$, the sliding motion consists of two phases:

The first phase: $x_{ei}(0) \rightarrow |x_{ei}| = \phi$, the first component of Equation (10) offers the role of providing a quick convergence rate and the second component has a smaller role.

$$\begin{aligned}\int_0^{t_1} dt &= \int_{\phi}^{x_{ei}(0)} \frac{1}{\frac{2\gamma_1}{1+e^{-\mu_1(|x_{ei}|-\phi)}}x_{ei}+\frac{2\gamma_2}{1+e^{\mu_2(|x_{ei}|-\phi)}}|x_{ei}|^{\alpha}}d(|x_{ei}|)\\ &< \int_{\phi}^{x_{ei}(0)} \frac{1}{\gamma_1|x_{ei}|}d(|x_{ei}|) = \frac{\ln(|x_{ei}(0)|)-\ln(\phi)}{\gamma_1}\end{aligned} \tag{13}$$

The second phase: $|x_{ei}| = \phi \rightarrow x_{ei} = 0$, the second component of Equation (10) offers a role greater than the first one.

$$\begin{aligned}\int_0^{t_2} dt &= \int_0^{\phi} \frac{1}{\frac{2\gamma_1}{1+e^{-\mu_1(|x_{ei}|-\phi)}}x_{ei}+\frac{2\gamma_2}{1+e^{\mu_2(|x_{ei}|-\phi)}}|x_{ei}|^{\alpha}}d(|x_{ei}|)\\ &< \int_0^{\phi} \frac{1}{\gamma_2|x_{ei}|^{\alpha}}d(|x_{ei}|) = \frac{1}{\gamma_2(1-\alpha)}|\phi|^{1-\alpha}\end{aligned} \tag{14}$$

The total time of the sliding motion phase is defined as:

$$T_s = t_1 + t_2 < \frac{\ln(|x_{ei}(0)|)-\ln(\phi)}{\gamma_1} + \frac{1}{\gamma_2(1-\alpha)}|\phi|^{1-\alpha} \tag{15}$$

The state variable of the dynamic in Equation (10) converges to sliding manifold ($s(0) \to 0$) within the defined time T_r, which was pointed out in [8]. Therefore, the total time for stability on the sliding manifold is computed as:

$$T \leq T_r + T_s \tag{16}$$

3.2. The Designed Control Methodology

Let us take the time derivative of Equation (8):

$$
\dot{s} = \dot{x}_{de} + \frac{2\gamma_1}{1+e^{-\mu_1(|x_e|-\phi)}} x_{de} + \frac{2\gamma_1\mu_1 x_{de}\mathrm{sgn}(x_e)e^{-\mu_1(|x_e|-\phi)}}{\left(1+e^{-\mu_1(|x_e|-\phi)}\right)^2} x_e
$$
$$
+ \frac{2\gamma_2\alpha}{1+e^{\mu_2(|x_e|-\phi)}}|x_e|^{\alpha-1}x_{de} - \frac{2\gamma_2\mu_2 x_{de}e^{\mu_2(|x_e|-\phi)}}{\left(1+e^{\mu_2(|x_e|-\phi)}\right)^2}|x_e|^{\alpha} \tag{17}
$$

With $\dot{x}_{de} = \ddot{x}_2 - \ddot{x}_d$, the time derivation of Equation (17) gets along with the system in Equation (5) as follows:

$$
\dot{s} = \Pi(x,t) + \Phi(x,t)u + D(\theta,\Delta u) - \ddot{x}_d + \frac{2\gamma_1}{1+e^{-\mu_1(|x_e|-\phi)}} x_{de} + \frac{2\gamma_1\mu_1 x_{de}\mathrm{sgn}(x_e)e^{-\mu_1(|x_e|-\phi)}}{\left(1+e^{-\mu_1(|x_e|-\phi)}\right)^2} x_e
$$
$$
+ \frac{2\gamma_2\alpha}{1+e^{\mu_2(|x_e|-\phi)}}|x_e|^{\alpha-1}x_{de} - \frac{2\gamma_2\mu_2 x_{de}e^{\mu_2(|x_e|-\phi)}}{\left(1+e^{\mu_2(|x_e|-\phi)}\right)^2}|x_e|^{\alpha} \tag{18}
$$

In order to facilitate controller design, there is the following assumption:

Assumption 1. *The lumped uncertain terms, $D(\theta,\Delta u) = [D_1(\theta,\Delta u),\ldots,D_n(\theta,\Delta u)]$, need to satisfy the following standard condition:*

$$\|D_i(\theta,\Delta u)\| \leq K_i|s_i|^{\frac{1}{2}}; i = 1,\ldots,n \tag{19}$$

where $K_i > 0$.

In order to achieve the stabilization target of the robot system, the following control action is proposed:

$$u = -\Phi^{-1}(x,t)\left(u_{eq} + u_r\right). \tag{20}$$

Here, it should be noted that the u_{eq} is designed as:

$$
u_{eq} = \Pi(x,t) - \ddot{x}_d + \frac{2\gamma_1}{1+e^{-\mu_1(|x_e|-\phi)}} x_{de} + \frac{2\gamma_1\mu_1 x_{de}\mathrm{sgn}(x_e)e^{-\mu_1(|x_e|-\phi)}}{\left(1+e^{-\mu_1(|x_e|-\phi)}\right)^2} x_e
$$
$$
+ \frac{2\gamma_2\alpha}{1+e^{\mu_2(|x_e|-\phi)}}|x_e|^{\alpha-1}x_{de} - \frac{2\gamma_2\mu_2 x_{de}e^{\mu_2(|x_e|-\phi)}}{\left(1+e^{\mu_2(|x_e|-\phi)}\right)^2}|x_e|^{\alpha} \tag{21}
$$

and u_r is designed as:

$$
u_r = \Sigma_1|s|^{\frac{1}{2}}\mathrm{sgn}(s) + \eta
$$
$$
\dot{\eta} = -\Sigma_2\mathrm{sgn}(s) \tag{22}
$$

where $\Sigma_1 = diag(\Sigma_{11},\ldots,\Sigma_{1n})$ and $\Sigma_2 = diag(\Sigma_{21},\ldots,\Sigma_{2n})$. Σ_{1i} and Σ_{2i} are assigned to satisfy the following relationship [8]:

$$
\begin{cases}
\Sigma_{1i} > 2K_i \\
\Sigma_{2i} > \Sigma_{1i}\dfrac{5K_i\Sigma_{1i}+4K_i^2}{2(\Sigma_{1i}-2K_i)}
\end{cases} ; i = 1,2,\ldots,n \tag{23}
$$

Based on those above statements, the following theorems are written to prove the stability problem.

Theorem 2. *Consider the robot system in Equation (1). If the designed torque actions are proposed for system in Equation (1) as Equations (20)–(22), then x_{ei} and x_{dei} stabilize to zero in finite time. That means that robot system in Equation (1) runs in a stable mode.*

Proof. Applying control torque in Equation (20)–(22) to Equation (19) gains:

$$\begin{cases} \dot{s} = D(\theta, \Delta u) - \Sigma_1 |s|^{1/2} \mathrm{sgn}(s) - \eta \\ \dot{\eta} = -\Sigma_2 \mathrm{sgn}(s) \end{cases} \tag{24}$$

Based on the assumption in Equation (19), and the selection condition of the sliding gains in Equation (23), it can be verified that the sliding manifold and its time derivative will converge to zero in finite time. Now, considering one of the elements in Equation (24) as follows:

$$\begin{cases} \dot{s}_i = D_i(\theta, \Delta u) - \Sigma_{1i} |s_i|^{1/2} \mathrm{sgn}(s_i) - \eta_i \\ \dot{\eta}_i = -\Sigma_{2i} \mathrm{sgn}(s_i) \end{cases} \tag{25}$$

The following Lyapunov function is defined for the system in Equation (25):

$$L_2 = \sigma^T \Gamma \sigma \tag{26}$$

Here, $\sigma = \left[s_i^{1/2}, \lambda_i \right]^T$, $\Gamma = \frac{1}{2} \begin{bmatrix} 4\Sigma_{2i} + \Sigma_{1i}^2 & -\Sigma_{1i} \\ -\Sigma_{1i} & 2 \end{bmatrix}$. If $\Sigma_{2i} > 0$, so, according to Rayleigh's inequality:

$$\lambda_{\min}(\Gamma)\|\sigma\|^2 \leq L_2 \leq \lambda_{\max}(\Gamma)\|\sigma\|^2 \tag{27}$$

with $\|\sigma\|^2 = |s_i| + \eta_i^2$.

The time derivation of Equation (26) is:

$$\dot{L}_2 = -\frac{1}{|s_i|^{1/2}} \sigma^T \Phi \sigma + \frac{1}{|s_i|^{1/2}} [D_i(\theta, \Delta u), 0] \Gamma \sigma \tag{28}$$

with $\Phi = \frac{\Sigma_{1i}}{2} \begin{bmatrix} 2\Sigma_{2i} + \Sigma_{1i}^2 & -\Sigma_{1i} \\ -\Sigma_{1i} & 1 \end{bmatrix}$.

Based on the assumption in Equation (19), we can gain:

$$\begin{aligned} \dot{L}_2 &\leq -\frac{1}{|s_i|^{1/2}} \sigma^T \widetilde{\Phi} \sigma \\ &\leq -\frac{1}{|s_i|^{1/2}} \lambda_{\min}\left(\widetilde{\Phi}\right)\|\sigma\|^2 \end{aligned} \tag{29}$$

where $\widetilde{\Phi} = \frac{\Sigma_{1i}}{2} \begin{bmatrix} 2\Sigma_{2i} + \Sigma_{1i}^2 - (4\Sigma_{2i} + \Sigma_{1i})K_i & -(\Sigma_{1i} + 2K_i) \\ -(\Sigma_{1i} + 2K_i) & 1 \end{bmatrix}$.

$\widetilde{\Phi}$ is selected to be greater than zero. Consequently, $\dot{L}_2 < 0$.

Applying the Equation (27) gives:

$$|s_i|^{1/2} \leq \|\sigma\| \tag{30}$$

It follows that:

$$\dot{L}_2 \leq v L_2^{1/2} \tag{31}$$

with $v = \frac{\lambda_{\min}\left(\widetilde{\Phi}\right)}{\lambda_{\max}^{1/2}(\Gamma)}$.

According to [8], s_i and $\dot{s}_i$ are equal to zero in finite-time ($t_{ri} = 2L_2^{1/2}(t = 0)/v$). Therefore, s and $\dot{s}$ equal to zero in finite time ($T_r = \max_{i=1,\dots,n}\{t_{ri}\}$) and both x_{ei} and x_{dei} also stabilize to equilibrium in finite time ($T \leq T_r + T_s$) under the control action in Equation (20)–(22).

4. Numerical Simulation Studies

In this numerical example, a 3-DOF PUMA560 robot manipulator (with the first three joints and the last three joints blocked) is adopted. The MATLAB/SIMULINK software (2019a MATLAB Version of The MathWorks, Inc. 3 Apple Hill Drive Natick, MA 01760 USA) was used for all computation, the sampling time was set to 10^{-3} s, and the solver ode3 was used. The kinematic description for the robot system is displayed in Figure 1. The design parameters and dynamic models of the robot system are referenced from the document [39]. There are many essential parameters of a robot that need to be presented. Therefore, to present briefly, the design parameters and dynamic models of the robot system are reported in [39].

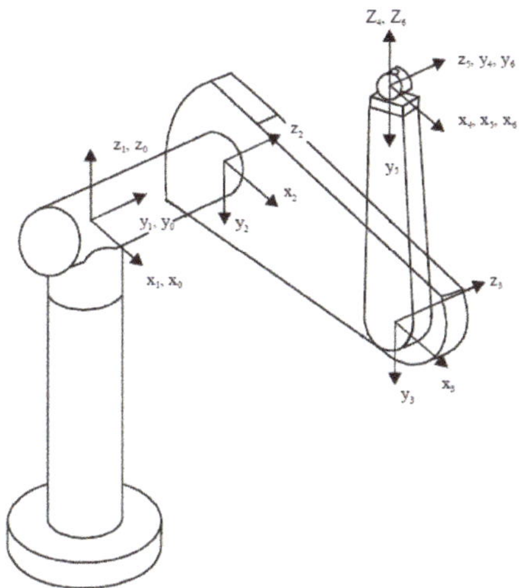

Figure 1. The kinematic description of 3-Degrees of Freedom (DOF) PUMA560 robot manipulator.

To explore the potential of our designed approach, the robot is controlled to follow the designated trajectory configuration at first. Later, its control performance is then evaluated and compared with the performance of different control algorithms, including SMC and NFTSMC. These control methods for comparison are briefly explained as follows:

The normal SMC [14] has the following control torque:

$$u = -\Phi(x,t)^{-1}\left[\Pi(x,t) + c\left(x_2 - \dot{x}_d\right) - \ddot{x}_d + (\Sigma + \xi)\mathrm{sgn}(s)\right] \tag{32}$$

where $s = x_{de} + cx_e$ is the linear sliding manifold, c is a positive constant.

Further, the NFTSMC [40] has the following control torque:

$$u(t) = -\Phi(x,t)^{-1}\left[\Pi(x,t) + \omega\frac{q}{l}x_{de}^{2-l/q} - \ddot{x}_d + (\Sigma + \xi)\mathrm{sgn}(s)\right] \tag{33}$$

where $s = x_e + \omega^{-1}x_{de}^{l/q}$ is a nonlinear sliding manifold.

The designed parameters of three control methodologies are given in Table 1.

Table 1. Parameter values for three different control methodologies.

Control Schemes	Parameters	Values
SMC	c	2
	ξ, Σ	$0.01, 20$
NFTSMC	l, q, ω	$5, 3, 2$
	ξ, Σ	$0.01, 20$
Proposed FTSMC	$\gamma_1, \gamma_2, \mu_1, \mu_2$	$1, 1, 1.2, 1.4$
	$\phi, \alpha, \Sigma_1, \Sigma_2$	$1, 0.6, 40, 50$

The designated trajectory configuration for position tracking when the robot manipulator operates:

$$\theta_d = \begin{cases} \theta_{d1} = 0.6 + \cos\left(\frac{t}{6\pi}\right) - 1 \\ \theta_{d2} = -0.6\sin\left(\frac{t}{6\pi} + 0.5\pi\right) - 1 \\ \theta_{d3} = 0.6 + \sin\left(0.2\frac{t}{\pi} + 0.5\pi\right) - 1 \end{cases} \tag{34}$$

Friction and disturbance models are hypothesized to analyze the strong capability of the designed FTSMC. It is not amenable to accurately calculate these friction and disturbance terms; therefore, the physical values of frictions and disturbances are not measured. Therefore, the following friction forces and disturbances were modeled, respectively:

$$f_r\left(\dot{\theta}\right) = \begin{cases} f_{r1} = 2.3\dot{\theta}_1 + 2.14\mathrm{sgn}\left(3\dot{\theta}_1\right) \\ f_{r2} = 4.5\dot{\theta}_2 + 2.35\mathrm{sgn}\left(2\dot{\theta}_2\right) \\ f_{r3} = 1.7\dot{\theta}_3 + 1.24\mathrm{sgn}\left(2\dot{\theta}_3\right) \end{cases} \tag{35}$$

$$\delta_d(t) = \begin{cases} \delta_{d1}(t) = 7.6\sin\left(\dot{\theta}_1\right) \\ \delta_{d2}(t) = 6.6\sin\left(\dot{\theta}_2\right) \\ \delta_{d3}(t) = 4.23\sin\left(\dot{\theta}_3\right) \end{cases} \tag{36}$$

To clearly present the results within the simulation period and to facilitate easier comparison, the averaged tracking error i:

$$E_j = \sqrt{\frac{1}{N}\sum_{i=1}^{N}\left(\|e_j(k)\|^2\right)}; \; j = 1, 2, 3 \tag{37}$$

where Z is the number of simulation steps.

To demonstrate the superiority of the designed controller, the average control error is calculated over two different simulation periods (10 s and 30 s).

The averaged tracking errors are reported in Table 2.

Table 2. The averaged tracking errors under the input of the controllers.

Error Control System	E_1 (10 s)	E_2 (10 s)	E_3 (10 s)	E_1 (30 s)	E_2 (30 s)	E_3 (30 s)
SMC	0.03102	0.01946	0.03113	0.01077	0.00671	0.01082
NFTSMC	0.02322	0.01284	0.02208	0.00774	0.00428	0.00737
Proposed Controller	0.01939	0.01037	0.01793	0.00646	0.00345	0.00597

The designated trajectory configuration and real trajectory under three control methods at the first three joints are displayed in Figure 2. It can be seen from Figure 2 that all three controllers appear to have a similar tracking control performance. However, they have different convergence times in the following order: the designed controller has the fastest convergence time among all three control methods, and NFTSMC has faster convergence time than the normal SMC.

Figures 3 and 4 show the position control errors and the velocity control errors, respectively. It can be seen from Figure 3 and Table 2, the position control errors of the designed control scheme are relatively small compared to those of the other control methods, in the order of $10^{-7}rad$. The position control errors of the NFTSMC are in the order of $10^{-6}rad$. SMC provides the largest position control errors of the three control methods, in the order of $10^{-4}rad$.

From Figure 4, it is seen that the designed control method also has the smallest velocity control errors among all the three control methods.

The control torque for all three control manners, including SMC, NFTSMC, and the designed FTSMC, are displayed in Figure 5. It can be recognized from Figure 5, SMC and NFTSMC have discontinuous control torque because of using the high-frequency control law. Meanwhile, the designed system has smooth control torque with a significant elimination of the chattering phenomena. To achieve this goal, the suggested controller applies STCL to substitute the high-frequency control law in removing chattering behavior.

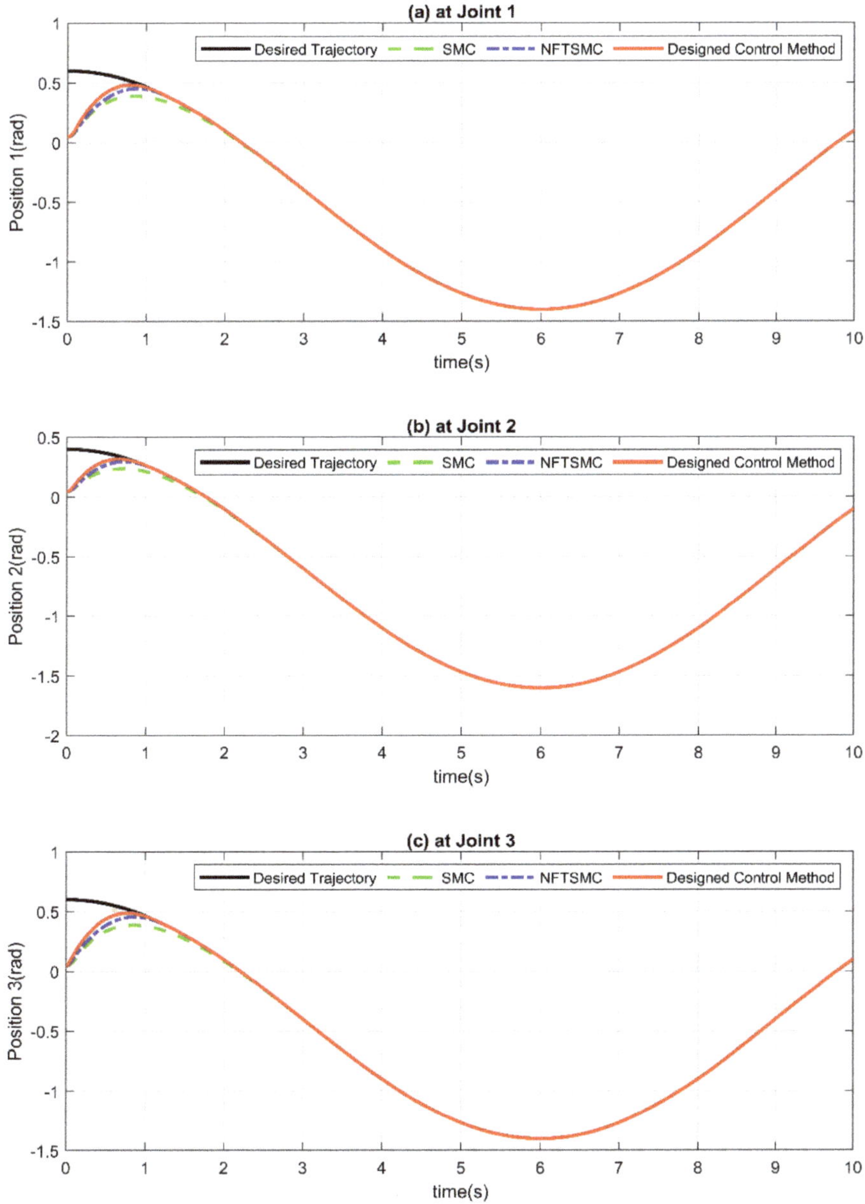

Figure 2. The designated trajectory configuration and real trajectory under three control methods: (**a**) at Joint 1, (**b**) at Joint 2, and (**c**) at Joint 3.

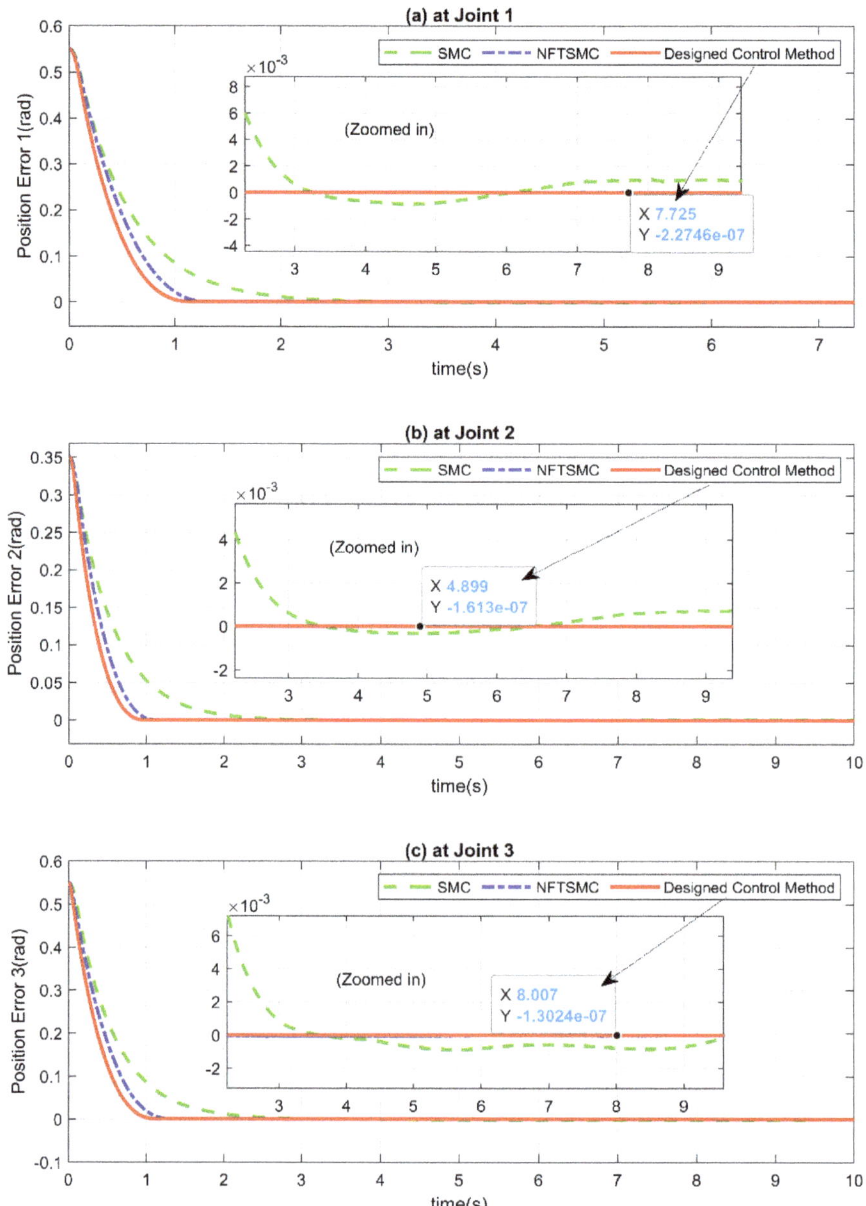

Figure 3. The position control errors: (**a**) at Joint 1, (**b**) at Joint 2, and (**c**) at Joint 3.

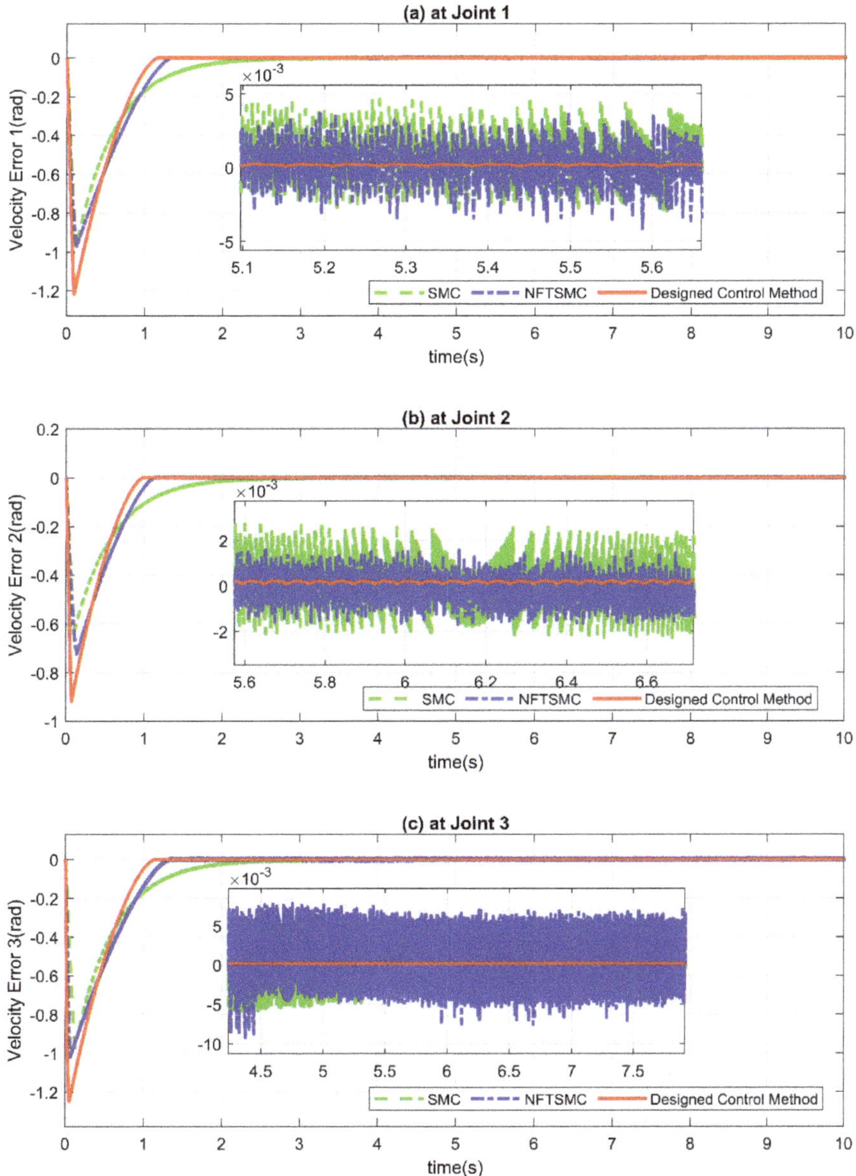

Figure 4. The velocity control errors: (**a**) at Joint 1, (**b**) at Joint 2, and (**c**) at Joint 3.

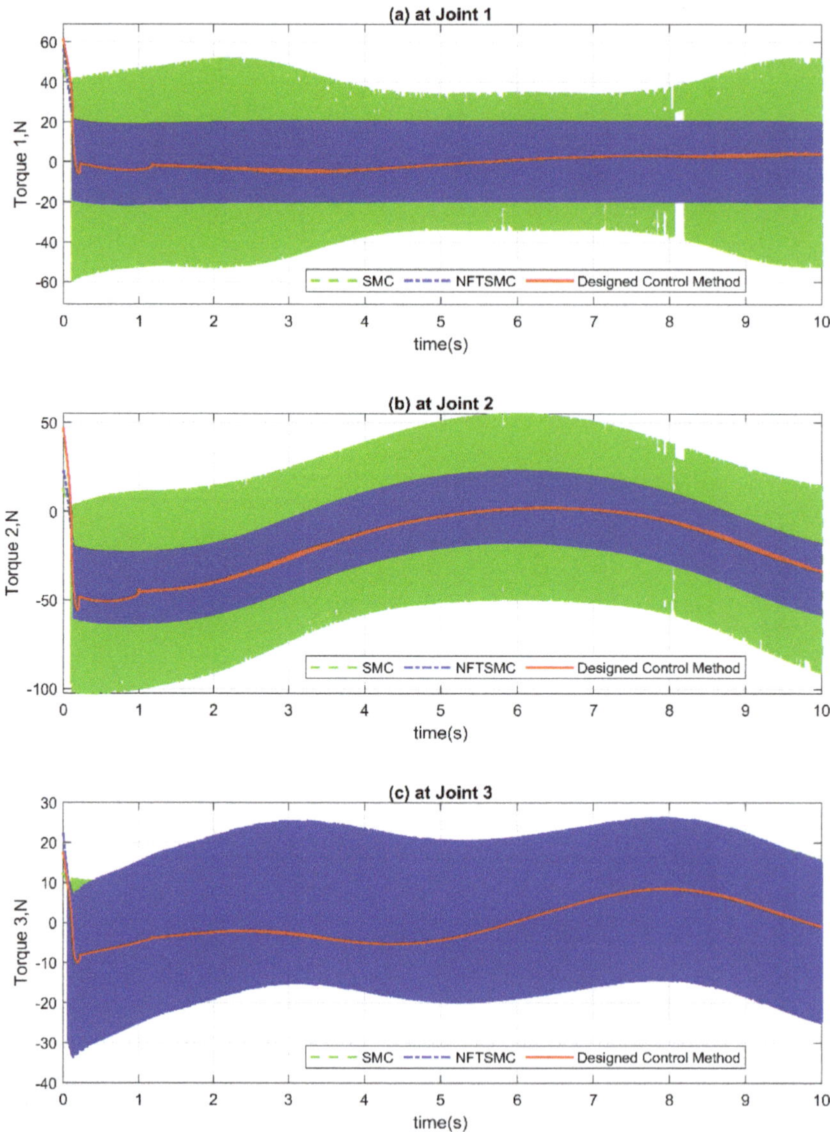

Figure 5. The control input signals: (**a**) Sliding Mode Control (SMC), (**b**) Non-singular Fast Terminal Sliding Mode Control (NFTSMC), and (**c**) designed Fast Terminal Sliding Mode Control (FTSMC).

Response time of the sliding mode manifolds, including SMC, NFTSMC, and designed FTSMC, are shown in Figure 6.

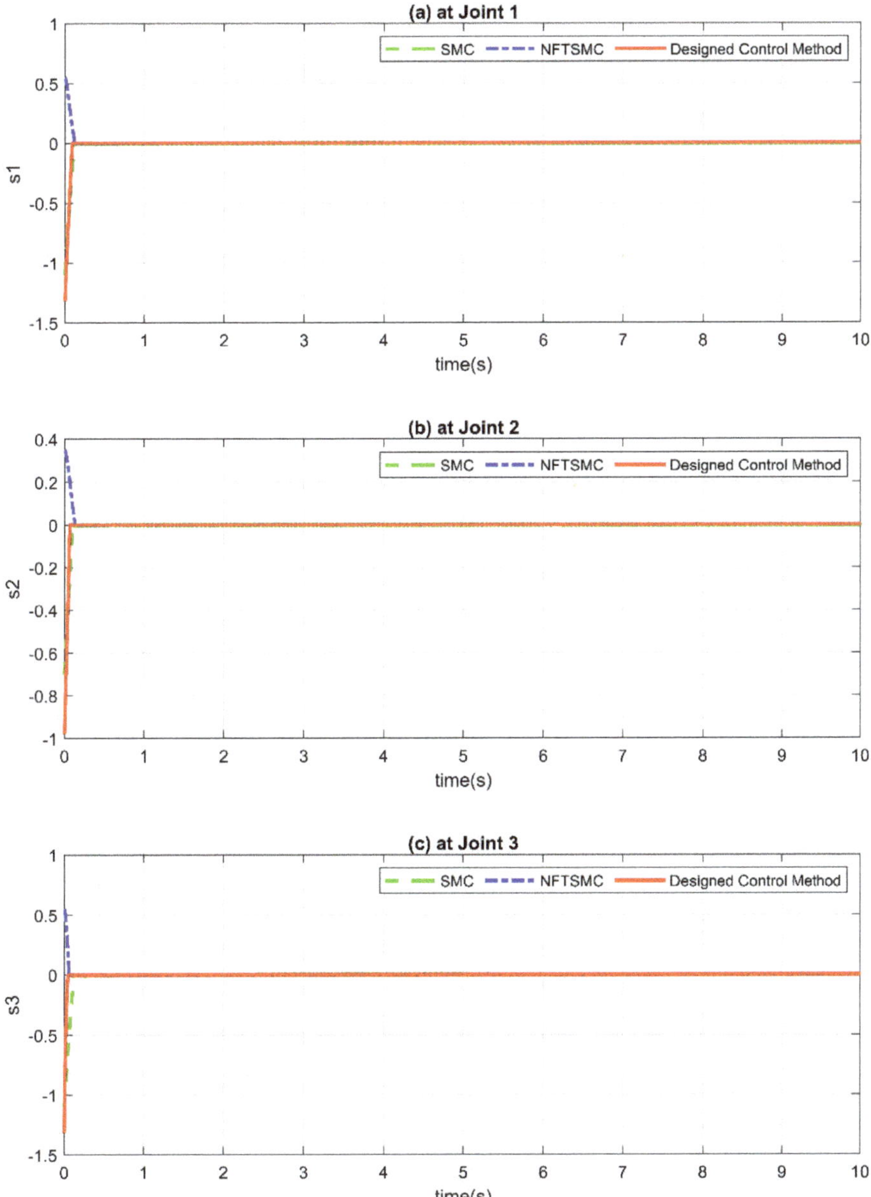

Figure 6. Response time of the sliding mode manifolds: (**a**) at Joint 1, (**b**) at Joint 2, and (**c**) at Joint 3.

5. Conclusions

This paper focuses on designing a novel FTSMC for robot manipulators. In the first step, the novel FTSMM is developed to enhance response capability, fast convergence time, uncertainties opposition, and especially, improve the accuracy of the tracking position. To alleviate unknown nonlinear parameters in the control system, STCL is then applied. Thanks to this valuable technique, exterior disturbances and nonlinear elements are compensated more rapidly and more correctly with the smooth control torque. Finally, combining STCL and our proposed sliding mode manifold, under the flexible controller, the stability and robustness of the control system are guaranteed with high-performance and limited chattering. To evaluate the efficiency, a simulation example is performed for the trajectory tracking control of a 3-DOF robotic manipulator.

From theoretical evidence, simulation results, and a comparison with SMC and NFTSMC, our proposed controller has some of the following contributions: (1) the proposed controller provides finite-time convergence and faster transient performance without singularity problem in controlling; (2) the proposed controller inherits the benefits of the FTSMC and CRCL in the characteristics of robustness towards the existing uncertainties; (3) a new FTSMM was proposed, and evidence of finite-time convergence was sufficiently proved; (4) the precision of the proposed controller was further improved in the trajectory tracking control; (5) the proposed controller shows the smoother control torque commands with lesser oscillation.

Author Contributions: Conceptualization, Q.V.D., A.T.V., and T.D.L.; methodology, Q.V.D. and T.D.L.; software, A.T.V.; validation, Q.V.D., A.T.V., and N.H.A.N.; formal analysis, Q.V.D., A.T.V., and H.-J.K.; investigation, T.D.L. and N.H.A.N.; resources, A.T.V. and T.D.L.; data curation, H.-J.K. and T.D.L.; writing—original draft preparation, Q.V.D., A.T.V., and N.H.A.N.; writing—review and editing, Q.V.D., H.-J.K., and N.H.A.N.; visualization, A.T.V.; supervision, T.D.L.; project administration, T.D.L. and Q.V.D.; funding acquisition, T.D.L. and Q.V.D. All authors have read and agreed to the published version of the manuscript.

Funding: This research is funded by Funds for Science and Technology Development of the University of Danang under project number B2019-DN02-52.

Conflicts of Interest: The authors declare no conflict of interest.

References

1. Shtessel, Y.; Edwards, C.; Fridman, L.; Levant, A. *Sliding Mode Control and Observation*; Springer: Berlin/Heidelberg, Germany, 2014.
2. Young, K.D.; Utkin, V.I.; Ozguner, U. A control engineer's guide to sliding mode control. *IEEE Trans. Control Syst. Technol.* **1999**, *7*, 328–342. [CrossRef]
3. Islam, S.; Liu, X.P. Robust sliding mode control for robot manipulators. *IEEE Trans. Ind. Electron.* **2011**, *58*, 2444–2453. [CrossRef]
4. Ferrara, A.; Incremona, G.P. Design of an integral suboptimal second-order sliding mode controller for the robust motion control of robot manipulators. *IEEE Trans. Control Syst. Technol.* **2015**, *23*, 2316–2325. [CrossRef]
5. Roopaei, M.; Jahromi, M.Z. Chattering-free fuzzy sliding mode control in MIMO uncertain systems. *Nonlinear Anal. Theory Methods Appl.* **2009**, *71*, 4430–4437. [CrossRef]
6. Utkin, V. Discussion aspects of high-order sliding mode control. *IEEE Trans. Automat. Contr.* **2016**, *61*, 829–833. [CrossRef]
7. Perruquetti, W.; Barbot, J.-P. *Sliding Mode Control in Engineering*; CRC Press: Boca Raton, FL, USA, 2002.
8. Moreno, J.A.; Osorio, M. A Lyapunov approach to second-order sliding mode controllers and observers. In Proceedings of the 2008 47th IEEE Conference on Decision and Control, Cancun, Mexico, 9–11 December 2008; pp. 2856–2861.
9. Qi, Z.; McInroy, J.E.; Jafari, F. Trajectory tracking with parallel robots using low chattering, fuzzy sliding mode controller. *J. Intell. Robot. Syst.* **2007**, *48*, 333–356. [CrossRef]
10. Vo, A.T.; Kang, H.-J. An Adaptive Neural Non-Singular Fast-Terminal Sliding-Mode Control for Industrial Robotic Manipulators. *Appl. Sci.* **2018**, *8*, 2562. [CrossRef]
11. Vo, A.T.; Kang, H.-J.; Le, T.D. An Adaptive Fuzzy Terminal Sliding Mode Control Methodology for Uncertain Nonlinear Second-Order Systems. In Proceedings of the International Conference on Intelligent Computing, Chennai, India, 2–3 February 2018; pp. 123–135.
12. Kamal, S.; Moreno, J.A.; Chalanga, A.; Bandyopadhyay, B.; Fridman, L.M. Continuous terminal sliding-mode controller. *Automatica* **2016**, *69*, 308–314. [CrossRef]
13. Moreno, J.A.; Negrete, D.Y.; Torres-González, V.; Fridman, L. Adaptive continuous twisting algorithm. *Int. J. Control* **2016**, *89*, 1798–1806. [CrossRef]

14. Edwards, C.; Spurgeon, S. *Sliding Mode Control: Theory and Applications*; CRC Press: Boca Raton, FL, USA, 1998.

15. Lee, H.; Kim, E.; Kang, H.-J.; Park, M. A new sliding-mode control with fuzzy boundary layer. *Fuzzy Sets Syst.* **2001**, *120*, 135–143. [CrossRef]

16. Mu, C.; Xu, W.; Sun, C. On switching manifold design for terminal sliding mode control. *J. Franklin Inst.* **2016**, *353*, 1553–1572. [CrossRef]

17. Tan, C.P.; Yu, X.; Man, Z. Terminal sliding mode observers for a class of nonlinear systems. *Automatica* **2010**, *46*, 1401–1404. [CrossRef]

18. Zhang, F. High-speed nonsingular terminal switched sliding mode control of robot manipulators. *IEEE/CAA J. Autom. Sin.* **2017**, *4*, 775–781. [CrossRef]

19. Vo, A.T.; Kang, H. An Adaptive Terminal Sliding Mode Control for Robot Manipulators with Non-singular Terminal Sliding Surface Variables. *IEEE Access* **2018**, *7*, 7801–8712. [CrossRef]

20. Vo, A.T.; Kang, H. A Chattering-Free, Adaptive, Robust Tracking Control Scheme for Nonlinear Systems with Uncertain Dynamics. *IEEE Access* **2019**, *7*, 10457–10466. [CrossRef]

21. Tuan, V.A.; Kang, H.-J. A New Finite-time Control Solution to The Robotic Manipulators Based on the Nonsingular Fast Terminal Sliding Variables and Adaptive Super-Twisting Scheme. *J. Comput. Nonlinear Dyn.* **2019**, *14*, 031002. [CrossRef]

22. Nojavanzadeh, D.; Badamchizadeh, M. Adaptive fractional-order non-singular fast terminal sliding mode control for robot manipulators. *IET Control Theory Appl.* **2016**, *10*, 1565–1572. [CrossRef]

23. Mobayen, S.; Baleanu, D.; Tchier, F. Second-order fast terminal sliding mode control design based on LMI for a class of non-linear uncertain systems and its application to chaotic systems. *J. Vib. Control* **2017**, *23*, 2912–2925. [CrossRef]

24. Boukattaya, M.; Mezghani, N.; Damak, T. Adaptive nonsingular fast terminal sliding-mode control for the tracking problem of uncertain dynamical systems. *ISA Trans.* **2018**, *77*, 1–19. [CrossRef]

25. Edelbaher, G.; Jezernik, K.; Urlep, E. Low-speed sensorless control of induction machine. *IEEE Trans. Ind. Electron.* **2006**, *53*, 120–129. [CrossRef]

26. Li, H.; Dou, L.; Su, Z. Adaptive nonsingular fast terminal sliding mode control for electromechanical actuator. *Int. J. Syst. Sci.* **2013**, *44*, 401–415. [CrossRef]

27. Wang, Y.; Chen, J.; Yan, F.; Zhu, K.; Chen, B. Adaptive super-twisting fractional-order nonsingular terminal sliding mode control of cable-driven manipulators. *ISA Trans.* **2019**, *86*, 163–180. [CrossRef] [PubMed]

28. Van, M.; Franciosa, P.; Ceglarek, D. Fault diagnosis and fault-tolerant control of uncertain robot manipulators using high-order sliding mode. *Math. Probl. Eng.* **2016**, *2016*, 7926280. [CrossRef]

29. Van, M.; Ge, S.S.; Ren, H. Robust fault-tolerant control for a class of second-order nonlinear systems using an adaptive third-order sliding mode control. *IEEE Trans. Syst. Man, Cybern. Syst.* **2017**, *47*, 221–228. [CrossRef]

30. Levant, A. Higher-order sliding modes, differentiation and output-feedback control. *Int. J. Control* **2003**, *76*, 924–941. [CrossRef]

31. Rubio-Astorga, G.; Sánchez-Torres, J.D.; Cañedo, J.; Loukianov, A.G. High-order sliding mode block control of single-phase induction motor. *IEEE Trans. Control Syst. Technol.* **2014**, *22*, 1828–1836. [CrossRef]

32. Feng, Y.; Han, F.; Yu, X. Chattering free full-order sliding-mode control. *Automatica* **2014**, *50*, 1310–1314. [CrossRef]

33. Feng, Y.; Zhou, M.; Zheng, X.; Han, F.; Yu, X. Full-order terminal sliding-mode control of MIMO systems with unmatched uncertainties. *J. Franklin Inst.* **2018**, *355*, 653–674. [CrossRef]

34. Song, Z.; Duan, C.; Wang, J.; Wu, Q. Chattering-free full-order recursive sliding mode control for finite-time attitude synchronization of rigid spacecraft. *J. Franklin Inst.* **2018**, *356*, 998–1020. [CrossRef]

35. Xiang, X.; Liu, C.; Su, H.; Zhang, Q. On decentralized adaptive full-order sliding mode control of multiple UAVs. *ISA Trans.* **2017**, *71*, 196–205. [CrossRef]

36. Li, H.; Wang, J.; Wu, L.; Lam, H.-K.; Gao, Y. Optimal Guaranteed Cost Sliding-Mode Control of Interval Type-2 Fuzzy Time-Delay Systems. *IEEE Trans. Fuzzy Syst.* **2018**, *26*, 246–257. [CrossRef]

37. Van, M. An Enhanced Robust Fault Tolerant Control Based on an Adaptive Fuzzy PID-Nonsingular Fast Terminal Sliding Mode Control for Uncertain Nonlinear Systems. *IEEE/ASME Trans. Mechatron.* **2018**, *23*, 1362–1371. [CrossRef]

38. Duc, T.M.; Van Hoa, N.; Dao, T.-P. Adaptive fuzzy fractional-order nonsingular terminal sliding mode control for a class of second-order nonlinear systems. *J. Comput. Nonlinear Dyn.* **2018**, *13*, 31004. [CrossRef]

39. Armstrong, B.; Khatib, O.; Burdick, J. The explicit dynamic model and inertial parameters of the PUMA 560 arm. In Proceedings of the 1986 IEEE International Conference on Robotics and Automation, San Francisco, CA, USA, 7–10 April 1986; Volume 3, pp. 510–518.

40. Feng, Y.; Yu, X.; Man, Z. Non-singular terminal sliding mode control of rigid manipulators. *Automatica* **2002**, *38*, 2159–2167. [CrossRef]

applied sciences

Article

Adaptive Feedforward Control of a Pressure Compensated Differential Cylinder

Konrad Johan Jensen *, Morten Kjeld Ebbesen and Michael Rygaard Hansen

Department of Engineering Sciences, University of Agder, 4879 Grimstad, Norway;
morten.k.ebbesen@uia.no (M.K.E.); michael.r.hansen@uia.no (M.R.H.)
* Correspondence: konrad.j.jensen@uia.no

Received: 29 September 2020; Accepted: 31 October 2020; Published: 5 November 2020

Abstract: This paper presents the design, simulation and experimental verification of adaptive feedforward motion control for a hydraulic differential cylinder. The proposed solution is implemented on a hydraulic loader crane. Based on common adaptation methods, a typical electro-hydraulic motion control system has been extended with a novel adaptive feedforward controller that has two separate feedforward states, i.e, one for each direction of motion. Simulations show convergence of the feedforward states, as well as 23% reduction in root mean square (RMS) cylinder position error compared to a fixed gain feedforward controller. The experiments show an even more pronounced advantage of the proposed controller, with an 80% reduction in RMS cylinder position error, and that the separate feedforward states are able to adapt to model uncertainties in both directions of motion.

Keywords: adaptive control; hydraulics; differential cylinder; feedforward; motion control

---

1. Introduction

For hydraulically actuated systems such as cranes, the hydraulic cylinder is the most common actuator since it can provide a linear motion with, generally speaking, a large force to volume ratio, a high efficiency and at a modest price. For systems which require a cylinder force in both directions, a double acting cylinder is needed, and the differential cylinder is an obvious choice due to its low cost and simple design. The main disadvantage is the difference in effective hydraulic area which leads to a jump in both velocity and force gain when changing sign of direction, i.e., around zero velocity.

For many hydraulic systems, the pressure compensated directional control valve is a practical choice due to the fact that it provides load independent flow control of the actuators. The pressure compensator senses the load pressure, and adjusts the pressure drop over the directional control valve to give a load independent flow. Since the velocity of the actuator is proportional to the hydraulic flow through the valve, this translates to load independent velocity control. For manually operated systems, the velocity control makes it easy for an operator to control systems that are subjected to large variations in external load.

For closed loop control systems, the load independent velocity control can be utilized in a control system using feedforward [1]. In this case, both a position reference and a velocity reference are generated in the control system. An example of a typical closed loop electro-hydraulic motion control system with feedforward is shown in Figure 1. The feedback controller uses the position reference and the measured cylinder position, whereas the feedforward controller uses the velocity reference. The pressure compensator is connected to a supply line which is shared with other actuators. The red dashed lines show the hydraulic pilot lines for the counterbalance valve and the pressure compensator.

It should be noted that feedforward control cannot be used alone. A feedback controller is also needed to help track the position reference, to eliminate steady state position error, and to counteract any drift. Normally the feedforward gain is based on system components, and is defined as the ratio of valve opening to actuator velocity. With this in mind, it follows that modeling errors and model uncertainties, in addition to external disturbances and system dynamics, may yield sub-optimal performance with a fixed feedforward gain.

This paper focuses on modeling and motion control of a hydraulic loader crane with pressure compensated differential cylinders. An adaptive feedforward controller is investigated to improve performance of the motion control system. Two different approaches to feedforward control have been implemented, the first is based on the MIT-rule [2], and the second is based on the sign-sign algorithm [3].

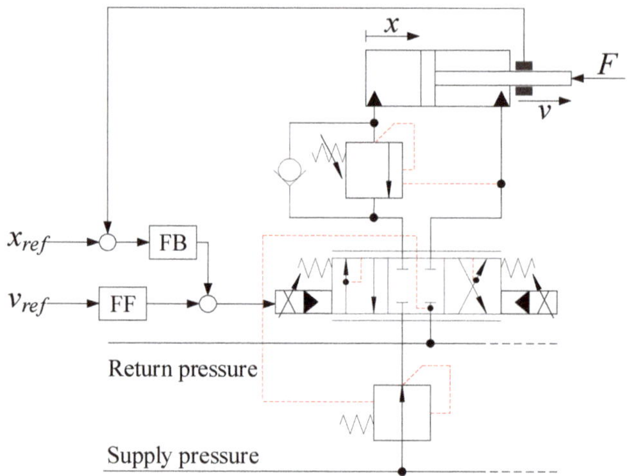

Figure 1. Electro-hydraulic motion control system with feedforward.

2. Background and Method

Adaptive systems have long been used for system identification and parameter estimation. One of the first methods is described in [4]. Another common method is the least mean squares algorithm, which was developed in [5]. An example of this is shown in Equations (1)–(3). Given the linear system:

$$Y = \theta^T \cdot X \tag{1}$$

$$E = Y - \hat{\theta}^T \cdot X \tag{2}$$

$$\dot{\hat{\theta}} = \gamma \cdot X \cdot E^T \tag{3}$$

where

Y = system output;
θ = system parameters;
X = system input;
E = estimation error;
$\hat{\theta}$ = estimated parameters;
γ = adaptation gain, constant.

The estimated parameters will converge towards the system parameters. The idea of using the sign function in the adaptive law comes from the sign-sign least mean squares algorithm, and was first introduced by [3]. Equation (3) then becomes:

$$\dot{\theta} = \gamma \cdot \text{sign}(X) \cdot \text{sign}(E^T) \tag{4}$$

By taking the sign of the estimation error and system input, the adaptation becomes insensitive to the magnitudes of E and X, and as such only the adaptation gain γ sets the adaptation speed.

The MIT rule is also used for adaptive control, and is described in [2]. A typical application is model reference adaptive control, shown in Figure 2. Based on the model output y_m, an additional control output $\hat{u}$ is multiplied with the command signal u_c to shape the plant output y. The equations for the model reference adaptive control is shown in Equations (5) and (6).

$$\dot{\hat{u}} = -\gamma \cdot y_m \cdot (y - y_m) \tag{5}$$

$$u = u_c \cdot \hat{u} \tag{6}$$

where

u = control output;

$\hat{u}$ = adaptive control output;

u_c = command signal;

y_m = model output;

y = plant output.

Early work in adaptive control can be found in [6–10]. Other work on adaptive control include [11] which investigates adaptive feedback and feedforward control of robot manipulators, Reference [12] which models and implements adaptive control of a flexible arm, and [13] which uses model reference adaptive control on linear time-varying plants. Adaptive fuzzy sliding mode control is investigated and implemented on an inverted pendulum in [14].

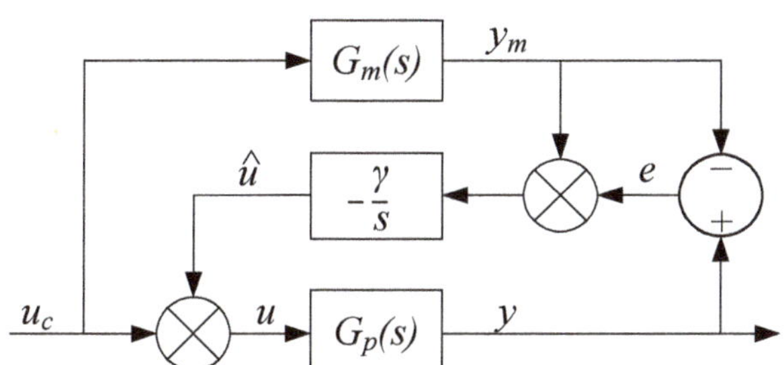

Figure 2. Model reference adaptive control based on MIT-rule.

Newer applications of adaptive control systems include adaptive friction compensation with an adaptive velocity estimator to compensate for the estimated non-linear friction force [15]. In [16], a fuzzy model reference adaptive control of an active magnetic bearing for a milling process is investigated to reduce the milling dynamics. Adaptive integral robust control of an electro-hydraulic servo system is investigated in [17], using parameter estimation and integral control to compensate for disturbances and plant uncertainties. Adaptive control of quadrotors is investigated in [18], which uses an cerebellar model arithmetic computer to adapt to model uncertainties and disturbances. In [19], adaptive control based on least-mean-fourth is implemented for a three-phase grid connected solar system, which is able to provide load balancing and power factor correction.

As for motion control of hydraulic systems, different approaches have previously been investigated, including vector control [20], pressure control [21,22], force control [23,24], and feedforward control [25].

To the knowledge of the authors, adaptive feedforward motion control of hydraulic cylinders has not previously been investigated, and this paper will focus on this novel concept.

In this paper, two adaptive controllers have been tested on a hydraulic differential cylinder and compared to a fixed gain feedforward controller. Based on a typical fixed gain feedforward controller, an adaptive controller can be made by extending it with the MIT rule. An illustration of a control system with feedforward with fixed gain is shown in Figure 3.

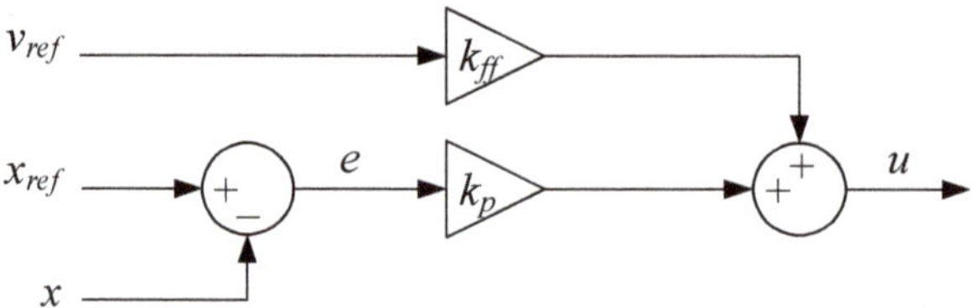

Figure 3. Feedforward with fixed gain.

Defining the position error e as the position reference x_{ref} minus the measured position x, the control output for this control system is given in Equation (7)

$$u = k_p \cdot e + k_{ff} \cdot v_{ref} \tag{7}$$

where

u = controller output;
k_p = proportional gain;
e = position error;
k_{ff} = feedforward gain;
v_{ref} = velocity reference.

Extending the traditional feedforward controller into an adaptive feedforward controller is done by replacing the fixed feedforward gain with the MIT-rule. An illustration of the adaptive feedforward scheme is shown in Figure 4.

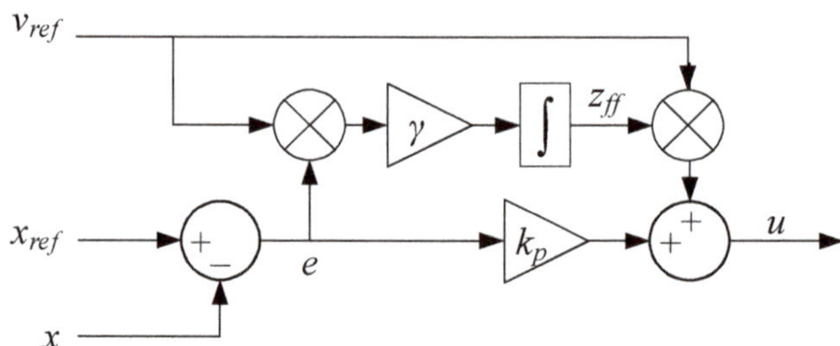

Figure 4. MIT-rule adaptive feedforward.

The MIT-rule adaptive feedforward controller uses the position error, the velocity reference, and the constant γ to update the feedforward gain. The update law and the control output for this adaptive control system is then given in Equations (8) and (9).

$$\dot{z}_{ff} = \gamma \cdot v_{ref} \cdot e \tag{8}$$
$$u = k_p \cdot e + z_{ff} \cdot v_{ref} \tag{9}$$

where

γ = adaptation gain;
z_{ff} = feedforward gain.

Extending this controller to use sign-sign is then straightforward. An illustration of the sign-sign adaptive feedforward scheme is shown in Figure 5.

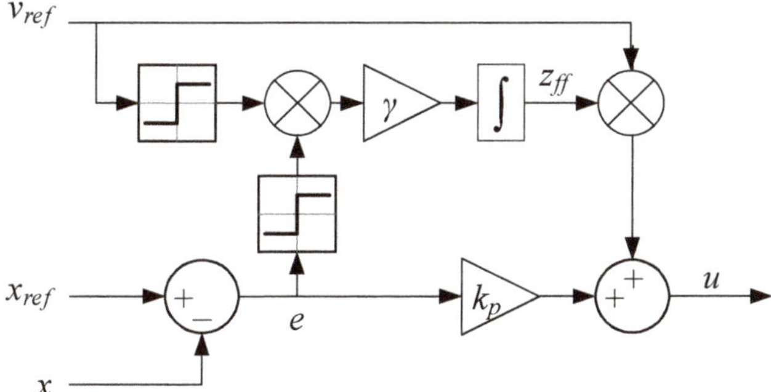

Figure 5. Sign-sign adaptive feedforward.

The update law and the control output for this adaptive control system is shown in Equations (10) and (11).

$$\dot{z}_{ff} = \gamma \cdot \text{sign}(v_{ref}) \cdot \text{sign}(e) \tag{10}$$

$$u = k_p \cdot e + z_{ff} \cdot v_{ref} \tag{11}$$

It should be noted that the sign function can produce unnecessary chattering when the input is oscillating around zero, due to the inherent discontinuity. Therefore the sign function has been replaced with the tanh function, shown in Equation (12).

$$\text{sign}(e) \approx \tanh(k \cdot e) \tag{12}$$

This gives a smooth output when the input is oscillating around zero. Increasing the parameter k gives a sharper rise and a closer approximation to $\text{sign}(e)$. Another advantage of using tanh is that the adaptation stops when the position error is zero. The parameter k has been set to $k = 100 \, \text{m}^{-1}$ and $k = 100 \, \text{s} \cdot \text{m}^{-1}$ for the position error and velocity reference, respectively.

3. Considered System

In this paper an 2020K4 loader crane made by HMF Group A/S, Højbjerg, Denmark has been used for experiments. An illustration of the crane is shown in Figure 6. This crane has two hydraulic differential cylinders: the main cylinder, and the knuckle cylinder. For this paper, the knuckle cylinder has been used for simulation and experiments, since it can experience both resistive and assistive loads in both directions of motion, equivalent to four quadrant operation. The relevant data for the knuckle cylinder is shown in Table 1, and the data for the knuckle boom is given in Figure 7 and Table 2.

Each actuator is controlled via a pressure compensated proportional directional control valve which ensures load independent flow control of the actuators. Counterbalance valves made by Oil Control S.p.A, Modena, Italy are also used for load holding, assisting in lowering of the booms, and pressure relief of pressure surges. An illustration of the hydraulic system for the knuckle cylinder is shown in Figure 8.

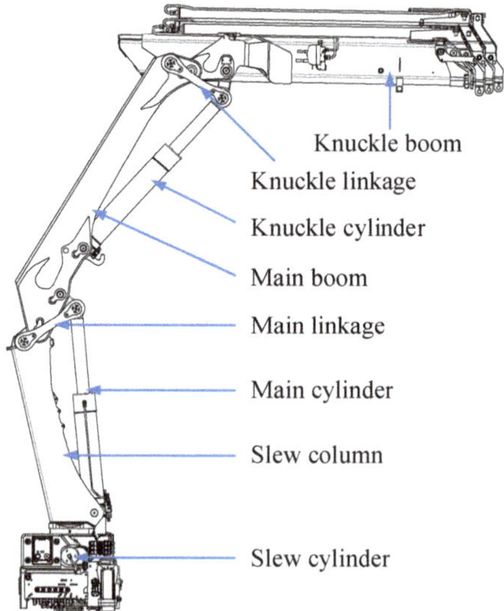

Figure 6. Illustration of the HMF 2020K4 loader crane.

Table 1. Knuckle cylinder data.

Name	Parameter	Value
Piston diameter	D_p	0.15 m
Piston area	A	0.0177 m^2
Rod diameter	D_r	0.1 m
Annulus area	A_a	0.0098 m^2
Piston area ratio	$\phi = \frac{A_a}{A}$	0.5556
Valve maximum flow	Q_{max}	40 L/min

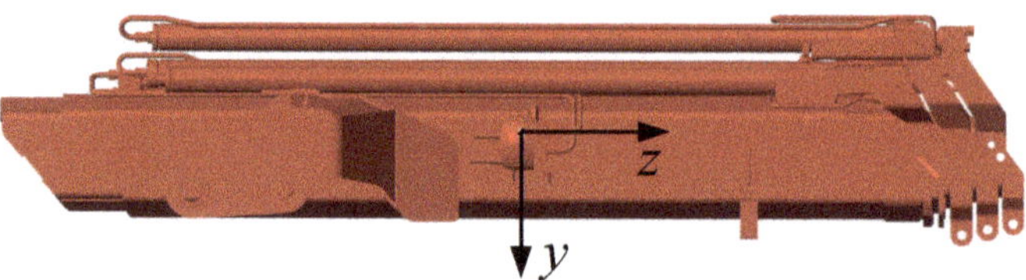

Figure 7. Knuckle boom center of mass.

Table 2. Knuckle boom data.

Name	Parameter	Value		
Mass	m_k	851.972 kg		
Inertia matrix	I_k	$\begin{bmatrix} 579.552 & 8.74629 & 11.5456 \\ 8.74629 & 573.285 & 0.174433 \\ 11.5456 & 0.174433 & 32.2491 \end{bmatrix}$ kg·m^2		

The control system is implemented on a CompactRIO 9075 controller made by National Instruments, Austin, TX, USA. The CompactRIO contains the reference generator and feedforward motion controllers. The block diagram of the connections is shown in Figure 9.

The CompactRIO communicates with a PC, sends control signals to the valves, and reads the sensors on the crane.

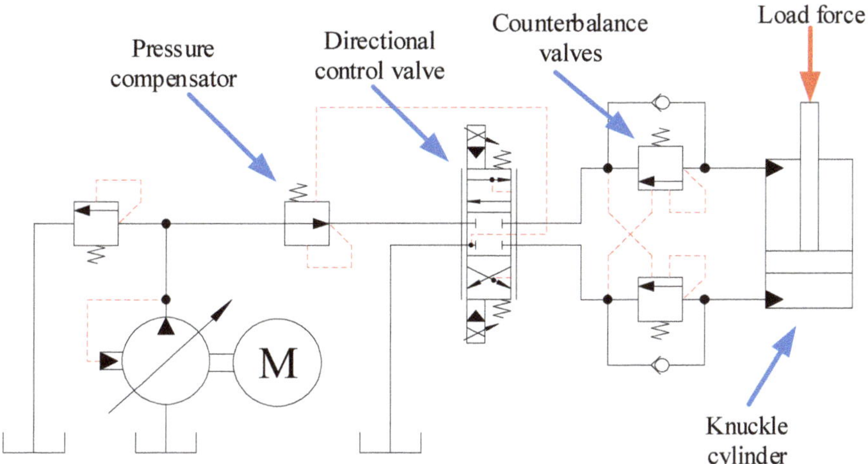

Figure 8. Hydraulic system for the knuckle cylinder.

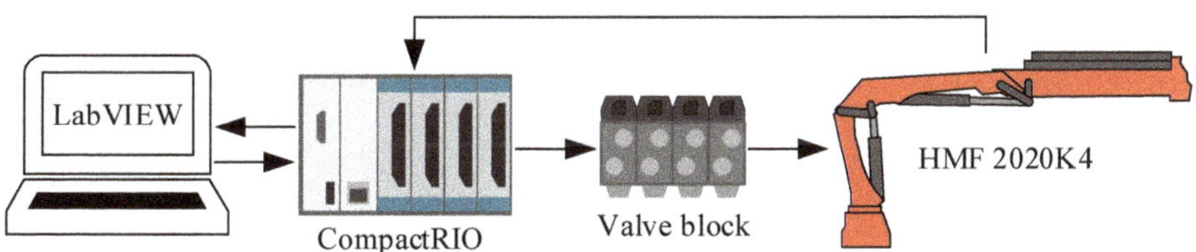

Figure 9. Connection between the crane and CompactRIO controller.

4. Modelling

A dynamic model of the crane has been made in Simscape™ by MathWorks®, Natick, MA, USA. 3D computer-aided design (CAD) models have been imported into the model using the Multibody library. The hydraulic circuit has been made using the hydraulic library of Simscape™. A picture of the CAD model is shown in Figure 10.

In the configuration shown in Figure 10, the knuckle cylinder experiences both resistive and assistive loads in both directions of motion when retracting fully, and extending back out again. The knuckle cylinder is controlled by a pressure compensated directional control valve, shown in Figure 11.

The pressure compensator ensures that there is a constant pressure drop over the directional control valve, which gives a load independent flow. The governing equations of the pressure compensator are given in Equations (13)–(15).

$$u_{pc} = \frac{p_{set} + p_{load} - p_p}{\Delta p} \tag{13}$$

$$p_{load} = \begin{cases} p_a & \text{if } u_{spool} \geq 0 \\ p_b & \text{otherwise} \end{cases} \tag{14}$$

$$Q_{pc} = k_{pc} \cdot u_{pc} \cdot \sqrt{p_i - p_p} \tag{15}$$

where

u_{pc} = opening of compensator, $0 \leq u_{pc} \leq 1$

p_p = compensated pressure at port p;

Δp = pressure difference between fully closed and fully open;

p_a = pressure at port a;

p_b = pressure at port b;

p_t = tank pressure;

p_{set} = spring pressure setting;

p_{load} = load pressure;

u_{spool} = position of the main spool, $-1 \leq u_{spool} \leq 1$;

Q_{pc} = flow in pressure compensator;

k_{pc} = flow gain of compensator;

p_i = compensator inlet pressure.

Figure 10. 3D view of the simulation model of the HMF 2020K4 in Simscape.

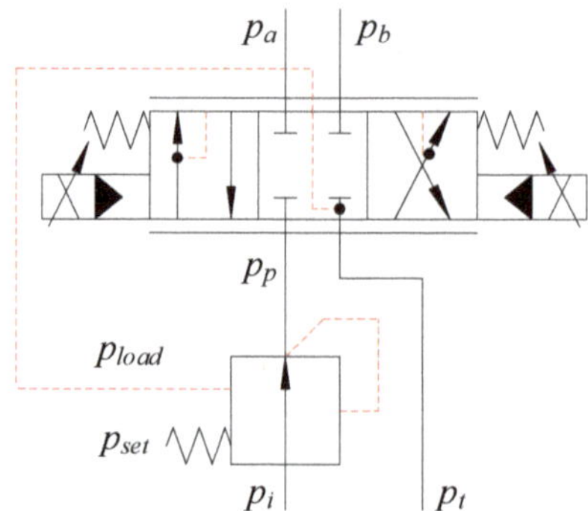

Figure 11. Hydraulic pressure compensated directional control valve for the knuckle cylinder.

The steady state of p_p is then given by Equation (16).

$$p_p = p_{load} + p_{set} \tag{16}$$

The sensing of the load pressures p_a and p_b ensures that the pressure drop over the directional control valve always equals p_{set}, and that the flow is load independent. This is shown in the orifice equation in Equation (17).

$$
\begin{aligned}
Q &= C_d \cdot A_d \cdot u_{spool} \cdot \sqrt{\frac{2}{\rho} \cdot (p_p - p_{load})} \\
&= C_d \cdot A_d \cdot u_{spool} \cdot \sqrt{\frac{2}{\rho} \cdot p_{set}} \\
&= Q_{max} \cdot u_{spool}
\end{aligned}
\tag{17}
$$

where

Q = flow in the valve;
C_d = discharge coefficient;
A_d = maximum discharge area;
ρ = mass density;
Q_{max} = maximum valve flow;

Double counterbalance valves are used on the knuckle cylinder. An illustration of the counterbalance valves is shown in Figure 12.

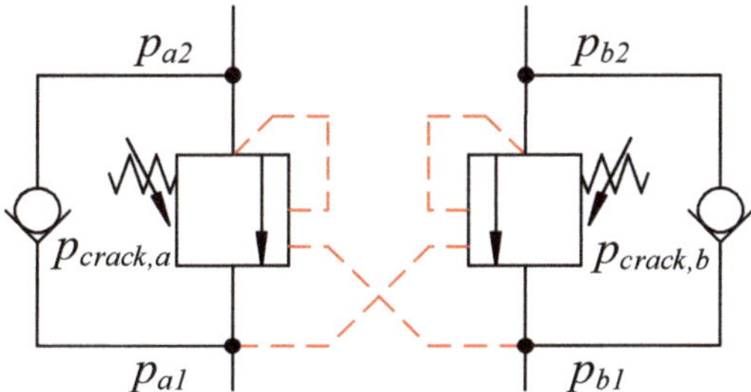

Figure 12. Double counterbalance valve.

The unitless openings of the counterbalance valves are calculated in Equations (18) and (19).

$$u_a = \frac{p_{a2} + \psi \cdot p_{b1} - p_{crack,a}}{\Delta p} \tag{18}$$

$$u_b = \frac{p_{b2} + \psi \cdot p_{a1} - p_{crack,b}}{\Delta p} \tag{19}$$

where

u_a = opening of valve a, $0 \le u_a \le 1$;
u_b = opening of valve b, $0 \le u_b \le 1$;
p_{a1} = pressure at valve a input side;
p_{a2} = pressure at valve a actuator side;
p_{b1} = pressure at valve b input side;
p_{b2} = pressure at valve b actuator side;

$p_{crack,a}$ = crack pressure of valve a;
$p_{crack,b}$ = crack pressure of valve b;
ψ = pilot area ratio;
Δp = pressure difference between fully closed and fully open.

When u_a and u_b are 0, the valves are closed. When they are 1, the valves are fully open. During assistive loads the valves tend to be somewhere between 0 and 1, meaning that they are throttling the flow. The dynamics of the valves are included as a time constant, since the valves have a finite bandwidth.

5. Adaptive Control Design

Since the actuator is a hydraulic differential cylinder, two separate states z_{ff}^+ and z_{ff}^- are used for out-stroke and in-stroke motion to handle model uncertainties both directions of motion. Consequently, both the feedforward control output and the update law for the two gains are only active during out-stroke or in-stroke motion respectively. To handle this, some switching logic is introduced based on the sign of the velocity reference. The block diagram for the differential MIT-rule adaptive feedforward is shown in Figure 13.

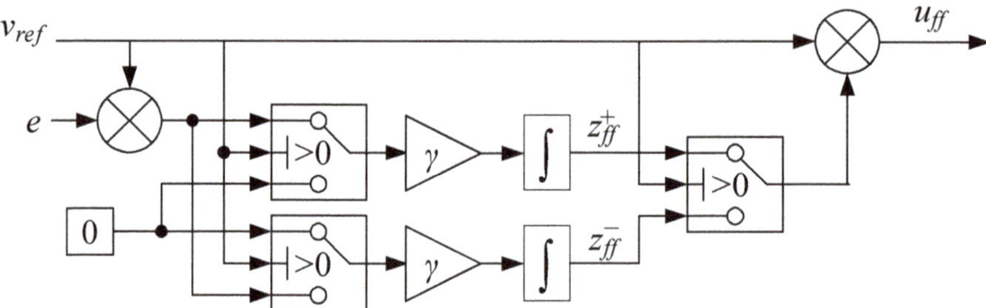

Figure 13. Differential MIT-rule adaptive feedforward.

The governing equations for the differential MIT-rule adaptive feedforward are shown in Equations (20)–(23).

$$\dot{z}_{ff}^+ = \begin{cases} \gamma \cdot v_{ref} \cdot e, & v_{ref} > 0 \\ 0, & \text{otherwise} \end{cases} \tag{20}$$

$$\dot{z}_{ff}^- = \begin{cases} 0, & v_{ref} > 0 \\ \gamma \cdot v_{ref} \cdot e, & \text{otherwise} \end{cases} \tag{21}$$

$$u_{ff} = \begin{cases} z_{ff}^+ \cdot v_{ref}, & v_{ref} > 0 \\ z_{ff}^- \cdot v_{ref}, & \text{otherwise} \end{cases} \tag{22}$$

$$u = k_p \cdot e + u_{ff} \tag{23}$$

where

z_{ff}^+ = out-stroke feedforward gain;
z_{ff}^- = in-stroke feedforward gain;
u_{ff} = feedforward controller output.

Extending the controller to sign-sign is straightforward. The block diagram for the differential sign-sign adaptive feedforward is shown in Figure 14.

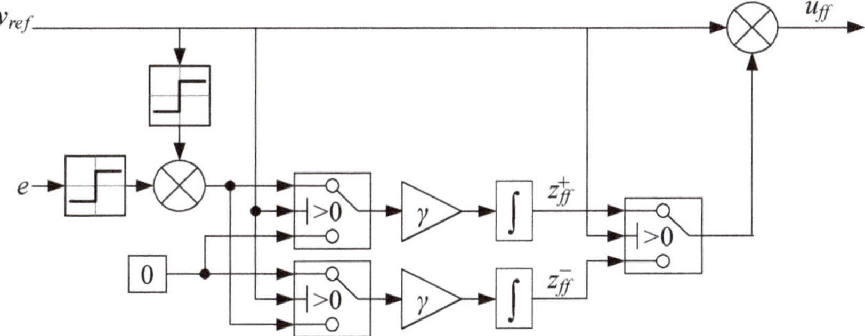

Figure 14. Differential sign-sign adaptive feedforward.

The governing equations for the differential sign-sign adaptive feedforward are shown in Equations (24)–(27).

$$\dot{z}_{ff}^{+} = \begin{cases} \gamma \cdot \text{sign}(v_{ref}) \cdot \text{sign}(e), & v_{ref} > 0 \\ 0, & \text{otherwise} \end{cases} \tag{24}$$

$$\dot{z}_{ff}^{-} = \begin{cases} 0, & v_{ref} > 0 \\ \gamma \cdot \text{sign}(v_{ref}) \cdot \text{sign}(e), & \text{otherwise} \end{cases} \tag{25}$$

$$u_{ff} = \begin{cases} z_{ff}^{+} \cdot v_{ref}, & v_{ref} > 0 \\ z_{ff}^{-} \cdot v_{ref}, & \text{otherwise} \end{cases} \tag{26}$$

$$u = k_p \cdot e + u_{ff} \tag{27}$$

6. Simulation Results

For the simulation, a point-to-point trapezoidal velocity path generator has been used as a reference. The point-to-point path generator has previously been developed in [26]. The path generator operates in actuator space, which eliminates the effects of the non-linearities between the hydraulic cylinder strokes and the joint angles in joint space. A path has been made such that the cylinder experiences both resistive and assistive loads in both directions of motion. The references for position and velocity are shown in Figure 15. The adaptation gain γ is different for the two controllers, due to the use of $\text{sign}(x)$, and has been experimentally set to $\gamma = 200 \text{ s} \cdot \text{m}^{-3}$ for the MIT-rule feedforward, and $\gamma = 0.1 \text{ m}^{-1}$ for the sign-sign feedforward. The unit is adapted accordingly to obtain the correct output.

The position error for the MIT-rule feedforward simulation is shown in Figure 16. The position error decreases towards a bounded error of ± 6 mm, which is shown with the dashed lines. The RMS error after convergence is 1.6 mm, showing high performance.

The states z_{ff} for the MIT-rule feedforward simulation are shown in Figure 17. The dashed lines show the theoretical values for a fixed feedforward gain. The states converge to values slightly larger than the theoretical ones. This small discrepancy can be attributed to the constant velocity reference and ramped position reference. When moving with a ramp position reference, there will always be a small constant position error without an integrator in the position controller. Having a slightly larger feedforward gain helps reducing this constant position error by giving the cylinder a small velocity boost. Since the position error is measured, the adaptive controller is able to adapt the feedforward gains to minimize the position error.

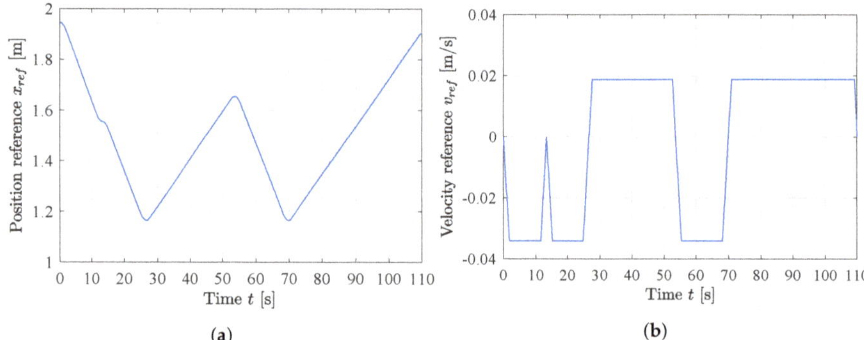

Figure 15. Point-to-point path references for simulation. (**a**) Position reference; (**b**) Velocity reference.

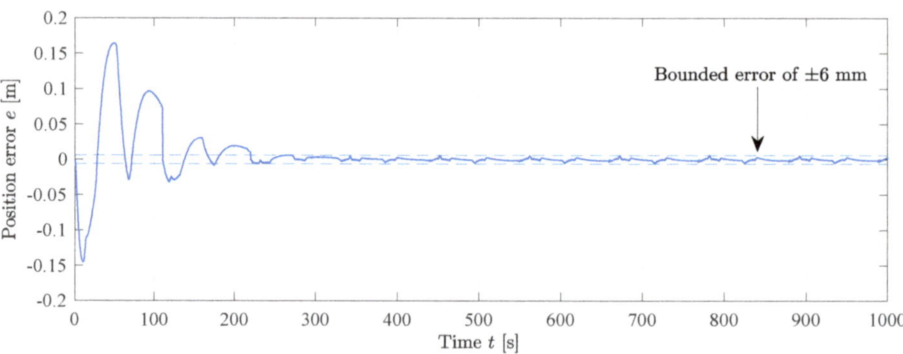

Figure 16. Cylinder position error during MIT-rule feedforward simulation, $\gamma = 200\,\text{s} \cdot \text{m}^{-3}$.

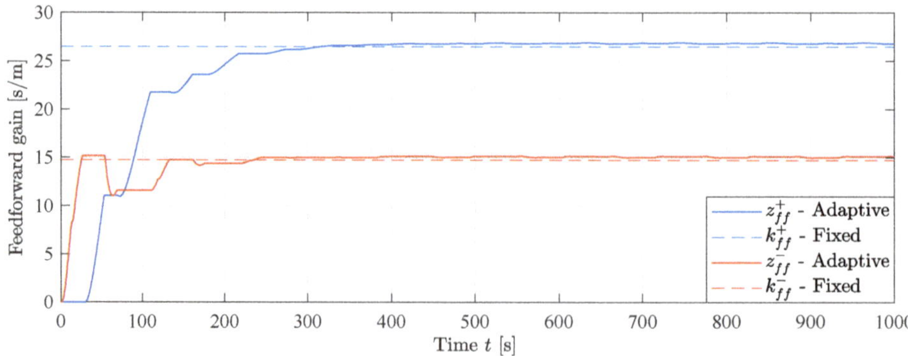

Figure 17. Feedforward states during MIT-rule feedforward simulation, $\gamma = 200\,\text{s} \cdot \text{m}^{-3}$.

Figure 18 shows the control signals u_{ff} and u_{fb} from the feedforward and feedback controller, respectively. Given that the total control signal $u = u_{fb} + u_{ff}$, it can be seen that the contribution from the feedforward controller clearly dominates, providing more than 95% at steady state.

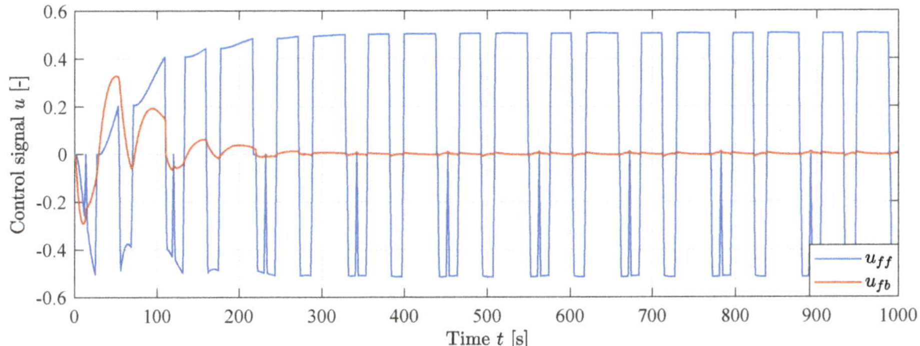

Figure 18. Control signals from feedforward and feedback during simulation, $\gamma = 200 \, \text{s} \cdot \text{m}^{-3}$.

The position error for the sign-sign feedforward simulation is shown in Figure 19. The same bounded error of ± 6 mm is shown with the dashed lines. The RMS error after convergence is 2.1 mm.

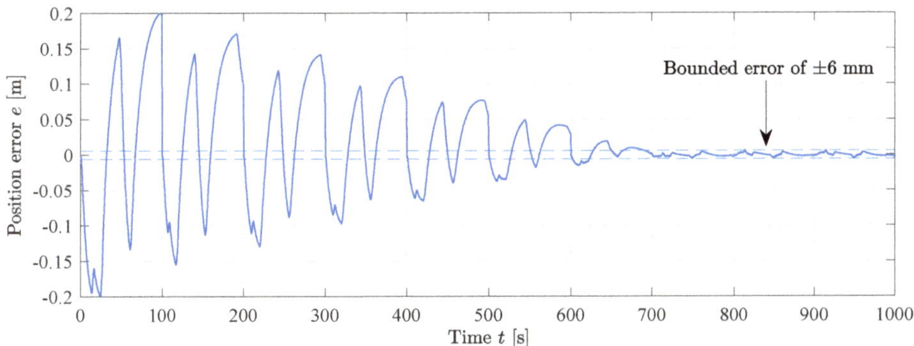

Figure 19. Cylinder position error during sign-sign feedforward simulation, $\gamma = 0.1 \, \text{m}^{-1}$.

The states z_{ff} for the sign-sign feedforward simulation are shown in Figure 20. The dashed lines show the theoretical values for a fixed feedforward gain. The same results can be seen here as with the MIT-rule, the states converge to values slightly larger than the theoretical ones, although convergence is slower with 700 s compared to 400 s.

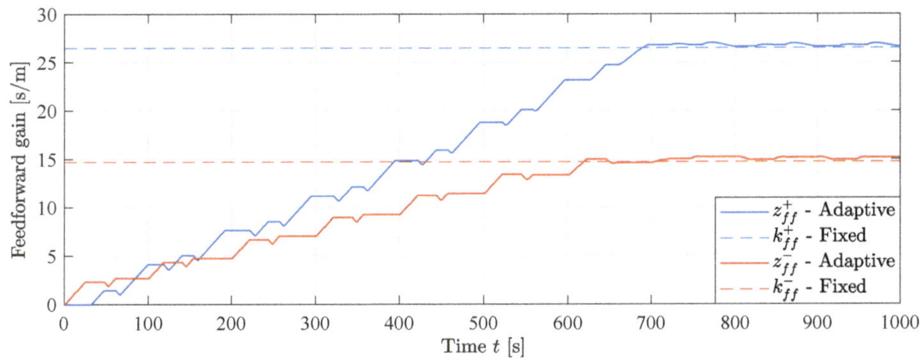

Figure 20. Feedforward states during sign-sign feedforward simulation, $\gamma = 0.1 \, \text{m}^{-1}$.

To show the difference in performance between the fixed gain controller and the adaptive controllers, a simulation with fixed gain feedforward has been made and compared with the MIT-rule feedforward at a simulation time where the states z_{ff} have converged, at $t = 800$ s. This is shown in Figure 21. It can be seen that the position error for the MIT-rule feedforward is lower compared to the fixed gain feedforward, showing that the MIT-rule feedforward controller outperforms the fixed gain controller even with an ideal model with correlation between cylinder velocity and feedforward gain.

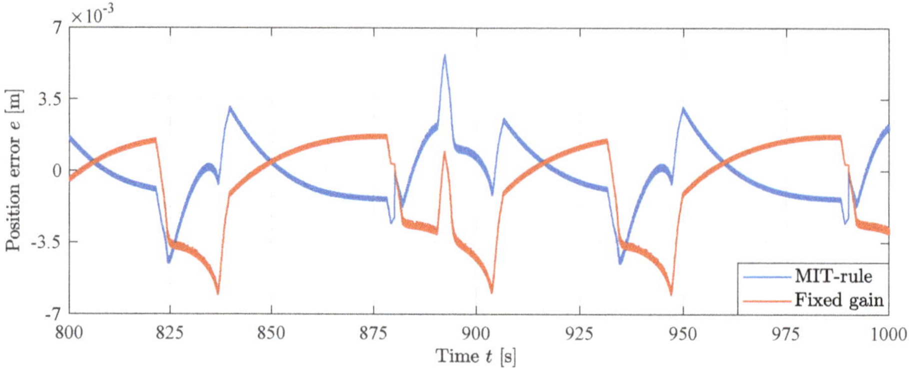

Figure 21. Position error comparison between MIT-rule and fixed gain feedforward in simulation.

The RMS position error for each controller after convergence of the states z_{ff} is shown in Table 3. Even though the fixed gain feedforward is based on an ideal model, the MIT-rule adaptive feedforward controller yields better position tracking with a 23% decrease in RMS position error. This shows the improved performance of the adaptive controller.

Table 3. Comparison of RMS position error after convergence in simulation.

	MIT-Rule	**Sign-Sign**	**Fixed Gain**
RMS error	1.6 mm	2.1 mm	2.1 mm

7. Experimental Results

The three controllers have been implemented on the CompactRIO controller in the laboratory. The control laws are implemented in discrete-time based on backward euler integration. A picture of the HMF 2020K4 loader crane in the laboratory is shown in Figure 22. The figure shows the crane in the starting position. During motion the knuckle boom is folded down.

There is some deadband in the valves on the HMF 2020K4 loader crane, and therefore deadband compensation has been implemented for the laboratory experiments. The identified deadbands for the knuckle boom valve are shown in Table 4.

Table 4. Identified deadbands for the knuckle boom valve.

Name	Parameter	Value
Out-stroke deadband	u^{+}	0.21
In-stroke deadband	u^{-}	-0.31

The equation for the deadband compensation is shown in Equation (28). By adding a small deadband $\tilde{u}$, it is ensured that the valve will be able to stay closed when no movement is needed.

$$\hat{u} = \begin{cases} u^+ + (1 - u^+) \cdot u, & u > \tilde{u} \\ u^- + (1 + u^-) \cdot u, & u < -\tilde{u} \\ 0, & \text{otherwise} \end{cases} \tag{28}$$

where

$\hat{u}$ = compensated control signal;
u = control signal;
u^+ = Out-stroke deadband;
u^- = In-stroke deadband;
$\tilde{u}$ = desired deadband, 0.001.

Figure 22. HMF 2020K4 loader crane in the laboratory.

The cylinder is running with a point-to-point path in actuator space equal to the simulations. The position error for the MIT-rule feedforward is shown in Figure 23. It is shown that the position error decreases towards a bounded error of ± 14 mm. The RMS error after convergence is 5.2 mm. The convergence of the position error is similar to the simulations, showing that the proposed adaptive controller is feasible in a real world scenario, albeit with slightly larger position error.

The states z_{ff} for the MIT-rule feedforward experiment are shown in Figure 24. The dashed lines show the theoretical values for a fixed feedforward gain. The states converge to values that differ from the theoretical ones. The state z_{ff}^+ is higher than the theoretical, while the state z_{ff}^- is lower. This means that there exist some model uncertainties that the controller is able to adapt to. In addition, the ratio of the feedforward gains differs from the cylinder area ratio ϕ, i. e. $\frac{z_{ff}^-}{z_{ff}^+} \neq \frac{A_a}{A}$, showing the importance of using two separate feedforward states. Since the two states are not mathematically linked by the cylinder area ratio ϕ, they are able to converge to values that minimizes position error in both directions of motion regardless of their ratio. This would not be possible if the traditional MIT-rule with a single state was used.

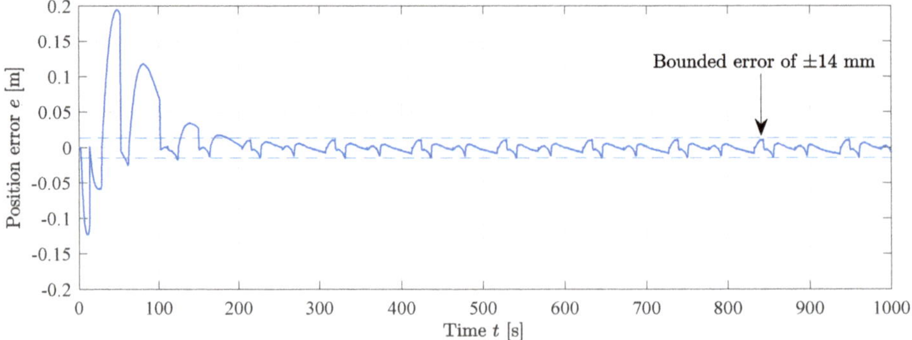

Figure 23. Position error during MIT-rule feedforward experiment, $\gamma = 200\,\text{s} \cdot \text{m}^{-3}$.

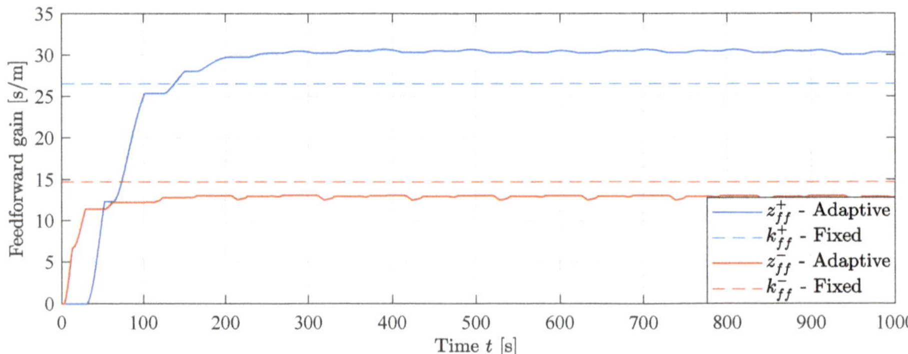

Figure 24. Feedforward states during MIT-rule feedforward experiment, $\gamma = 200\,\text{s} \cdot \text{m}^{-3}$.

The position error for the sign-sign feedforward is shown in Figure 25. The same bounded error of ± 14 mm is shown. The RMS error after convergence is 5.3 mm.

The states z_{ff} for the sign-sign feedforward experiment are shown in Figure 26. Similar results can be seen here as with the MIT-rule, the states converge to values that differ from the theoretical ones. The dashed lines show the theoretical values for a fixed feedforward gain. The convergence is slower than the MIT-rule feedforward, and even though convergence speed is not critical, it may be a minor disadvantage compared to the MIT-rule feedforward.

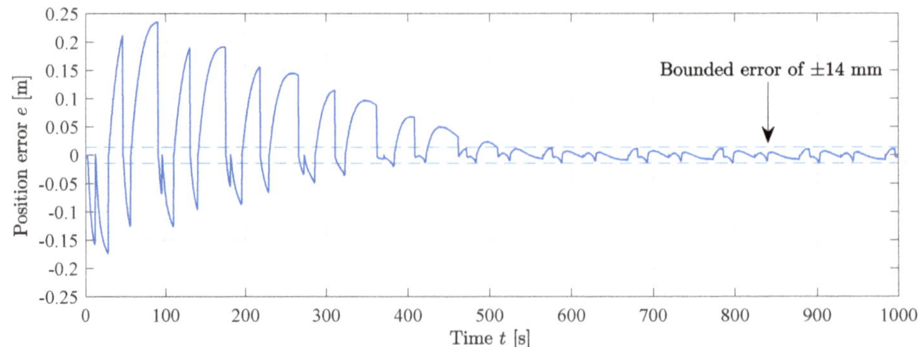

Figure 25. Position error during sign-sign feedforward experiment, $\gamma = 0.1\,\text{m}^{-1}$.

The same comparison as in the simulations is made in the laboratory. An experiment with fixed gain feedforward has been made and compared with the MIT-rule feedforward at a time where the states z_{ff} have converged, at $t = 800$ s. Figure 27 shows the difference in performance between the fixed gain controller and the adaptive controller, where the position error for the MIT-rule feedforward is significantly lower compared to the fixed gain feedforward.

The RMS position error for each controller after convergence of the states z_{ff} is shown in Table 5. The two adaptive feedforward controllers yield excellent performance with an 80% decrease in RMS position error.

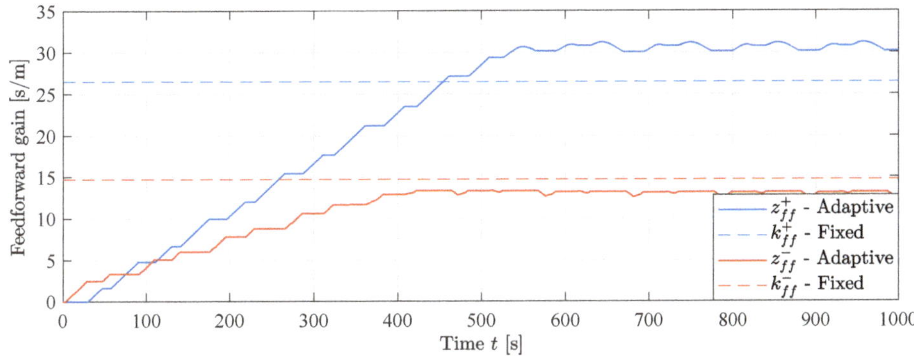

Figure 26. Feedforward states during sign-sign feedforward experiment, $\gamma = 0.1 \text{ m}^{-1}$.

In general, the RMS position errors are slightly larger than in the simulations, but this is expected and can be attributed to the unmodeled flexibility of the crane, and other unmodeled dynamics. However, the advantage of the adaptive feedforward controller is clear. The independent adaptation of the out-stroke and in-stroke states z_{ff}^{+} and z_{ff}^{-} provides significantly increased performance on a physical system with model uncertainties.

Table 5. Comparison of RMS position error after convergence in experiment.

	MIT-Rule	Sign-Sign	Fixed Gain
RMS error	5.2 mm	5.3 mm	24.9 mm

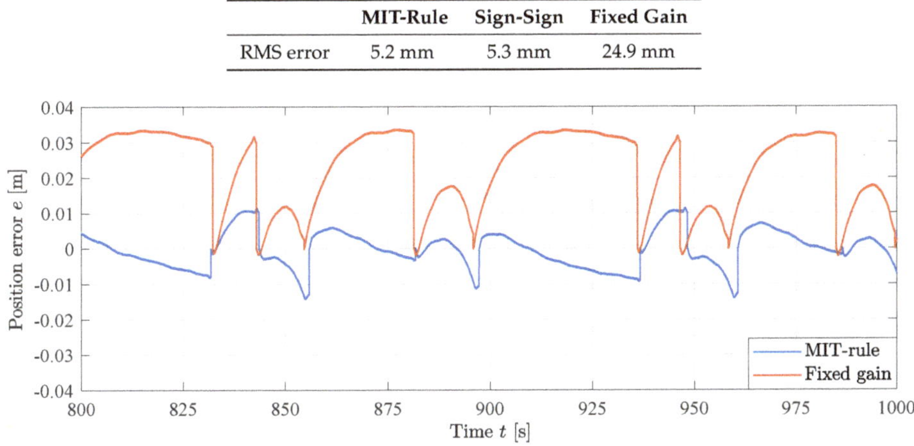

Figure 27. Position error with fixed and adaptive gains.

8. Conclusions

In this paper two adaptive feedforward motion controllers are designed, simulated, evaluated, implemented and experimentally verified on a loader crane with hydraulic differential cylinders. The controllers are based on common and proven adaptation methods to extend a typical electro-hydraulic motion control system into a novel adaptive feedforward motion controller. One of the challenges associated with a differential cylinder, namely the jump in both velocity and force gain when changing sign of direction, is solved by creating two separate feedforward states for out-stroke and in-stroke motion of the hydraulic differential cylinder, respectively. This separation makes the controller able to adapt to model uncertainties where the ratio between the in-stroke and out-stroke feedforward gains is not equal to the cylinder area ratio ϕ. Adaptation of the feedforward states only occurs when the hydraulic cylinder is moving in the direction of motion associated with the feedforward state.

Simulation results show high performance with good position tracking and that the states z_{ff} converge to values slightly higher than the theoretical ones. The cylinder position error is lowest for the MIT-rule controller with an RMS error of 1.6 mm, and shows faster convergence than the sign-sign controller. Compared to a fixed gain feedforward controller, where the gain is equal to the ratio of valve opening to cylinder velocity, the RMS error is reduced by 23%, showing the improved performance of the novel adaptive feedforward controllers.

Experiments in the laboratory show even better results than in the simulations. The adaptive feedforward controllers converge and show good position tracking, while the MIT-rule feedforward converges faster than the sign-sign feedforward. Compared to a fixed gain feedforward, the RMS position error is reduced by 80% to 5.2 mm for the MIT-rule. The results show the feasabillty of the novel adaptive feedforward controllers on a physical system. In addition, the differential structure of the controllers shows its advantage, as the ratio of the feedforward states converges to values different than the cylinder area ratio ϕ, showing the excellent performance of the adaptive feedforward controller and its capability of handling model uncertainties in both directions of motion.

Future work may include stability analysis of the adaptive controllers, since the feedforward gains are dependent on feedback of the cylinder position error e. The effects of the adaptation gain γ may also be investigated to see if there exists an upper boundary where the system becomes unstable.

Author Contributions: Conceptualization, K.J.J., M.K.E. and M.R.H.; methodology, K.J.J.; software, K.J.J.; validation, K.J.J.; formal analysis, K.J.J.; investigation, K.J.J.; data curation, K.J.J.; writing–original draft preparation, K.J.J.; writing–review and editing, K.J.J., M.K.E. and M.R.H.; visualization, K.J.J.; supervision, M.K.E. and M.R.H. All authors have read and agreed to the published version of the manuscript.

Funding: This research was funded by the Norwegian Ministry of Education and Research grant number 155597. The APC was funded by the University of Agder.

References

1. Bak, M.K.; Hansen, M.R. Analysis of Offshore Knuckle Boom Crane—Part Two: Motion Control. *Model. Identif. Control.* **2013**, *34*, 175–181. [CrossRef]
2. Mareels, I.M.; Anderson, B.D.; Bitmead, R.R.; Bodson, M.; Sastry, S.S. Revisiting the Mit Rule for Adaptive Control. *IFAC Proc. Vol.* **1987**, *20*, 161–166. [CrossRef]
3. Lucky, R.W. Techniques for adaptive equalization of digital communication systems. *Bell Syst. Tech. J.* **1966**, *45*, 255–286. [CrossRef]
4. Widrow, B. Adaptive sampled-data systems. *IFAC Proc. Vol.* **1960**, *1*, 433–439. [CrossRef]
5. Widrow, B.; Hoff, M.E. Adaptive Switching Circuits. *1960 IRE Wescon Conv. Rec.* **1960**, 96–104.
6. Unbehauen, H. Theory and Application of Adaptive Control. *IFAC Proc. Vol.* **1985**, *18*, 1–17. [CrossRef]
7. Truxal, J.G. Adaptive control. *IFAC Proc. Vol.* **1963**, *1*, 386–392. [CrossRef]
8. Strietzel, R.; Töpfer, H. Feedforward Adaption to Control Processes in Chemical Engineering. *IFAC Proc. Vol.* **1985**, *18*, 115–120. [CrossRef]

9. M'Saad, M.; Duque, M.; Landau, I. Robust LQ Adaptive Controller for Industrial Processes. *IFAC Proc. Vol.* **1985**, *18*, 91–97. [CrossRef]

10. Unbehauen, H.D. Adaptive Systems for Process Control. *IFAC Proc. Vol.* **1986**, *19*, 15–23. [CrossRef]

11. Oh, B.; Jamshidi, M.; Seraji, H. Two Adaptive Control Structures of Robot Manipulators. *IFAC Proc. Vol.* **1989**, *22*, 371–377. [CrossRef]

12. Van den Bossche, E.; Dugard, L.; Landau, I. Adaptive Control of a Flexible Arm. *IFAC Proc. Vol.* **1987**, *20*, 271–276. [CrossRef]

13. Tsakalis, K.; Ioannou, P. Adaptive control of linear time-varying plants. *Automatica* **1987**, *23*, 459–468. [CrossRef]

14. Hušek, P. Adaptive fuzzy sliding mode control for uncertain nonlinear systems. *IFAC Proc. Vol.* **2014**, *47*, 540–545. [CrossRef]

15. Sato, K.; Tsuruta, K. Adaptive Friction Compensation for Linear Slider with adaptive differentiator. *IFAC Proc. Vol.* **2010**, *43*, 467–472. [CrossRef]

16. Lee, R.M.; Chen, T.C. Adaptive Control of Active Magnetic Bearing against Milling Dynamics. *Appl. Sci* **2016**, *6*, 52. [CrossRef]

17. Yang, G.; Yao, J.; Le, G.; Ma, D. Adaptive integral robust control of hydraulic systems with asymptotic tracking. *Mechatronics* **2016**, *40*, 78–86. [CrossRef]

18. Nicol, C.; Macnab, C.; Ramirez-Serrano, A. Robust adaptive control of a quadrotor helicopter. *Mechatronics* **2011**, *21*, 927–938. [CrossRef]

19. Agarwal, R.K.; Hussain, I.; Singh, B. LMF-Based Control Algorithm for Single Stage Three-Phase Grid Integrated Solar PV System. *IEEE Trans. Sustain. Energy* **2016**, *7*, 1379–1387. [CrossRef]

20. Krus, P.; Palmberg, J.O. Vector Control of a Hydraulic Crane. *Int. Off Highw. Powerpl. Congr. Expo.* **1992**. [CrossRef]

21. Sørensen, J.K.; Hansen, M.R.; Ebbensen, M.K. Boom Motion Control Using Pressure Control Valve. In Proceedings of the 8th FPNI Ph.D Symposium on Fluid Power, Lappeenranta, Finland, 11–13 June 2014. [CrossRef]

22. Sørensen, J.K.; Hansen, M.R.; Ebbesen, M.K. Load Independent Velocity Control on Boom Motion Using Pressure Control Valve. In Proceedings of the Fourteenth Scandinavian International Conference on Fluid Power, Tampere, Finland, 20–22 May 2015.

23. Beiner, L. Identification and Control of a Hydraulic Forestry Crane. *Mechatronics* **1997**, *7*, 537–547. [CrossRef]

24. Mattila, J.; Virvalo, T. Energy-efficient Motion Control of a Hydraulic Manipulator. In Proceedings of the 2000 IEEE International Conference on Robotics and Automation, San Francisco. CA, USA, 24–28 April 2000. [CrossRef]

25. Zhang, Q. Hydraulic Linear Actuator Velocity Control Using a Feedforward-plus-PID Control. *Int. J. Flex. Autom. Integr. Manuf.* **1999**, *7*, 277–292.

26. Jensen, K.J.; Kjeld Ebbesen, M.; Rygaard Hansen, M. Development of Point-to-Point Path Control in Actuator Space for Hydraulic Knuckle Boom Crane. *Actuators* **2020**, *9*, 27. [CrossRef]

Publisher's Note: MDPI stays neutral with regard to jurisdictional claims in published maps and institutional affiliations.

 *applied sciences*

Article

Modeling and Control of a Cable-Suspended Sling-Like Parallel Robot for Throwing Operations [†]

Deng Lin [1], Giovanni Mottola [2], Marco Carricato [2],* and Xiaoling Jiang [1],*

[1] Faculty of Mechanical Engineering and Automation, Zhejiang Sci-Tech University, Hangzhou 310018, China; 201810501007@mails.zstu.edu.cn
[2] Department of Industrial Engineering, University of Bologna, Viale Risorgimento 2, 40136 Bologna, Italy; giovanni.mottola3@unibo.it
* Correspondence: marco.carricato@unibo.it (M.C.); jxl1212@zstu.edu.cn (X.J.); Tel.: +39-051-209-3443 (M.C.)
† This paper is an extended version of our paper published in Dynamically-Feasible Trajectories for a Cable-Suspended Robot Performing Throwing Operations. Presented at 23rd CISM IFToMM Symposium on Robot Design, Dynamics and Control, Sapporo, Japan, 20–24 September 2020.

Received: 3 November 2020; Accepted: 15 December 2020; Published: 18 December 2020

Featured Application: Sling-type cable-suspended robots could be applied for gaming and entertainment purposes: the robot could be used, for instance, to throw a ball towards human players and catch it back. The advantage of cable robots is their large workspace, which can span across a playing field. Moreover, an ideal application for a sling-like CSPR would be automatic waste handling, where there are no strict requirements on positioning accuracy nor a risk of damaging the objects being sorted.

Abstract: Cable-driven parallel robots can provide interesting advantages over conventional robots with rigid links; in particular, robots with a cable-suspended architecture can have very large workspaces. Recent research has shown that dynamic trajectories allow the robot to further increase its workspace by taking advantage of inertial effects. In our work, we consider a three-degrees-of-freedom parallel robot suspended by three cables, with a point-mass end-effector. This model was considered in previous works to analyze the conditions for dynamical feasibility of a trajectory. Here, we enhance the robot's capabilities by using it as a sling, that is, by throwing a mass at a suitable time. The mass is carried at the end-effector by a gripper, which releases the mass so that it can reach a given target point. Mathematical models are presented that provide guidelines for planning the trajectory. Moreover, results are shown from simulations that include the effect of cable elasticity. Finally, suggestions are offered regarding how such a trajectory can be optimized.

Keywords: cable-driven parallel robots; cable-suspended robots; dynamic workspace; trajectory planning; throwing robots; casting robot

---

1. Introduction

Cable-Suspended Parallel Robots (CSPRs) are a class of parallel robots where the end-effector (EE) is supported by n extendable cables and in which the gravity force is used to keep cable tensions $\tau_1, \ldots, \tau_n$ positive. In this work, we focus in particular on robots that employ as many (taut) cables as the number of Degrees-of-Freedom (DoFs) of the EE.

Such robots are often considered to move quasi-statically, that is, slowly enough that inertial effects can be disregarded. If a CSPR operates in static or quasi-static conditions, the EE must remain within the Static Equilibrium Workspace (SEW), namely, the set of poses where the EE can be in static equilibrium [1]. However, considering dynamic trajectories has become more and more important over the years. In [2], the authors defined the Dynamic Workspace (DW) as the set of EE poses that can be reached with positive cable tensions by taking advantage of inertia forces, with the EE moving in non-static conditions (nonzero velocity or acceleration): since the SEW is a subset of the DW, this expands the potential applications of CSPRs.

For a dynamic trajectory to be feasible, the cable tensions must be positive at any instant, as the EE moves along the trajectory. This condition can be verified numerically, by solving the inverse dynamics problem at each time step; however, this approach can be computationally demanding, to the point that it is difficult to apply in real-time conditions. Another possibility is to use analytic methods: thus, several authors proposed trajectories defined by harmonic motions where the frequency must lie within a feasible range [3–9] that can be directly computed from the trajectory parameters. One of the first works in this field was [8], where the author considered several types of parametric dynamic trajectories; this work was extended and generalized in [7], where the authors considered elliptic paths in space, including circles and straight lines as special cases. Later works found similar results for cable robots having rotational DoFs [6] or overconstrained cable-driven parallel robots (CDPRs) [9]. Finally, it was proved in [10,11] that the feasibility conditions found for the 3-cable, 3-DoF spatial robot with point-mass EE (considered in [3–5,7,8]) can be extended to a 6-cable, 3-DoF spatial robot with purely translational motion and a finite-size EE; this is a more realistic architecture for practical applications (for instance, a gripper can be attached on the EE [12] to carry an object).

Operations that require dynamic manipulation of an object, such as throwing and catching, were attempted by both rigid-link robots and CSPRs. These techniques can be used not only to enlarge the workspace, but also to reduce operation times, which provides greater application possibilities. For rigid-link robots, some of the first works in this field were [13,14], where the authors considered a one-joint manipulator performing throwing and batting operations on an object in the plane, and proved that pose and velocity could be completely controlled (within a subset of the state space) by underactuated manipulation. Other works on 1-DoF manipulators with similar dynamic motions were reported in [15], where the authors presented a 1-DoF robot prototype for tossing a ball towards a target, and [16], where a control strategy was proposed for a similar prototype that launched an object towards a target point. More recently, a concept similar to that presented in [13,14] was applied to a 1-DoF robot capable of juggling a ball in three dimensions in an open-loop configuration [17]. For a general survey of dynamic manipulation using only unilateral constraints (which includes throwing movements), we refer the reader to [18].

Throwing can also be performed by more complex architectures, such as serial robotic arms: an early work is [19], where a 2-DoF robot moving in a vertical plane along an elliptical path was used to throw a ball and catch it back using visual feedback. For similar 2-link, underactuated robot arms, other authors considered the problem of optimizing the pitching trajectory [20,21], while in [22] a control strategy for the throwing motion was proposed and applied to the Pendubot prototype. Regarding higher-DoF serial systems, a serial-link robot launching a ball towards a target was explored in [23], where the authors reported a success rate of 40%; later, the authors of [24] reported up to 85% accuracy by using both a physical simulation and the predictions of a deep-learning network. A 6-DoF serial arm was used for dynamic release and catching of a ball (in a "dribbling" motion) by predicting its trajectory through a model of the free-flight phase using data from a visual tracking system [25]. A more general trajectory design approach for n-link manipulators was introduced in [26]: the algorithm considers both kinematic and dynamic constraints, and finds a solution (if any) with a known probability. Finally, the sensitivity of the launch trajectory was considered in [27], where the authors showed how to find the optimal launch trajectory in order to minimize the variance of the landing-point position.

A number of works by Frank et al. [28] considered the possibility of throwing an object from a gripper; the proposed application was rapid transportation of parts in a manufacturing system. In [29], this idea was developed specifically for throwing and catching cylinder-shaped objects. The position of the launched object in midair was observed by a visual tracking system [30].

A slightly different approach, compared to throwing robots, was used in so-called "casting" robots, proposed in [31]: here the authors design a planar, underactuated 2-DoF robot whose EE is a gripper attached through an extendable cable to a rigid-link mechanism. This design, further developed in [32], was used to prove the possibility of reaching a target point far from the robot base, by swinging the rigid mechanism and then launching the EE by releasing the string at a suitable time. Later, the same authors proposed in [33] the idea of controlling the EE motion in midair, by controlling the tension of the string to which the EE is attached. In [34], this approach was extended to a 3-DoF spatial parallel robot; later on, in [35], a casting robot was proposed using a serial robot arm mounted on a mobile robot. In [36], a robotic arm launched a probe to a target location while the probe orientation was kept fixed during the launch motion. A system that launched an anchor ball attached to a cable was considered in [37], together with a ropeway robot moving along the launched cable after its anchor had been fixed at the target position. Notably, in a casting robot, the cable may interact not only with the launched part, but also with the environment, for instance by wrapping itself around a pole [38]. A concept closely correlated to casting was explored in [39], where the authors considered a shooting robot that catapulted an object connected to an elastic cable with an impulsive force; an advantage of this approach is the ability to reach "blind spots" behind objects. A potential drawback of casting manipulators, where the gripper detaches from the robot, is that they require long cables that, especially in cluttered areas, can be difficult to safely reel back.

In this work, we focus on a 3-DoF CSPR with a point-mass EE attached to three cables and carrying a gripper from which a mass, modeled as a frictionless point object, must be thrown towards a given target by opening the gripper at a suitable instant. The obtained results can be readily extended to a 3-DoF CSPR with a finite-size EE such as the one in [10], as long as some conditions on the robot architecture are respected. The CSPR with parallelogram-type actuation in [10] can perform purely-translational motion: this is useful, since keeping a constant orientation of the EE helps in controlling the launch motion of the object.

Throwing robots were proposed in the literature for application in industrial environments, for instance to quickly transport small parts [28,40], to collect small objects into bins efficiently [24] or to store objects in warehouses [15,16]; for a survey of throwing as a transportation and sorting method for objects with mass up to 100 g, see [40]. In particular, an ideal application for a sling-like CSPR would be automatic waste handling, where there are no strict requirements on positioning accuracy nor a risk of damaging the objects being sorted: while the concept of sorting waste through throwing robots was already explored with both serial [41] or parallel [42] architectures, the application to CSPRs is new, to the best of our knowledge. Throwing robots were also applied for entertainment purposes [19,25], for instance in juggling [17], therefore we also envision an application for our work to a robotic game-playing platform: the robot could be used to launch a ball towards players and catch it back from them, taking advantage of the large DW of CSPRs to move across a playing field. Another interesting application would be in hostile environments, such as space [34,35] or disaster-stricken areas [33,36,37], for example to launch a grappling device [37] or a penetrator [36] (with an attached wire) towards a target point in order to retrieve samples or to attach a tether. Finally, we also suggest that our robot could be used in a casting configuration for fly-fishing purposes, by throwing a hook in a body of water.

In this paper, we begin by defining the dynamic model of the robot (Section 2) as it moves along elliptical paths according to the conditions for dynamic feasibility defined in [4,7]. In Section 3 we find the ballistic trajectory of the launched mass, from which we define conditions so that it can reach the target point. Then, in Section 4 the trajectory thus found is optimized according to two objective functions, through the use of a genetic algorithm. A more complete model of the robot, that takes into account cable elasticity, is presented in Section 5: this is especially relevant considering the discontinuity in cable tensions before and after the launch instant. Finally, in Section 6, we summarize our work and present possible directions for further research.

2. Kinematic and Dynamic Model

In our work, we consider a 3-DoF robot suspended by three cables such as the one shown in Figures 1 and 2; the cables are modeled as inextensible, massless and having fixed cable exit points A_i. The points A_i have position vectors $\mathbf{a}_i = A_i - O$ with respect to the fixed reference frame (on the robot base) having its origin in O. The EE is modeled as a point P having mass m and position $\mathbf{p} = P - O = [x, y, z]^T$. One can write the inverse kinematic equations (which provide the cable lengths ρ_i as a function of $\mathbf{p}$) as

$$\rho_i = \sqrt{(\mathbf{p} - \mathbf{a}_i)^T (\mathbf{p} - \mathbf{a}_i)}, \quad i = 1, 2, 3. \tag{1}$$

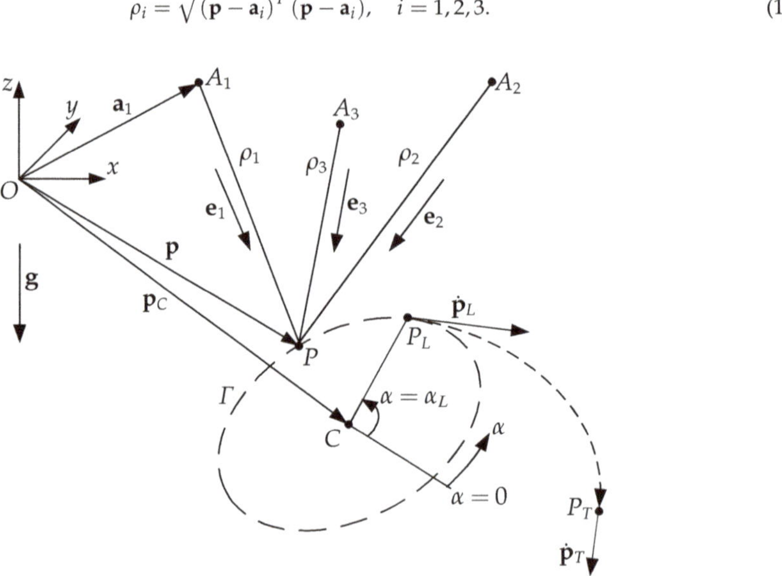

Figure 1. A 3-DoF (Degrees-of-Freedom) spatial Cable-Suspended Parallel Robot (CSPR) launching an object from the launch point P_L to the target point P_T.

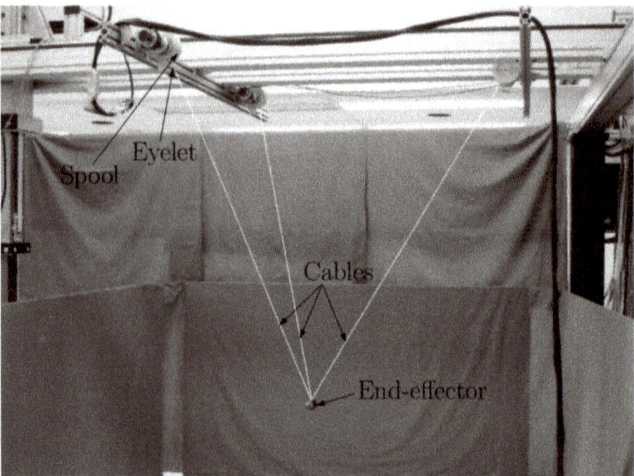

Figure 2. An example 3-DoF, 3-cable spatial CSPR (image taken with permission from [43]).

We define the unit vectors $\mathbf{e}_i$ directed as the ith cable and oriented towards the EE as

$$\mathbf{e}_i = \frac{\mathbf{p} - \mathbf{a}_i}{\rho_i}, \quad i = 1, 2, 3. \tag{2}$$

We obtain the dynamic model of the robot by writing the force equilibrium equation for the EE:

$$m\mathbf{g} - \sum_{i=1}^{3} \tau_i \mathbf{e}_i = m\ddot{\mathbf{p}} \tag{3}$$

In Equation (3), τ_i denotes the tensile force in the ith cable, $\mathbf{g} = -[0, 0, g]^T$ is the gravity vector, and g is the gravitational acceleration. It is convenient to refer to the cable tensions per unit mass $\mu_i = \tau_i/m$, as the EE mass varies before and after launch. We now define the vector $\boldsymbol{\mu}$ and the structure matrix $\mathbf{M}$ as

$$\boldsymbol{\mu} = \begin{bmatrix} \mu_1 \\ \mu_2 \\ \mu_3 \end{bmatrix} = \frac{1}{m} \begin{bmatrix} \tau_1 \\ \tau_2 \\ \tau_3 \end{bmatrix} = \frac{1}{m}\boldsymbol{\tau}, \quad \mathbf{M} = \begin{bmatrix} \mathbf{e}_1 & \mathbf{e}_2 & \mathbf{e}_3 \end{bmatrix} \tag{4}$$

After some algebraic manipulation, Equation (3) can be solved (in matrix form) for $\boldsymbol{\mu}$ as

$$\boldsymbol{\mu} = \mathbf{M}^{-1} \left(\mathbf{g} - \ddot{\mathbf{p}} \right) \tag{5}$$

where matrix $\mathbf{M}$ is invertible if and only if P does not lie on the plane through points A_1, A_2 and A_3 [4]. In the following, for simplicity we assume that the EE mass varies instantaneously from m_1 (before the launch) to m_2 (after the launch) and that this happens by simply releasing the launched object, without introducing external forces on the EE; with the previous assumptions of point-mass EE and inextensible cables, this implies that the tensions $\boldsymbol{\tau}$ instantaneously vary at launch. For a more comprehensive treatment of variable-mass systems, we refer the reader to [44].

The robot is fully actuated, that is, it has as many controlled DoFs as actuated cables: therefore, any launch trajectory could in principle be used in order to reach a desired launch condition (defined by a launch point and launch velocity), provided that said trajectory is feasible. We assume that P moves along an elliptical path Γ, so that we may usefully apply many results from previous works [5,7,8], especially in regards to determining trajectory feasibility by means of analytical methods. The position $\mathbf{p}$ of the EE is then defined by

$$\mathbf{p} = \mathbf{p}_C + \mathbf{b} \begin{bmatrix} \sin(\omega t) \\ \cos(\omega t) \end{bmatrix}, \quad \mathbf{p}_C = \begin{bmatrix} x_C \\ y_C \\ z_C \end{bmatrix}, \quad \mathbf{b} = \begin{bmatrix} x_{rs} & x_{rc} \\ y_{rs} & y_{rc} \\ z_{rs} & z_{rc} \end{bmatrix} \tag{6}$$

where $\mathbf{p}_C = C - O$ denotes the position of the center C of Γ, while the entries in matrix $\mathbf{b}$ are path parameters that define the shape and orientation of the trajectory. The motion amplitude along the x axis is given by $x_A = \sqrt{x_{rs}^2 + x_{rc}^2}$, thus the minimum and maximum values of the x coordinate as P moves along Γ are $x_{min} = x_C - x_A$ and $x_{max} = x_C + x_A$; similar extrema can be found for the motion along axes y and z.

3. Launch Trajectory Planning

A mass is launched at the launch instant $t = t_L$ from the EE and we require it to pass through a given target point P_T, which has position $\mathbf{p}_T = [x_T, y_T, z_T]^T$ (see Figure 1). The position $\mathbf{p}_L$ at launch point P_L of the EE and its launch velocity $\dot{\mathbf{p}}_L$ can be obtained from Equation (6) as:

$$\mathbf{p}_L = \begin{bmatrix} x_L \\ y_L \\ z_L \end{bmatrix} = \mathbf{p}_C + b \begin{bmatrix} \sin\alpha_L \\ \cos\alpha_L \end{bmatrix}, \quad \dot{\mathbf{p}}_L = \begin{bmatrix} \dot{x}_L \\ \dot{y}_L \\ \dot{z}_L \end{bmatrix} = \omega b \begin{bmatrix} \cos\alpha_L \\ -\sin\alpha_L \end{bmatrix} \tag{7}$$

where $\alpha_L = \omega t_L$ is the launch angle (using α as the path coordinate along Γ); note that $\mathbf{p}_L$ and $\dot{\mathbf{p}}_L$ define the initial conditions for the motion of the mass after launch.

The trajectory planning of the launched mass requires some simplifying assumptions. Most authors [13–17,19,21,25–27,31–34,36–39] consider a simplified model where, during the free-flight phase, the only force acting is gravity; under these assumptions, it can be readily shown that the center of mass of the launched object will move along a parabolic arc in a vertical plane. Considering the effect of air drag leads to a more realistic model; however, the dynamic problem including the air-drag model does not have a closed-form solution. Therefore, we will disregard this effect, which is acceptable if the object being launched is sufficiently small and heavy.

With these assumptions, gravity is the only external force acting during the free-flight motion and the equation of ballistic flight can be written as:

$$\mathbf{p}(t) = \frac{1}{2}\mathbf{g}\,(t - t_L)^2 + \dot{\mathbf{p}}_L\,(t - t_L) + \mathbf{p}_L \tag{8}$$

If we denote by Δt_{LT} the flight time in which the mass goes from P_L to P_T, then $\mathbf{p}_T = \mathbf{p}\,(t_L + \Delta t_{LT})$, and thus:

$$x_T = \dot{x}_L \Delta t_{LT} + x_L \tag{9}$$

$$y_T = \dot{y}_L \Delta t_{LT} + y_L \tag{10}$$

$$z_T = \dot{z}_L \Delta t_{LT} - \frac{1}{2}g\Delta t_{LT}^2 + z_L \tag{11}$$

Equations (9) and (10) yield

$$\frac{x_T - x_L}{y_T - y_L} = \frac{\dot{x}_L}{\dot{y}_L} \tag{12}$$

By substituting Equation (7) into Equation (12) and simplifying, we obtain

$$\begin{aligned} &\left[(x_T - x_C)\,y_{rs} - (y_T - y_C)\,x_{rs}\right]\cos\alpha_L \\ &- \left[(x_T - x_C)\,y_{rc} - (y_T - y_C)\,x_{rc}\right]\sin\alpha_L + x_{rs}y_{rc} - x_{rc}y_{rs} = 0 \end{aligned} \tag{13}$$

which, using the tangent half-angle substitution, can be rewritten as a 2nd-degree equation in $t_n = \tan\left(\frac{\alpha_L}{2}\right)$, namely:

$$Gt_n^2 + Ht_n + I = 0 \tag{14}$$

where G, H and I are known coefficients that depend on the geometry of ellipse Γ and the coordinates of the target point P_T, namely:

$$\begin{aligned} G &= x_{rs}y_{rc} - x_{rc}y_{rs} - x_Ty_{rs} + x_Cy_{rs} + x_{rs}y_T - x_{rs}y_C \\ H &= 2\left(-x_Ty_{rc} + x_Cy_{rc} + x_{rc}y_T - x_{rc}y_C\right) \\ I &= x_{rs}y_{rc} - x_{rc}y_{rs} + x_Ty_{rs} - x_Cy_{rs} - x_{rs}y_T + x_{rs}y_C \end{aligned} \tag{15}$$

Solving Equation (14) yields two values for t_n:

$$t_{n,1} = \frac{-H + \sqrt{H^2 - 4GI}}{2G}, \quad t_{n,2} = \frac{-H - \sqrt{H^2 - 4GI}}{2G} \tag{16}$$

and thus two values for α_L:

$$\alpha_{L1} = 2\arctan\left(t_{n,1}\right), \quad \alpha_{L2} = 2\arctan\left(t_{n,2}\right) \tag{17}$$

Two possible launch positions $\mathbf{p}_{L1}$ and $\mathbf{p}_{L2}$ are obtained substituting the two values of α_{Li} in Equation (7); it is worth pointing out that neither α_L nor $\mathbf{p}_L$ depend on ω.

Substituting the components of either possible solution for $\mathbf{p}_L$ and $\dot{\mathbf{p}}_L$ into Equations (9) and (11), we get two nonlinear equations in the two unknowns ω and Δt_{LT}, which are the last parameters that need to be computed:

$$x_T - x_L = \omega\left(x_{rs}\cos\alpha_{Li} - x_{rc}\sin\alpha_{Li}\right)\Delta t_{LT} \tag{18}$$

$$z_T - z_L = \omega\left(z_{rs}\cos\alpha_{Li} - z_{rc}\sin\alpha_{Li}\right)\Delta t_{LT} - \frac{1}{2}g\Delta t_{LT}^2 \tag{19}$$

for $i = 1, 2$. Equation (18) can be rewritten as

$$\omega = \frac{x_T - x_L}{\left(x_{rs}\cos\alpha_{Li} - x_{rc}\sin\alpha_{Li}\right)\Delta t_{LT}} \tag{20}$$

Substituting Equation (20) into Equation (19), we find

$$\Delta t_{LT} = \sqrt{\frac{2}{g}\left[\frac{\left(x_T - x_L\right)\left(z_{rs}\cos\alpha_{Li} - z_{rc}\sin\alpha_{Li}\right)}{x_{rs}\cos\alpha_{Li} - x_{rc}\sin\alpha_{Li}} - \left(z_T - z_L\right)\right]} \tag{21}$$

Substituting Equation (21) into Equation (20), two values of ω can be found. Thus, we obtain the values of ω and Δt_{LT} that ensure that the launched mass reaches the target point. The method above gives two solutions for α_L and thus two solutions for the launch trajectory.

A necessary and sufficient condition for Equation (21) to have solutions is

$$\frac{\left(x_T - x_L\right)\left(z_{rs}\cos\alpha_{Li} - z_{rc}\sin\alpha_{Li}\right)}{x_{rs}\cos\alpha_{Li} - x_{rc}\sin\alpha_{Li}} - \left(z_T - z_L\right) > 0 \tag{22}$$

We offer the conjecture (verified on a large number of numerical simulations) that Equation (22) holds if and only if the target point lies below the plane containing the ellipse Γ.

4. Optimal Trajectory

4.1. Optimization Method

In Section 3, we showed how to find, for a given trajectory Γ and target point P_T, two possible launch positions $\mathbf{p}_{L1}$ and $\mathbf{p}_{L2}$. In this Section, instead, we seek an optimal trajectory Γ_{opt} (for a given P_T), that is, one having the lowest "cost" (which can be defined according to users' requirements) while respecting desired constraints. Since an elliptical path is completely defined by vector $\mathbf{p}_C$ and matrix $\mathbf{b}$, while the remaining trajectory parameters α_L and ω can be computed from Equations (17) and (20), we look for the optimal choice for $\mathbf{p}_C$ and $\mathbf{b}$, which minimizes a given objective function F.

In a throwing operation, we may be interested in limiting both the energy required for throwing and the motion time of the launched object. Reducing the motion time provides obvious advantages in terms of productivity; however, we noted empirically that minimizing Δt_{LT} alone tends to produce unacceptable solutions, as the velocities involved and the corresponding cable tensions may become excessively high. It thus appears sensible to evaluate the energy E required for the launch motion, as E depends on both tensions and velocities. Therefore, we include the ballistic flight time Δt_{LT} and the total energy E required for launch in a vector-valued function F with two optimization objectives:

$$F = \begin{bmatrix} E \\ \Delta t_{LT} \end{bmatrix} \tag{23}$$

Some sensible constraints may be set on the velocity $\dot{\mathbf{p}}_T = [\dot{x}_T, \dot{y}_T, \dot{z}_T]^T = \mathbf{n}_T \|\dot{\mathbf{p}}_T\|$ at the target point: here, we choose to assign the direction $\mathbf{n}_T = [n_{Tx}, n_{Ty}, n_{Tz}]^T$ and a maximum magnitude $\dot{p}_{max}$, such that $\|\dot{\mathbf{p}}_T\| \leq \dot{p}_{max}$.

We note that, by differentiating Equation (8) with respect to time, $\dot{\mathbf{p}}_L$ can be derived from $\dot{\mathbf{p}}_T$ as

$$\begin{aligned} \dot{x}_L &= \dot{x}_T \\ \dot{y}_L &= \dot{y}_T \\ \dot{z}_L &= \dot{z}_T + g\Delta t_{LT} \end{aligned} \tag{24}$$

The flight time Δt_{LT} can be found from Equation (21), but it is convenient to express it as a function of the optimization parameters. By introducing $\Delta xy_{LT} = \sqrt{(x_T - x_L)^2 + (y_T - y_L)^2}$ and $n_{T,xy} = \sqrt{n_{Tx}^2 + n_{Ty}^2}$, the flight time can be expressed from Equations (9) and (10) as

$$\Delta t_{LT} = \frac{\Delta xy_{LT}}{\|\dot{\mathbf{p}}_T\| n_{T,xy}} \tag{25}$$

The energy required for launch is the sum of a kinetic and a potential term due to gravity. In order to simplify the trajectory planning, we choose to start the EE from rest and let it reach the dynamical conditions of trajectory Γ through transition trajectories [7] with progressively increasing motion amplitudes. With this choice, the initial rest position is at the center P_C of Γ. The kinetic energy at start is zero, while the kinetic energy E_{KL} at launch is

$$E_{KL} = \frac{1}{2}m\left(\dot{x}_L^2 + \dot{y}_L^2 + \dot{z}_L^2\right) \tag{26}$$

Substituting Equations (24) and (25) in Equation (26) yields, after rearrangement

$$E_{KL} = \frac{1}{2}m\|\dot{\mathbf{p}}_T\|^2 + \frac{n_{Tz}}{n_{T,xy}}mg\Delta xy_{LT} + \frac{1}{2}mg^2\frac{\Delta xy_{LT}^2}{\|\dot{\mathbf{p}}_T\|^2 n_{T,xy}^2} \tag{27}$$

The variation of the potential energy as the EE moves from center point P_C to launch point P_L is

$$\Delta E_P = E_{PL} - E_{PC} = mg\left(z_L - z_C\right) \tag{28}$$

As the EE moves from P_C to P_L, the total energy provided by the motors is then at least

$$\Delta E_L = E_{KL} - E_{KC} + E_{PL} - E_{PC} = E_{KL} + \Delta E_P \tag{29}$$

Another possible approach is to consider the energy variation ΔE_{max} from P_C to the point on Γ that has the maximum value of ΔE, instead of the energy variation from P_C to P_L: ΔE_{max} can be of interest for the optimization, as the motors will have to provide that amount of energy to the EE. The total energy variation at any point on Γ (with respect to the start point P_C) can be expressed as

$$\Delta E = \frac{1}{2}m\|\dot{\mathbf{p}}\|^2 + mg\left(z - z_C\right) \tag{30}$$

and thus, by differentiating with respect to time

$$\frac{d}{dt}\Delta E = m\dot{p}\ddot{p} + mg\dot{z} = 0 \tag{31}$$

Here, $\dot{p} = \|\dot{\mathbf{p}}\|$ and $\ddot{p} = \|\ddot{\mathbf{p}}\|$, while $\dot{z}$ is the component of $\dot{\mathbf{p}}$ along the z axis. Substituting Equation (6) into Equation (31) and introducing $s = \tan\left(\frac{\omega t}{2}\right)$ yields

$$c_0 + c_1 s + c_2 s^2 + c_3 s^3 + c_4 s^4 = 0 \tag{32}$$

where

$$
\begin{aligned}
c_0 &= g\left(z_{rc} + z_{rs}\right) \\
c_1 &= 2\omega^2\left(x_{rc}^2 + y_{rc}^2 + z_{rc}^2 - x_{rs}^2 - y_{rs}^2 - z_{rs}^2\right) - 2gz_{rc} \\
c_2 &= 4\omega^2\left(x_{rc}x_{rs} + y_{rc}y_{rs} + z_{rc}z_{rs}\right) - 2gz_{rc} \\
c_3 &= 2\omega^2\left(x_{rs}^2 + y_{rs}^2 + z_{rs}^2 - x_{rc}^2 - y_{rc}^2 - z_{rc}^2\right) - 2gz_{rc} \\
c_4 &= g\left(z_{rc} - z_{rs}\right)
\end{aligned}
\tag{33}
$$

Equation (32) is a fourth-degree polynomial equation, which can be solved either analytically or numerically for the unknown variable s; from the four roots of s we get four possible values of $\alpha = \omega t$ and, thus, four local extrema of $\Delta E(\alpha)$, the largest of which is the maximum energy ΔE_{max} along the trajectory.

In defining our objective function (see Equation (23)), E can be defined as either the maximum energy ΔE_{max} seen above or as ΔE_L in Equation (29), according to the designer's preference. In the following, we use the definition in Equation (29), for simplicity.

From Equation (8), one can see that, when $\Delta_{xy,LT}$ and $\|\dot{\mathbf{p}}_T\|$ are determined, $\mathbf{p}_L$ can also be found from $\mathbf{p}_T$ and $\mathbf{n}_T$; then, one can also find $\dot{\mathbf{p}}_L$ by using Equation (24). We can show, from Equations (25) and (27)–(29), that the time to reach P_T and the launch energy do not depend on x_C and y_C; the same is also true if we consider the maximal energy obtained from Equation (30) and the solution of Equation (32). Thus, we may assume the coordinates x_C and y_C to be given, while we seek to optimize the starting height z_C. The optimal path Γ_{opt} is uniquely defined from $\mathbf{p}_C$, $\mathbf{p}_L$, the direction of $\dot{\mathbf{p}}_L$ and the linear eccentricity ε of Γ_{opt}; ε is defined as the distance from the center of the ellipse to either of its two foci. The frequency ω can then be found from Γ_{opt}, $\mathbf{p}_L$ and $\|\dot{\mathbf{p}}_L\|$. The optimal choice for $\Delta_{xy,LT}$, $\|\dot{\mathbf{p}}_T\|$, z_C and ε is the one that minimizes F.

When optimizing the objective function F, we may also wish to set some constraints on the trajectory. First, we limit the EE motion within a rectangular cuboid workspace defined as

$$
\begin{aligned}
X_{min} &< x < X_{max} \\
Y_{min} &< y < Y_{max} \\
Z_{min} &< z < Z_{max}
\end{aligned}
\tag{34}
$$

In a practical application, we also have to limit the torques and velocities at the robot's actuators: these are proportional, respectively, to the cable tensions and cable velocities, if the robot uses cable winches controlled by rotary motors (as is common for CDPRs [34]). So we set constraints directly on the tensions and velocities, as

$$
\begin{aligned}
\max\{\tau_1, \tau_2, \tau_3\} &< \tau_{max} \\
\max\{\dot{\rho}_1, \dot{\rho}_2, \dot{\rho}_3\} &< \dot{\rho}_{max}
\end{aligned}
\tag{35}
$$

where τ_{max} and $\dot{\rho}_{max}$ depend on the robot's motors; for instance, denoting the maximum torque and maximum angular speed of the motors as, respectively, T_M and Ω_M, and the radius of the winches that actuate the cables as R, one has $\tau_{max} = T_M/R$ and $\dot{\rho}_{max} = \Omega_M R$ (assuming that the motors and the winches of all cables are identical). The cable tensions in Equation (35) are found from Equations (4) and (5), while the velocities are found by differentiating Equation (1) with respect to time.

Finally, when a CSPR moves along a dynamical trajectory, cable tensions must also be ensured to be always positive. For a given Γ, the feasible range $[\omega_{min}, \omega_{max}]$ for the trajectory frequency ω can be calculated by the method presented in [7]. If (and only if) the value of ω is in this range, then cable tensions τ_i are positive both before and after launch, since the method in [7] does not depend on the EE mass m, which in this case changes instantaneously at $t = t_L$. Therefore, the optimization problem also requires the constraint

$$\omega_{min} < \omega < \omega_{max} \tag{36}$$

4.2. Numerical Example

We now present a numerical example (in the following, time, lengths and forces are given in seconds, meters and newtons, respectively). We set

$$\mathbf{p}_T = \begin{bmatrix} 2 \\ 5 \\ -5 \end{bmatrix}, \quad \mathbf{n}_T = \begin{bmatrix} 1 \\ 4 \\ -8 \end{bmatrix}, \quad \mathbf{a}_1 = \begin{bmatrix} 2 \\ 1 \\ 1 \end{bmatrix}, \quad \mathbf{a}_2 = \begin{bmatrix} -3 \\ -2 \\ 0 \end{bmatrix}, \quad \mathbf{a}_3 = \begin{bmatrix} -1 \\ 3 \\ -1 \end{bmatrix} \tag{37}$$

and the maximum magnitude of the velocity at the target point as $\dot{p}_{max} = 40$. For a realistic choice of motors and winch radius, we set $\tau_{max} = 8$ and $\dot{\rho}_{max} = 7$. The mass m of the EE before and after launch is $m_1 = 1$ and $m_2 = 0.6$, respectively.

We seek the optimal trajectory under the following workspace constraints:

$$\begin{aligned} X_{min} &= -4, & X_{max} &= 4 \\ Y_{min} &= -5, & Y_{max} &= 5 \\ Z_{min} &= -5, & Z_{max} &= 0 \end{aligned} \tag{38}$$

We also assign the x and y coordinates of trajectory center as $x_C = -1$, $y_C = 1$.

Several algorithms may be used to solve the optimization problem: we chose to use a genetic algorithm with the MATLAB function `gamultiobj`, setting an initial population of 1500 individuals and a maximum of 5000 generations. The algorithm converges within the specified number of generations with a function tolerance of 10^{-5}; all solutions also respect the constraints set in Equations (34)–(36). The result of the optimization is a Pareto front that displays the relationship between the first and the second element in the vector-valued function F. In Figure 3, every point represents a potential solution on the Pareto front for the two objectives to be minimized E and Δt_{LT}. As expected, we can see that we cannot minimize both objectives at the same time.

We also show an example solution from the Pareto front, defined by

$$\mathbf{p}_L = \begin{bmatrix} 1.954 \\ 0.815 \\ -1.693 \end{bmatrix}, \quad \mathbf{p}_C = \begin{bmatrix} -1 \\ 1 \\ -2.43 \end{bmatrix}, \quad \mathbf{b} = \begin{bmatrix} -2.94 & 0.599 \\ -0.599 & -2.02 \\ -1.01 & -0.547 \end{bmatrix}, \quad \omega = 1.964 \tag{39}$$

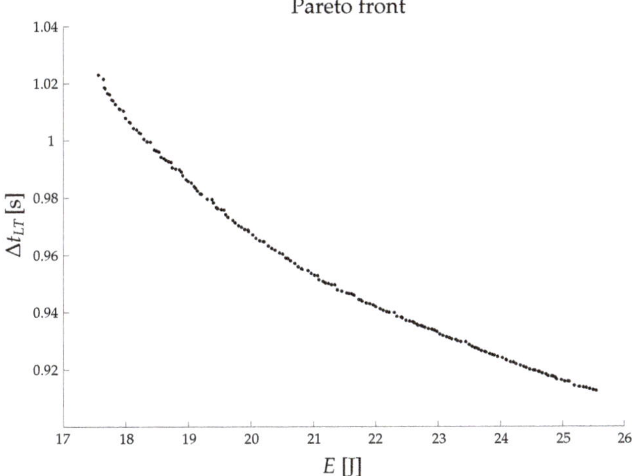

Figure 3. Pareto front for the two objectives E and Δt_{LT}.

Figure 4a,b show the optimal trajectory and the corresponding cable tensions; we notice that the cable tensions are discontinuous at the launch instant, as the EE mass m suddenly changes.

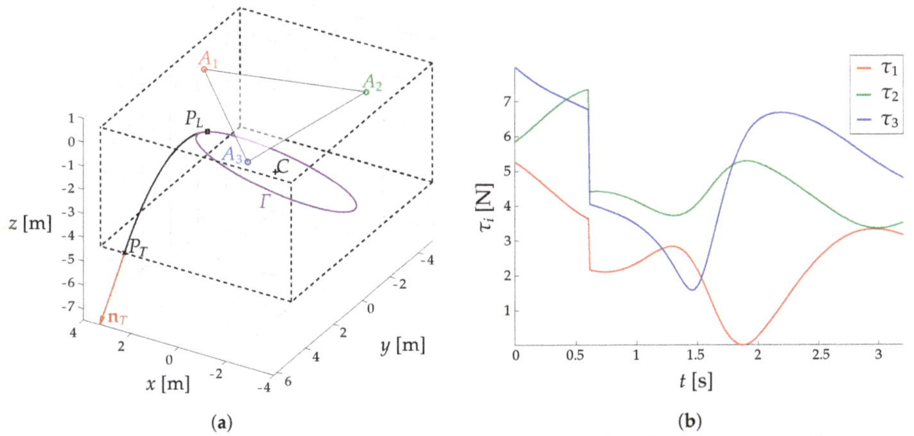

(a) (b)

Figure 4. (a) Optimal trajectory: 3D view of the path Γ_{opt}. (b) The cable tensions along Γ_{opt}.

5. Cable Elasticity

So far we have considered cables to be inextensible. With this model, it can be clearly seen from Figure 4b that, at the launch instant, tensions suddenly change due to the instantaneous variation of the EE mass m. Though inextensible-cable models can be very useful for predicting the dynamic feasibility of a trajectory for a CDPR [3–10], tension discontinuity is unrealistic. Thus, we consider now a more realistic model where cables behave as linear elastic springs: by this second approach, tension discontinuities disappear, but the sudden variation of the gravity load applied on the EE causes oscillations.

We introduce $\mathbf{p}_I(t) = [x_I(t), y_I(t), z_I(t)]^T$ and $\mathbf{p}_A(t) = [x_A(t), y_A(t), z_A(t)]^T$ as the positions of the EE as it moves along the ideal (Γ_I) and the actual (Γ_A) trajectory, respectively: that is, Γ_I is the trajectory defined in the planning phase considering ideal inextensible cables, while Γ_A is the trajectory obtained when elastic effects are introduced. Accordingly, cable lengths become

$$\rho_{i,A} = \|\mathbf{p}_A - \mathbf{a}_i\|, \quad \rho_{i,I} = \|\mathbf{p}_I - \mathbf{a}_i\|, \quad i = 1, 2, 3 \tag{40}$$

We also define $\Delta\rho_i = \rho_{i,A} - \rho_{i,I}$. We use the following linear elastic model:

$$\tau_{i,A} = \begin{cases} \frac{K_0}{\rho_{i,I}} \Delta\rho_i, & \Delta\rho_i \geq 0 \\ 0, & \Delta\rho_i < 0 \end{cases}, \quad i = 1, 2, 3 \tag{41}$$

where K_0 is the stiffness of a unit length of cable (assumed equal for all cables).

The cable-vector directions are

$$\mathbf{e}_{i,A} = \frac{\mathbf{p}_A - \mathbf{a}_i}{\rho_{i,A}}, i = 1, 2, 3 \tag{42}$$

and the actual structure matrix is

$$\mathbf{M}_A = \begin{bmatrix} \mathbf{e}_{1,A} & \mathbf{e}_{2,A} & \mathbf{e}_{3,A} \end{bmatrix} \tag{43}$$

Since any real system is affected by energy dissipation, damping needs to be taken into account to obtain a realistic model. A complete model of all dissipative forces in a real system is beyond the scope of the present work; therefore, for simplicity we group all damping effects in a linear term, as

$$\mathbf{F}_d = K_d \dot{\mathbf{p}}_A \tag{44}$$

where K_d is the damping coefficient and $\dot{\mathbf{p}}_A$ is the velocity of the EE as it moves along the actual trajectory. This dissipative force could be due, for instance, to air drag. So, the general equilibrium equation for the elastic model is

$$\mathbf{M}_A \boldsymbol{\tau}_A = m \left(\mathbf{g} - \ddot{\mathbf{p}}_A \right) - \mathbf{F}_d \tag{45}$$

where $\boldsymbol{\tau}_A = [\tau_{1,A}, \tau_{2,A}, \tau_{3,A}]^T$.

We can then combine Equations (41), (44) and (45) to obtain a set of ordinary differential equations. For given initial values of $\mathbf{p}_A(t)$ and $\dot{\mathbf{p}}_A(t)$, this set can be numerically solved to yield the actual trajectory and actual cable tensions.

To provide an example, we use the optimal trajectory parameters found in Section 4.2, as expressed in Equations (37) and (39); here, the robot starts moving from rest at C with a transition trajectory, reaches the target trajectory Γ and finally performs the launch. The other parameters are set as $K_0 =$ 10,000 N and $K_d = 0.01$ Ns/m; the value of K_0 is obtained considering polymer cables having diameter ≈ 1 mm (using the data presented in [45] for reference), while the value of K_d is only a reasonable parameter for simulation. The results are shown in Figures 5–7.

In Figure 5, the continuous lines represent the actual trajectory, while the dashed lines are the ideal trajectory. In Figure 7, the black crosses are the target point's coordinates along the coordinate axes. The continuous line is the actual trajectory when considering an elastic-cable model. The two figures clearly show that even when considering elastic cables the launched object can be thrown acceptably close to the target point. We can also see in Figure 6 that the cable tensions are continuous before and after launch, although they briefly reach zero with the elastic model. Significant tension oscillations are triggered at the launch instant $t = t_L$, which can lead to losing tension in some cables: this could affect the robot stability after launch.

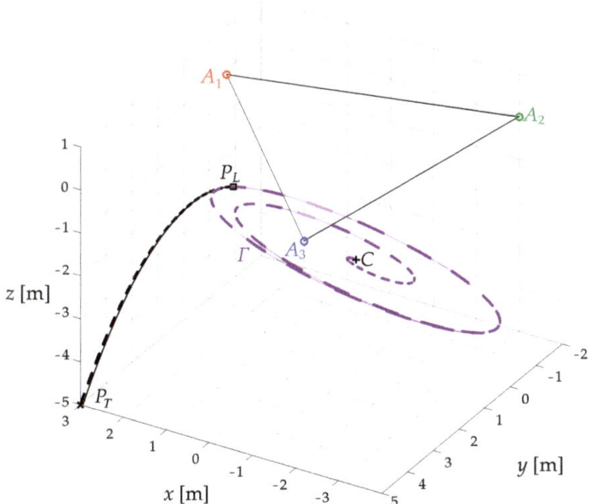

Figure 5. Ideal trajectory (dashed line) and actual trajectory (continuous).

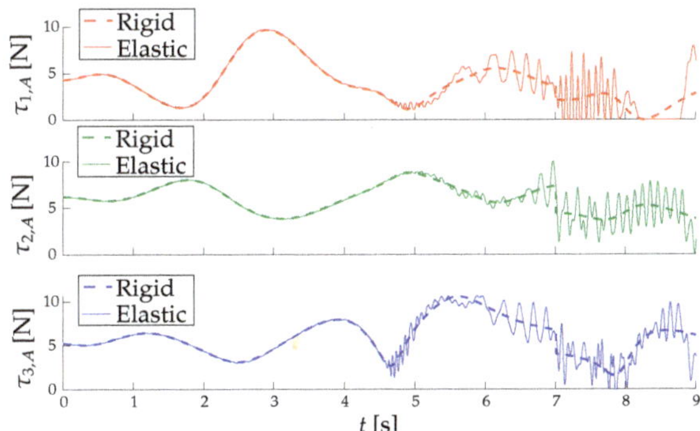

Figure 6. Plots of the cable tensions over time, with the inextensible model (dashed line) and elastic model (continuous).

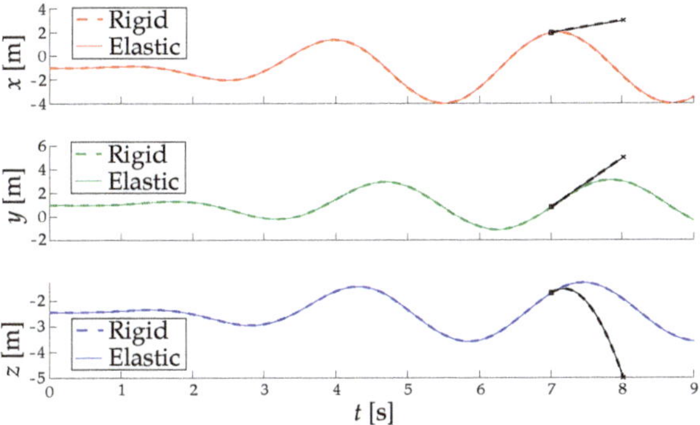

Figure 7. Coordinates of the end-effector (EE) (in color) and of the launched object (black) with both models, along the three axes.

6. Conclusions

In this paper, we considered general elliptical trajectories for a 3-cable suspended robot with a point-mass EE and we showed how these trajectories can be applied for launching a mass from the EE to a target point. We provided analytical conditions for the object to reach its target and we optimized the motion in terms of two parameters of interest (namely, energy consumption and total flight time) by using a genetic algorithm. Constraints on the cable tensions were set to guarantee that the trajectories are feasible, both before and after launch. Notably, the results presented for trajectory planning can be extended to a 6-cable suspended robot with parallelogram-type actuation [10,11]. We also verified the motion feasibility by numerical simulations including a model of cable elasticity.

The goal of our work is to expand the capabilities of CDPRs in terms of the types of motion that can be achieved and the total workspace that can be reached. Directions for future work include:

- optimizing the robot trajectory in order to reduce the discontinuity (in the rigid-body model) of cable tensions at the launch instant, which may cause oscillations after launch;
- studying tension fluctuations that are triggered, in the elastic model, even before the launch (see Figure 6);
- analyzing the sensitivity of the final position of the launched object to errors in the actual launch position and velocity with respect to their ideal values, similarly to the works presented in [11,27];
- considering the effect of air drag; the trajectory of a point-mass object in free-flight through a viscous medium is described by differential equations that generally do not allow closed-form solutions, but some authors [29] propose tractable models that approximate the real solution;
- experimentally validating our results by tests on a prototype (currently under development).

Author Contributions: Conceptualization, D.L., G.M. and M.C.; software, D.L. and G.M.; methodology, D.L. and G.M.; writing—original draft preparation, D.L.; writing—review and editing, D.L., G.M., M.C. and X.J.; visualization, D.L. and G.M.; supervision, M.C. and X.J.; project administration, M.C. and X.J.; funding acquisition, X.J. All authors have read and agreed to the published version of the manuscript.

Funding: This research was funded by the National Natural Science Foundation of China (NSFC) under grants 51525504 and 51805486.

Acknowledgments: The authors wish to thank Qinchuan Li for his support throughout the project. We also thank Edoardo Idà for his advice regarding CSPRs and their actuators.

Abbreviations

The following abbreviations are used in this manuscript:

CDPR	Cable-Driven Parallel Robot
CSPR	Cable-Suspended Parallel Robot
EE	End-Effector
DoF	Degree of Freedom
SEW	Static Equilibrium Workspace
DW	Dynamic Workspace

References

1. Riechel, A.T.; Ebert-Uphoff, I. Force-feasible workspace analysis for underconstrained point-mass cable robots. In Proceedings of the IEEE 2004 ICRA, New Orleans, LA, USA, 26 April–1 May 2004; Volume 5, pp. 4956–4962. [CrossRef]
2. Barrette, G.; Gosselin, C.M. Determination of the dynamic workspace of cable-driven planar parallel mechanisms. *ASME J. Mech. Des.* **2005**, *127*, 242–248. [CrossRef]

3. Zhang, N.; Shang, W. Dynamic trajectory planning of a 3-DOF under-constrained cable-driven parallel robot. *Mech. Mach. Theory* **2016**, *98*, 21–35. [CrossRef]

4. Zhang, N.; Shang, W.; Cong, S. Geometry-based trajectory planning of a 3-3 cable-suspended parallel robot. *IEEE Trans. Robot.* **2016**, *33*, 484–491. [CrossRef]

5. Zhang, N.; Shang, W.; Cong, S. Dynamic trajectory planning for a spatial 3-DoF cable-suspended parallel robot. *Mech. Mach. Theory* **2018**, *122*, 177–196. [CrossRef]

6. Jiang, X.; Gosselin, C.M. Dynamically feasible trajectories for three-DOF planar cable-suspended parallel robots. In Proceedings of the ASME 2014 IDETC/CIE, Buffalo, NY, USA, 17–20 August 2014; Volume 5A, p. V05AT08A085. [CrossRef]

7. Mottola, G.; Gosselin, C.M.; Carricato, M. Dynamically feasible periodic trajectories for generic spatial three-degrees-of-freedom cable-suspended parallel robots. *ASME J. Mech. Rob.* **2018**, *10*, 031004. [CrossRef]

8. Gosselin, C.M. Global planning of dynamically feasible trajectories for three-DOF spatial cable-suspended parallel robots. In *Cable-Driven Parallel Robots*; Mechanisms and Machine Science; Bruckmann, T., Pott, A., Eds.; Springer: Berlin/Heidelberg, Germany, 2013; Volume 12, pp. 3–22. [CrossRef]

9. Shao, Z.; Li, T.; Tang, X.; Tang, L.; Deng, H. Research on the dynamic trajectory of spatial cable-suspended parallel manipulators with actuation redundancy. *Mechatronics* **2018**, *49*, 26–35. [CrossRef]

10. Mottola, G.; Gosselin, C.M.; Carricato, M. Dynamically feasible motions of a class of purely-translational cable-suspended parallel robots. *Mech. Mach. Theory* **2019**, *132*, 193–206. [CrossRef]

11. Mottola, G.; Gosselin, C.M.; Carricato, M. Effect of actuation errors on a purely-translational spatial cable-driven parallel robot. In Proceedings of the IEEE 9th Annual International Conference on CYBER Technology in Automation, Control, and Intelligent Systems (CYBER), Suzhou, China, 29 July–2 August 2019; pp. 701–707. [CrossRef]

12. Saber, O.; Zohoor, H. Workspace analysis of a cable-suspended robot with active-passive cables. In Proceedings of the ASME 2013 IDETC/CIE, Portland, OR, USA, 4–7 August 2013; Volume 6A, p. V06AT07A071. [CrossRef]

13. Mason, M.T.; Lynch, K.M. Dynamic manipulation. In Proceedings of the IEEE/RSJ 1993 IROS, Yokohama, Japan, 26–30 July 1993; Volume 1, pp. 152–159. [CrossRef]

14. Lynch, K.M.; Mason, M.T. Dynamic nonprehensile manipulation: controllability, planning, and experiments. *Int. J. Rob. Res.* **1999**, *18*, 64–92. [CrossRef]

15. Tabata, T.; Aiyama, Y. Tossing manipulation by 1 degree-of-freedom manipulator. In Proceedings of the IEEE/RSJ 2001 IROS, Maui, HI, USA, 29 October–3 November 2001; Volume 1, pp. 132–137, [CrossRef]

16. Miyashita, H.; Yamawaki, T.; Yashima, M. Control for throwing manipulation by one joint robot. In Proceedings of the IEEE 2009 ICRA, Kobe, Japan, 12–17 May 2009; pp. 1273–1278. [CrossRef]

17. Reist, P.; D'Andrea, R. Design and analysis of a blind juggling robot. *IEEE Trans. Robot.* **2012**, *28*, 1228–1243. [CrossRef]

18. Ruggiero, F.; Lippiello, V.; Siciliano, B. Nonprehensile dynamic manipulation: A survey. *IEEE Robot. Autom. Lett.* **2018**, *3*, 1711–1718. [CrossRef]

19. Sakaguchi, T.; Fujita, M.; Watanabe, H.; Miyazaki, F. Motion planning and control for a robot performer. In Proceedings of the IEEE 1993 ICRA, Atlanta, GA, USA, 2–6 May 1993; Volume 3, pp. 925–931. [CrossRef]

20. Mettin, U.; Shiriaev, A.S.; Freidovich, L.B.; Sampei, M. Optimal ball pitching with an underactuated model of a human arm. In Proceedings of the IEEE 2010 ICRA, Anchorage, AL, USA, 3–7 May 2010; pp. 5009–5014. [CrossRef]

21. Yedeg, E.L.; Wadbro, E. State constrained optimal control of a ball pitching robot. *Mech. Mach. Theory* **2013**, *69*, 337–349. [CrossRef]

22. Shoji, T.; Katsumata, S.; Nakaura, S.; Sampei, M. Throwing motion control of the springed Pendubot. *IEEE Trans. Control Syst. Technol.* **2013**, *21*, 950–957. [CrossRef]

23. Senoo, T.; Namiki, A.; Ishikawa, M. High-speed throwing motion based on kinetic chain approach. In Proceedings of the IEEE/RSJ 2008 IROS, Nice, France, 22–26 September 2008; pp. 3206–3211. [CrossRef]

24. Zeng, A.; Song, S.; Lee, J.; Rodriguez, A.; Funkhouser, T. TossingBot: learning to throw arbitrary objects with residual physics. *IEEE Trans. Robot.* **2020**, *36*, 1307–1319. [CrossRef]

25. Bätz, G.; Lee, K.K.; Wollherr, D.; Buss, M. Robot basketball: A comparison of ball dribbling with visual and force/torque feedback. In Proceedings of the IEEE 2009 ICRA, Kobe, Japan, 12–17 May 2009; pp. 514–519. [CrossRef]

26. Sintov, A.; Shapiro, A. A stochastic dynamic motion planning algorithm for object-throwing. In Proceedings of the IEEE 2015 ICRA, Seattle, WA, USA, 26–30 May 2015; pp. 2475–2480. [CrossRef]

27. Okada, M.; Pekarovskiy, A.; Buss, M. Robust trajectory design for object throwing based on sensitivity for model uncertainties. In Proceedings of the IEEE 2015 ICRA, Seattle, WA, USA, 26–30 May 2015; pp. 3089–3094. [CrossRef]

28. Frank, H.; Wellerdick-Wojtasik, N.; Hagebeuker, B.; Novak, G.; Mahlknecht, S. Throwing objects—A bio-inspired approach for the transportation of parts. In Proceedings of the IEEE 2006 ROBIO, Kunming, China, 17–20 December 2006; pp. 91–96. [CrossRef]

29. Frank, T.; Janoske, U.; Mittnacht, A.; Schroedter, C. Automated throwing and capturing of cylinder-shaped objects. In Proceedings of the IEEE 2012 ICRA, Saint Paul, MN, USA, 14–18 May 2012; pp. 5264–5270. [CrossRef]

30. Barteit, D.; Frank, H.; Kupzog, F. Accurate prediction of interception positions for catching thrown objects in production systems. In Proceedings of the IEEE 2008 6th INDIN, Daejeon, Korea, 13–16 July 2008; pp. 893–898. [CrossRef]

31. Arisumi, H.; Kotoku, T.; Komoriya, K. A study of casting manipulation (swing motion control and planning of throwing motion). In Proceedings of the IEEE/RSJ 1997 IROS, Grenoble, France, 11 September 1997; Volume 1, pp. 168–174. [CrossRef]

32. Arisumi, H.; Kotoku, T.; Komoriya, K. Swing motion control of casting manipulation. *IEEE Contr. Syst. Mag.* **1999**, *19*, 56–64. [CrossRef]

33. Arisumi, H.; Yokoi, K.; Komoriya, K. Casting manipulation—Midair control of a gripper by impulsive force. *IEEE Trans. Robot.* **2008**, *24*, 402–415. [CrossRef]

34. Fagiolini, A.; Torelli, A.; Bicchi, A. Casting robotic end-effectors to reach far objects in space and planetary missions. In Proceedings of the 9th ESA Workshop Advanced Space Technologies for Robotics and Automation, ASTRA 2006, Noordwijk, The Netherlands, 28–30 November 2006.

35. Fagiolini, A.; Belo, F.A.W.; Catalano, M.G.; Bonomo, F.; Alicino, S.; Bicchi, A. Design and control of a novel 3D casting manipulator. In Proceedings of the IEEE 2010 ICRA, Anchorage, AL, USA, 3–7 May 2010; pp. 4169–4174. [CrossRef]

36. Arisumi, H.; Otsuki, M.; Nishida, S. Launching penetrator by casting manipulator system. In Proceedings of the IEEE/RSJ 2012 IROS, Vilamoura, Portugal, 7–12 October 2012; pp. 5052–5058. [CrossRef]

37. Tsukagoshi, H.; Watari, E.; Fuchigami, K.; Kitagawa, A. Casting device for search and rescue aiming higher and faster access in disaster site. In Proceedings of the IEEE/RSJ 2012 IROS, Vilamoura, Portugal, 7–12 October 2012; pp. 4348–4353. [CrossRef]

38. Hill, L.; Woodward, T.; Arisumi, H.; Hatton, R.L. Wrapping a target with a tethered projectile. In Proceedings of the IEEE 2015 ICRA, Seattle, WA, USA, 26–30 May 2015; pp. 1442–1447. [CrossRef]

39. Hatakeyama, T.; Mochiyama, H. Shooting manipulation inspired by chameleon. *IEEE Trans. Mechatron.* **2013**, *18*, 527–535. [CrossRef]

40. Pongratz, M.; Pollhammer, K.; Szep, A. KOROS initiative: automatized throwing and catching for material transportation. In *Leveraging Applications of Formal Methods, Verification, and Validation (ISoLA 2011)*; Communications in Computer and Information Science; Hähnle, R., Knoop, J., Margaria, T., Schreiner, D., Steffen, B., Eds.; Springer: Berlin/Heidelberg, Germany, 2011; Volume 336, pp. 136–143. [CrossRef]

41. Chen, H.; Zhang, B.; Fuhlbrigge, T. Robot throwing trajectory planning for solid waste handling. In Proceedings of the IEEE 9th Annual International Conference on CYBER Technology in Automation, Control, and Intelligent Systems (CYBER), Suzhou, China, 29 July–2 August 2019; pp. 1372–1375. [CrossRef]

42. Raptopoulos, F.; Koskinopoulou, M.; Maniadakis, M. Robotic pick-and-toss facilitates urban waste sorting. In Proceedings of the IEEE 16th International Conference on Automation Science and Engineering (CASE), Hong Kong, China, 20–21 August 2020; pp. 1149–1154. [CrossRef]

43. Jiang, X. Dynamic Trajectory Planning and Synthesis for Fully Actuated Cable-Suspended Parallel Robots. Ph.D. Thesis, Université Laval, Québec, QC, Canada, 2017.

44. Cveticanin, L. Dynamics of Bodies with Time-Variable Mass. In *Mathematical and Analytical Techniques with Applications to Engineering*; Springer: Cham, Switzerland, 2016. [CrossRef]

45. Pott, A. Cable-driven parallel robots: theory and application, In *Springer Tracts in Advanced Robotics*; Springer: Cham, Switzerland, 2018; Volume 120. [CrossRef]

Publisher's Note: MDPI stays neutral with regard to jurisdictional claims in published maps and institutional affiliations.

Article

Performance Investigation of Integrated Model of Quarter Car Semi-Active Seat Suspension with Human Model

Saransh Jain [1], Shubham Saboo [1], Catalin Iulian Pruncu [2,3,*] and Deepak Rajendra Unune [1,4]

[1] Department of Mechanical-Mechatronics Engineering, The LNM Institute of Information Technology, Jaipur 302031, India; saransh.j1997@gmail.com (S.J.); 16ume048@lnmiit.ac.in (S.S.); deepunune@gmail.com (D.R.U.)

[2] Department of Mechanical Engineering, Imperial College London, Exhibition Rd., London SW7 2AZ, UK

[3] Department of Mechanical Engineering, School of Engineering, University of Birmingham, Birmingham B15 2TT, UK

[4] Department of Materials Science and Engineering, INSIGNEO Institute for in silico Medicine, University of Sheffield, Sheffield S1 3JD, UK

* Correspondence: c.pruncu@imperial.ac.uk

Received: 24 March 2020; Accepted: 29 April 2020; Published: 2 May 2020

Abstract: In this paper, an integrated model of a semi-active seat suspension with a human model over a quarter is presented. The proposed eight-degrees of freedom (8-DOF) integrated model consists of 2-DOF for the quarter car model, 2-DOF for the semi-active seat suspension and 4-DOF for the human model. A magneto-rheological (MR) damper is implemented for the seat suspension. The fuzzy logic-based self-tuning (FLST) proportional–integral–derivative (PID) controller allows to regulate the controlled force on the basis of sprung mass velocity error and its derivative as input. The controlled force is tracked by the Heaviside step function which determines the supply voltage for the MR damper. The performance of the proposed integrated model is analysed, in-terms of human head accelerations, for several road profiles and at different speeds. The performance of the semi-active seat suspension is compared with the traditional passive seat suspension to validate the effectiveness of the proposed integrated model with a semi-active seat suspension. The simulation results show that the semi-active seat suspension improves the ride comfort significantly by reducing the head acceleration effectively compared to the passive seat suspension.

Keywords: semi-active seat suspension; integrated model; control; fuzzy logic-based self-tuning; PID

---

1. Introduction

Vibration created due to the unevenness of the road profile causes serious problems for the driver and the passengers. Especially, the vibration transmitted from the road to the human body at a low frequency causes damage to the spine in the long run. Vibrations create digestion problems, fatigue and discomfort while driving. Therefore, the seat suspension is mainly used to improve the driver's comfort and reduce the intensity of the vibration transmitting to the human body.

A seat suspension system is broadly classified into three types, i.e., the passive seat suspension, semi-active seat suspension and active seat suspension. In the passive suspension, the ride comfort cannot be achieved without compromising the road-handling capabilities, as mentioned by Qin et al. [1]. The fixed characteristics of the passive seat suspension decrease the versatility and prevent the damper from varying according to the working conditions. Unaune et al. [2] revealed the design constraints associated with a passive suspension system using the quarter car suspension model simulations in MATLAB. A passive seat suspension also has similar design constraints and requires an optimum design of parameter-bound work within a limited frequency range [3]. The active and semi-active seat suspension outperforms the passive suspension and overcomes the inadequacies of the passive seat suspension. The active seat suspension is comprised of an actuator which generates the force regulated by using the control law driven by signals measured with the help of the sensors. The controller controls the actuators to generate an adequate amount of force or dissipate the energy from the system. In spite of the great accuracy and the performance of the active seat suspension, disadvantages such as a high capital and running costs, a complicated design and the power requirements make it less popular and redundant for commercial vehicle application [4]. The semi-active seat suspension was used widely because of its characteristics such as the semi-active suspension that does not require high power, while varying the damping force is one of the key features. Another crucial feature of the semi-active suspension is that it can even work as the passive suspension when the controller fails [5]. Therefore, the semi-active seat suspension has the features of both the active and passive seat suspensions. The entirely new segment of the smart dampers provides wide applications in the industries owing to their versatility and promising performance as compared to the conventional oil-filled dampers. The magneto-rheological (MR) and electro-rheological (ER) dampers belong to the segment of smart dampers, as they can give a quick response and can change the damping in just a few milliseconds, as per real-time requirements [6,7]. MR dampers are highly reliable and can easily be maintained [8]. Due to these reasons, semi-active seat suspension is considered in this work.

The performance of semi-active seat suspension has been studied by many researchers in the past. Choi and Han [9] showed the performance of a robust sliding mode control on the electrorheological seat suspension with the human model. Bae et al. [10] implemented the integrated semi-active seat suspension with the MR damper. One of the authors of this study has previously investigated the performance of semi-active suspension systems with different control strategies such as skyhook, groundhook and hybrid and reported the suitability of the control strategies for the desired application [11]. Advanced and intelligent controls, e.g., fuzzy logic control (FLC), artificial neural networks (ANN), adaptive neuro-fuzzy inference system (ANFIS), evolving radial basis function networks and so forth have attracted the minds of researchers to controlled suspensions, owing to their better control performance [12]. Mulla and Unune [13] compared FLC and ANFIC control strategies with a traditional skyhook control for a semi-active suspension under International Organization for Standardization (ISO) road disturbance.

Singh and Aggarwal [14] analysed the passenger seat vibration of the semi-active quarter car model with a hybrid Fuzzy-PID control. They reported that the proposed Fuzzy-PID approach offered a better performance in terms of passenger seat acceleration and displacement. Typical studies report the study of the active and semi-active seat suspension separately from the active and semi-active suspension of the vehicle. Bhattacharjee et al. [15] applied the chaotic fruit fly algorithm for the tuning of PID. Choi and Han [9] and Bae and Kang [16] have studied the seat suspension with the human model without considering the vehicle suspension model. Gohari and Tahmasebi [17] implemented the active control on the seat suspension while analysing the performance with various road disturbances, but only the seat suspension was considered. Swethamarai and Lakshmi [18] implemented a Fuzzy-PID controller for the quarter car active suspension system with 3-DOF and reported that the Fuzzy-PID control reduced the driver's body acceleration better than Fuzzy and PID controls, thus improving the ride comfort. However, they did not consider the seat suspension and human model in detail and they implemented Fuzzy-PID control for active suspension. Metered and Šika [19] proposed a semi-active MR damper for the truck seat suspension. They used FLC to calculate the desired force of the MR damper, and a significant improvement in the ride comfort was observed. However, the approach of considering only the seat suspension with the human model while analysing the ride comfort is debatable. The study of an integrated model of the seat suspension, integrated with both the vehicle suspension and human model, is more pragmatic. In the integrated model, a semi-active seat suspension is considered to be mounted over the quarter car suspension model along and the human model. Rare work is available which considers the integrated model of the seat suspension with the human model and the vehicle suspension. Du et al. [20] proposed the integrated model of the seat suspension with the human model. Rajendrian and Lakshmi [21] implemented the fractional order terminal sliding mode controller (SMC) to reduce the head acceleration when considering the half car suspension model. Nevertheless, the analysis was done using active control in Reference [20,21]. As previously mentioned, the semi-active seat suspensions are the more preferred choice, as compared to the active seat suspensions. This encouraged current work to implement the integrated semi-active seat suspension with the human model.

The ride comfort of the rider is typically determined in terms of the acceleration of the human body in the vertical direction [20]. To achieve a better ride comfort performance, an integrated semi-active seat suspension had been modelled including a 2-DOF semi-active seat suspension which is mounted over the 2-DOF quarter-car model and 4-DOF human model. The integrated model of the semi-active seat suspension is more practical because the road disturbances are altered by the vehicle suspension before coming to the cabin floor. Previous work which considered the direct supply of the road input to the cabin floor for analysing the performance of the seat suspension seems to be inappropriate and may provide inaccurate results. Wang et al. [22] implemented the active seat suspension, however, they considered the human body and seat as a single entity. Similarly, Sathishkumar et al. [23] considered only the seat while analysing the effect of semi-active force control in their semi-active seat suspension model. Thus, it further becomes important to integrate the human body model to get more accurate results, instead of just analysing the active or semi-active seat suspension.

From the literature review, it was observed that rare work is reported on the study of the integrated model of the semi-active seat suspension, consisting of both the vehicle suspension and the human models. Further, no work is available that studies the semi-active suspension system with such an integrated model. Further, the implementation of ISO road profiles while investigating such systems is missing. Thus, in this work, an integrated model of seat suspension comprising the semi-active quarter car suspension and human model is proposed. The MR damper, governed by the fuzzy logic-based self-tuning PID controller (FLST-PID), is implemented in the seat suspension. The FLST-PID determines the desired force and then it is compared with the force generated by the MR damper, and on the basis of the comparison when the MR damper, is switched either on or off. The FLST-PID controller is designed under the constraints of the suspension deflection limitation. Numerical simulations are performed to evaluate the performance of the integrated model on the various types of road profiles including the bump and ISO road profiles of different grades such as C, D and E-grades. Finally, the effectiveness of the semi-active suspension with the MR dampers is compared with the passive suspension on the bump and ISO road profile with different speeds that vary from 30 km/h to 120 km/h.

2. Integrated Seat Suspension and MR Damper Model

The proposed integrated semi-active seat suspension model, as shown in Figure 1, consists of the quarter-car suspension model and the semi-active seat suspension with the MR damper mounted over it (see, Figure 1a). The human model with 4-DOF was also taken into account. The human body was modelled as a 4-DOF system consisting of thighs, lower torso, high torso and head (see Figure 1b). The arms and legs were integrated with the upper torso and thighs. In this paper, only the vertical motion of the system was considered for analysis.

The notations used for the integrated model are mentioned in Table 1. The values of the parameters of the system for the numerical simulation were taken from Reference [18].

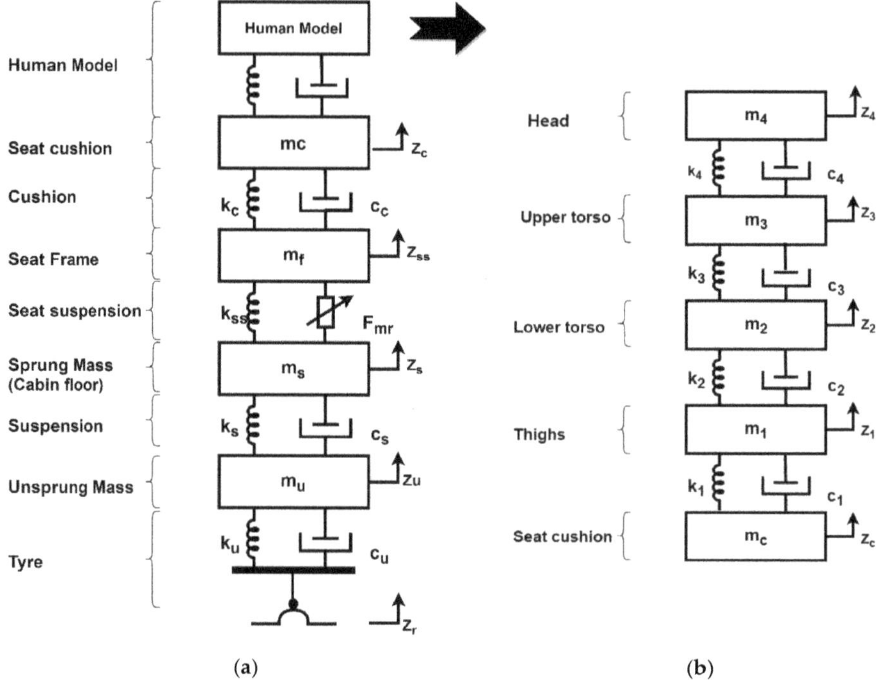

Figure 1. Integrated model of quarter car semi-active seat suspension with the human body model: (**a**) integrated quarter car semi-active seat suspension model; (**b**) human body model.

Table 1. Parameters of integrated seat suspension system.

Symbol	Parameter	Value
c_u	Damping of tire	0 (Ns/m)
c_s	Damping of suspension	2000 (Ns/m)
F_{mr}	MR damper force	-
c_c	Damping of seat cushion	200 (Ns/m)
c_1	Damping of buttocks and thighs	2064 (Ns/m)
c_2	Damping of lumber spine	4585 (Ns/m)
c_3	Damping of thoracic spine	4750 (Ns/m)
c_4	Damping of cervical spine	400 (Ns/m)
k_u	Tyre stiffness	180,000 (N/m)
k_s	Sprung mass stiffness	10,000 (N/m)
k_{ss}	Seat suspension stiffness	31,000 (N/m)
k_c	Seat cushion stiffness	18,000 (N/m)
k_1	Buttocks and thigh stiffness	90,000 (N/m)
k_2	Lumber spine stiffness	162,800 (N/m)
k_3	Thoracic spine stiffness	183,000 (N/m)
k_4	Cervical spine stiffness	310,000 (N/m)

Table 1. *Cont.*

Symbol	Parameter	Value
m_u	Unsprung mass	20 (kg)
m_s	Sprung mass	300 (kg)
m_f	Seat frame	15(kg)
m_c	Seat cushion	1(kg)
m_1	Thighs	12.78 (kg)
m_2	Lower torso	8.62 (kg)
m_3	Upper torso	28.49 (kg)
m_4	Head	5.31 (kg)
z_r	Road displacement input	-
z_u	Unsprung mass displacement	-
z_s	Sprung mass displacement	-
z_{ss}	Seat frame displacement	-
z_c	Seat cushion displacement	-
z_1	Thighs displacement	-
z_2	Lower torso displacement	-
z_3	Upper torso displacement	-
z_4	Head displacement	-

The dynamical equations of the integrated model were derived using Newton's second law of motion.

$$m_u \ddot{z}_u = -k_u(z_u - z_r) - c_u(z_u - z_r) + k_s(z_s - z_u) + c_s(z_s - z_u) \tag{1}$$

$$m_s \ddot{z}_s = -k_s(z_s - z_u) - c_s(z_s - z_u) + k_{ss}(z_{ss} - z_s) + F_d \tag{2}$$

$$m_f \ddot{z}_{ss} = -k_{ss}(z_{ss} - z_s) - F_d + k_c(z_c - z_{ss}) + c_c(z_c - z_f) \tag{3}$$

$$m_c \ddot{z}_c = -k_c(z_c - z_{ss}) - c_c(z_c - z_f) + k_1(z_1 - z_c) + c_1(z_1 - z_c) \tag{4}$$

$$m_1 \ddot{z}_1 = -k_1(z_1 - z_c) - c_1(z_1 - z_c) + k_2(z_2 - z_1) + c_2(z_2 - z_1) \tag{5}$$

$$m_2 \ddot{z}_2 = -k_2(z_2 - z_1) - c_2(z_2 - z_1) + k_3(z_3 - z_2) + c_3(z_3 - z_2) \tag{6}$$

$$m_3 \ddot{z}_3 = -k_3(z_3 - z_2) - c_3(z_3 - z_2) + k_4(z_4 - z_3) + c_4(z_4 - z_3) \tag{7}$$

$$m_4 \ddot{z}_4 = -k_4(z_4 - z_3) - c_4(z_4 - z_3) \tag{8}$$

Equations (1) and (2) represent the dynamics of the quarter car model. The dynamics of the seat suspension are represented using Equations (3) and (4). Human body dynamics are represented by Equations (5) to (8). Here, F_d is the force generated by the damper. The damper force is given by Equation (9):

$$F_d = \begin{cases} c_{ss}(z_{ss} - z_s) & \text{(Passive suspension)} \\ F_{mr} & \text{(semi - active suspension)} \end{cases} \tag{9}$$

In this model, the viscous damping, c_{ss}, was considered as zero while analysing the performance of the MR damper. The MR damper generates the controlled force, F_{mr}. However, to compare the performance of the semi-active seat suspension with the passive seat suspension, a non-zero value of c_{ss} was considered only while analysing the passive seat suspension.

2.1. Dynamic Model of MR Damper

The MR damper model has been studied and applied by several researchers in the past. In this work, the Modified Bouc–Wen model of the MR damper is implemented. This model was adopted from Karkoub and Zribi [24]. The force generated by the MR damper depends on the time history of the voltage in the coil of the damper, i.e., the supply voltage, and it also depends on the displacement of the damper, z_r:

$$z_r = z_{ss} - z_s \tag{10}$$

The FLST-PID determines the supply voltage v of the MR damper. The force generated by the MR damper was computed by solving the Modified Bouc–Wen model using the voltage v and the z_r. The MR damper force has been computed by Equation (11):

$$F_{mr} = c_1 y + k_1 (z_r - x_0) \tag{11}$$

where, k_1 represents the accumulated stiffness as a spring with a constant value c_1 and is used to simulate the hysteresis loop at low frequencies.

$$y = \frac{1}{c_0 + c_1} \{ \alpha z + c_0 x + k_0 (z_r - y) \} \tag{12}$$

In Equation (12), c_0 denotes the damping at the higher velocities. To control the stiffness at the large velocities another spring with the constant k_0 was introduced in Equation (12). Furthermore, x_0 denotes the initial displacement. The internal moments in the MR damper is signified by Equation (12). The hysteresis loop of the MR damper is presented by z, which is known as the evolutionary variable. The behaviour of the evolutionary variable is governed by Equation (13):

$$z = -\gamma |z_r - y| |z|^{n-1} z - \beta (z_r - y) |z|^n + A(z_r - y) \tag{13}$$

The shape of the hysteresis loop can be synchronized by γ, β, n and A. The force generated by the MR damper was controlled by the input u. The control input is associated with the amount of voltage supply v by the following equations, Equations (14)–(17).

$$\alpha = \alpha(u) = \alpha_a + \alpha_b u \tag{14}$$

$$c_1 = c_1(u) = c_{1a} + c_{1b} u \tag{15}$$

$$c_0 = c_0(u) = c_{0a} + c_{0b} u \tag{16}$$

$$u = -\eta(u - v) \tag{17}$$

The effect of the sensor delays is not considered in this work.

2.2. MR Damper Controller

The force generated by the MR damper 'F_{mr}' depends on the control input u which is related to the supply voltage v. The supply voltage was controlled by the Heaviside step function adopted from Dyke et al. [25]. The governing equation of the Heaviside step function is given by Equation (18):

$$v = V_{max} H((F_c - F_{mr}) F_{mr}) \tag{18}$$

$$F_c = K_p e(t) + K_i \int e(t) dt + K_d \frac{d}{dt} e(t) \tag{19}$$

Here, V_{max} in Equation (18) is the saturation voltage of the coil in the MR damper and H is the Heaviside step function. The value of the supply voltage v depends on F_c, i.e., the desired controlled force, which is estimated by the FLST-PID and the force produced by the MR damper. In Equation

(19), K_p, K_i *and* K_d are the PID gain which were determined through fuzzy logic. e is the sprung mass velocity error. Equation (19) allows to calculate the controlled force F_c (desired force) which feeds into Equation (18). Equation (18) permits to determine the input voltage of the MR damper.

When $F_c > F_{mr}$, it means that the requirement of the control force F_c (desired force) is greater than the force produced by the MR damper F_{mr}. Therefore, the voltage must be increased to its maximum value to satisfy the requirement. When $F_c = F_{mr}$, the control force (desired force) is equal to the MR damper force F_{mr}. There should be no change in the voltage to maintain the voltage. In cases other than those mentioned above, the voltage remains zero. In this case, there is no requirement for activating the MR damper.

If the force produced by the MR damper is equal to the desired control force, then the supply voltage should remain constant [24].

If the magnitude of the MR damper force is less than the desired force and if both forces are in the same direction, then the applied voltage is the maximum to increase the force generated by the MR damper. Otherwise, the supply voltages are sets to zero [24]. The parameters of the MR damper model used for the numerical simulation were taken from References [19,26] and are mentioned in Table 2.

Table 2. Parameters of the magneto-rheological damper.

Symbol	Parameter	Value
η	Frequency	190 (/s)
c_{0a}	Viscous damping at high damper velocity	784 (Ns/m)
c_{0b}	Viscous damping at high damper velocity	1803 (Ns/m)
k_0	Stiffness at high damper velocity	3610 (N/m)
c_{1a}	Viscous damping at low damper velocity	14,649 (Ns/m)
c_{1b}	Viscous damping at low damper velocity	34,622 (Ns/Vm)
k_1	Accumulator stiffness	840 (N/m)
x_0	Effect of MR damper accumulator	0 (m)
α_a	Coefficient of stiffness	12,441 (Nm)
α_b	Coefficient of stiffness	38,430 (N/Vm)
γ	Characteristics constant of Bouc-Wen model	136,320 (m^{-2})
β	Characteristics constant of Bouc-Wen model	2,059,020 (m^{-2})
δ	Characteristics constant of Bouc-Wen model	58
n	Characteristics constant of Bouc-Wen model	2

3. Controller Design for the Semi-Active Seat Suspension

To determine the desired force required to minimize the transmissibility of the vibration to the human body for a comfortable ride, a proper controller design is essential. Therefore, in this work, the self-tuning PID controller based on fuzzy logic has been implemented. The PID controller is one of the most popular and highly used control designs in the industry. The performance of the PID controller is based on the optimal values of gains. However, as the operating conditions change and the system parameters change, the PID controllers show unwanted overshoots, and this sometimes leads to a slow response when subjected to external disturbances [19]. To overcome such a drawback, the FLST-PID has been implemented in this work. The FLST-PID tunes the gain on the basis of fuzzy logic, which provides the advantage of the auto-tuning of the proportional, integral and derivative gains for the PID controller for a better performance when subjected to varying external disturbances [27]. This control takes the error and error derivative as the input and determines suitable values of the proportional gain (K_p), integral gain (K_i) and derivative gain (K_d) as the output [27,28]. The values of K_p, K_i and K_d were tuned by the fuzzy logic. The block diagram of the FLST-PID is shown in Figure 2.

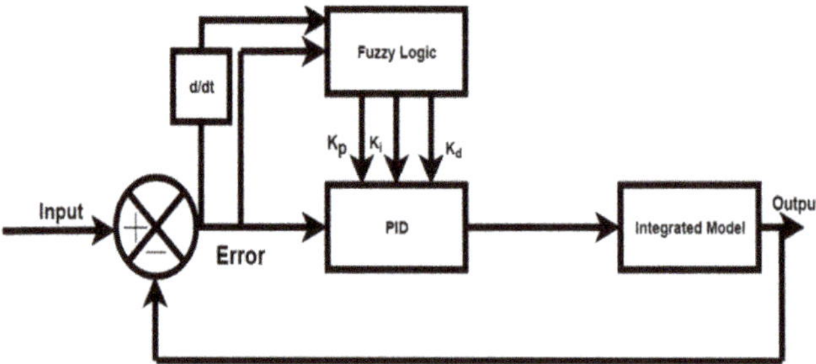

Figure 2. Block diagram of Fuzzy Logic based Self-Tuning Proportional-Integral-Differential Controller.

The state variables, sprung mass velocity and accelerations used in this work are measured by using velocity sensors or accelerometers. The sprung mass velocity and acceleration are measurable typically and, hence, considered as the input variables. Therefore, the sprung mass velocity (z_s) error, i.e., $(\dot{z}_s\text{-}0)$, is denoted as 'V_e', and the error of derivative $(\ddot{z}_s\text{-}0)$ is denoted as 'C_e', i.e., the sprung mass acceleration $(\ddot{z}_s)$, were chosen as inputs. For the fuzzy logic design, the linguistic variables of the input membership functions were taken as: Negative Large (NL), Negative Small (NS), Zero (ZE), Positive Small (PS) and Positive Large (PL). Similarly, the linguistic variables of the output membership functions were taken as: Positive Very Small (PVS), Positive Small (PS), Positive Medium Small (PMS), Positive Medium (PM), Positive Medium Large (PML), Positive Large (PL) and Positive Very Large (PVL). Membership functions were chosen on the basis of the literature [14,18,27,28], the trial and error method and the experience of the authors and are shown in Figure 3.

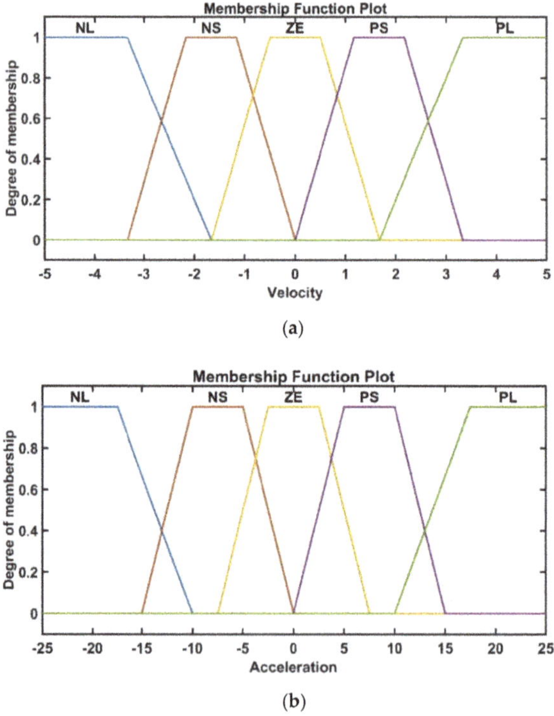

Figure 3. *Cont.*

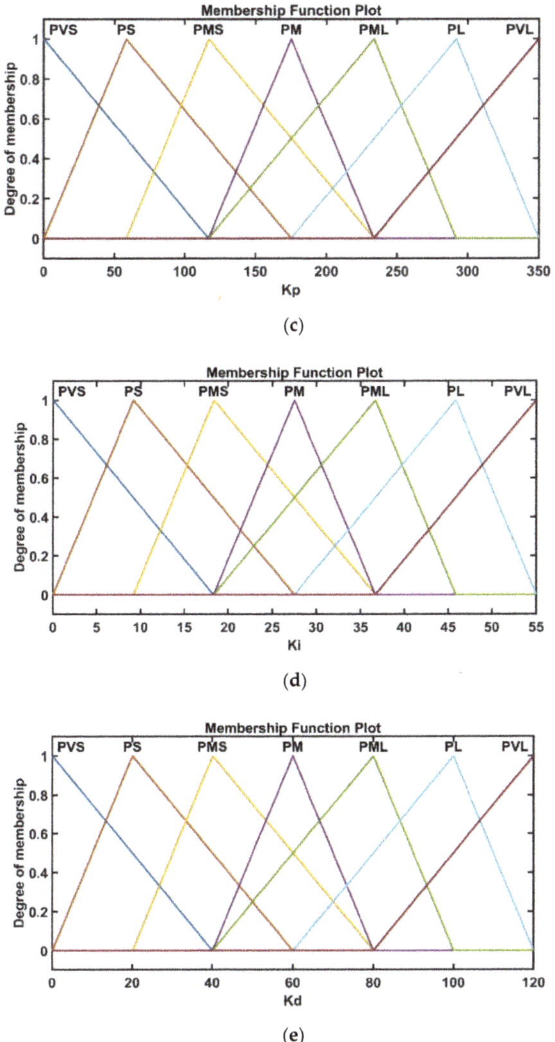

Figure 3. Membership functions for fuzzy logic: (**a**) membership function for variable sprung mass velocity; (**b**) membership function for acceleration; (**c**) membership function of proportional gain; (**d**) membership function of integral gain; (**e**) membership function of deferential gain.

The triangular membership functions were considered for the input as well as the output variables. The fuzzy rules for tuning and gain were based on the error and the derivative of the error. The output was computed by the centre average defuzzification method. For instance, in the rules of K_p, provided in Table 3, if the velocity error V_e is in the region of NL and its change C_e is also in the NL region, then, according to the fuzzy mapping rule, the output will be in the PVL region. In other words, for very large NE, there must be a PL value of K_p to minimize it. For Ki and K_d similar criteria logic was used to determine the respective gains. Hence, as the velocity error or the acceleration error changes its region, at the same time, the region of output K_p, Ki and K_d change according to the rules described in Tables 3–5. Furthermore, as the input changes, the region of the value of all three gain values change according to the fuzzy rule. The PID gains change according to instantaneous error values, therefore, this is known as a self-tuning PID controller based on fuzzy logic.

Table 3. Fuzzy rules for calculating K_p.

V_e/C_e	NL	NS	ZE	PS	PL
NL	PVL	PVL	PVL	PVL	PVL
NS	PML	PML	PML	PML	PML
ZE	PVS	PVS	PS	PMS	PMS
PS	PML	PML	PML	PM	PM
PL	PVL	PVL	PVL	PVL	PVL

Note: *Ve*—velocity error; *Ce*—change in velocity error (i.e., acceleration error); NL—Negative Large; NS—Negative Small; ZE—Zero; PS—Positive Small; PL—Positive Large; PVS—Positive Very Small; PMS—Positive Medium Small; PM—Positive Medium; PML—Positive Medium Large; PVL—Positive Very Large.

Table 4. Fuzzy rules for calculating K_i.

V_e/C_e	NL	NS	ZE	PS	PL
NL	PM	PM	PM	PM	PM
NS	PMS	PMS	PMS	PMS	PMS
ZE	PS	PS	PVS	PS	PS
PS	PMS	PMS	PMS	PMS	PMS
PL	PM	PM	PM	PM	PM

Table 5. Fuzzy rules for calculating K_d.

V_e/C_e	NL	NS	ZE	PS	PL
NL	PM	PM	PM	PM	PM
NS	PMS	PMS	PMS	PMS	PMS
ZE	PS	PS	PVS	PS	PS
PS	PMS	PMS	PMS	PMS	PMS
PL	PM	PM	PM	PM	PM

The centre average method, also known as the centroid method, determines the crisp value of the output, taking into consideration, in a weighted manner, all the influences obtained from the rules fired by the particular state of the inputs at a certain moment. This method is adopted from the mechanics and are specific to calculate the abscissa of the centroid [29]. Hence, as the input variables change, then the output variables change according to the rule shown in Table 3. From these rules, the values of the PID gains were computed. The rules for tuning K_p, are shown in Table 3.

The rules for tuning K_i are shown in Table 4.

The rules for tuning K_d are shown in Table 5.

The semi-active control of the integrated seat suspension model with FLST-PID and the Heaviside function's block diagram is shown in Figure 4.

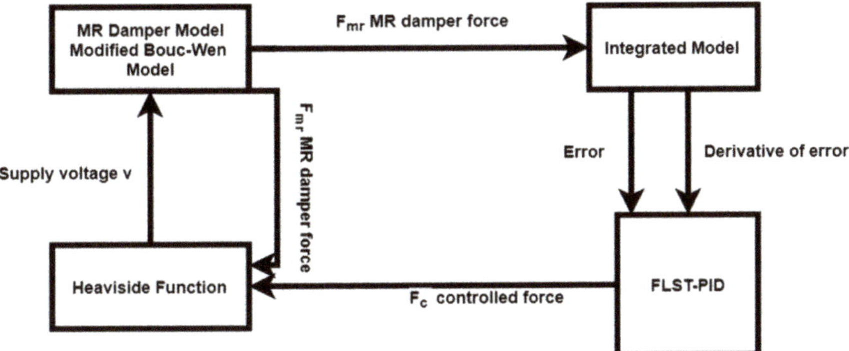

Figure 4. Semi-active control system for integrated seat suspension.

4. Numerical Simulation

To validate the effectiveness of the integrated semi-active seat suspension, numerical simulations were performed for the different road profiles. The numerical simulations were computed using MATLAB. The ode45 was used as the solver with a step size of 0.04 s. The performance of the semi-active and passive seat suspension was compared on the bump profile and the ISO road profiles of the different grades such as the C, D and E-grade with the varying speed from 30 km/h, 60km/h and 90 km/h. The bump profile applied to the vehicle's wheels were determined by using Equation (20):

$$z_r(t) = \begin{cases} \frac{a}{2}\left(1 - \cos\left(\frac{2\pi v_0}{l}t\right)\right), & 0 \leq t \leq \frac{l}{v_0} \\ 0, & t \geq \frac{l}{v_0} \end{cases} \tag{20}$$

More information on Equation (19) can be found in Reference [30].

The bump profile applied to the vehicle's wheels are shown in Figure 5. The bump road profiles at 30km/h, 60 km/h and 90 km/h are shown in Figure 5a,c,e, respectively.

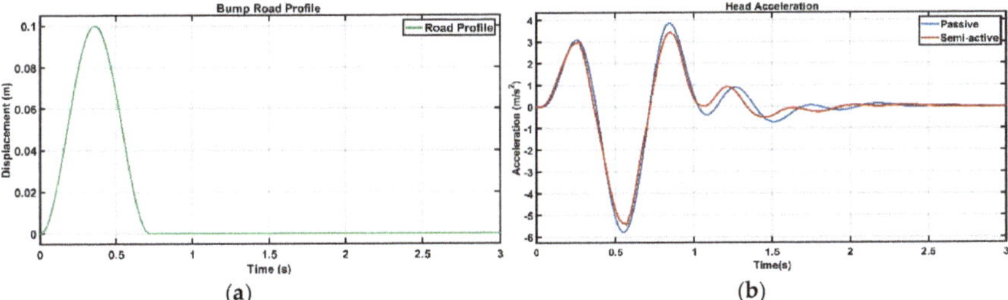

Figure 5. *Cont.*

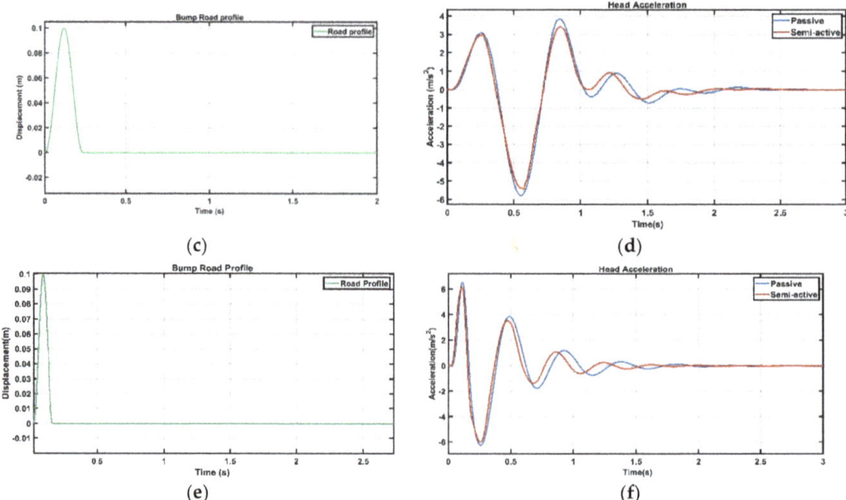

Figure 5. Bump road profile and head acceleration at different speeds: (**a**) bump road profile at a speed of 10 Km/h; (**b**) head acceleration comparison for 10km/h bump input; (**c**) bump road profile at a speed of 30 Km/h; (**d**) head acceleration comparison for 30km/h bump input; (**e**) bump road profile at a speed of 50 Km/h; (**f**) head acceleration comparison for 50km/h bump input.

In Equation (20), a is the amplitude of the bump, which is 0.1 m, the velocity (v_0) of the vehicle which is taken as 30 km/h. The driver's head acceleration response for the bump input at different velocities for semi-active and passive seat suspension is shown in Figure 5b,d,f. In Figure 5, passive represents the passive seat suspension without any controlled damper, whereas semi-active represents the semi-active seat suspension with the MR damper controlled by FLST-PID. For instance, the semi-active seat suspension reduces the peak-to-peak value of the head acceleration by 7.68%, as compared to that of the passive seat suspension only, improving the ride comfort in case of a bump road profile at 30 kmph vehicle speed. Although this percentage reduction is value is not very large; such an observation is due to the inclusion of the vehicle suspension model in the proposed integrated seat suspension model. In this study, we give the bump profile with a 0.1 m amplitude road input to the wheels of the vehicle and suspension damper, suppressing the vibrations at first, and then the displacement of the sprung-mass Z_s actually serves the input to the seat suspension system. Much previous work reports road input directly applied to the seat suspension as the disturbance and this may lead to larger values of percentage reduction of the head acceleration of the semi-active seat suspension, as compared to that of thee passive system. Therefore, the proposed approach of the integrated seat suspension with a vehicle suspension and human model looks more pragmatic. Gad et al. [31] proposed a semi-active MR seat suspension system with a human body model without considering the vehicle suspension in their model. They simulated their model for bump road input and used a fractional-order proportional-integral-derivative (FOPID) with a genetic algorithm controller for their semi-active seat suspension system. They reported that the peak-to-peak value of the human head acceleration came out at 7.9442 m/s². We simulated our model with the same input as that of Gad et al. and found the peak-to-peak value of the head acceleration to be 3.523 m/s² (see Figure 6) and, thus, show a reduction of ~55% (both studies use semi-active suspension models). This result shows the effectiveness of the proposed model as well as the controller implemented. The model parameters for both studies are same, however, they did not consider vehicle suspension in their study. This justifies the need to consider integrated seat suspension modelling.

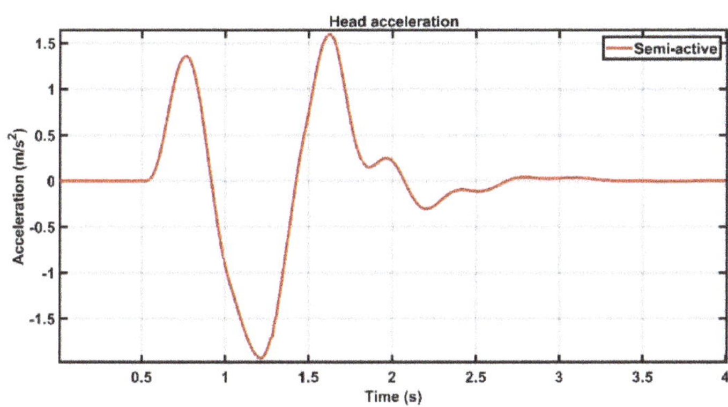

Figure 6. Human head acceleration for the bump profile with an amplitude of 0.07 m and speed of 3.04 km/h).

Table 6 shows the comparison of the performance of passive and semi-active seat suspension for the bump road for the peak-to-peak values of the human body parts, in other words, the head, upper torso, lower torso and things. Table 6 also shows the percentage improvement in the peak-to-peak accelerations owing to a semi-active seat suspension, as compared to the passive system. Overall, it was found that a semi-active seat suspension reduces the acceleration values and, thus, improves the ride comfort.

Table 6. Peak-to-peak value of acceleration of body parts.

Bump Profile Speed	Body Part	Peak-to-Peak Acceleration (m/s^2)		Improvement (%)
		Passive	Semi-Active	
10 Km/h	Head	9.631	8.873	7.87
	Upper Torso	9.641	8.877	7.92
	Lower Torso	9.969	8.926	10.46
	Thigh	10.234	9.324	8.89
30 Km/h	Head	15.74	14.53	7.68
	Upper Torso	15.66	14.47	7.59
	Lower Torso	14.93	13.76	7.83
	Thigh	14.86	13.30	10.49
50 Km/h	Head	12.83	11.88	7.40
	Upper Torso	12.71	11.64	8.41
	Lower Torso	12.04	11.00	8.56
	Thigh	12.22	10.96	10.31

The force generated by the MR damper is shown in Figure 7. It can be noted that the maximum force generated by the MR damper is 500 N, which is under the constraint of the limited seat suspension stroke. This constraint is quite practical to consider in the design, otherwise the suspension might collide with the end's stop [32].

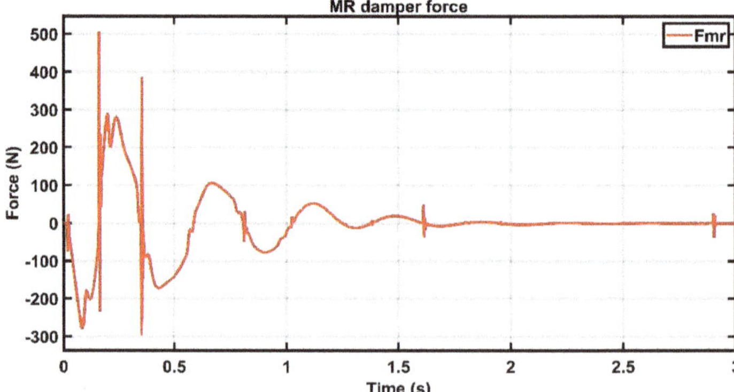

Figure 7. Force generated by MR damper.

From Figure 8 it can be observed that the stroke length for the MR damper is within the range of ±20 mm which is acceptable for the seat suspension [19].

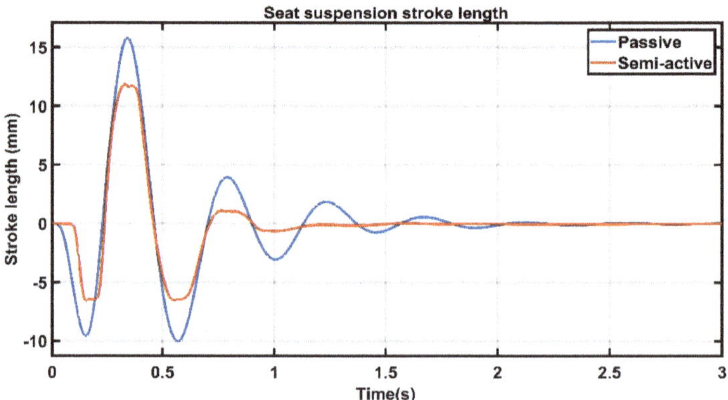

Figure 8. Seat suspension stroke length.

The calculated values of the PID gains with the time for the bump input are shown in Figure 9.

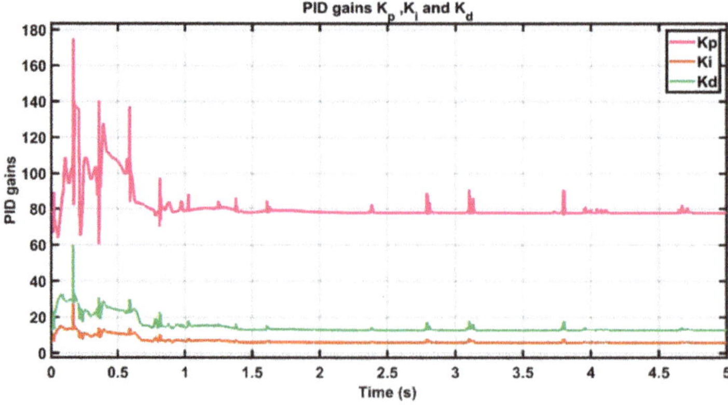

Figure 9. Variation of proportional–integral–derivative gains.

The FLST-PID can easily be implemented to design the suspension module and can be integrated with the electronic control unit (ECU) of the car which is responsible for other electronic controls in the vehicles. The three phases of fuzzy logic, i.e., fuzzification, fuzzy inference engine and defuzzification, can be embedded into the circuit. The suspension model receives the feedback from the accelerometer or velocity sensors fitted at the sprung mass and the seat of the drivers. Thus, for FLST-PID, the inputs coming from such sensors can be directed to the fuzzification board. Then, in the fuzzy inference engine, the fuzzy logic rules that depend on the analogue signals' input voltages ranging from − 5 V to + 5 V can be implemented, generating an analogue voltage distribution on the signal lines. Such a voltage distribution is characterized by the membership functions of the individual inference result. The membership function of the final fuzzy inference result can be applied to the buffer amplifier array (input stage of defuzzifier), through a data bus of analogue signal lines (rules) and, then, the outputs fed to the two resistor arrays, one of which produces a weighted sum of the membership function, and the other, a normal sum of two output data (scalar values). These outputs are then transferred to the analogue divider to produce a centre of gravity of the membership function applied to this block. Finally, the defuzzifier board groups output signals from the rule boards to determine the PID gains and compute the controlled force. By doing an if–else computation, the Heaviside function controller accordingly provides the signal to the power transistors whether to switch the solenoid of the MR Damper on and off. Thus, the FLST-PID can compute the amount of the force required to minimize the vibration. For further reading, please refer to Reference [33].

To further validate the performance of the semi-active seat suspension, a random road profile was used as the input. Here, to check the effectiveness of the controller, the different speeds of the vehicle, i.e., 30 km/h, 60 km/h, 90 km/h and 120 km/h for each grade of the road were considered.

For a constant speed in the time domain, the power spectral is the white noise signal and the spectral density of the signal is given by Equation (21):

$$k = 4\pi^2 n_0^2 G_q(n_0)v \tag{21}$$

According to the ISO 2631 standard, the value of $G_q(n_0)$ for the C-grade = 64 ×10^{-6} m^3, D-grade = 256 ×10^{-6} m^3 and for the E-grade = 1024 ×10^{-6} m^3, n_0 is the space frequency which is taken as 0.1 m^{-1} and v is the velocity in m/s [30]. The random road profile can be generated using Equation (22):

$$q(t) = \sqrt{k} \int_0^t w(t) \tag{22}$$

Here, $w(t)$ is the white noise signal, t denotes the time. One sample of the random road profile of the E-grade at 120 km/h is shown in Figure 10.

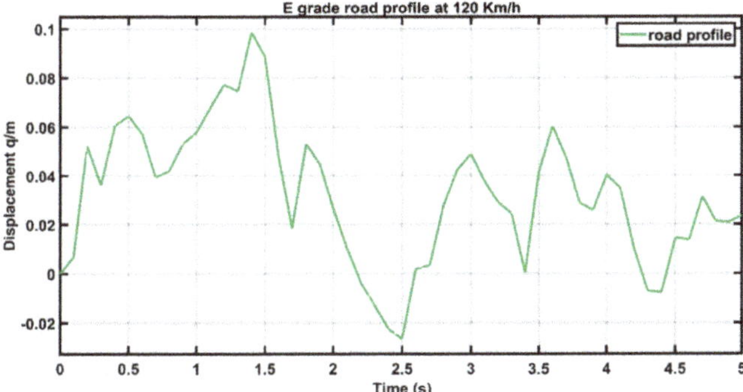

Figure 10. Random road profile of E-grade at 120 Km/h.

The performance of the passive seat suspension and the proposed integrated semi-active seat suspension on the random road profile has been compared on the basis of peak-to-peak value of the head acceleration. The speed of the vehicle was varied from 30 to 120 Km/h with an interval of 30 km/h. The performance comparison of integrated semi-active seat suspension with passive seat suspension in terms of peak-to-peak value of head acceleration for C-grade, D-grade and E-grade road profiles at different vehicle speeds is shown in Figure 11.

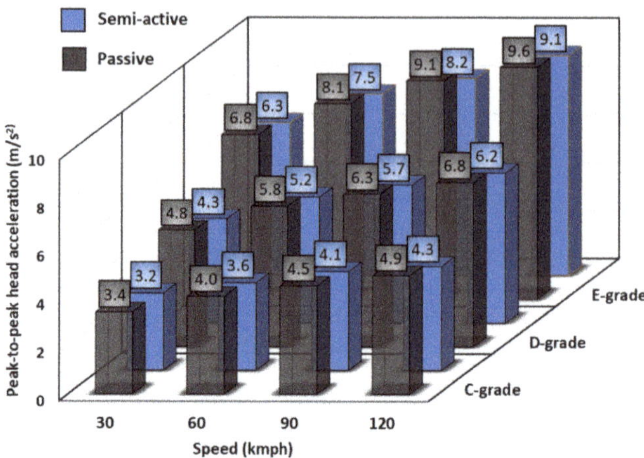

Figure 11. Comparison of peak-to-peak value of head acceleration for semi-active seat suspension and passive seat suspension for different grades of road profiles.

It can be observed from Figure 11 that the semi-active seat suspension outperforms the passive seat suspension in terms of the peak to the peak value of the head acceleration at considered vehicle speed values. From these results, it is reasonable to claim that the semi-active set suspension improves the ride comfort by reducing the head acceleration as compared with the passive seat suspension.

5. Conclusions

In this work, an integrated semi-active seat suspension mounted over a quarter car with the human model was implemented. The human model was broken down into 4-DOF including thighs, lower torso, upper torso, and head. The MR damper was implemented for the semi-active seat suspension and the controller force was determined by the fuzzy logic-based self-tuning PID controller. The Heaviside step function was used to track the desired damper force and supply the required voltage to the MR damper. The performance of the semi-active seat suspension and the passive seat suspension was compared using different road profiles such as bump input and the ISO road profile with the different grades and the vehicle's speed.

The simulation results show that the MR damper seat suspension can significantly attenuate the vibration transmitting to the human head. For the bump road input profile, the semi-active seat suspension shows 9.49% reduction in the peak-to-peak value of head acceleration, as compared to that of passive seat suspension, thus, improving the ride comfort. Similarly, for the C-grade road profile there is an average of 8.3% reduction in the peak-to-peak values of head acceleration at different speeds. Furthermore, for the D and E-grade road profiles, the peak-to-peak head accelerations were reduced by 7.8% and 6.5%, respectively. A notable reduction in peak-to-peak values of head accelerations were observed for the proposed integrated semi-active suspension, as compared to the passive seat suspension. In future work, the effectiveness of the proposed integrated semi-active seat suspension model on the ride quality, in terms of seat effective amplitude transmissibility, vibration dose value, and crest factor, are required to be investigated considering the frequency domain analysis. Further, the effect of sensor delays in the proposed models' performance also needs to be investigated.

Author Contributions: Conceptualization, S.J. and D.R.U..; methodology, S.J.; software, S.J. and S.S.; formal analysis, S.J. and C.I.P.; investigation, S.J. and D.R.U.; data curation, S.J. and S.S.; writing—original draft preparation, S.J.; writing—review and editing, D.R.U. and C.I.P.; visualization, C.I.P.; supervision, D.R.U; All authors have read and agreed to the published version of the manuscript.

Funding: This research received no external funding.

Conflicts of Interest: The authors declare no conflict of interest.

Abbreviations

MR	Magneto-Rheological;
ER	Electro-rheological;
DOF	Degrees of Freedom;
FLST-PID	Fuzzy Logic based Self-Tuning Proportional Integral Differential;
ISO	International Organization for Standardization;
PID	Proportional Integral Differential;
K_p	Proportional gain;
K_i	Integral gain;
K_d	Differential gain;
ECU	Electronic Control Unit;
V_e	Velocity error;
C_e	Change in velocity error (i.e., acceleration error);
NL	Negative Large;
NS	Negative Small;
ZE	Zero;
PS	Positive Small;
PL	Positive Large;
PVS	Positive Very Small;
PMS	Positive Medium Small;
PM	Positive Medium;
PML	Positive Medium Large;
PVL	Positive Very Large.

References

1. Qin, Y.; Dong, M.; Langari, R.; Gu, L.; Guan, J. Adaptive hybrid control of vehicle semiactive suspension based on road profile estimation. *Shock. Vib.* **2015**. [CrossRef]
2. Unune, D.R.; Pawar, M.J.; Mohite, S.S. Ride analysis of quarter vehicle model. In Proceedings of the International Conference of Industrial Engineering (ICIE), National Institute of Technology, Surat, India, 17–19 November 2011; pp. 17–19.
3. Maciejewski, L.; Meyer, T.; Krzyzynski, I. Modelling and multi-criteria optimisation of passive seat suspension vibro-isolating properties. *J. Sound Vib.* **2009**, *324*, 520–538. [CrossRef]
4. Gao, H.; Zhao, Y.; Sun, W. Input-delayed control of uncertain seat suspension systems with human-body model. *IEEE Trans. Control. Syst. Technol.* **2009**, *18*, 591–601. [CrossRef]
5. Ata, W.G.; Salem, A.M. Semi-active control of tracked vehicle suspension incorporating magnetorheological dampers. *Veh. Syst. Dyn.* **2017**, *55*, 626–647. [CrossRef]
6. Park, J.H.; Park, O.O. Electrorheology and magnetorheology. *Korea Austral. Rheol. J.* **2011**, *13*, 13–17.
7. Kolekar, S.; Venkatesh, K.; Oh, J.S.; Choi, S.B. Vibration Controllability of Sandwich Structures with Smart Materials of Electrorheological Fluids and Magnetorheological Materials: A Review. *J. Vib. Eng. Technol.* **2019**, 1–19. [CrossRef]
8. Zhu, S.; Tang, L.; Liu, J.; Tang, X.; Liu, X. A novel design of magnetorheological damper with annular radial channel. *Shock Vib.* **2016**. [CrossRef]

9.   Choi, S.B.; Han, Y.M. Vibration control of electrorheological seat suspension with human-body model using sliding mode control. *J. Sound. Vib.* **2007**, *303*, 391–404. [CrossRef]

10.   Bai, X.X.; Jiang, P.; Qian, L.J. Integrated semi-active seat suspension for both longitudinal and vertical vibration isolation. *J. Intell. Mater. Sys. Struct.* **2017**, *28*, 1036–1049. [CrossRef]

11.   Mulla, A.; Unune, D.R.; Jalwadi, S.N. Performance analysis of skyhook, groundhook and hybrid control strategies on semiactive suspension system. *Int. J. Curr. Eng. Technol.* **2014**, *3*, 265–269.

12.   Soliman, A.; Kaldas, M. Semi-active suspension systems from research to mass-market—A review. *J. Low Freq. Noise Vib. Act. Control.* **2019**. [CrossRef]

13.   Mulla, A.A.; Unune, D.R. *Investigation on the Performance of Quarter-Car Semi-Active Suspension with Skyhook, Fuzzy Logic, Adaptive Neuro-Fuzzy Inference System Control Strategies for ISO-Classified Road Disturbance*; SAE Technical Paper 2020-01-5040; SAE: Warrendale, PA, USA, 2020; In Press. [CrossRef]

14.   Singh, D.; Aggarwal, M.L. Passenger seat vibration control of a semi-active quarter car system with hybrid Fuzzy–PID approach. *Int. J. Dyn. Control.* **2017**, *5*, 287–296. [CrossRef]

15.   Bhattacharjee, V.; Chatterjee, D.; Karabasoglu, O. Hybrid Control Strategy for a Semi Active Suspension System Using Fuzzy Logic and Bio-Inspired Chaotic Fruit Fly Algorithm. *CoRR* **2017**. abs/1703.08878. Available online: http://arxiv.org/abs/1703.08878 (accessed on 12 March 2020).

16.   Bae, J.J.; Kang, N. Development of a Five-Degree-of-Freedom Seated Human Model and Parametric Studies for Its Vibrational Characteristics. *Shock Vibr.* **2018**. [CrossRef]

17.   Gohari, M.; Tahmasebi, M. Active off-road seat suspension system using intelligent active force control. *J. Low Freq. Noise Vibr. Act. Control.* **2015**, *34*, 475–489. [CrossRef]

18.   Swethamarai, P.; Lakshmi, P. Design and Implementation of Fuzzy-PID Controller for an Active Quarter Car Driver Model to minimize Driver Body Acceleration. In Proceedings of the IEEE International Systems Conference (SysCon), Orlando, FL, USA, 8–11 April 2019; pp. 1–6. [CrossRef]

19.   Metered, H.; Šika, Z. Vibration control of a semi-active seat suspension system using magnetorheological damper. In Proceedings of the 2014 IEEE/ASME 10th International Conference on Mechatronic and Embedded Systems and Applications (MESA), Senigallia, Italy, 10–12 September 2014; pp. 1–7. [CrossRef]

20.   Du, H.; Li, W.; Zhang, N. Integrated seat and suspension control for a quarter car with driver model. *IEEE Trans. Vehic. Technol.* **2012**, *61*, 3893–3908. [CrossRef]

21.   Rajendiran, S.; Lakshmi, P. Performance Analysis of Fractional Order Terminal SMC for the Half Car Model with Random Road Input. *J. Vibr. Eng. Technol.* **2019**, 1–11. [CrossRef]

22.   Wang, Y.; Li, S.; Cheng, C.; Su, Y. Adaptive control of a vehicle-seat-human coupled model using quasi-zero-stiffness vibration isolator as seat suspension. *J. Mech. Sci. Technol.* **2018**, *32*, 2973–2985. [CrossRef]

23.   Sathishkumar, P.; Jancirani, J.; John, D. Reducing the seat vibration of vehicle by semi active force control technique. *J. Mech. Sci. Technol.* **2014**, *28*, 473–479. [CrossRef]

24.   Karkoub MAZribi, M. Active/semi-active suspension control using magnetorheological actuators. *Int. J. Syst. Sci.* **2006**, *37*, 35–44. [CrossRef]

25.   Dyke, S.J.; Spencer, B.F., Jr.; Sain, M.K.; Carlson, J.D. Modeling and control of magnetorheological dampers for seismic response reduction. *Smart Mater. Struct.* **1996**, *5*, 565. [CrossRef]

26.   Lai, C.Y.; Liao, W.H. Vibration control of a suspension system via a magnetorheological fluid damper. *J. Vib. Control.* **2002**, *8*, 527–547. [CrossRef]

27.   Khodadadi, H.; Ghadiri, H. Self-tuning PID controller design using fuzzy logic for half car active suspension system. *Int. J. Dyn. Control.* **2018**, *6*, 224–232. [CrossRef]

28.   Soyguder, S.; Karakose, M.; Alli, H. Design and simulation of self-tuning PID-type fuzzy adaptive control for an expert HVAC system. *Expert Syst. Appl.* **2009**, *36*, 4566–4573. [CrossRef]

29.   Rutkowska, D. *Neuro-Fuzzy Architectures and Hybrid. Learning*; Physica-Verlag: Heidelberg, Germany; New York, NY, USA, 2002. [CrossRef]

30.   Du, H.; Li, W.; Zhang, N. Vibration control of vehicle seat integrating with chassis suspension and driver body model. *Adv. Struct. Eng.* **2013**, *16*, 1–9. [CrossRef]

31.   Gad, S.; Metered, H.; Bassuiny, A.; Abdel Ghany, A. Multi-objective genetic algorithm fractional-order PID controller for semi-active magnetorheologically damped seat suspension. *J. Vibr. Control.* **2017**, *23*, 1248–1266. [CrossRef]

32. Wang, J.; Qiang, B. Road simulation for four-wheel vehicle whole input power spectral density. *AIP Conf. Proc.* **2017**, *1839*, 020147. [CrossRef]
33. Yamakawa, T. Electronic circuits dedicated to fuzzy logic controller. *Sci. Iran.* **2011**, *18*, 528–538. [CrossRef]

Article

Study on the Vibration Active Control of Three-Support Shafting with Smart Spring While Accelerating over the Critical Speed

Miao-Miao Li *, Liang-Liang Ma, Chuan-Guo Wu and Ru-Peng Zhu

National Key Laboratory of Science and Technology on Helicopter Transmission, Nanjing University of Aeronautics and Astronautics, Nanjing 210016, China; maliangl@nuaa.edu.cn (L.-L.M.); wuchuanguo@nuaa.edu.cn (C.-G.W.); rpzhu@nuaa.edu.cn (R.-P.Z.)
* Correspondence: limiaomiao@nuaa.edu.cn; Tel.: +86-1590-516-8902

Received: 30 July 2020; Accepted: 31 August 2020; Published: 2 September 2020

Abstract: Smart Spring is a kind of active vibration control device based on piezoelectric material, which can effectively suppress the vibration of the shaft system in an over-critical state, and the selection of control strategy has great influence on the vibration reduction effect of the Smart Spring. In this paper, the authors investigate the control of the over-critical vibration of the transmission shaft system with Smart Spring, based on the ADAMS and MATLAB joint simulation method. Firstly, the joint simulation model of three-support shafting with Smart Spring is established, and the over-critical speed simulation analysis of the three-support shafting under the fixed control force of the Smart Spring is carried out. The simulation results show that the maximum vibration reduction rate is 71.6%. The accuracy of the joint simulation model is verified by the experiment of the three-support shafting subcritical vibration control. On this basis, a function control force vibration control strategy with time-varying control force is proposed. By analyzing the axis orbit of the shafting, the optimal fixed control force at different speeds is obtained, the control force function is determined by polynomial fitting, and the shafting critical crossing simulation under the function control force is carried out. The simulation results show that the displacement response of the shafting under the function control force is less than that under the fixed control force in the whole speed range.

Keywords: multi-support shaft system vibration control; combined simulation; transverse bending vibration; Smart Spring

---

1. Introduction

The multi-support shafting is a typical rotor system [1]. When the shafting is accelerated and over-critical, the vibration is intensified, which restricts the performance of the tail drive system. Excessive vibration may lead to the failure of system components, reduce the service life of equipment, waste energy, generate excessive noise, damage the health of staff, and even cause major accidents resulting in heavy property loss and casualties. Therefore, it is necessary to suppress the bending vibration of the shafting in the over-critical state [2]. At present, vibration control methods include mainly the following [3]: dynamic balance, passive control, and active control. Active control vibration reduction technology attenuates vibration by actively applying control force. The commonly used active shock absorbers are electromagnetic damper [4], squeeze film damper [5], magnetorheological damper, electrorheological damper, and piezoelectric actuator. Among them, the piezoelectric actuator has the advantages of simple structure, light weight, high control accuracy, and fast response; it has a good application prospect in the field of vibration control, but its actuation displacement is small. A.B. Palazzolo et al. were the first to study the vibration control of the piezoelectric actuator in a rotor system. By controlling the output force of the piezoelectric actuator to change the supporting parameters of the bearing pedestal, the vibration of the rotor was reduced [6,7]. The advantages and disadvantages of active damping feedback control and active stiffness feedback control were analyzed. The experimental results showed that the speed feedback control method can effectively suppress the rotor shaft system [8]. The Smart Spring [9] is an active vibration damping device based on a piezoelectric actuator, which is composed of a main support and an auxiliary support. The main support is the elastic support of shafting itself. By controlling the actuating force of the piezoelectric ceramic actuator on the auxiliary support, positive pressure is generated between friction elements on the main support and the auxiliary support, and vibration energy is consumed through dry friction between friction plates.

According to the research on the vibration reduction technology of Smart Spring, Nitzsche of Carlton University of Canada has done a lot of work on the application of the Smart Spring to helicopter blade vibration reduction. In 2012, Nitzsche et al. applied a set of semi-active Smart Spring systems to the vibration reduction of a helicopter rotor system. The control scheme selected the state switching, and the vibration reduction performance of the Smart Spring was verified by experiments [10]. In 2013, Nitzsche et al. improved the active variable pitch pull rod APL (Active pitch link) of the damping device and verified the damping performance of the improved APL in the rotating tower test. The results showed that the APL effectively attenuated the vibration response of the blade and reduced the transmission force [11]. Other scholars on Nitzsche's team have also done a lot of work on the application of Smart Springs to helicopter blade damping. Afagh et al. placed the Smart Spring mechanism on the load transfer path of the blade to achieve the purpose of vibration reduction and studied the stability of the blade in the Elastohydrodynamic state [12]. Coppotelli et al. studied the dynamic characteristics of the Smart Spring installed on the blade of a non-rotating helicopter, analyzed the influence of the Smart Spring on the modal characteristics, and finally established the finite element model of the blade [13]. Some scholars from the Aeronautical Research Institute of Canada also studied the Smart Spring damping technology. Chen Yong et al. established a mathematical model under harmonic excitation and designed an adaptive notch algorithm to reduce vibration on the DSP (Digital Signal Processor) platform. Wind tunnel tests were carried out to verify the damping effect of the Smart Spring [14]. Wickramasinghe et al. proved the ability of the Smart Spring to control the dynamic impedance characteristics of the structure through experimental research, and the vibration reduction performance was proved by the wind tunnel test [15]. Grewal et al. designed a control scheme to make the stiffness of the Smart Spring device change continuously and compared it with the control effect of the state switching control algorithm [16]. To sum up, the research into Smart Spring has been mainly on the aspect of torsional vibration control of the helicopter blade, and the research into Smart Spring vibration reduction applied to multi-support shafting has been mainly dynamic modeling and characteristic analysis. Ni De [17] carried out dynamic analysis on the shaft system with Smart Spring, studied the design method of intelligent spring parameters, and analyzed the vibration control effect of the intelligent spring in multi-support shafting. Peng Bo [18] made a preliminary exploration of the multi-span shafting vibration reduction of the intelligent spring, established the multi-span shafting dynamic model of the intelligent spring, analyzed the influence of intelligent spring control force and its parameters on vibration reduction performance, and optimized the parameters of the intelligent spring with a genetic algorithm.

At present, there is a lack of theoretical guidance and related experimental verification for the research on using Smart Spring to reduce the vibration of tail drive shafting. The application of Smart Spring to the vibration reduction of the tail drive shaft system needs further research. In this paper, the control strategy of exerting fixed control force and function control force is put forward. By using the ADAMS and MATLAB joint simulation method, the research on over-critical vibration control of three-support shafting with Smart Spring support is carried out, and the accuracy of the joint simulation method is verified by experiments.

2. Research on Vibration Reduction of the Shaft System with Fixed Control Force Applied by Smart Spring

2.1. Establishment of Joint Simulation Model of Three-Support Shafting Based on ADAMS and MATLAB

The double-disk three-support shafting with Smart Spring was taken as the research object, as shown in Figure 1. The three-support shafting consisted of motor, diaphragm coupling, transmission shaft, rigid disk, elastic support, and Smart Spring pair support. The Smart Spring consisted of basic support, auxiliary support, and a piezoelectric ceramic actuator (PZTA). The simplified principle of Smart Spring support is shown in Figure 2a. The radial stiffness and damping of the basic support are k and c, and the radial stiffness and damping of the active support are k' and c', m' is the equivalent mass of the PZTA and other components connected with the active support. F_d is the sliding friction force caused by the control force N_t between the friction pair of the basic support and the auxiliary support. The lower end of the Smart Spring support was fixed on the foundation, and the upper end bore the support reaction force Ft from the shaft section. The radial displacement of the main support is u and that of the auxiliary support is u'. There was an initial clearance between the main support and the auxiliary support. The structure diagram of Smart Spring is shown in Figure 2b.

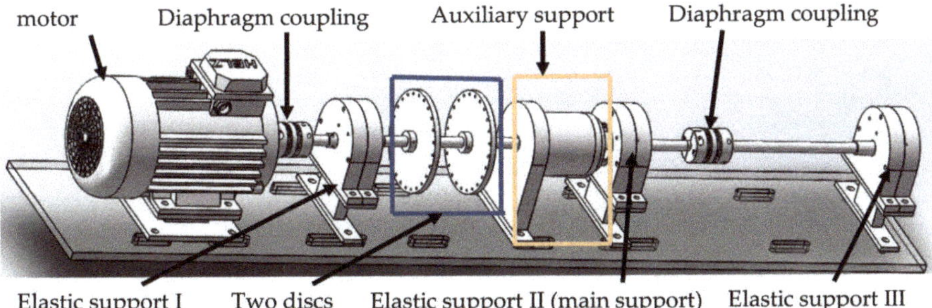

Figure 1. Three-dimensional model of three-support shafting.

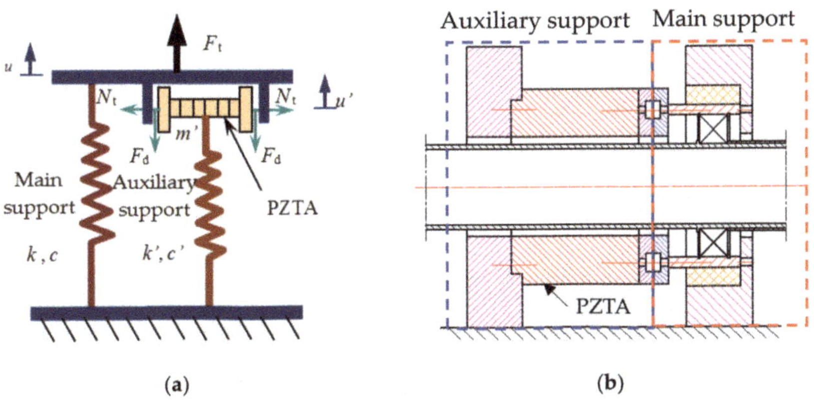

Figure 2. (**a**) Principle sketch and (**b**) structure diagram of Smart Spring.

The shaft system was a typical variable cross-section continuous rotor system with disk rotor, shaft sleeve, and coupling. Based on the finite element method, the variable cross-section of the three-support shafting was truncated and discretized. As shown in Figure 3, taking no account of the influence of the coupling on the bending vibration of the shaft system in discrete processing, the three-support shafting was discretized into disk, shaft section, and bearing seat, in turn, while the two sections of shaft were regarded as a single optical axis. The dynamic model of three-support shafting with double disks was established. The allowable degrees of freedom in the model were x-axis rotation and the translation of Y and Z. The basic parameters are shown in Table 1.

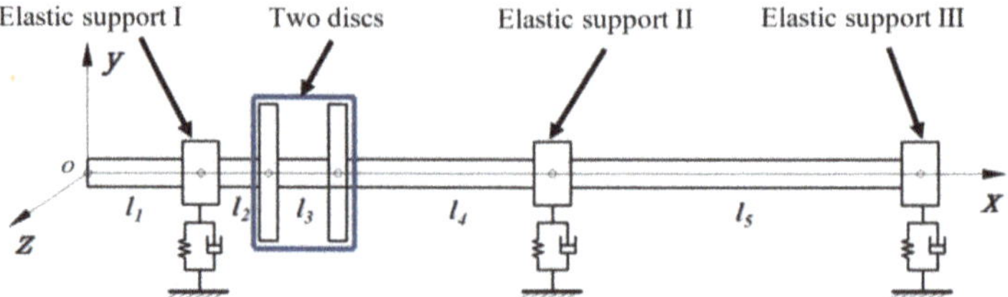

Figure 3. Dynamic model of three-support shafting with double disks.

Table 1. Basic parameters of three-support shafting.

Parameter Name	Value/Unit	Parameter Name	Value/Unit
The density of the shaft ρ	$7850/(\mathrm{kg\cdot m^{-3}})$	Elastic modulus E	$2 \times 10^{11}/\mathrm{Pa}$
Shaft radius r	7.5/mm	Disc radius R	75/mm
Length of shaft l_1	120/mm	Disc width b	8/mm
Length of shaft l_2	70/mm	Support stiffness k_b	$1.7 \times 10^{5}/(\mathrm{N\cdot m^{-1}})$
Length of shaft l_3	80/mm	Support damping c_b	$60/(\mathrm{N\cdot s\cdot m^{-1}})$
Length of shaft l_4	270/mm	Unbalance magnitude e_0	6.3×10^{-5} kg·m
Length of shaft l_3	420/mm	Auxiliary support stiffness k_a	6×10^{5} N·m^{-1}

According to the dynamic model of three-support shafting, the virtual prototype model of shafting was established in ADAMS simulation software, as shown in Figure 4. Combined with the actual working state of the shafting, the constraint conditions of the virtual prototype model of the shafting were set. The eccentricity was added to the flexible shaft to simulate the imbalance of the shaft, and the spring with stiffness and damping was used to simulate the elastic support. A rotation drive was added to the left end face of the flexible shaft. The rotation drive adopted the mode of constant acceleration, and the acceleration was set as 20π rad/s^2. In order to realize the contact and action between the friction disks of the main support and the auxiliary support of the Smart Spring, as shown in Figure 5, two friction disks were established on the virtual prototype model to the equivalent mass of the main and auxiliary supports, and the control force and contact pair were added to the friction disk. Furthermore, the friction force was linearized, which was equal to the product of the friction coefficient and control force after linearization.

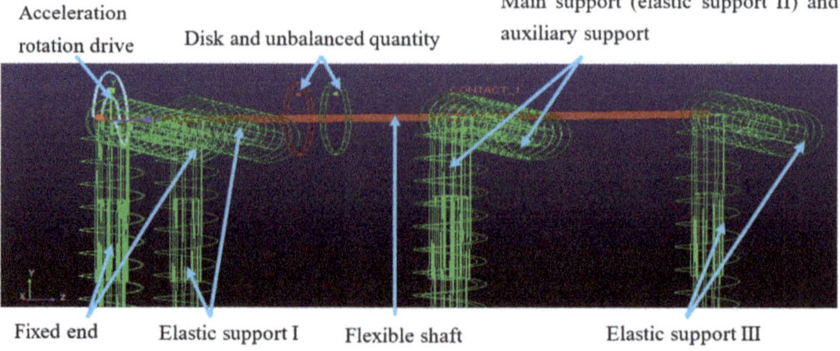

Figure 4. Virtual prototype model of double-disk shafting.

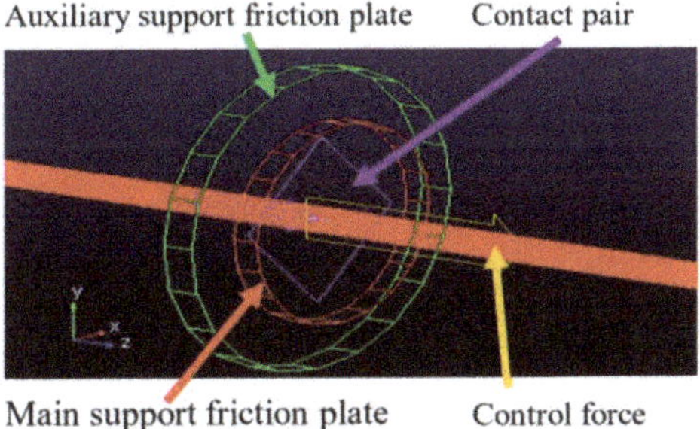

Figure 5. Realization of the Smart Spring function.

ADAMS software focuses on the mechanical dynamics simulation, while MATLAB focuses on the control system simulation. Using the ADAMS and MATLAB joint simulation method, based on the ADAMS/control module, we imported the mechanical system simulation model from ADAMS directly to MATLAB without tedious derivation and the listing of a large number of equations to describe the law of the control system, which greatly simplified the workload of modeling. At the same time, we added the complex control directly to the ADAMS model, which can simulate the whole system at one time, and, even if there are problems, they can be easily solved. The ADAMS virtual prototype model was imported into MATLAB, and the open-loop control joint simulation model of double-disk shafting was obtained, as shown in Figure 6. The joint simulation model consisted of the ADAMS module, input module, and output module. The input module was the control force of the actuator, and the control force was input into ADAMS module. The two output modules were the displacement responses in X direction and Y direction at the middle position of the shafting.

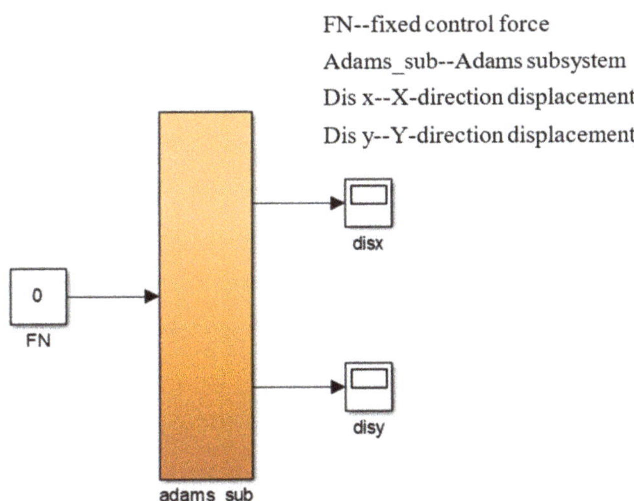

Figure 6. Joint simulation model of open-loop control.

2.2. Analysis of Simulation Results of Shafting Over-Critical

The mid-point of support I and support II was selected as the measuring point, and fixed control forces of 100 N, 200 N, 300 N, 425 N, 800 N, and 1000 N were applied respectively to simulate the shaft system passing through the critical point, and the displacement response at the measuring point was obtained. Due to the isotropy of the rotor, the shafting had similar displacement response curves in X and Y directions. Therefore, the displacement response curves in the Y direction were selected for analysis.

Figure 7 shows the over-critical bending vibration response of a double-disk three-support shafting with Smart Spring support under different fixed control forces. It can be seen from Figure 7 that the application of fixed control force can significantly suppress the critical response peak value and the critical speed shifts to the right. When the fixed control force was less than 200 N, the vibration response decreased in the whole speed range. When t < 6 s, i.e., the speed was lower than 3600 r/min, the vibration could be attenuated by applying 0–1000 N fixed control force. However, when t > 7 s, i.e., the rotating speed was greater than 4200 r/min, and the fixed control force was greater than 200 N, the vibration was increased by applying the fixed control force.

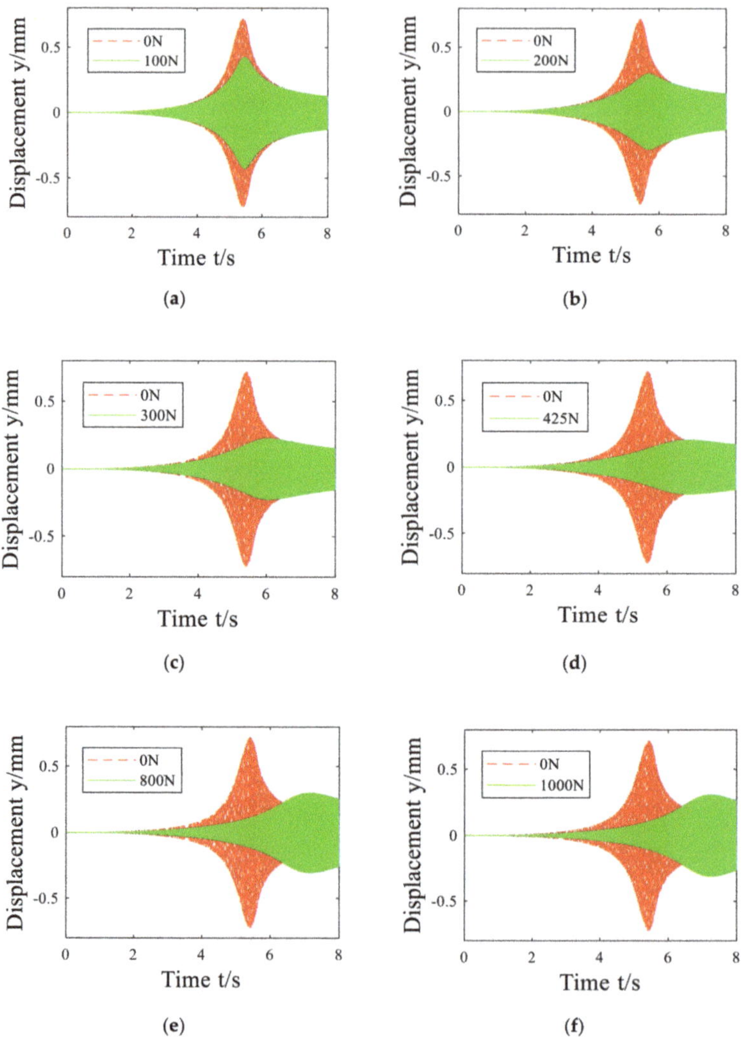

Figure 7. Response of the shaft system under (**a–f**) fixed control forces, respectively 100 N, 200 N, 300 N, 425 N, 800 N and 1000 N of the Smart Spring.

Figure 8 shows the relationship between the critical response peak value and the fixed control force. The results showed that with the increase of the fixed control force, the peak value of the over-critical response first decreased and then increased. When the fixed control force was 0 N, the peak value was 0.722 mm; when the fixed control force was 425 N, the peak value was the minimum, which was 0.205 mm. At this time, the vibration reduction rate was the largest, reaching 71.6%. Therefore, the control force of the Smart Spring was not better when greater, but there was an optimal fixed control force. Figure 9 shows that with the increase of the fixed control force, the first-order critical speed of the double-disk shaft system gradually increased, and tended to be stable when the control force was greater than 700 N, which was close to 4350 r/min, that is, the damping control gradually transited to the stiffness control.

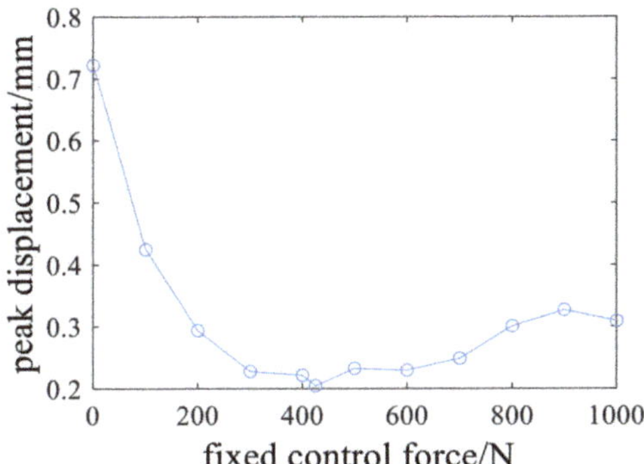

Figure 8. Relationship between peak value of over-critical response of shafting and fixed control force.

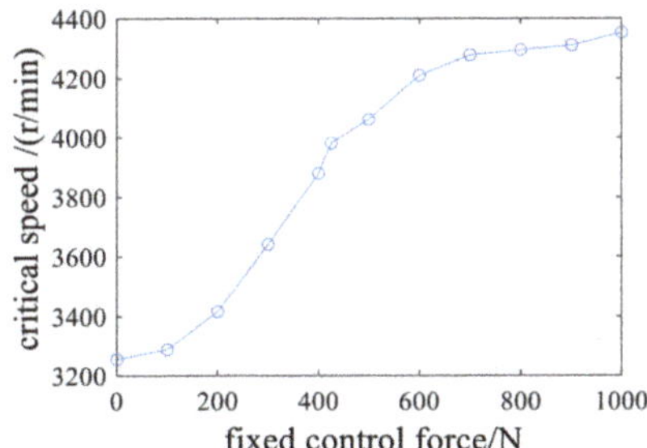

Figure 9. Relationship between first critical speed and fixed control force of shafting.

3. Experimental Verification of Shaft System Over-Critical Vibration Control

3.1. Design of Test System

The test system consisted of the shafting test bed, executive system, data acquisition system, and control system. The actuator system consisted of the piezoelectric ceramic controller and piezoelectric actuator. The data acquisition system included sensor, data collector DH5922N and its supporting software, DHDAS dynamic signal acquisition and analysis system. The control system included controller AD5436 and its supporting software. The layout of the test platform is shown in Figures 10 and 11. In the experiment, the output voltage of the piezoelectric ceramic controller was increased by 15 V, and the corresponding shaft system over-critical response was recorded, and the test results were analyzed.

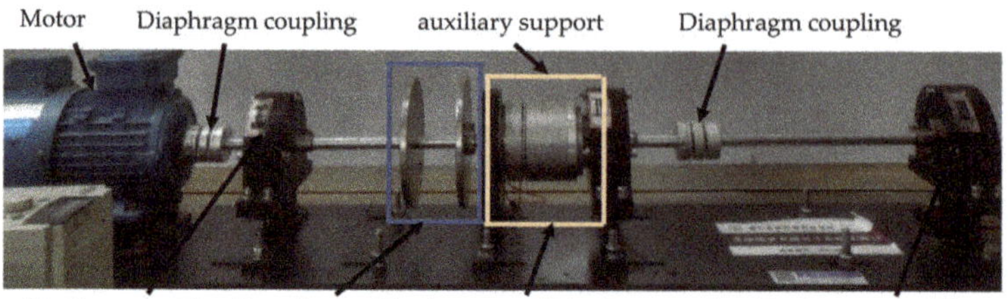

Figure 10. Shafting test bench.

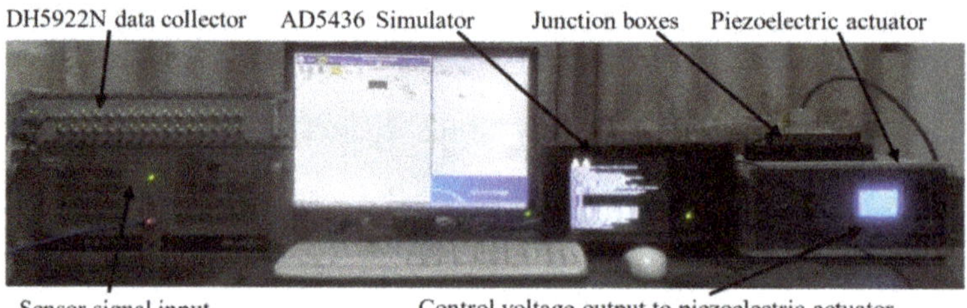

Figure 11. Layout of test data acquisition device.

When the voltage was applied, the PZTA extended to both sides to make the main and auxiliary friction surfaces of the Smart Spring support tight. If the control voltage of the PZTA was increased after jacking, the support of the Smart Spring pair deformed in the axial direction. The axial displacement x_a of the support under different loading voltages (0 V–150 V, step size 15 V) was measured by an eddy current sensor, which was the deformation x_a of the Smart Spring support. Assuming that the axial stiffness k_a of the auxiliary support remained constant with the increase of the force, then:

$$N(t) = k_a \times x_a. \tag{1}$$

In Equation (1) k_a is the axial stiffness of the auxiliary support and x_a is the axial displacement of the auxiliary support.

As shown in Figure 12, the relationship between control force and control voltage was not a precise linear relationship, and there was a certain hysteresis effect. When the control voltage increased from 0 V to 150 V at 15 V intervals and then decreased to 0 V at 15 V intervals, the curves of the control force could not coincide, that is, the piezoelectric ceramics had a hysteresis effect. When the control voltage was 150 V, the maximum control force was 447 N.

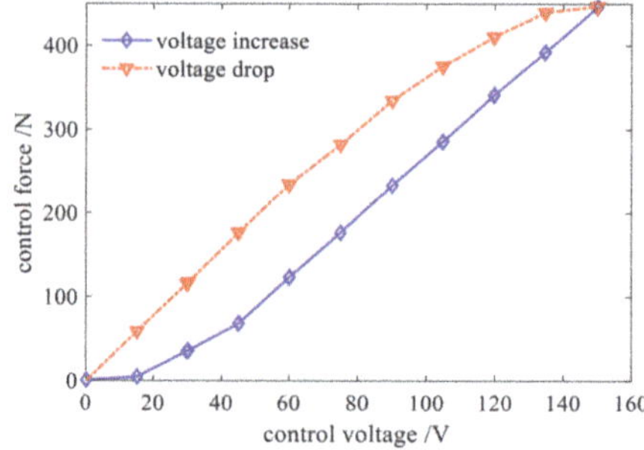

Figure 12. Relationship between control force and control voltage of the Smart Spring support.

3.2. Analysis of Experiment Result

Figure 13 shows the critical time domain response of shafting under various control voltages. It can be seen from the figure that after the control voltage was applied, the peak value of the shaft acceleration over-critical value decreased obviously. With the increase of the control voltage, the vibration peak value decreased more obviously when the shafting was over-critical.

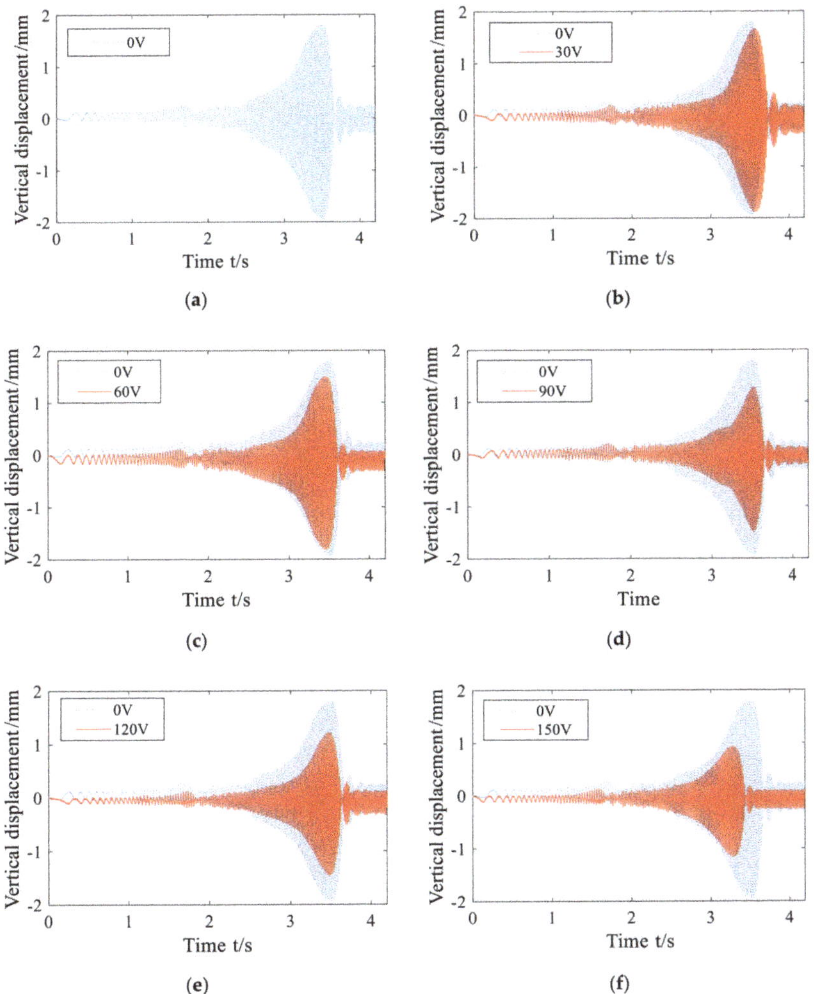

Figure 13. Accelerated over-critical vibration response of shafting under (**a**–**f**) control voltages, respectively 0 V, 30 V, 60 V, 90 V, 120 V and 150 V.

The relationship between the peak value of shaft acceleration and the control voltage under each control voltage is shown in Figure 14. It can be seen from the figure that the vibration reduction performance of the Smart Spring support was good, and the maximum vibration reduction rate reached 44.2% with the change trend of the vibration peak value with the control voltage.

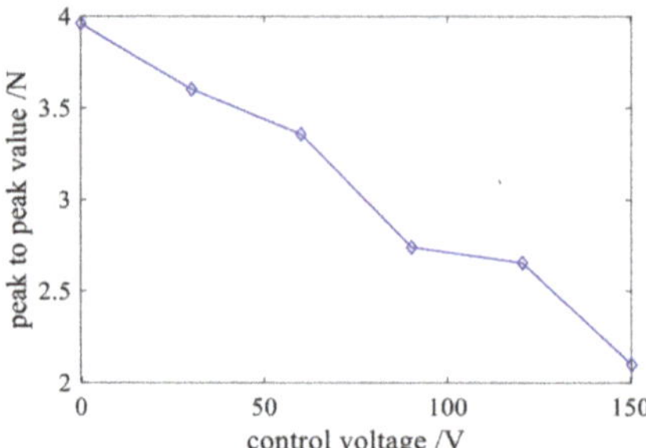

Figure 14. Relationship between peak and peak value of shaft system over-critical vibration response and control voltage.

Comparing Figures 7 and 14, the change trend of the first section of the two figures was consistent, and the displacement response decreased with the increase of the control parameters. However, because of the small axial stiffness of the bearing pedestal and the limited displacement of the piezoelectric actuator, the optimal control parameter in Figure 14 appeared at the maximum value, and the maximum control force was 447 N, so the piezoelectric actuator failed to exert its maximum capacity.

Figure 15 shows the comparison results of the shafting vibration damping ratio under the different control voltages/control forces obtained by test and simulation analyses. It can be seen from the figure that under the trans-critical state of the shafting, the variation trends of the damping rate obtained by the test and simulation were the same, which indicated the accuracy of the joint simulation model. When the fixed control force of the intelligent spring increased from 100 N to 200 N, the vibration reduction rate of the simulation results increased by 21 percentage points, and the vibration reduction rate of the test results increased by 15 percentage points; when the control force of the intelligent spring increased from 200 N to 300 N, the vibration reduction rate of the simulation results and that of the test results increased by three percentage points. The difference between the experimental damping rate and the simulated damping rate was due to the error of the test bench and the limitation of the control force of the piezoelectric actuator. For example, there was a certain gap between the angular acceleration of the motor and the simulated angular acceleration, the axial stiffness of the bearing pedestal was too small, and the displacement of the piezoelectric ceramic actuator was limited.

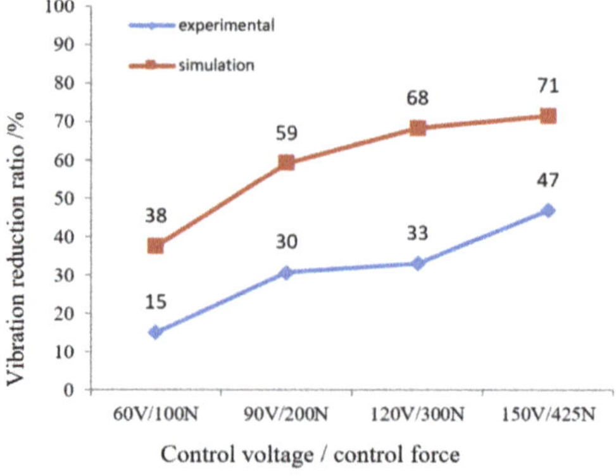

Figure 15. Vibration reduction rate of shafting under different control voltages/forces.

4. Research on Vibration Reduction of Shafting with Function Control Force Exerted by Smart Spring

It can be seen from the previous analysis results that the vibration damping rate of the shafting was different at different speeds when the Smart Spring applied fixed control force. In order to obtain better control effect of shaft system over-critical vibration, a vibration suppression method with function control force was proposed. In other words, the function control force that changes with acceleration and time was applied to the Smart Spring. Therefore, after establishing the joint simulation model of shafting, the relationship between control force F_N and time was studied to determine the optimal control force under constant operating speed, and then the control force function was determined to carry out joint simulation analysis on the critical open-loop control of the Smart Spring applying function control force.

4.1. Establishment of Simulation Model of Three-Support Shafting with Smart Spring

This section takes the three-support shafting test bed shown in Figure 16 as the research object, establishes the virtual prototype model and imports it into MATLAB to obtain the joint simulation model of three-support shafting with Smart Spring.

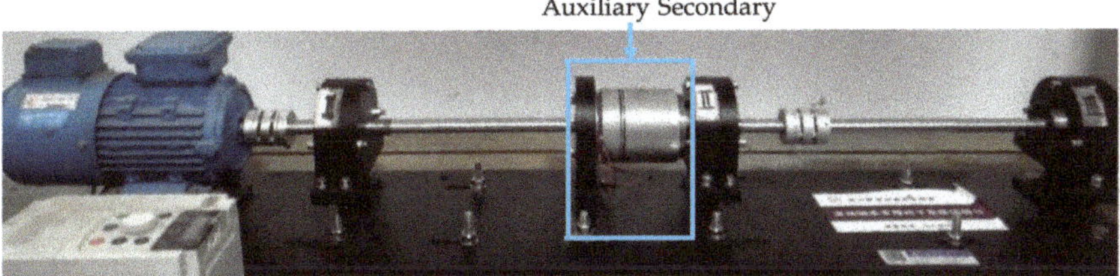

Figure 16. Three-support shafting test bench with Smart Spring pair support.

4.2. Determination of Optimal Fixed Control Force at Constant Speed

The joint simulation model under constant speed as shown in Figure 17 was established. In order to realize the constant speed operation of the shafting, a state variable was added to the input module to input the constant speed. In the speed range of 0–8400 r/min, the simulation of optimal fixed control force at constant speed was carried out with the interval of 600 r/min. The axis orbit of the shafting under different fixed control forces was obtained by simulation, and the control force with the minimum radius of the axis trajectory was selected as the optimal control force F_{op}.

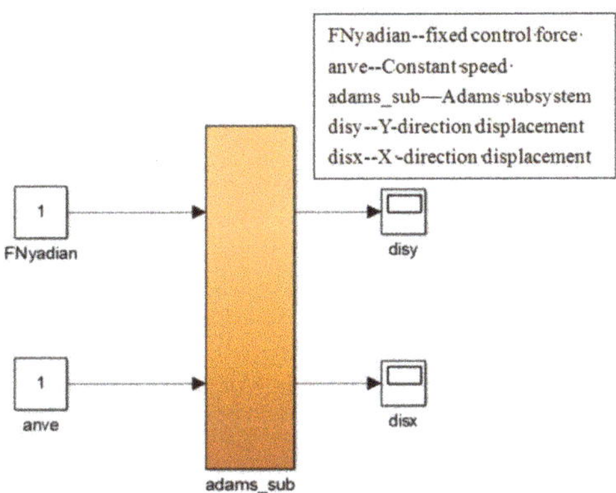

Figure 17. Joint simulation model of applying function control force at constant speed.

Figure 18 shows the axis orbit of the steady-state response under each fixed control force when the shaft speed was 6000 r/min. It can be seen from the figure that the axis trajectories of the steady-state response were all circular. With the increase of the control force, the radius of the steady-state response also changed, which indicated that different fixed control forces had different suppression effects on the steady-state response of constant speed operation.

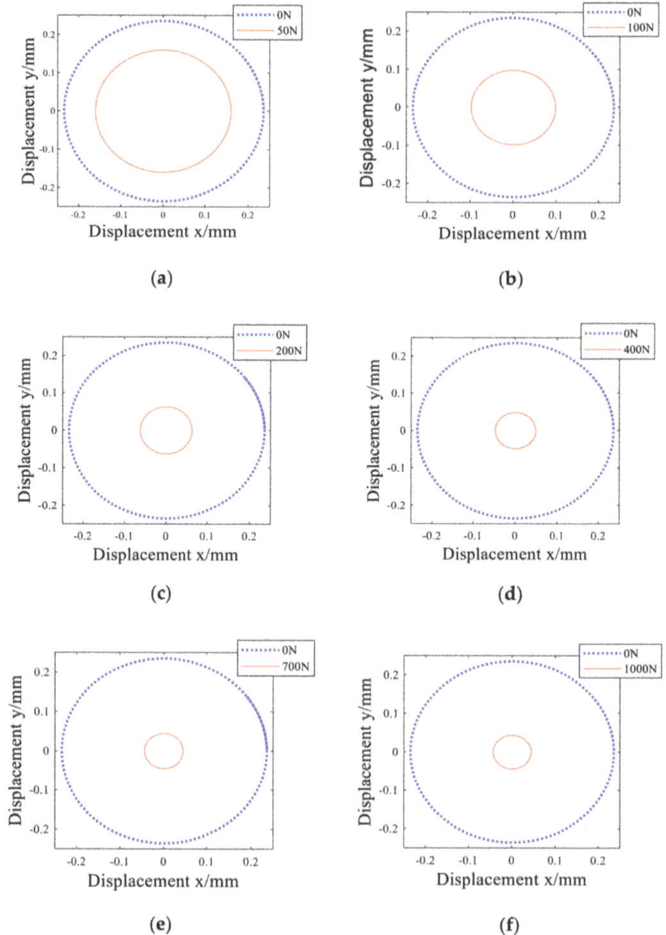

Figure 18. Steady-state response center orbit under (**a**–**f**) control forces at 6000 r/min, respectively 50 N, 100 N, 200 N, 400 N, 700 N and 1000 N.

Since the optimal control forces in the speed range of 0–6000 r/min were all 1000 N, it was no longer necessary to analyze the case where the shaft speed was lower than 6000 r/min. Figure 19 shows the relationship between the fixed control force of the Smart Spring and the steady response of the steady-state response at the speeds of 6000 r/min, 6600 r/min, 7200 r/min, 7800 r/min, and 8400 r/min. It can be seen from the figure that with the increase of speed, the optimal control force F_{op} and the maximum damping rate of constant speed operation became smaller and smaller. When the speed was 6000 r/min, the optimal control force F_{op} was 1000 N, and the maximum vibration reduction rate was about 80%; when the speed was 6600 r/min, the optimal control force F_{op} was 650 N, and the maximum vibration reduction rate was about 58%; when the speed was 7200 r/min, the optimal control force F_{op} was 200 N, and the maximum damping rate was about 17%; when the speed was 8400 r/min, the optimal control force F_{op} was 25 N, and the maximum damping rate was close to zero.

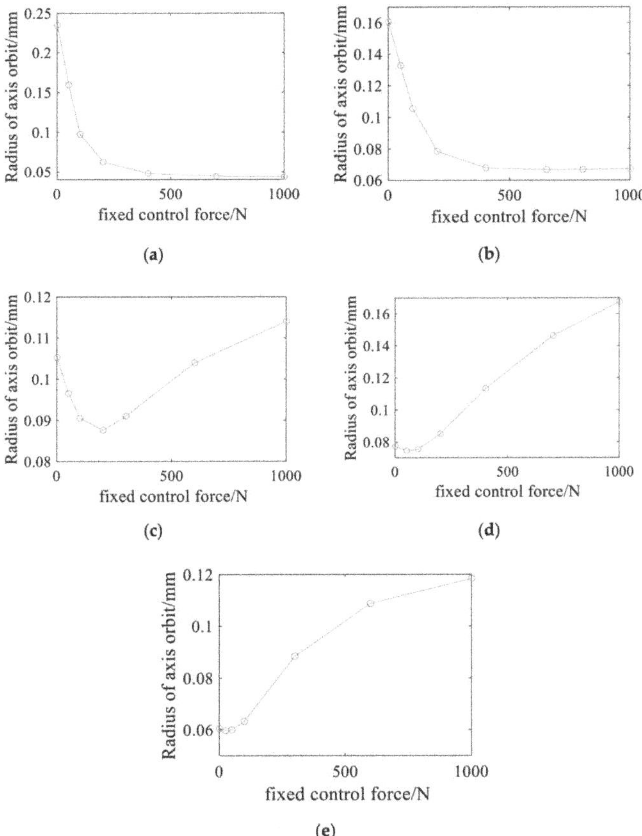

Figure 19. Relationship between the radius of the axis orbit and the fixed control force under (**a**–**e**) speeds, respectively 6000 r/min, 6000 r/min, 6600 r/min, 7200 r/min, 7800 r/min and 8400 r/min.

4.3. Polynomial Fitting of Function Control Force

According to the above analysis, the required control force is related to acceleration and time, so that

$$F_{\mathrm{N}}(t) = a_0 + a_1 t + a_2 t^2 + \cdots + a_m t^m. \tag{2}$$

When the shaft system was accelerated with an angular acceleration of 20π rad/s^2, the times to reach 6000 r/min, 6600 r/min, 7200 r/min, 7800 r/min, and 8400 r/min were 10 s, 11 s, 12 s, 13 s, and 14 s, respectively, and the optimal control forces were 1000 N, 650 N, 200 N, 50 N and 25 N, respectively. Equation (2) is used to fit the functional relationship between control force F_N and acceleration and time. The goal of fitting is to minimize the square sum D of the upper deviation, that is, Equation (3) is the minimum.

$$D = \sum_{i=1}^{N} \delta_i{}^2 = \sum_{i=1}^{N} \left[F_{\mathrm{N}}(t_i) - F_{\mathrm{OP}_i} \right]^2. \tag{3}$$

In Equation (3), the optimal control force is $F_{\mathrm{OP}i}$, the fixed control force is $F_{\mathrm{N}}(t_i)$ and the variance is δ.

According to the least square theorem, if $N > m + 1$, then there is a unique set of polynomial coefficients $a_0, a_1 \ldots a_m$, which minimize the value of Equation (3), and

$$x_m = \left(A^{\mathrm{T}}A\right)^{-1}A^{\mathrm{T}}b. \tag{4}$$

In Equation (4), $x_m = [a_0\, a_1\, a_2 \cdots a_m]^{\mathrm{T}}$, $b = [F_{\mathrm{OP}_1}\, F_{\mathrm{OP}_2} \cdots F_{\mathrm{OP}_N}]^{\mathrm{T}}$, the value of matrix A is as follows:

$$A = \begin{bmatrix} 1 & t_1 & t_1^2 & \cdots & t_1^m \\ 1 & t_2 & t_2^2 & \cdots & t_2^m \\ \vdots & \vdots & \vdots & & \vdots \\ 1 & t_N & t_N^2 & \cdots & t_N^m \end{bmatrix}.$$

By substituting the time and fixed control force values corresponding to speeds of 6000 r/min, 6600 r/min, 7200 r/min, 7800 r/min, and 8400 r/min into Equation (4), the results show that $a_0 = 143{,}280$, $a_1 = -57{,}510$, $a_2 = 7687.2$, $a_3 = -342.14$, that is, the fitting function of optimal control force and time for 10 s $\leq$ t $\leq$ 14 s is as follows:

$$F_{\mathrm{N}}(t) = 143{,}280 - 57{,}510t + 7687.2t^2 - 342.14t^3. \tag{5}$$

According to Equation (5), the curve shown in Figure 20 is obtained.

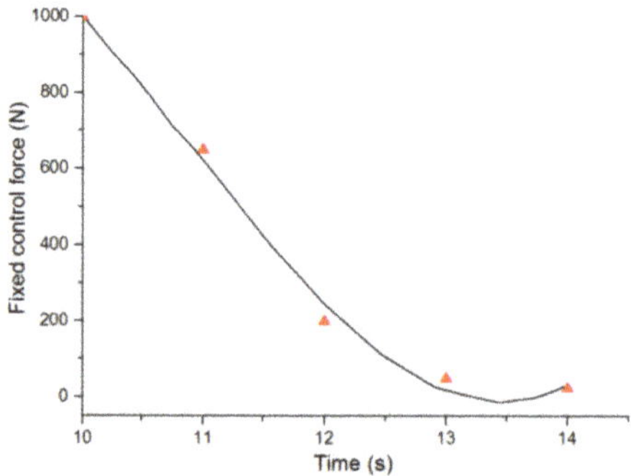

Figure 20. Control force fitting curve for 10 s $\leq t \leq$ 14 s.

According to the simulation results of the shaft center trajectory under constant speed, the optimal control force of the shafting was 1000 N in the speed range of 0–6000 r/min, i.e., within 0–10 s, the function control force expression of the optimal control force and time is as follows:

$$F_{\mathrm{N}}(t) = \begin{cases} 1000, & 0 \leq t \leq 10 \\ 143{,}280 - 57{,}510t - 7687.2t^2 - 342.14t^3, & 10 < t \leq 14 \end{cases}. \tag{6}$$

4.4. Analysis of Simulation Results of Supercritical

4.4.1. Analysis of Simulation Results of Over-Critical with Fixed Control Force

The midpoint of the central line between support I and support II was selected as the measuring point, and the displacement response at this point was measured. Figure 21 shows the critical bending vibration response of a three-support shafting with Smart Spring support under different fixed control forces. The fixed control forces applied were 50 N, 100 N, 150 N, 200 N, 250 N, 300 N, 600 N, and 1000 N, respectively. It can be seen from the figure that the application of fixed control force significantly inhibited the peak value of critical response, and the critical speed obviously shifted to the right; when the fixed control force was not greater than 100 N, the vibration response in the whole speed range of 0–8400 r/min (i.e., 0 s < t < 14 s) decreased; when the speed was lower than 7200 r/min (i.e., t < 12 s), the vibration was attenuated by applying 0–1000 N fixed control force, but when the rotating speed was less than 7200 r/min (when the speed was higher than 7800 r/min (i.e., t > 13 s)) and the fixed control force was greater than 150 N, the vibration increased when the fixed control force was applied.

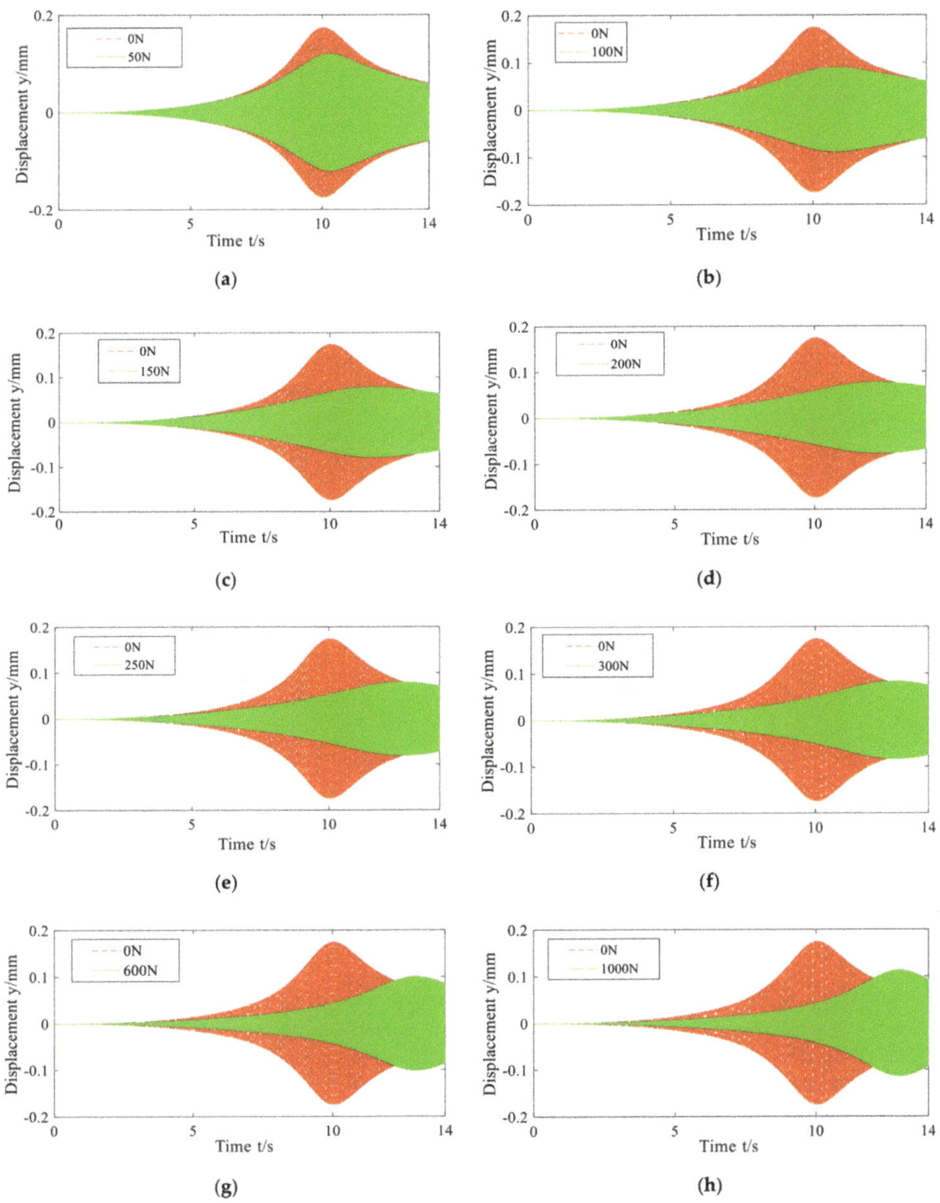

Figure 21. Response of shaft system under (**a**–**h**) fixed control forces, respectively 50 N, 100 N, 150 N, 200 N, 250 N, 300 N, 600 N and 1000 N.

Figure 22 shows the relationship between the peak displacement response of the measuring point and the fixed control force when the shafting was over-critical. It can be seen from the figure that with the increase in fixed control force, the peak value of critical response first decreased and then increased gradually. When the control force was 0 N, the peak value was 0.1749 mm; when the control force was 200 N, the peak value was the minimum, which was 0.0777 mm, and the maximum vibration reduction rate was 55.6%.

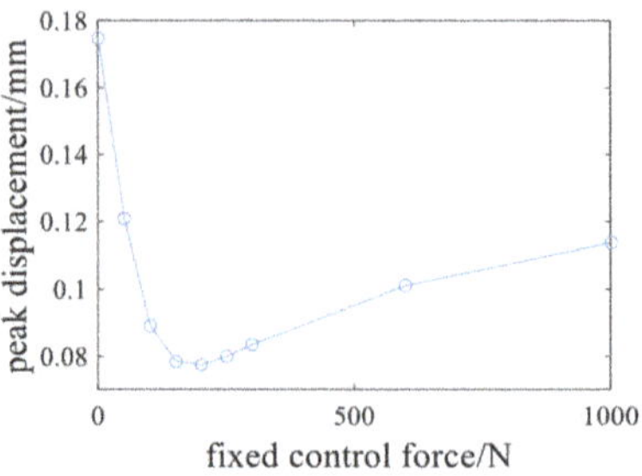

Figure 22. Relationship between peak displacement of measurement point and fixed control force.

As shown in Figure 23, the first critical speed increased gradually and tended to be stable after the control force was greater than 600 N, which was close to 7800 r/min, that is, the damping control gradually transited to the stiffness control.

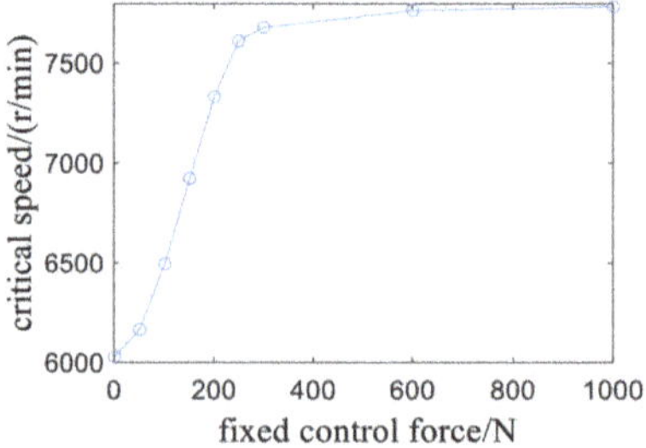

Figure 23. Relationship between first critical speed and fixed control force.

It can be concluded from Figures 22 and 23 that the Smart Spring support has good vibration reduction performance. By applying fixed control force, the bending vibration response of three-support shafting accelerating through critical can be greatly attenuated, but the greater the control force, the better. The optimal fixed control force is 200 N, and the vibration response of shafting is the minimum.

4.4.2. Analysis of Simulation Results of Applying Function Control Force Over-Critical

According to the determined function control force, the joint simulation model established in Figure 16 was modified, and then the function control force was applied to the model. A variable F_N of function control force was established in the MATLAB workspace. F_N contained two columns of data, the first column was time and the second column was the corresponding control force. The two output state variables are the displacement in X direction and Y direction at the midpoint of elastic supports I and II.

By comparing the vibration analysis results of the three-support shafting of the Smart Spring without control force, fixed control force, and function control force, Figures 24 and 25 are obtained. It can be seen from the figure that the critical peak value of the function control force and the best fixed control force was 200 N, but in the whole speed range, the displacement response of the function control force was less than that of the fixed control force of 200 N, which achieved a better vibration reduction effect. Therefore, for the open-loop control of shaft system acceleration over-critical vibration, the vibration reduction effect of Smart Spring with function control force is the best.

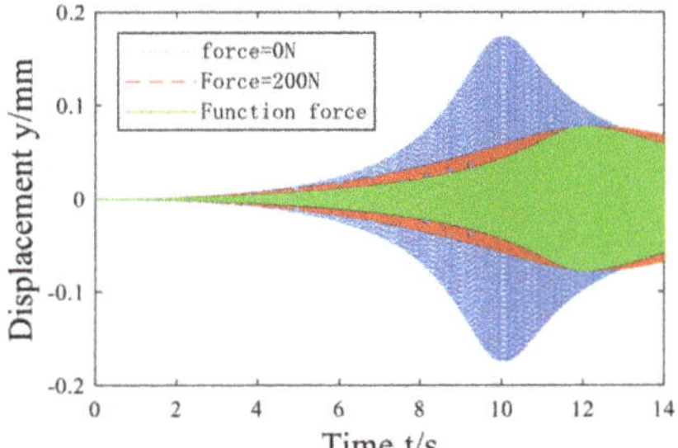

Figure 24. Response curves of the shaft system under different control methods.

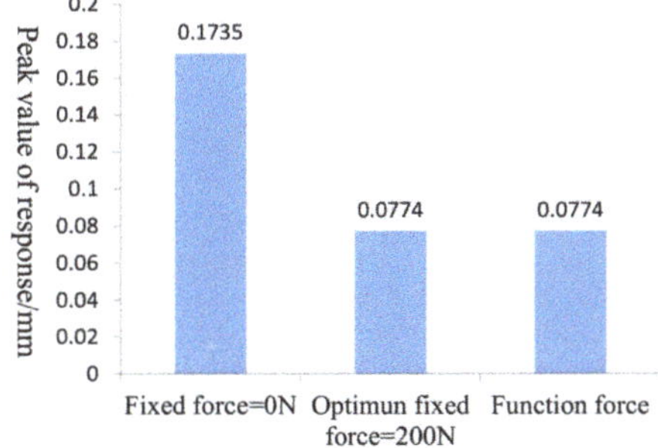

Figure 25. Critical vibration peak value of shafting under different control methods.

5. Conclusions

Previous studies have proved that constant control force exerted by Smart Spring is effective for vibration control of shafting. However, there are few researches on the improvement of control strategy. Based on ADAMS and MATLAB simulation software, this paper analyzes the lateral bending vibration control of the shaft system under critical state, and improves the control strategy of the intelligent spring and proposes a vibration suppression method based on function control force. The main conclusions are as follows:

(1) In this paper, the authors established a joint simulation model for the double-disk three-support shafting with Smart Spring, and carried out the open-loop control simulation and experimental study of the shafting over-critical vibration. The simulation results showed that the fixed control force exerted by the Smart Spring can greatly reduce the displacement response of the double-disk shaft system accelerating through the critical value. However, the fixed control force is not better when greater, but there is an optimal fixed control force making the vibration control effect the best. The optimal fixed control force is 425 N, and the vibration reduction rate is the maximum, which is 71.6%. The double-disk shafting test system was designed and tested. The test results showed that when the maximum control voltage is 150 V, the vibration reduction rate is the highest, reaching 44.2%, which verified that the Smart Spring support has good control effect on the lateral bending vibration of the three-support shafting when the acceleration is over-critical, and further proved the rationality of the joint simulation method.

(2) The open-loop control of the spring system was carried out by the joint control function of the spring system. The simulation results showed that with the increase of fixed control force,

the first critical speed gradually increases and tends to be stable, and the peak value decreases first and then increases. The optimal fixed control force is 200 N, of which the maximum vibration reduction rate reaches 55.6%. Compared with the fixed control force, the displacement response of the function control force is significantly less than that of the fixed control force of 200 N, so it has a better vibration reduction effect.

Author Contributions: Conceptualization, M.-M.L. and L.-L.M.; methodology, M.-M.L.; software, M.-M.L.; validation, M.-M.L., L.-L.M. and C.-G.W.; data curation, L.-L.M.; writing—original draft preparation, L.-L.M.; writing—review and editing, M.-M.L.; supervision, C.-G.W.; project administration, R.-P.Z.; funding acquisition, R.-P.Z. All authors have read and agreed to the published version of the manuscript.

Funding: This work was supported by the National Natural Science Foundation of China (Grant No. 51775265), and the National Key Laboratory of Science and Technology on Helicopter Transmission (Nanjing University of Aeronautics and Astronautics) (Grant No. HTL-A-19G09).

Conflicts of Interest: The authors declare no conflict of interest.

References

1. Meng, G. Retrospect and Prospect to the Research on Rotor Dynamics. *Shock Noise* **2002**, *15*, 1–9.
2. Peng, B. Research on Multi-span Shaft Dynamics and Vibration Reduction via Smart Spring Support. Master's Thesis, Nanjing University of Aeronautics and Astronautics, Nanjing, China, 2017.
3. Li, W.Z.; Wang, L.Q. Review on Vibration Control Technology of High-speed Rotor System. *J. Mech. Strength* **2005**, *27*, 44–49.
4. Zhang, T.; Meng, G.; Zhang, Z.X. Active Control of the Rotor System by Electromagnetic Bearing-Extrudod Oil Film Damper. *J. Xi'an Pet. Inst. Nat. Sci. Ed.* **2002**, *6*, 80–83.
5. Xing, J. Research on Active Vibration Control of Rotor System Using Magnetorheological Fluid Damper. Master's Thesis, Beijing University of Chemical Technology, Beijing, China, 2015.
6. Qu, W.Z.; Sun, J.C.; Qiu, Y. Active control of vibration using a fuzzy control method. *J. Sound Vib.* **2004**, *275*, 917–930.
7. Ishimatsu, T.; Shimomachi, T.; Taguchi, N. Active vibration control of flexible rotor using electromagneticdamper. In Proceedings of the IECON'91: 1991 International Conference on Industrial Electronics, Control and Instrumentation, Kobe, Japan, 28 October 1991; IEEE: Piscataway, NJ, USA.
8. Palazzolo, A.B.; Jagannathan, S.; Kascak, A.F.; Montague, G.T.; Kiraly, L.J. Hybrid active vibration control of rotor bearing system using piezoelectric actuators. *J. Vib. Acoust.* **1993**, *115*, 111–119. [CrossRef]
9. De, N. Research on Design Method of Smart Spring Damping System and Its Application in the Drive Shaft System. Ph.D. Thesis, Nanjing University of Aeronautics and Astronautics, Nanjing, China, 2015.
10. Nitzsche, F. The use of smart structures in the realization of effective semi-active Control systems for vibration reduction. *J. Braz. Soc. Mech. Sci. Eng.* **2012**, *XXXIV*, 371–377. [CrossRef]
11. Nitzsche, F.; Feszty, D.; Grappasonni, C.; Coppotelli, G. Whirl-tower Open-loop Experiments and Simulations with an Adaptive Pitch Link Device for Helicopter Rotor Vibration. In Proceedings of the Aiaa/asme/asce/ahs/asc Structures, Structural Dynamics, & Materials Conference, Boston, MA, USA, 8–11 April 2013; pp. 1–13.
12. Afagh, F.F.; Nitzsche, F.; Morozova, N. Dynamic modelling and stability of hingeless helicopter blades with a smart spring. *Aeronaut. J.* **2004**, *108*, 369–377. [CrossRef]
13. Coppotelli, G.; Marzocca, P.; Ulker, F.D.; Campbell, J.; Nitzsche, F. Experimental Investigation on Modal Signature of Smart Spring/Helicopter Blade System. *J. Aircr.* **2008**, *45*, 1373–1380. [CrossRef]
14. Chen, Y. Development of the smart spring for active vibration control of helicopter blades. *J. Intell. Mater. Syst. Sructures* **2004**, *15*, 37–47.
15. Wickramasinghe, V. Experimental evaluation of the smart spring impedance control approach for adaptive vibration suppression. *J. Intell. Mater. Syst. Sructures* **2008**, *19*, 171–179. [CrossRef]
16. Cavalini, A.A., Jr. Vibration attenuation in rotating machines using smart spring mechanism. In *Mathematical Problems in Engineering*; Hindawi Publishing Corporation: London, UK, 2011; pp. 1–14.

17. NI, D.; Zhu, R.-P. Influencing factors of vibration suppression performance for a smart spring device. *J. Vib. Shock* **2012**, *31*, 87–98.
18. Peng, B.; Zhu, R.; Li, M. Bending Vibration Suppression of a Flexible Multispan Shaft Using Smart Spring Support. *Shock Vib.* **2017**, *2017*, 1–12. [CrossRef]

Article

Trajectory Optimization of Industrial Robot Arms Using a Newly Elaborated "Whip-Lashing" Method

Rabab Benotsmane [1], László Dudás [1] and György Kovács [2,*

[1] Department of Information Technology, University of Miskolc, 3515 Miskolc, Hungary;
    iitrabab@uni-miskolc.hu (R.B.); iitdl@uni-miskolc.hu (L.D.)
[2] Institute of Logistics, University of Miskolc, 3515 Miskolc, Hungary
*   Correspondence: altkovac@uni-miskolc.hu

Received: 12 November 2020; Accepted: 1 December 2020; Published: 3 December 2020

Abstract: The application of the Industry 4.0's elements—e.g., industrial robots—has a key role in the efficiency improvement of manufacturing companies. In order to reduce cycle times and increase productivity, the trajectory optimization of robot arms is essential. The purpose of the study is the elaboration of a new "whip-lashing" method, which, based on the motion of a robot arm, is similar to the motion of a whip. It results in achieving the optimized trajectory of the robot arms in order to increase velocity of the robot arm's parts, thereby minimizing motion cycle times and to utilize the torque of the joints more effectively. The efficiency of the method was confirmed by a case study, which is relating to the trajectory planning of a five-degree-of-freedom RV-2AJ manipulator arm using SolidWorks and MATLAB software applications. The robot was modelled and two trajectories were created: the original path and path investigate the effects of using the whip-lashing induced robot motion. The application of the method's algorithm resulted in a cycle time saving of 33% compared to the original path of RV-2AJ robot arm. The main added value of the study is the elaboration and implementation of the newly elaborated "whip-lashing" method which results in minimization of torque consumed; furthermore, there was a reduction of cycle times of manipulator arms' motion, thus increasing the productivity significantly. The efficiency of the new "whip-lashing" method was confirmed by a simulation case study.

Keywords: manipulator arm; trajectory optimization; "whip-lashing" method; reduction of cycle time; trajectory planning; SolidWorks and MATLAB software applications

---

1. Introduction

Nowadays, the modern industry in all sectors is facing a new revolution known as Industry 4.0 [1,2], where many challenges and requirements are taken into consideration with the aim of building smart factories that combine flexibility and ability concepts [3,4] by developing a new paradigm based on the latest technologies, where automation and network systems present the efficient keys for realizing the new industrial revolution [5].

Recently, industrial robotics has become an important solution used in different sectors due to the advantages guaranteed by industrial robots [6] as manipulator arms and parallel robots represented with higher precision and higher productivity. This optimizes the lead time of the production process [7].

Especially with technological developments, the manipulator arm, for example, presents the most often used tool in the production sector [8], where it can cooperate with its environment and work safety [9,10]. In addition to the ability to pick huge products and control itself in a flexible and smart way, the physical structure of the manipulator arm regroups two essential parts [11]. These parts are the serial links articulated as an arm and the end-effector, which can be reconfigurable according to the task [12,13]. The motion planning of a manipulator arm is always based on the degree of freedom characterized by the joints placed in each link, where the number of the degree of freedom limits the workspace and defines the redundancy of the robot. The control of manipulator robots can be studied in two directions, depending on the requirements needed to achieve it [14]: (1) task execution, where the process is based on pick and place, welding, and painting; (2) path planning execution, where the process is based on the trajectory of the end effector, depending on each joint connected to the link of the robot arm.

In the area of robotics, the trajectory planning of manipulator arms represents an essential field for focus. The execution of a robot arm's defined task optimizes its trajectory, which can guarantee many benefits such as a reduced cycle time and energy consumption, as well as increased productivity. Basically, the main objective in the trajectory planning field is to compute the desired points that represent the reference input data for the controller of a robot using mathematical techniques [15,16]. The motion executed from the reference inputs always represents two categories known as forward and inverse kinematics: (1) in free space based on the joint angles, where the motion is limited by the structure constraints, i.e., velocity, torque, and workspace limits; or (2) in task space based on the position and the orientation of the end-effector, where it depends on precision and avoiding obstacles [17]. The approaches generally used are polynomial interpolation function, the bang-bang law, the trapezoid law, etc. [18].

Over the years, researchers studied this field deeply by proposing many methods and solutions to solve trajectory problems for industrial robots [19–22]. The definition of the optimality concept is divided in many directions. Some scientists focus on a time-optimal trajectory to increase productivity [23,24], while others work on the smoothness of trajectories [25,26], taking into account reducing cycle time by implementing fast trajectories combined with optimal jerk values in order to reduce the excitation of the resonant frequencies and limit the vibrations of the mechanical system [27–29]. From the literature, a basic approach is known for generating a trajectory using splines [30,31], where the virtual points are required to ensure the continuity of the trajectory from the starting point to the endpoint. The development of this approach motivated the authors of this study to apply an improved technique in the aspect of motion optimality using B-spline interpolation, based on the calculation of inverse of Jacobian matrix.

Regarding time optimality, an approach was proposed for a hyper-redundant robot taking into account the obstacles located in a 3D workspace [32,33]. It aims to minimize the cycle time during the execution of required tasks, regarding trajectory optimization for robots in terms of energy consumption and minimizing joint torque. Other researchers described a new scheme to determine the trajectory of a redundant robot arm with the purpose of minimizing the total energy consumption [34]. In order to optimize both the energy consumption and the time required for executing a trajectory, many researchers have elaborated new methods based on a fuzzy logic algorithm, a genetic algorithm, or an ant colony algorithm [34,35]. By using a genetic algorithm, a contribution was proposed to optimize the torque applied at the joints of the robot [36,37]. We can also cite the second contribution for the same target, which uses a unified quadratic-programming-based dynamic system [38], as well as the role of neural networks for the optimized dynamics of redundant robots [39]. Most of the literature in motion planning features deals with point-to-point applications in free space without any obstacles, where the starting and ending points of the end-effector are predefined.

The main purpose of this literature analysis was to guarantee a deeper understanding of the path planning field so that researchers could find an optimal solution without any constraints. Further, time and energy consumption presents the most important factors for evaluation [40].

The purpose of this study is to elaborate upon a new "whip-lashing" method that aims to realize an optimized trajectory for the five-degree-of-freedom RV-2AJ robot arm, i.e., to generate a path for a manipulator arm without any design constraints. This newly elaborated method seeks to minimize the cycle time of the trajectory with constrained torque values applied in joints for executing smooth motions. This new method originates from the motion of a whip analogue applied for a RV-2AJ robot arm. First, we introduce the methodology of this newly elaborated method, identifying all the steps required using SolidWorks and MATLAB software applications. After, the main features of whip-lashing are introduced and the Section 4 presents a concrete simulation executed for both paths in order to compare the normal motion of the arm to a whip lashing motion, based on cycle time. The final section presents the results of the joints torque variation according to the cycle time calculated in the previous section in order to show the reader the effect of the whip-lashing motion on the cycle time and the torque consumption.

The main value of the research is that a manipulator arm can be treated as a whip in certain conditions, which can guarantee running an improved path that results in reduced cycle time. The proof of this novelty is presented by applying the real parameters of an RV-2AJ arm to simulation tools.

2. Materials and Methods: Modelling the Robot Arm and the Original and Improved Trajectories

The idea of trajectory improvement is taken from the motion of a whip. When applying a huge force at the handle, it will propagate along the whip's length, with a wave running alongside the whip transferring the energy from the handle to the tip. In order to realize this idea and check its real effect on the cycle time of robot arm motion, SolidWorks [41] and MATLAB [42] software applications were used in the dynamic motion simulations. For comparison, an "original" path with simple angle interpolation at the joints was also simulated. The two will be compared.

2.1. Dynamic Analysis of RV-2AJ Robot Arm

The dynamic analysis of RV-2AJ arm requires the computation of the inertia matrix for each joint. This computation is performed using SolidWorks software by drawing the robot model and applying its real geometric and mass measurements. SolidWorks offers the possibility to transform all of the robot parameters in a URDF (Unified Robotic Description Format) file; this file can be used for the simulation in MATLAB software using Robotic Toolbox [43]. Figure 1 presents the CAD model of RV-2AJ arm in SolidWorks environment and the creating of its URDF file for MATLAB software. The URDF file allows us to import the dynamic parameters of RV-2AJ arm into a MATLAB environment, where we can visualize the robot arm and calculate its forward and inverse kinematics using the Robotic Toolbox that provides the following:

- RigidBodyTree (RBT) object,
- Home configuration function,
- Inverse Kinematic solver.

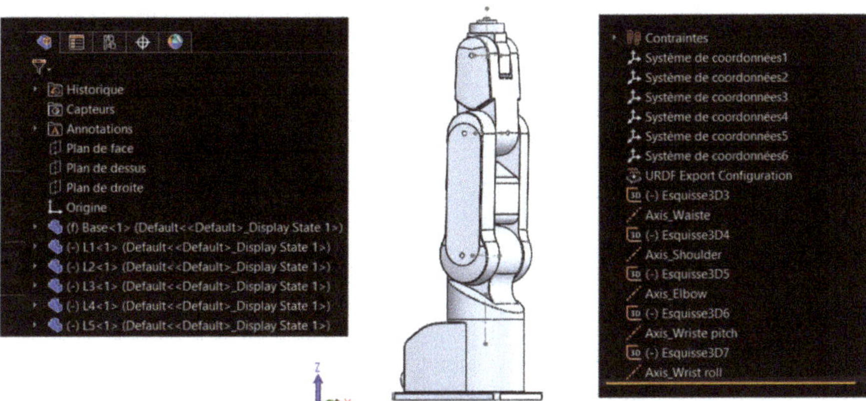

Figure 1. RV-2AJ model in SolidWorks and the created URDF (Unified Robotic Description Format) file for the CAD model of the RV-2AJ arm.

Figure 2 visualizes the imported structure of RV-2AJ arm in the MATLAB environment and the script code used for it. The visualization of the robot arm uses the RV-2AJ.URDF definition file and the "show (robot, Qhome)" command performs the visualization, as can be followed in the MATLAB code.

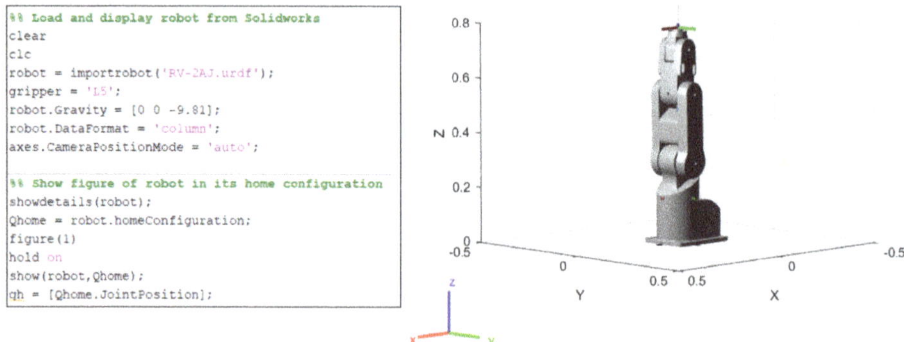

Figure 2. Visualization of RV-2AJ arm in the MATLAB environment.

For finding better or "quasi-optimal" solution and proving the effectiveness of the newly elaborated whip-lashing method, two paths were generated with the same starting and ending points but with different internal motions. During the application of the method both paths require the rotation of second–third–fourth joints, whereas the first and the fifth articulation are not used.

2.2. The Original Path

The original trajectory as a usual interpolated path is presented in Figure 3 as a continuous red arc starting from start point S and ending in final point E. In this path the RV-2AJ robot arm executes its motion by computing the angle steps for each internal point dividing the start-end angle of every joint with the number of points minus one step.

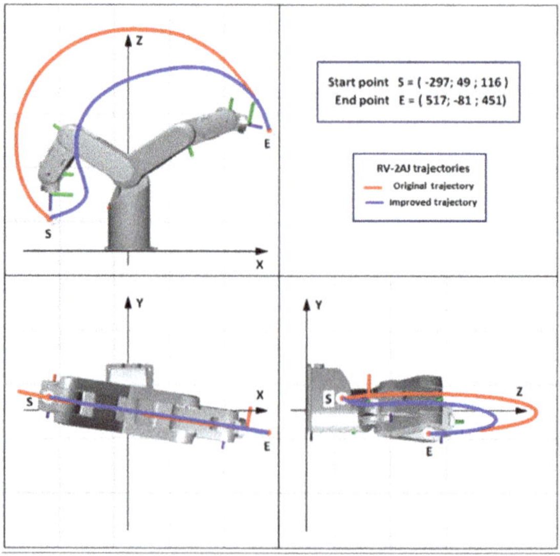

Figure 3. The trajectories of the RV-2AJ arm in three views.

2.3. The Improved Path

The suggested trajectory—named "improved path" given in blue in three views (front, top, and left side views) with the robot arm in Figure 3—is based on the newly elaborated special method that imitates the natural motion of a whip. The improved path aims to decrease the cycle time for RV-2AJ arm's movement from the starting point S to the final point E. This method is based on principle of the whip-lashing motion that determines the torques applied in closing and opening of joints of the arm, which will be discussed in the further sections.

3. Results: Newly Elaborated "Whip-Lashing" Method

For centuries whips have been considered instruments used to direct animals or torture slaves. Nowadays, this instrument has become an artform for some nations. In the last few years, researchers have been interested in the phenomenon executed by whips, which is characterized by the crack sound and the velocity that can reach supersonic speed [44,45]. Figure 4 presents the basic elements of a simple whip.

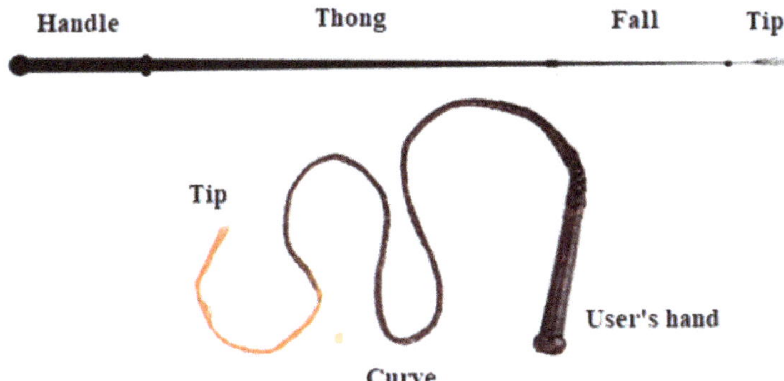

Figure 4. Basic elements of a whip.

When we give a force for a whip with the "handle" to move up to down and then stop, that produces kinetic energy. This energy will be transferred to the end of whip "tip" due to $p = m \times v$, where the mass (m) decreases and the velocity (v) increases.

The result of this is a sonic boom produced by a crack that is caused when some section of the whip moves faster than the speed of sound. The motion of a whip can include three types: a half wave, a full wave and a loop. Figures 5 and 6 present the transfer direction of momentum along the segments of whip and the diagrams describe the change of velocity, mass, kinetic energy and torque as a function of time. The input values used in the diagrams in Figure 6 were taken from Krehl et al. [46]. The subfigures show the following functions:

(1). The velocity diagram shows the changing of the velocity of the tip during a whip-lashing cycle.
(2). The mass diagram shows the changing of the mass of that part of the whip that has the most the kinetic energy during the whip-lashing as the wave impulse goes along the whip length.
(3). The kinetic energy diagram corresponds to the work of the whip user who moves the whip and conveys motion energy to the whip with his hand.
(4). The torque diagram shows the torque value changing in the wrist joint during the whip-lashing.

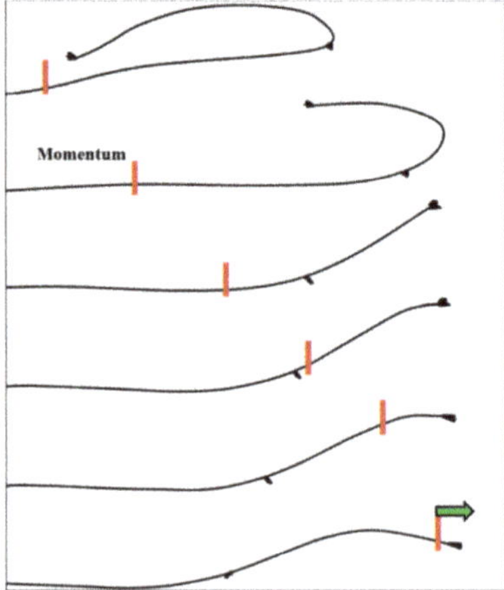

Figure 5. The motion of a whip.

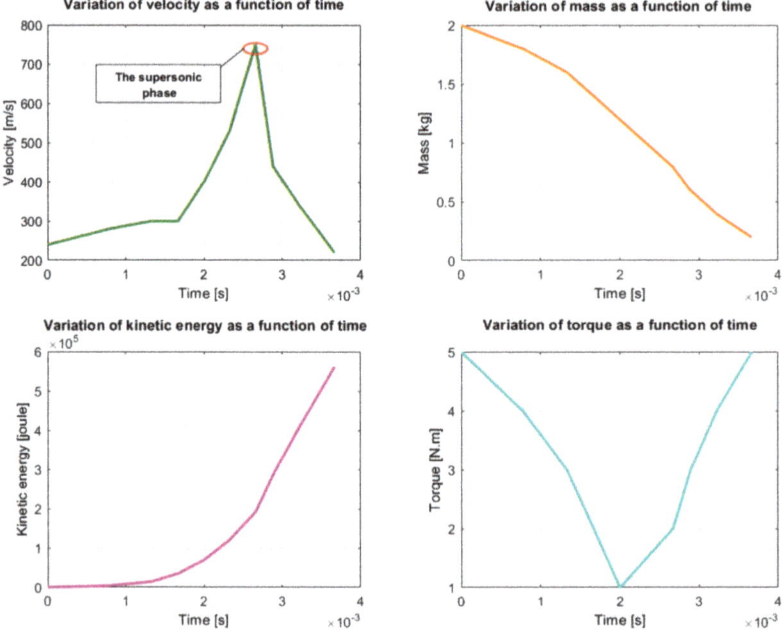

Figure 6. The variation of velocity, mass, kinetic energy, and torque as a function of time.

As we mentioned earlier, the motion of a whip is based on the momentum conservation from the beginning of the whip to the arrival to the tip, where—before cracking—many parameters vary, such as the moving mass along the whip which will be decreased and produce high velocity, presenting the supersonic boom. This phenomenon results in diminution in the torque and an increase in kinetic energy, as Figure 5 shows. A similar effect of conserving momentum can be seen in case of a figure skater doing a spin, when the skater closes his/her hands to his/her body, thus decreasing the rotational inertia and increasing the rotational velocity. Borrowing this characteristic of the motion from whip and considering the robot arm as a similar shape, the increase in the velocity of the robot parts will decrease the cycle time, so the motion of the robot arm can be used like in the mentioned whip or skater examples. The whip analogue fits the robot arm better because the arm is similar to a whip. It has many conjoined arm elements that become slimmer from the basement to the gripper.

3.1. Modelling of RV-2AJ Arm Motion as Whip-Lashing in MATLAB Software

Based on the brief analysis on the motion of the whip and considering the structural similarity to the manipulator arm, the suggestion of a whip-analogous motion and the resulted in trajectory of the gripper seems to be rational. In this article, we applied the same principle to the RV-2AJ arm, where the robot acted as a whip and the gripper executed the trajectory between two points faster than it otherwise would. The whip-lashing motion of the robot arm can be followed through the sub-figures of Figure 7. When the second joint rotation stopped, the momentum WAS transferred to the third and fourth joints. As the moving mass became smaller, the rotational velocity increased at the last links of the robot arm.

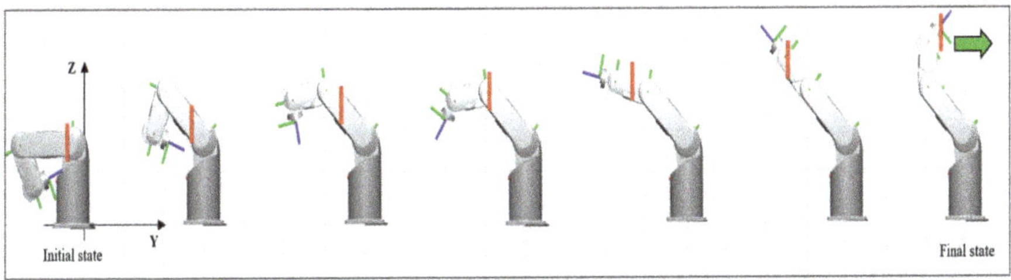

Figure 7. The motion of RV-2AJ arm.

The first step of operation of RV-2AJ arm started with the heavier segment, where the movement was realized by the torque applied at the second joint, at the "closer part to the base". In addition, the rotational inertia decreased when the motion of the links reached the smallest segment "end effector". As the motion proceeded the speedy "closer part to the base" decelerated and the outer joints opened. Finally, the "closer part to the base" stopped and the other joints finish the motion speeding up the motion of the "end effector".

In order to prove the efficiency of whip-lashing method, an iterative algorithm was created to execute the two trajectories (original and improved, see Sections 2.2 and 2.3) and calculated the minimal cycle time of each, trying to apply the maximum allowed torques for the joints.

The core of the algorithm inputted a given larger expected cycle time and determined the joint torque functions along the trajectory. If no torque maximum reaches the allowed torque limit for the given joint, the expected cycle time was decreased by a time step and the core process was repeated. If the expected cycle time was too small and the robot could only execute the trajectory with one or more joint torques exceeding the torque limit, then the decreasing time step was halved and the process applied a back step. This was repeated with a smaller time decrease. At the end of this successive approximation, the minimum cycle time that utilized the torque limit—usually this happens for one joint only—was obtained.

3.2. Elaboration of the Cycle Time Minimization (CTM) Algorithm

The organigram presented in Figure 8 contains different blocs (A–E) that define the following calculations:

- **A**: Filling up the **TM** torques matrix with the **T** torque vectors for every i-th trajectory point.
- **B**: Copying the i-th **T** torque vector in the i-th column of the **TM** torques matrix.
- **C**: Determining the $\mathbf{mT}[j]$ maximum torque for the j-th joint.
- **D**: Checking if any joint torque maximum $\mathbf{mT}[\,j\,]$ exceeds the allowed torque $\mathbf{aT}[\,j\,]$ for the j-th joint.
- **E**: Refinement of time step ts if necessary and continuing iteration, or finishing if time step ts goes below ending time step ets.

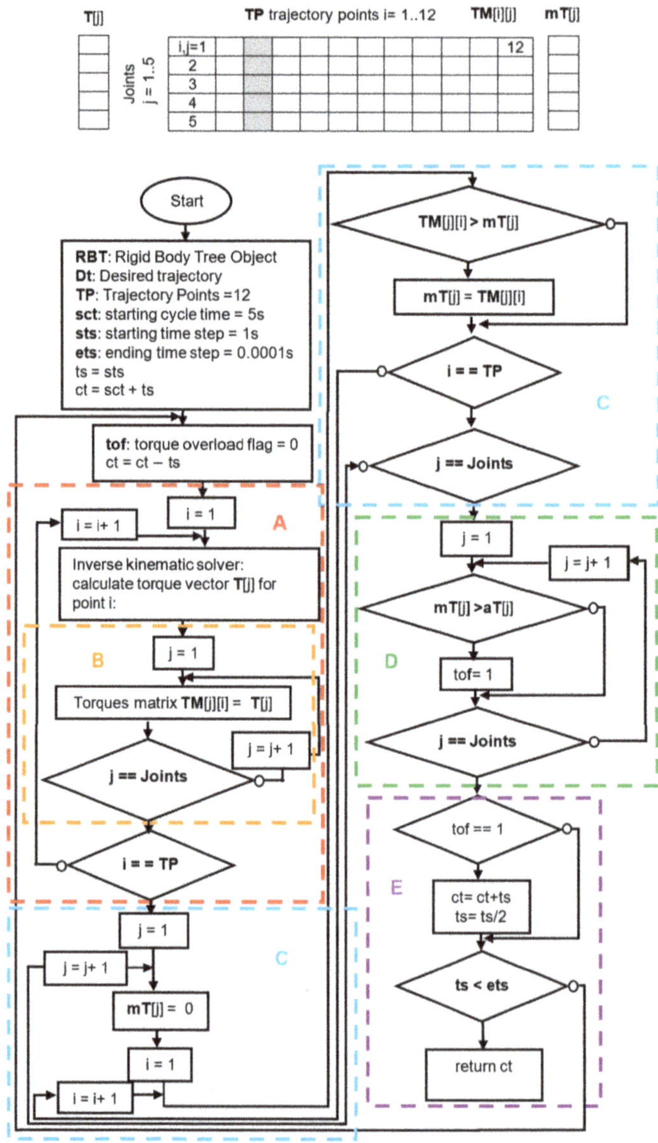

Figure 8. Cycle time minimization algorithm. A–E stand for different blocs that define different calculations.

By running the algorithm for the original and improved trajectories, the two minimum cycle times were determined and compared. In Figure 8, the elaborated cycle time minimization algorithm is introduced in detail.

The algorithm aimed to calculate the cycle time *ct* of the trajectory. It was made using a successive approximation algorithm as mentioned before. The concrete parameters of the algorithm are the next. The number of trajectory points $TP = 12$, where the index of a specified point is $i = 1, \dots , 12$. The number of RV-2AJ arm joints is 5, where the index of a specified joint is $j = 1, \dots , 5$. In the algorithm two conditions should be fulfilled:

- $ts \leq ets$ (*ts*—time step; *ets*—ending time step),
- $\mathbf{mT}[j] \leq \mathbf{aT}[j]$ ($\mathbf{mT}[j]$—the joints' torque maximums; $\mathbf{aT}[j]$—allowed torque for every j-th joint).

The algorithm aimed to determine the following data:

The $\mathbf{T}[j]$ torque vector is calculated for every i-th trajectory point and copied into the $\mathbf{TM}[j][i]$ torque matrix into the i-th column.

The $\mathbf{T}$ torque vector is determined by the "inverse Dynamics MATLAB Robotics System Toolbox function" of the MATLAB Robotic Toolbox. Then, the maximum of every j-th row of $\mathbf{TM}$ torque matrix is determined to the j-th cell of the $\mathbf{mT}[j]$ maximum torque vector. Then the $\mathbf{mT}[j] \leq \mathbf{aT}[j]$ condition setting the *tof* torque overload flag selects between cycle time decreasing or time step refinement and back stepping.

4. Discussion: Trajectory Optimization's Results of RV-2AJ Arm Using the Application of CTM Algorithm

The two trajectories are defined by positioning the same starting and ending points and 10 desired inner points that differ for each path using the spline function of MATLAB script to define the continuous path between the points. Figures 9–12 visualize the code script results for both trajectories (the original and the improved). The static Figures 10 and 12 cannot make the motion of the robot arms felt.

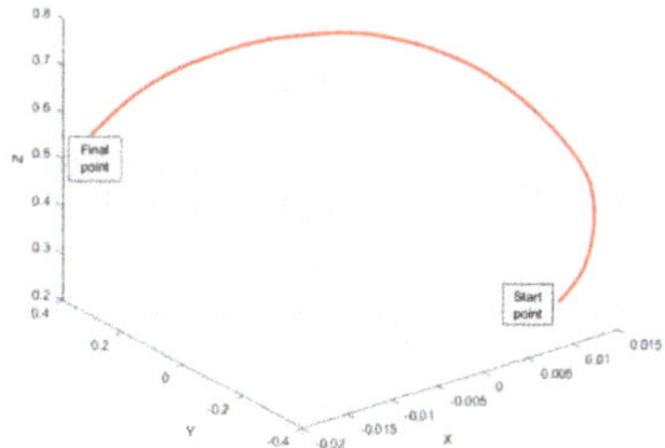

Figure 9. Original trajectory visualization.

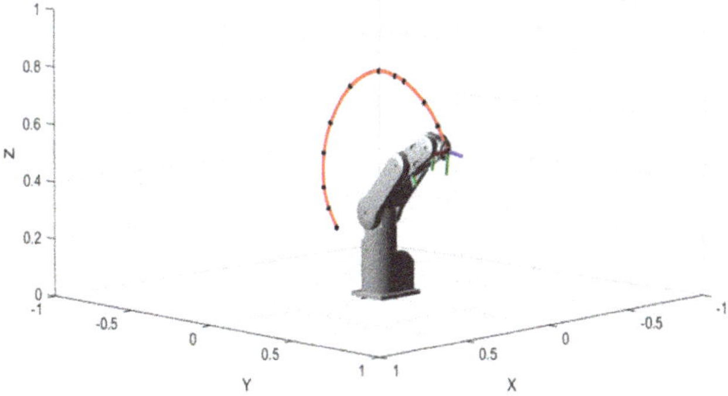

Figure 10. Visualization of the execution of the original trajectory by the RV-2AJ arm.

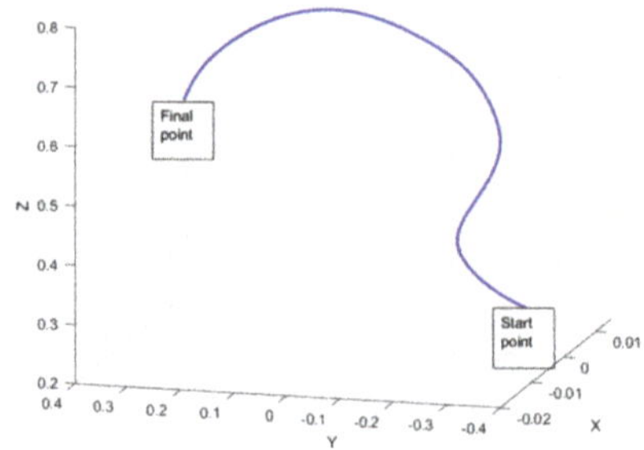

Figure 11. Improved trajectory visualization.

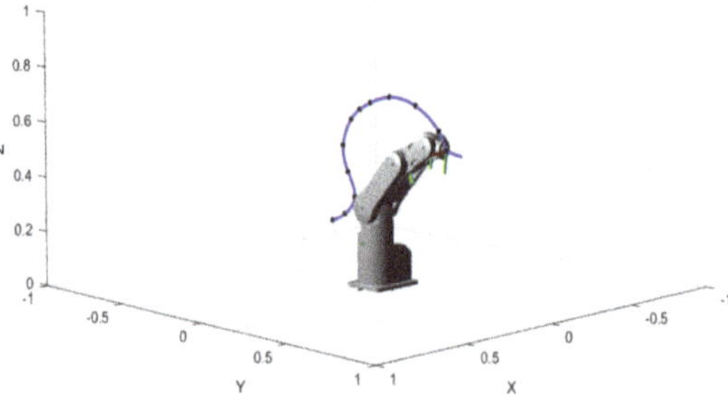

Figure 12. Visualization of the execution of the improved trajectory by RV-2AJ arm.

The execution of script code for both trajectories resulted in the value of cycle time ct for each path, as presented in Figure 13, uniting the two results with drawing manipulation to make comparable the relation between the two trajectories and robot arm motions alike. First, a starting large cycle time $ct = 5$ [s] was entered as input, then the algorithm was run till we obtained a new cycle time described as the searched cycle time sct when the torque limiting condition became unsatisfied, the algorithm continued to iterate the new values for cycle time ct and time step ts around the searched cycle time value till the value of tuned time step became smaller than the value of ending time step ets. For the original path, the minimum cycle time was $ct = 2.82$ [s], while for the improved path, the minimum cycle time was $ct = 1.86$ [s]. Comparing the two results, we proved that the improved trajectory, which is described as a whip motion analogue, utilized the maximum allowed torques with shorter cycle time than the original trajectory.

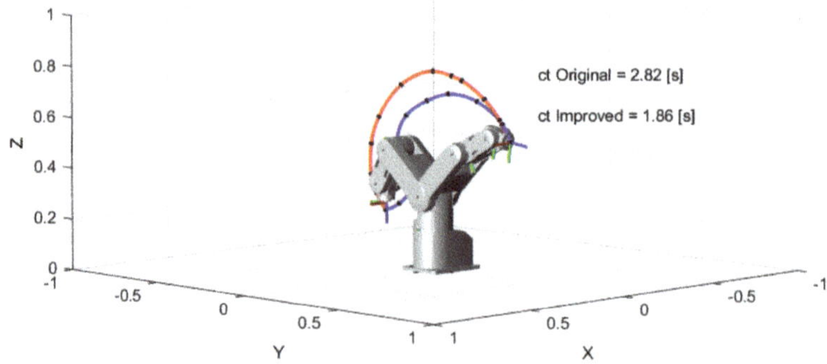

Figure 13. The original and the improved scenarios of RV-2AJ robot arm.

Modelling of RV-2AJ Arm Motion as Whip-Lashing in MATLAB Software

After studying the cycle time variation, the analysis of torque change effect regarding the both trajectories are discussed in this section. To perform the simulation in the Simulink environment, we imported the RV-2AJ body structure XML file into MATLAB. Then, we configured each link and joint in the structure to receive the vector of positions as input blocks that represent the original and the improved path, as well as to calculate the torques applied for providing the series of such positions.

Figures 14 and 15 present the block system needed to determine the finishing cycle time *ct* and torque vectors **T** calculated during the *ct* minimizing process, where the blocks "Improved Trajectory" and "Original Trajectory" contain spline function for each joint specified by angles vector that should RV-2AJ arm execute according to each position. "RV-2AJ arm block" contains the mechanical properties of the robot arm.

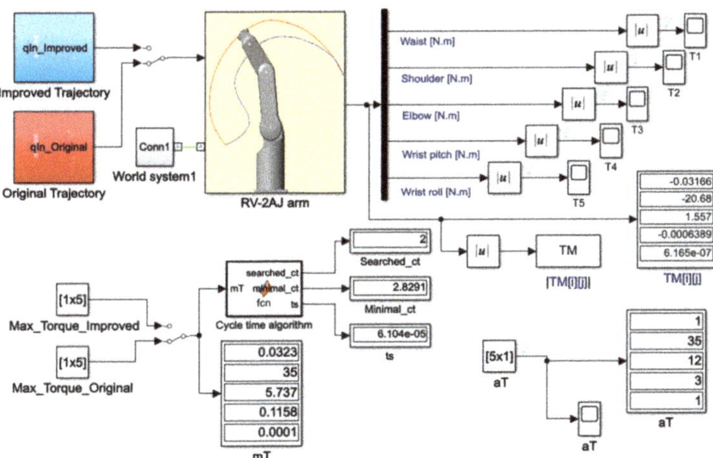

Figure 14. The block system scheme of RV-2AJ arm of the original trajectory.

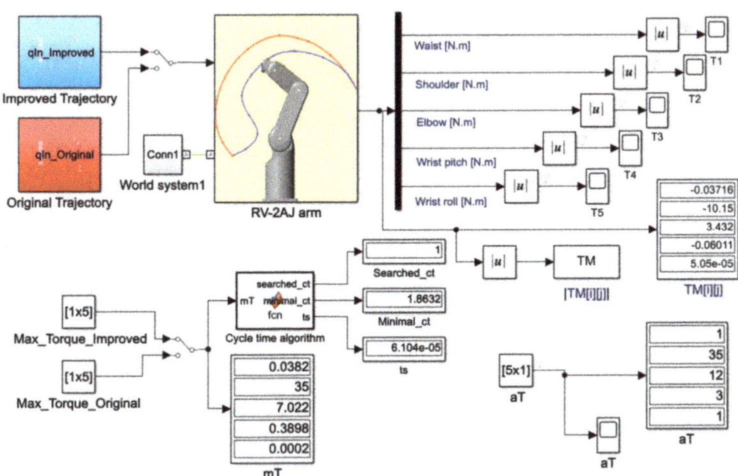

Figure 15. The block system scheme of RV-2AJ arm of the improved trajectory.

Using Scope Box, we visualized the torque diagram of each working joint (second–third–fourth, *j* = 2, 3, 4) for the original and improved paths (Figure 16).

In this section, we investigated the variation effect of the second joint (shoulder) and third joint (elbow) regarding the allowable torque values **aT** for those joints. As can be observed in Figure 16, the variation of torque values for the two paths for a starting cycle time $ct = 5$ [s] and $ts = 1$ [s] describes the behavior of the RV-2AJ arm.

In Figure 16, it can be seen that the torques of the fourth joint are close to zero due to the weight in this segment, which is very small compared to the other segments, which hold other segments. The comparison between the two paths was based on shoulder, which presents the second joint. It was clear that in the first iteration where $ct = 5$ [s], the torque curve for the original path reached higher peak values than the improved path, where the maximum torque value for the original path was **mT** = 31 Nm and for the improved path was **mT** = 27 Nm. We also observed an overshoot in the beginning stage for both curves, which is explained as the gravity effect on the mechanical structure.

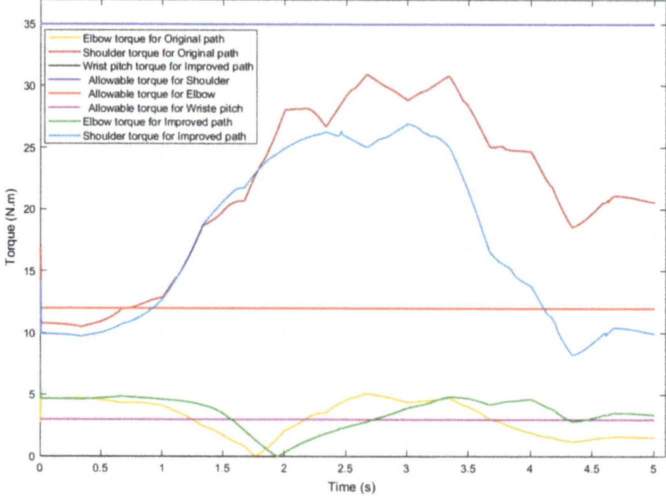

Figure 16. Variation of joint torques according to the two paths, $ct = 5$ [s].

As mentioned before, we started within a large cycle time $ct = 5$ [s] and within $ts = 1$ as a first iteration. The algorithm of minimization of cycle time started to calculate maximum torque for such a cycle time. If the maximum torque was always below the allowable torque for every joint, then the cycle time ct decreased by the time step ts.

Figure 17 presents the variation of joints torques according to the original path for different cycle times $ct = 5, 4, 3, 2$ [s] to see the behavior of torque curves and compare their peak values in each iteration, in order to observe when the torque peak of an iteration exceeds the allowable torque value **aT**.

Regarding the second joint (shoulder), in the interval 0–5 [s] we observed that the decrease of cycle time resulted in an increase of torque value for the original trajectory, where in $ct = 2$ [s] the maximum torque value of the original trajectory exceeded the allowable torque value with **mT** = 42 [N·m]. Therefore, this value $ct = 2$ [s] was stored as the searched cycle time value sct for the original trajectory and according to this value we calculated a new time step which was smaller than the first one, so the algorithm iterated the cycle time value more precisely.

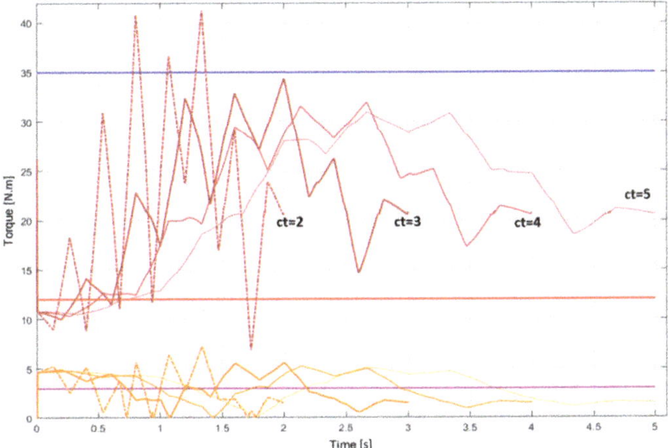

Figure 17. Variation of joint torques—original path—for different cycles.

As presented in Figure 18, the searched cycle time *sct* = 2 [s] obtained for the original path minimized the possible range to find the optimal maximum torque with optimal new cycle time. As a result, the optimal curve for shoulder was the red curve with *ct* = 2.82 [s]; this curve was obtained after the necessary iterations, starting from the first one, then minimizing the torque till obtaining the optimal curve, the "fourth iteration" m and because we had a condition to stop iteration when the time step *ts* was smaller than the ending time step *ets*; therefore, the algorithm completed the execution and found a better solution with *ct* = 2.82 [s], as presented in the first section.

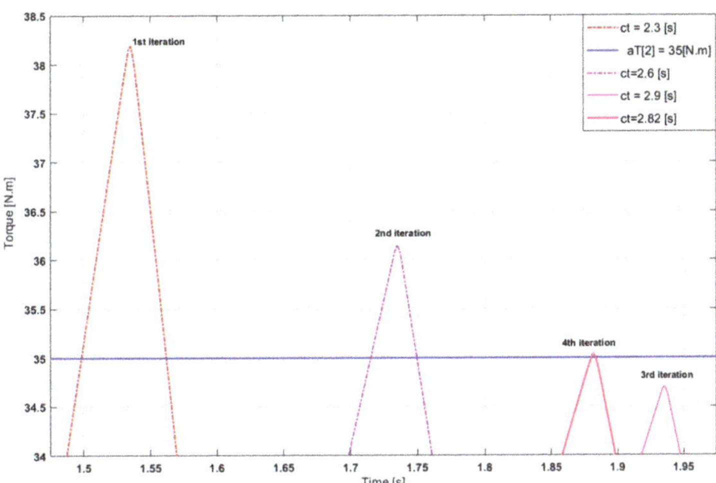

Figure 18. Different iterations of cycle time minimization algorithm for the original path around the searched cycle time *sct* = 2 [s].

Figure 19 presents the variation of joint torques for the improved path for different cycle times *ct* = 5, 4, 3, 2, 1 [s] by executing different iterations of cycle times *ct* in order to observe when the torque peak of an iteration exceeded the allowable torque value **aT**. For the improved path at *ct* = 1 [s], the maximum torque **mT** = 58 [N·m] exceeded the allowable torque value **aT**. Therefore, the searched cycle time for the improved path was *sct* = 1 [s].

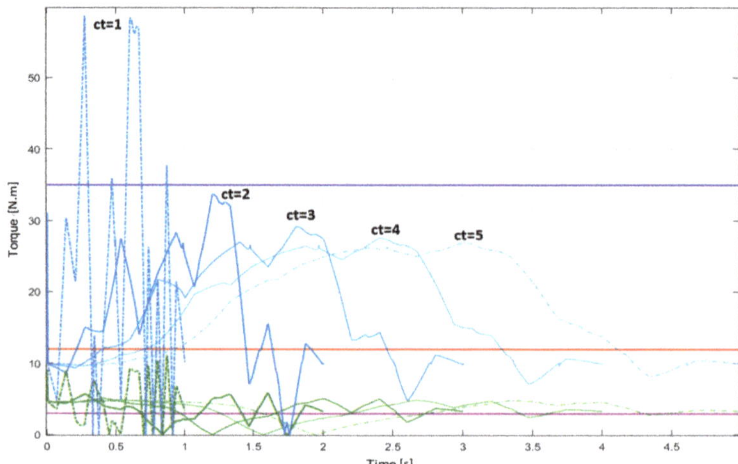

Figure 19. Variation of joints torques according to Improved path for different cycle times $ct = 5, 4, 3, 2, 1$ [s].

Figure 20 shows the execution of different iterations around the searched cycle time value obtained $sct = 1$ [s] from the results of Figure 19 to find the optimal curve. From the diagram, after 4 iterations around the searched cycle time sct = 1 [s], the algorithm achieved the optimal curve with $ct = 1.86$ [s].

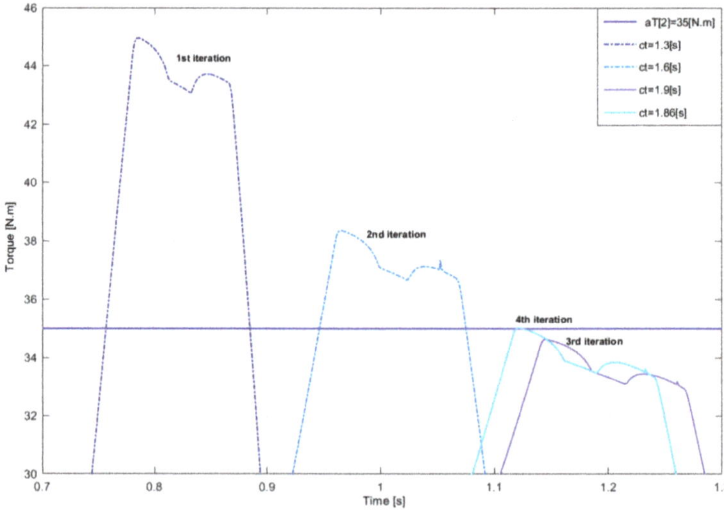

Figure 20. Different iterations of the cycle time minimization algorithm for improved path around the searched cycle time $sct = 1$ [s].

Based on the demonstration of diagrams and the comparison between the multi graphs of each path for the second joint, it was clear that the original path exceeded the allowable torque **aT** optimally with $ct = 2.82$ [s], unlike the improved path, which exceeded optimally **aT** with $ct = 1.86$ [s], as presented in Figure 21. Consequently, we proved that the improved path consumed 33% less time than the original path, which verified the concept of optimization.

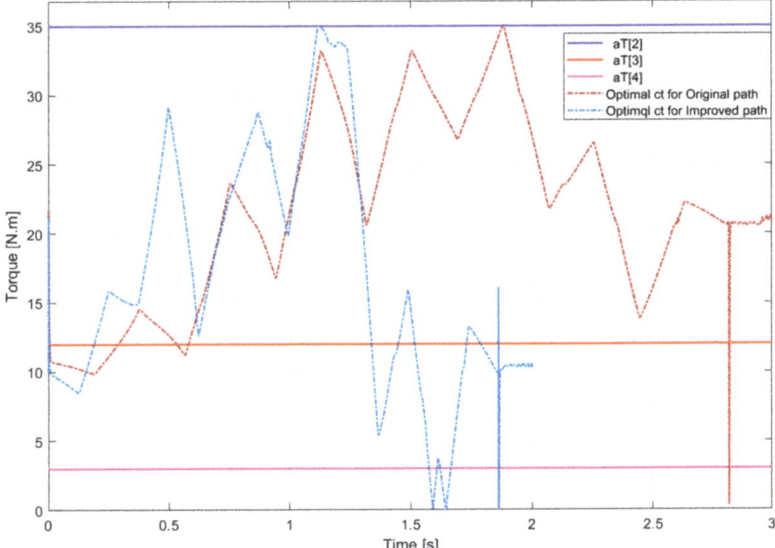

Figure 21. Optimal cycle time values for the original and improved paths.

5. Conclusions

The trajectory improvement of industrial robot arms plays an important role in the Industry 4.0 concept, since the trajectory optimization of robot arms reduces cycle times and energy consumption; furthermore, it increases productivity.

At first, a newly elaborated "whip-lashing" method was introduced for the trajectory planning of a robot arm. The idea of the "whip-lashing" method is that the motion and the shape of a whip are similar to the motion and shape of a robot arm, and it results in increasing the velocity of the robot arm's parts during operation; therefore, it reduces the cycle time. With the "whip-lashing" method, the optimized trajectory of the robot arm can be achieved in order to minimize the cycle times of manipulator arms' motion and utilize the torque of joints more effectively.

In the second part of the article, a case study was introduced to confirm the efficiency of the practical applicability of the "whip-lashing" method. In order to realize the idea of the "whip-lashing" method and check its real effect on the cycle time of robot arm motion, SolidWorks and MATLAB software applications were used in the dynamic motion simulations.

In the case study, the trajectory planning of a five-degrees-of-freedom RV-2AJ manipulator arm was described using SolidWorks and MATLAB software applications. At first, the RV-2AJ robot was modelled by the application of SolidWorks software. Two trajectories of the investigated manipulator arm were created using the application of MATLAB software and via the newly elaborated cycle time minimization algorithm. This was done in order to compare the original and the improved paths. It was found that application of the cycle time minimization algorithm resulted in a 33% shorter cycle time compared to the original path of RV-2AJ robot arm.

The main added value of the study is the elaboration and implementation of the newly elaborated "whip-lashing" method, which resulted in reduction of cycle times of manipulator arms' motion, thus increasing productivity significantly. The efficiency of the new "whip-lashing" method was confirmed by a simulation case study.

The application of this newly elaborated method can provide many advantages in industrial applications in the immediate future, where the robot arms will be designed as the shape of a whip; furthermore, the robot arms will be manufactured by usage of lightweight materials instead of recently used traditional metals. These innovative solutions (application of "whip-lashing" method and lightweight materials) will result in more flexible robot arms that can achieve motion with higher speeds without consuming higher energy, by the application of the momentum conservation law. This law can also be used by the existing rotation joint of robot arms where the motion can be achieved in a plain to increase the speed of the robot arm and decrease the motion time. This conception requires the application of new robot controllers and robot simulation software. It can be concluded that the application of the newly elaborated "whip-lashing" method results in achieving the optimized trajectory of the robot arms in order to increase velocity of the robot arm's parts, thereby minimizing motion cycle times and to utilize the torque of the joints more effectively. Consequently, the productivity will be increased significantly.

In the following research, the aim was to find a quasi-optimal trajectory between two given points. The searching for the quasi-optimum solution used the Tabu search method.

Author Contributions: Conceptualization, R.B., L.D., and G.K.; literature review and data collection, R.B.; methodology and software, R.B. and L.D.; formal analysis, R.B., L.D., and G.K.; writing—original draft preparation, R.B.; writing—review and editing, R.B., L.D., and G.K.; visualization, R.B.; supervision, L.D. and G.K.; project administration, L.D. and G.K. All authors have read and agreed to the published version of the manuscript.

Funding: This research and the APC were funded by the Stipendium Hungaricum Scholarship Program launched in 2013 by the Hungarian Government based on bilateral educational cooperation agreements signed between the Ministries responsible for education in the sending countries and Hungary.

Conflicts of Interest: The authors declare no conflict of interest.

References

1. Benotsmane, R.; Dudás, L.; Kovács, G. Collaborating robots in Industry 4.0 conception. In Proceedings of the XXIII International Conference on Manufacturing, IOP Conference Series: Materials Science and Engineering, Kecskemét, Hungary, 7–8 June 2018; pp. 1–9.
2. Benotsmane, R.; Dudás, L.; Kovács, G. Survey on new trends of robotic tools in the automotive industry. In *Vehicle and Automotive Engineering 3*; VAE 2020. Lecture Notes in Mechanical Engineering; Springer: Singapore, 2021; pp. 443–457.
3. Dima, I.C.; Kot, S. Capacity of production. In *Industrial Production Management in Flexible Manufacturing Systems*, 1st ed.; Dima, I.C., Ed.; Book News Inc.: Portland, OR, USA, 2013; pp. 40–67.
4. Kovács, G. Combination of Lean value-oriented conception and facility layout design for even more significant efficiency improvement and cost reduction. *Int. J. Prod. Res.* **2020**, *58*, 2916–2936. [CrossRef]
5. Benotsmane, R.; Kovács, G.; Dudás, L. Economic, social impacts and operation of smart factories in Industry 4.0 focusing on simulation and artificial intelligence of collaborating robots. *Soc. Sci.* **2019**, *8*, 143. [CrossRef]
6. Delgado, S.D.R.; Kostal, P.; Cagánová, D.; Cambál, M. On the possibilities of intelligence implementation in manufacturing: The role of simulation. *Appl. Mech. Mater.* **2013**, *309*, 96–104. [CrossRef]
7. Yildirim, C.; Sevil Oflaç, B.; Yurt, O. The doer effect of failure and recovery in multi-agent cases: Service supply chain perspective. *J. Serv. Theory Pract.* **2018**, *28*, 274–297. [CrossRef]
8. Gilchrist, A. *Industry 4.0: The Industrial Internet of Things*; Apress: Bangkok, Thailand, 2014; pp. 1–12.
9. El Zoghby, N.; Loscri, V.; Natalizio, E.; Cherfaoui, V. Robot cooperation and swarm intelligence. In *Wireless Sensor and Robot Networks: From Topology Control to Communication Aspects*, 1st ed.; Mitton, N., Simplot-Ryl, D., Eds.; World Scientific Publishing Company: Lille, France, 2014; pp. 168–201.
10. Alessio, C.; Maratea, M.; Mastrogiovanni, N.; Vallati, M. On the manipulation of articulated objects in human-robot cooperation scenarios. *Robot. Auton. Syst.* **2018**, *109*, 139–155.
11. Benotsmane, R.; Dudás, L.; Kovács, G. Trial—and—error optimization method of pick and place task for RV-2AJ robot arm. In *Vehicle and Automotive Engineering 3*; VAE 2020. Lecture Notes in Mechanical Engineering; Springer: Singapore, 2021; pp. 458–467.
12. Yim, M.; Shen, W.; Salemi, B.; Rus, D.; Moll, M.; Lipson, H.; Klavins, E.; Chirikjian, G.S. Modular self-reconfigurable robot systems. *IEEE Robot. Autom. Mag.* **2007**, *14*, 43–52. [CrossRef]
13. Koren, Y.; Heisel, U.; Jovane, F.; Moriwaki, T.; Pritschow, G.; Ulsoy, G.; Van Brussel, H. Reconfigurable manufacturing systems. *CIRP Ann. Manuf. Technol.* **1999**, *48*, 527–540. [CrossRef]

14. Lewis, F.L.; Abdallah, C.T.; Dawson, D.M.; Lewis, F.L. *Robot Manipulator Control: Theory and Practice*, 2nd ed.; Marcel Dekker: New York, NY, USA, 2004.

15. Wissama, K.; Etienne, D. *Modélisation Identification et Commande des Robots*, 2nd ed.; Harmes: Lavoisier, France, 1999.

16. Liu, X.; Qiu, C.; Zeng, Q.; Li, A. Kinematics analysis and trajectory planning of collaborative welding robot with multiple manipulators. *Procedia CIRP* **2019**, *81*, 1034–1039. [CrossRef]

17. Benotsmane, R.; Kacemi, S.; Benachenhou, M.R. Calculation methodology for trajectory planning of a 6 axis manipulator arm. *Ann. Fac. Eng. Hunedoara Int. J. Eng.* **2018**, *3*, 27–32.

18. Coiffet, P. *Les robots: Modélisation et Commande*, 1st ed.; Hermes Science Publications: Paris, France, 1986.

19. Kim, H.; Hong, J.; Ko, K. Optimal design of industrial manipulator trajectory for minimal time operation. *KSME J.* **1990**, *4*. [CrossRef]

20. Straka, M.; Žatkovič, E.; Schréter, R. Simulation as a means of activity streamlining of continuously and discrete production in specific enterprise. *Acta Logist.* **2014**, *1*, 11–16. [CrossRef]

21. Doan, Q.V.; Vo, A.T.; Le, T.D.; Kang, H.-J.; Nguyen, N.H.A. A novel fast terminal sliding mode tracking control methodology for robot manipulators. *Appl. Sci.* **2020**, *10*, 3010. [CrossRef]

22. Joo, S.-H.; Manzoor, S.; Rocha, Y.G.; Bae, S.-H.; Lee, K.-H.; Kuc, T.-Y.; Kim, M. Autonomous navigation framework for intelligent robots based on a semantic environment modeling. *Appl. Sci.* **2020**, *10*, 3219. [CrossRef]

23. Kim, J.; Kim, S.-R.; Kim, S.-J.; Kim, D.-H. A practical approach for minimum-time trajectory planning for industrial robots. *Ind. Robot Int. J.* **2010**, *37*, 51–61. [CrossRef]

24. Perumaala, S.; Jawahar, N. Synchronized trigonometric S-curve trajectory for jerk-bounded time-optimal pick and place operation. *Int. J. Robot. Autom.* **2012**, *27*, 385–395. [CrossRef]

25. Avram, O.; Valente, A. Trajectory planning for reconfigurable industrial robots designed to operate in a high precision manufacturing industry. *Procedia CIRP* **2016**, *57*, 461–466. [CrossRef]

26. Macfarlane, S.; Croft, E.A. Jerk-bounded manipulator trajectory planning: Design for real-time applications. *IEEE Trans. Robot. Autom.* **2003**, *19*, 42–52. [CrossRef]

27. Gasparetto, A.; Lanzutti, A.; Vidoni, R.; Zanotto, V. Experimental validation and comparative analysis of optimal time-jerk algorithms for trajectory planning. *Robot. Comput. Integr. Manuf.* **2012**, *28*, 164–181. [CrossRef]

28. Liu, H.; Xiaobo, L.; Wenxiang, W. Time-optimal and jerk-continuous trajectory planning for robot manipulators with kinematic constraints. *Robot. Comput. Integr. Manuf.* **2013**, *29*, 309–317. [CrossRef]

29. Martínez, J.R.G.; Reséndiz, J.R.; Prado, M.Á.M.; Miguel, E.E.C. Assessment of jerk performance s-curve and trapezoidal velocity profiles. In Proceedings of the XIII International Engineering Congress, Universidad Autónoma de Queretaro, Santiago de Queretaro, Mexico, 15–19 May 2017; pp. 1–7.

30. Fang, Y.; Hu, J.; Liu, W.; Shaw, Q.; Qi, J.; Peng, Y. Smooth and time-optimal S-curve trajectory planning for automated robots and machines. *Mech. Mach. Theory* **2019**, *137*, 127–153. [CrossRef]

31. Zheng, K.; Hu, Y.; Wu, B. Trajectory planning of multi-degree-of-freedom robot with coupling effect. *Mech. Sci. Technol.* **2019**, *33*, 413–421. [CrossRef]

32. Gasparetto, A.; Zanotto, V. A new method for smooth trajectory planning of robot manipulators. *Mech. Mach. Theory* **2007**, *42*, 455–471. [CrossRef]

33. Xidias, E.K. Time-optimal trajectory planning for hyper-redundant manipulators in 3D workspaces. *Robot. Comput. Integr. Manuf.* **2018**, *50*, 286–298. [CrossRef]

34. Hirakawa, A.; Kawamura, A. Trajectory planning of redundant manipulators for minimum energy consumption without matrix inversion. In Proceedings of the International Conference on Robotics and Automation, Albuquerque, NM, USA, 25 April 1997; Volume 3, pp. 2415–2420.

35. Baghli, F.Z.; El Bakkali, L.; Yassine, L. Optimization of arm manipulator trajectory planning in the presence of obstacles by ant colony algorithm. *Procedia Eng.* **2017**, *181*, 560–567. [CrossRef]

36. Saramago, S.F.P.; Steffen, V. Optimization of the trajectory planning of robot manipulators taking into account the dynamics of the system. *Mech. Mach. Theory* **1998**, *33*, 883–894. [CrossRef]

37. Devendra, G.; Manish, K. Optimization techniques applied to multiple manipulators for path planning and torque minimization. *Eng. Appl. Artif. Intell.* **2002**, *15*, 241–252.

38. Zhang, Y.; Shuzhi, S.G.; Tong, H.L. A unified quadratic-programming-based dynamical system approach to joint torque optimization of physically constrained redundant manipulators. *IEEE Trans. Syst.* **2014**, *34*, 2126–2132. [CrossRef]

39. Ding, H.; Li, Y.F.; Tso, S.K. Dynamic optimization of redundant manipulators in worst case using recurrent neural networks. *Mech. Mach. Theory* **2000**, *35*, 55–70. [CrossRef]
40. Saxena, P.; Stavropoulos, P.; Kechagias, J.; Salonitis, K. Sustainability assessment for manufacturing operations. *Energies* **2020**, *13*, 2730. [CrossRef]
41. Onwubolu, G. *A Comprehensive Introduction to SolidWorks*; SDC Publications: Mission, KS, USA, 2013.
42. Perutka, K. *MATLAB for Engineers—Applications in Control, Electrical Engineering, IT and Robotics*; Intech: Rijeka, Croatia, 2011.
43. Corke, P. *Robotics, Vision & Control: Fundamental Algorithms in MATLAB*, 2nd ed.; Springer: Victoria, Australia, 2017.
44. Goriely, A.; McMillen, T. Shape of a Cracking Whip. *Phys. Rev. Lett.* **2002**, *88*, 244301. [CrossRef]
45. Henrot, C. Characterization of Whip Targeting Kinematics in Discrete and Rhythmic Tasks. Bachelor's Thesis, MIT, Cambridge, MA, USA, 23 June 2016.
46. Krehl, P.; Engemann, S.; Schwenkel, D. The puzzle of whip cracking—Uncovered by a correlation of whip-tip kinematics with shock wave emission. *Shock Waves* **1998**, *8*, 1–9. [CrossRef]

Publisher's Note: MDPI stays neutral with regard to jurisdictional claims in published maps and institutional affiliations.

Article

A Novel Kinematic Directional Index for Industrial Serial Manipulators

Giovanni Boschetti

Department of Management and Engineering, University of Padova, 36100 Vicenza, Italy;
giovanni.boschetti@unipd.it; Tel.: +39-0444-99-8748

Received: 25 July 2020; Accepted: 26 August 2020; Published: 27 August 2020

Abstract: In the last forty years, performance evaluations have been conducted to evaluate the behavior of industrial manipulators throughout the workspace. The information gathered from these evaluations describes the performances of robots from different points of view. In this paper, a novel method is proposed for evaluating the maximum speed that a serial robot can reach with respect to both the position of the robot and its direction of motion. This approach, called Kinematic Directional Index (KDI), was applied to a Selective Compliance Assembly Robot Arm (SCARA) robot and an articulated robot with six degrees of freedom to outline their performances. The results of the experimental tests performed on these manipulators prove the effectiveness of the proposed index.

Keywords: directional index; serial robot; performance evaluation; kinematics

---

1. Introduction

Performance evaluation can provide useful information in the design of robots. For this reason, the first performance indexes were introduced as early as 1982. Manipulability [1], condition number [2], minimum singular value, kinematic isotropy [3], conditioning indexes [4] and dexterity analyses [5] provide a variety of information about a robot's performance throughout its workspace. Manipulability allows the robot to be kept far from the kinematic singularity while the condition number (and usually its reverse) is used to achieve a fast measurement of the workspace isotropy, i.e., the ratio between the maximum and the minimum performance of the robot.

These indexes, as with many others, are based on the Jacobian matrix of the velocity kinematic problem. The use of these approaches brought up some problems [6] that have been resolved with the evolution of these methods.

A modern performance index must provide information in a summarized form, such as a value, on the behavior that the manipulator can have in each position that it can reach within its workspace. The key features for a performance index are:

- To be homogeneous/independent of the unit of measurement [7–12].
- To be independent of the reference system [13,14].
- Providing increasingly accurate information [15–18].

A first sample of homogeneous indexes has been proposed in [7], where the main concept consists of directly relating the Cartesian and the actuator velocities. Another approach suggested introducing a novel dimensionally homogeneous Jacobian matrix [8]. In these methods, the velocities of three points of the flange have been considered and related with the motor velocities. A different approach was proposed in [10] where a proper performance index was developed to gather interesting information on a robot's behavior using a non-homogeneous Jacobian matrix.

An interesting approach was proposed in [13,14], where a motion/force transmission index was introduced. In these works, the use of proper virtual coefficients instead of the Jacobian matrix allowed introducing a frame-free performance index. This means that the evaluation is completely independent from the choice of the absolute reference frame in which it is computed.

Recent performance indexes [15–18] give more information compared to previous ones and are usually not affected by the adopted units of measurement and the choice of the reference frame.

Once a performance index can guarantee the above-mentioned features, it can be adopted for several purposes. In recent years, performance evaluation has been mainly conducted for the following uses:

- Optimization of manipulator design [19–21].
- Comparison between different robot architectures [22,23].
- Optimization of robot trajectory planning [24–27].
- Optimization of task locating (or robot positioning) [28,29].

The proposed performance index has been conceived to give useful information for the latter issue, i.e., finding the best location for a generic task, given the direction of the robot's main movements. The first performance index that took the direction of motion into account with respect to the actuator's movements was introduced in [30] and extended in [28]. Such an index, called a directional selective index (DSI), gives accurate information about the performance of parallel robots and allows finding the regions of the workspace where a parallel robot achieves its maximum and minimum performance. However, the DSI formulation cannot be easily extended to serial robots. For this reason, a novel approach, called the kinematic directional index (KDI), has been introduced in this paper. This method allows analyzing the behavior of a serial robot, taking into account the position of the robot and the direction of motion. As such, the index can be adopted for several purposes. In this paper, a performance analysis will be performed, by means of KDI, with two main goals:

- Finding the direction of maximum velocity with respect to a point.
- Finding the area of maximum velocity with respect to a direction of interest.

These two main analyses can provide very useful information for robot programming and the definition of the trajectory planning of industrial tasks. Performing the movements along the suggested directions can drastically reduce the time taken to complete a task.

The main motivations for this work are:

- Defining the novel KDI.
- Presenting two applications of the KDI.
- Proving the effectiveness of the KDI by exploiting two industrial robots.

The paper is organized as follows: in Section 2, the KDI is defined and formulated. In Section 3, two robots are presented for KDI computation and for the experimental tests. In the first part of Section 4, the KDI is computed at a point in the workspace in any direction of the horizontal plane, while in the second part, the index is computed along a direction of interest in the whole horizontal plane. In Section 5, the robot performances have been verified experimentally and are compared with the KDI. The strong correlation between the KDI values and the experimental results proves the effectiveness of the proposed index. Finally, in Section 6, the conclusions are addressed, and further research directions are given.

2. The KDI Performance Index

The KDI performance index allows evaluating the behavior of a serial manipulator in terms of linear velocity. For this purpose, the serial robot's kinematics are considered. As mentioned above, this index is based on the analysis of the Jacobian matrix J. As such, the analyzed problem can be identified by the forward velocity kinematic equation defined in Equation (1):

$$\dot{x} = J\dot{q} \tag{1}$$

where $\dot{x}$ and $\dot{q}$ are the velocity vectors in the Cartesian space and the joint space, respectively.

Since the KDI aims to determine the region in which a robot reaches its maximum translational velocity, only the translational part of the Jacobian matrix has been considered; hence, the motion of the wrist and its motors are not taken into account.

Once the direction of interest (i.e., the direction of motion) is defined, the velocities along the axes that are normal for the direction of interest are set as null values. Without loss of generality, it is possible to consider the direction of interest along the x-axis, and the velocity in this direction can be defined as follows in Equation (2):

$$
\left\{ \begin{array}{c} \dot{x} \\ \vdots \\ 0 \end{array} \right\} = \left[\begin{array}{cccc} j_{1,1} & j_{1,2} & \cdots & j_{1,m} \\ \vdots & \vdots & \ddots & \vdots \\ j_{n,1} & j_{n,2} & \cdots & j_{n,m} \end{array} \right] \left\{ \begin{array}{c} \dot{q}_1 \\ \vdots \\ \dot{q}_m \end{array} \right\} \tag{2}
$$

where n is the number of degrees of freedom in the Cartesian space and m is the number of actuators that defines the joint space.

Let us define R as the rotation matrix that aligns the x-axis to the direction of interest (d), J_R as the Jacobian matrix rotated by the matrix R, and $\dot{d}$ as the speed along d. In this way, in Equation (3), the velocity along the direction of interest is identified.

$$
\left\{ \begin{array}{c} \dot{d} \\ \vdots \\ 0 \end{array} \right\} = \left[\begin{array}{ccc} r_{1,1} & \cdots & r_{1,n} \\ \vdots & \ddots & \vdots \\ r_{n,1} & \cdots & r_{n,n} \end{array} \right] \left[\begin{array}{cccc} j_{1,1} & j_{1,2} & \cdots & j_{1,m} \\ \vdots & \vdots & \ddots & \vdots \\ j_{n,1} & j_{n,2} & \cdots & j_{n,m} \end{array} \right] \left\{ \begin{array}{c} \dot{q}_1 \\ \vdots \\ \dot{q}_m \end{array} \right\} = \left[\begin{array}{cccc} j_{R\,1,1} & j_{R\,1,2} & \cdots & j_{R\,1,m} \\ \vdots & \vdots & \ddots & \vdots \\ j_{R\,n,1} & j_{R\,n,2} & \cdots & j_{R\,n,m} \end{array} \right] \left\{ \begin{array}{c} \dot{q}_1 \\ \vdots \\ \dot{q}_m \end{array} \right\} \tag{3}
$$

The value of KDI is identified by the letter K and is defined as the maximum value taken by $\dot{d}$ as highlighted in Equation (4):

$$
K = \max\left(\dot{d}\right) \tag{4}
$$

When a robot moves at its maximum speed, some joint motors are also working at their maximum speed. Since Equation (3) is a linear system, it is possible to state that the minimum number of motors in this condition is equal to $m - n + 1$. For example, a solution can be found by means of a linear programming problem. However, most industrial robots are non-redundant, so a proper solution for these robots is proposed in detail. In this case, the number of degrees of freedom (n) is equal to the number of motors (m) $(n = m)$. Hence, the Jacobian matrix is a square matrix as defined in Equation (5).

$$
\left\{ \begin{array}{c} \dot{d} \\ \vdots \\ 0 \end{array} \right\} = \left[\begin{array}{ccc} j_{R\,1,1} & \cdots & j_{R\,1,n} \\ \vdots & \ddots & \vdots \\ j_{R\,n,1} & \cdots & j_{R\,n,n} \end{array} \right] \left\{ \begin{array}{c} \dot{q}_1 \\ \vdots \\ \dot{q}_n \end{array} \right\} \tag{5}
$$

It is important to highlight that the maximum velocity in the direction of interest is achieved when at least one motor reaches its maximum velocity. Therefore, in order to find the solution to this problem, the velocity in the direction of interest $(\dot{d})$ can be set to a fictitious value of "1" and the fictitious motor velocities can be easily computed since the following problem is defined by a linear system (Equation (6)) where the number of unknowns is equal to the number of equations.

$$
\left\{ \begin{array}{c} 1 \\ \vdots \\ 0 \end{array} \right\} = J_R \left\{ \begin{array}{c} \dot{q}_1 \\ \vdots \\ \dot{q}_n \end{array} \right\} \tag{6}
$$

The motor that limits the robot's performance is that which is nearest to its maximum velocity. In order to easily find this motor (p) and its ratio with respect to its maximum velocity, let us define, by means of Equation (7), the maximum value achieved by the ratios between each joint speed and its maximum.

$$\frac{1}{K} = \frac{\dot{q}_p}{\dot{q}_{p,max}} = \max\left(\frac{\dot{q}_1}{\dot{q}_{1,max}}, \ldots, \frac{\dot{q}_n}{\dot{q}_{n,max}}\right) \tag{7}$$

1/K is the maximum ratio between a joint speed and its maximum velocity, and this value is given by the joint p whose speed is the nearest to its maximum. This value can be also greater than 1, which can only happen because the solution to the problem is proposed using fictitious velocities. Nevertheless, this eventuality will not affect the validity of the solution.

Here, K identifies the KDI and also the maximum velocity that the robot can reach in the chosen direction of interest. This can be easily understood multiplying the velocities of Equation (5) by K and achieving Equation (8).

$$\left\{\begin{array}{c} K \\ \vdots \\ 0 \end{array}\right\} = J_R \left\{\begin{array}{c} K\dot{q}_1 \\ \vdots \\ K\dot{q}_n \end{array}\right\} \tag{8}$$

The value of K represents the maximum velocity of the robot, since $K\dot{q}_p$ is equal to the maximum value ($\dot{q}_{p,max}$), while the velocities of the other joints are below their maximum values.

From the definition of the KDI, it is possible to observe that it is not affected by the choice of the unit of measurement since only the value of K is used to compare the performances at different points in the workspace. Moreover, the index is also "frame free" because it does not depend on the use of an absolute reference frame but only on the robot configuration and on the direction of interest.

3. System Layout for KDI Validation

Two industrial manipulators were chosen to verify the accuracy of the KDI: an Adept Cobra 600 and an Adept Viper 650 robots (manufactured by Adept Technology, Pleasanton, CA, USA). This choice was made by looking for the most common kinematic architectures for serial robots. The first robot is a Selective Compliance Assembly Robot Arm (SCARA), while the second one is an articulated robot with six degrees of freedom (see Figure 1). In the SCARA robot, the index is used only in the horizontal plane since the vertical movements depend only on a prismatic joint whose performance does not change throughout the workspace; in the second case, the index can be useful along any direction of the workspace. However, in order to achieve comparable results, the performance of the articulated robot will also be investigated in a horizontal plane.

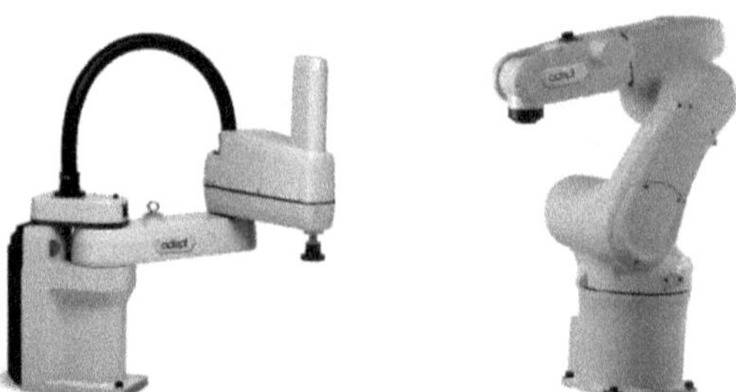

Figure 1. The Selective Compliance Assembly Robot Arm (SCARA) robot (**left**) and the articulated robot (**right**) exploited for the experimental validation of the Kinematic Directional Index (KDI) index.

3.1. SCARA Robot

The first test was performed on a horizontal plane by means of a SCARA robot. The horizontal movements depended on the motion of the first two links as represented in Figure 2, where the first two links are depicted in blue, and their reference frames are represented in violet together with the absolute frame.

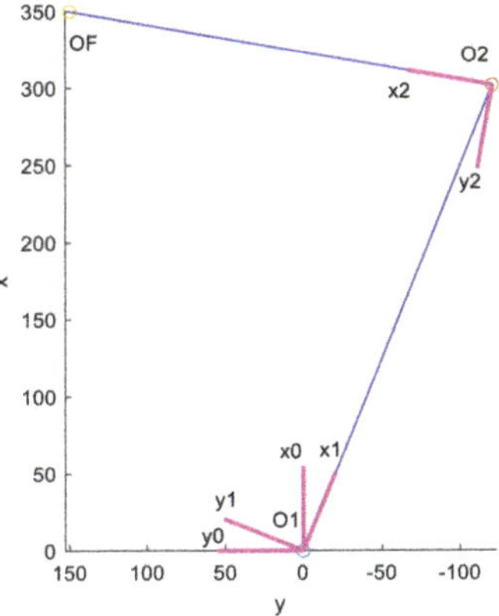

Figure 2. The kinematic scheme of the SCARA robot.

In Table 1, the complete Denavit Hartenberg table is defined in order to highlight the terms that are used for the computation of the Jacobian matrix of this robot.

Table 1. Denavit Hartenberg table of the Selective Compliance Assembly Robot Arm (SCARA) robot.

i	$T_{i,i-1}$	α_{i-1}	a_{i-1}	θ_i	d_i
1	$T_{1,0}$	0	0	θ_1	0
2	$T_{2,1}$	0	L_1	θ_2	0
3	$T_{3,2}$	π	L_2	0	d_3
4	$T_{4,3}$	0	0	θ_4	0

The projection in the horizontal plane of a SCARA manipulator allows highlighting the lengths of the two main links: $L_1 = 325$ mm and $L_2 = 275$ mm. The end-effector horizontal speed is given in function of the actuator speeds by means of the Jacobian matrix defined in Equation (9).

$$\mathbf{J}_S = \left[\begin{array}{cc} z_1 \times (\mathbf{OF} - \mathbf{O}_1) & z_2 \times (\mathbf{OF} - \mathbf{O}_2) \end{array} \right] \tag{9}$$

In this case, it is possible to easily calculate the Jacobian Matrix as follows in Equation (10):

$$\mathbf{J}_S = \left[\begin{array}{cc} -L_1 \sin(\theta_1) - L_2 \sin(\theta_{1,2}) & -L_2 \sin(\theta_2) \\ L_1 \cos(\theta_1) + L_2 \cos(\theta_{1,2}) & L_2 \cos(\theta_2) \end{array} \right] \tag{10}$$

where the expression $\theta_{i,j} = \theta_i + \theta_j$ has been used.

3.2. Articulated Robot

Figure 3 depicts the kinematic scheme of the articulated robot, without taking its wrist into account. The links that are involved in the translation of the wrist center are depicted in blue and their reference frames are represented in violet together with the absolute frame. The origin of the wrist is also highlighted (OF). In Table 2, all the kinematic parameters are highlighted by means of the Denavit Hartenberg table.

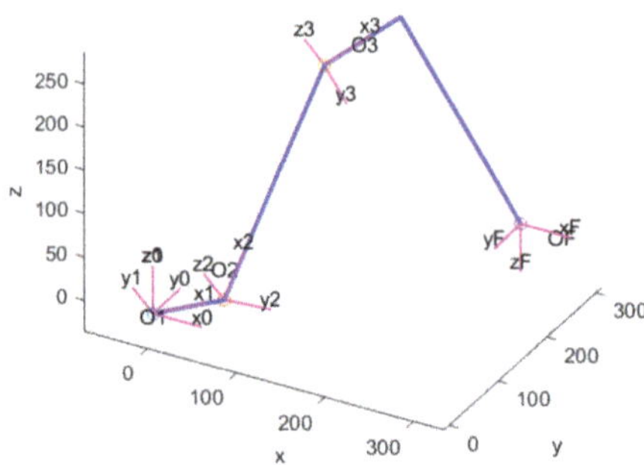

Figure 3. The kinematic scheme of the articulated robot.

Table 2. Denavit Hartenberg table of the articulated robot.

i	$T_{i,i-1}$	α_{i-1}	a_{i-1}	θ_i	d_i
1	$T_{1,0}$	0	0	θ_1	0
2	$T_{2,1}$	$-\pi/2$	a_1	θ_2	0
3	$T_{3,2}$	0	L_1	θ_3	0
4	$T_{4,3}$	$-\pi/2$	a_3	θ_4	L_2
5	$T_{5,4}$	$\pi/2$	0	θ_5	0
6	$T_{6,5}$	$\pi/2$	0	θ_6	0

Where the lengths of the links are $L_1 = 270\ mm$ and $L_2 = 295$ mm, while the non-null offset between the kinematic axes are $a_1 = 75$ mm and $a_3 = 90$ mm.

Using these parameters, the Jacobian matrix of Equation (11) can be computed at each point of the workspace.

$$J_A = \left[\ z_1 \times (OF - O_1) \quad z_2 \times (OF - O_2) \quad z_3 \times (OF - O_3)\ \right] \tag{11}$$

This Jacobian matrix relates the velocity of the wrist center with the ones of the joints.

4. Performance Investigation

Given the proper Jacobian matrix, it is possible to compute the KDI in any point of the workspace and in any direction. To highlight the usefulness of the KDI, two different performance investigations were proposed. The first one consisted of finding the maximum velocity that the robot can reach in a point for each direction of motion. The second one pointed out the areas where the robot reached its maximum velocity with respect to a direction of interest.

4.1. Performance Investigation in a Point

Without loss of generality, the KDI for a SCARA robot is plotted at the point P (350, 150) along any direction of the horizontal plane (see Figure 4). The performance along the x-axis is highlighted with a red line.

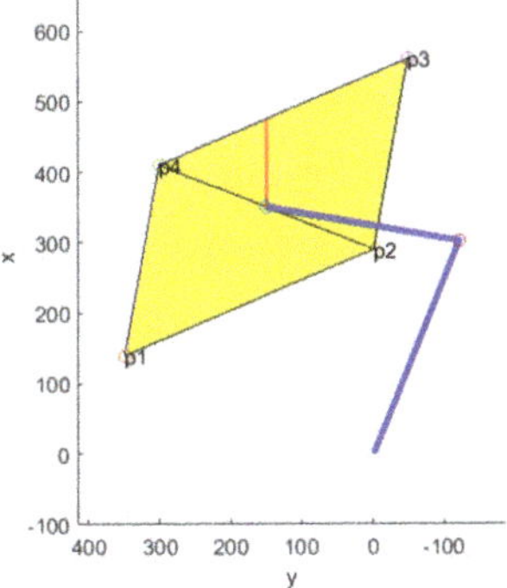

Figure 4. KDI computed for the SCARA robot in point P (350, 150) along any horizontal direction.

As mentioned above, when the maximum speed along any direction of a robot is achieved, at least one motor reaches its maximum speed. Therefore, given a generic point of the workspace, if this type of information is needed, the computation of the KDI can be greatly simplified. In fact, the maximum speed (absolute or relative) is reached when both motors reach their maximum (or minimum) speed. As such, the directions of the maximum Cartesian speed can be calculated by using a proper matrix that contains the four possible combinations of maximum and minimum speed. By multiplying such a matrix with the Jacobian matrix, the maximum speed direction can be identified by Equation (12).

$$P = J_S \begin{bmatrix} \dot{q}_{1,max} & -\dot{q}_{1,max} & -\dot{q}_{1,max} & \dot{q}_{1,max} \\ \dot{q}_{2,max} & \dot{q}_{2,max} & -\dot{q}_{2,max} & -\dot{q}_{2,max} \end{bmatrix} \tag{12}$$

where P is a matrix made up of the four virtual points that represent the vertexes of the parallelogram that delimits the robot speed in the horizontal plane as depicted in Figure 4.

By following the same approach with the articulated robot, the maximum and minimum values of the three joint speeds allowed defining the eight vertexes of a cuboid as depicted in Figure 5. Given these vertexes, the maximum speed of the robot could be gathered in any direction of the workspace. In Figure 5, for example, the performance of the robot along the x-axis is highlighted with a red line.

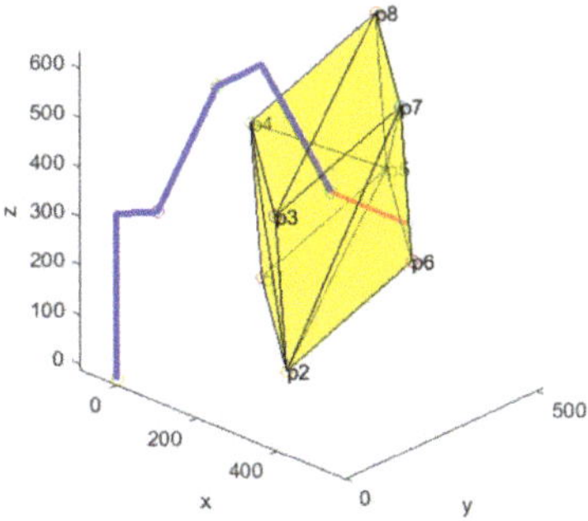

Figure 5. KDI computed for the articulated robot in point P (250, 300, 350) along any direction.

4.2. Performance Investigation in the Workspace

As mentioned above, the performance was evaluated in a horizontal plane. For each robot, a proper set of points was defined, and a direction of interest was chosen. The SCARA set of points was given by nine points along any radial direction of the workspace with an angular step of ten degrees.

Since the articulated robot was also investigated in a horizontal plane, the plane with the highest reach distance was chosen (i.e., where the wrist was at the same height as the shoulder). For this test, the set was given by seven points along any radial direction with an angular step of ten degrees. The points that are outside the workspace were considered. In Figure 6, the two sets of points are depicted together with the workspace limits.

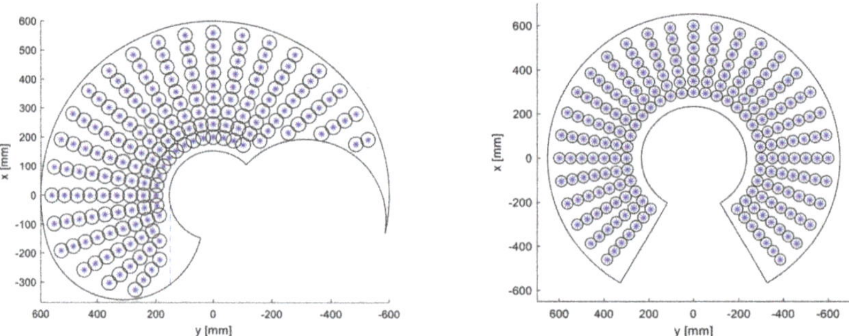

Figure 6. Point sets in which the experimental analyses were performed for the SCARA robot (**left**) and the articulated robot (**right**).

For the SCARA robot, the KDI index was computed along the direction of the y-axis (i.e., with a rotation of 90 degrees about the z-axis by means of the matrix $\boldsymbol{R}$). The KDI was computed for the articulated robot with a rotation of 45 degrees about the z-axis. The KDI was normalized and plotted in Figure 7 for both robots. The yellow areas indicate the regions where the robots achieved their best performance, while the blue areas show where the robots were slower.

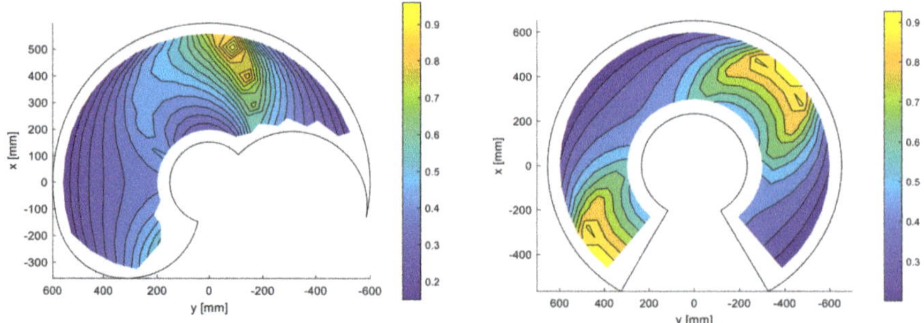

Figure 7. KDI computed for SCARA (**left**) and articulated (**right**) robots.

5. Experimental Setup and Performance Analysis in the Workspace

In order to experimentally verify where the robots can achieve the maximum velocity, the end effector of each robot was equipped with an accelerometer, as shown in Figure 8. The chosen device was a piezometric accelerometer, DeltaTron Type 4508001 (manufactured by Hottinger Brüel & Kjær, Nærum, Denmark), which did not require additional hardware for signal conditioning. The data were acquired by means of an LMS Pimento analyzer (manufactured by LMS International, Leuven, Belgium) at an acquisition frequency of 1000 Hz. The software supplied with the analyzer allowed directly computing the speed of the investigated movements.

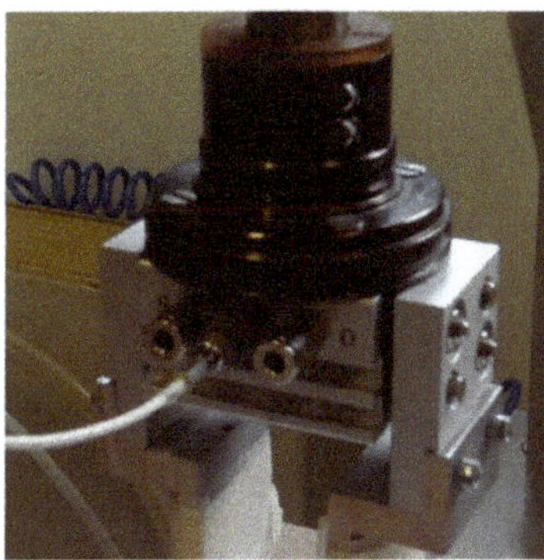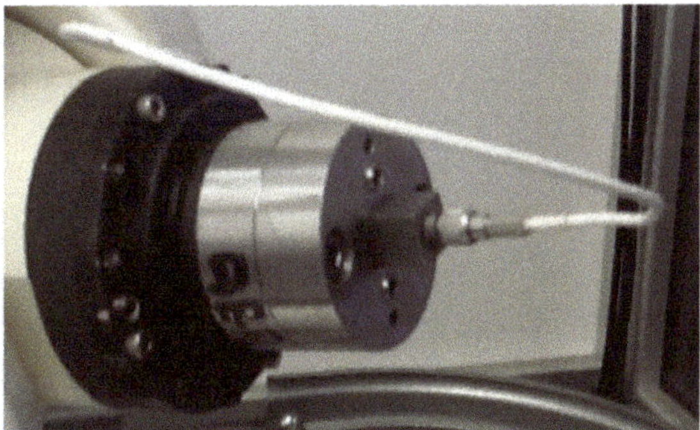

Figure 8. The end effectors of the robots equipped with accelerometers.

The experimental analysis was performed with the same set of points defined above and illustrated in Figure 6, in order to compare the results of the investigation with the values taken by the KDI. At each point, a back-and-forth movement with a length of 60 mm in the direction of interest was repeated five times so that the center of the displacement (where the maximum speed was reached) coincided with the investigated point. Using a data acquisition system, the signals of the accelerometers were gatherd, and the velocity of the robots was computed. In order to reduce the effect of measuring errors, the maximum experimentally measured speed was computed by calculating the average between the peaks of each repeated movement. In this way, the maximum speed reached in each point of the workspace could be collected. Such results have been normalized since the target of the work is to highlight the region of maximum speed without making performance comparisons between different robots. The experimentally gathered normalized performances are plotted in Figure 9 for both robots.

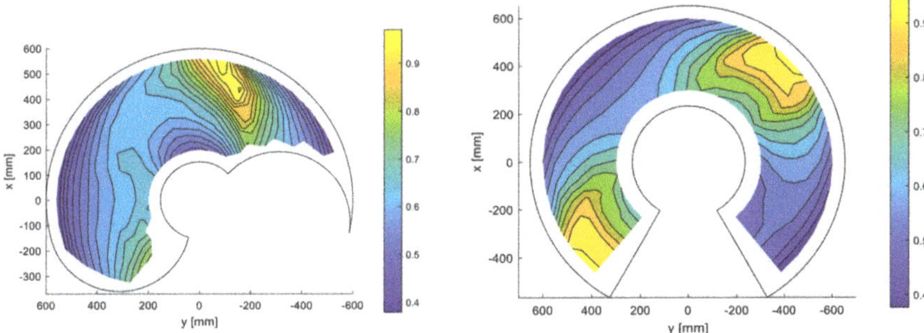

Figure 9. Normalized maximum speeds of the SCARA (**left**) and the articulated (**right**) robots.

It is now possible to appreciate that the values of the KDI perfectly matched the behavior of the robots; the best performance region identified by the KDI has a good correspondence with the measured region. Moreover, the trend of performance given by the KDI met the speed variation throughout the workspace. In fact, the comparison between Figures 7 and 9 highlights that the KDI can be adopted as a useful tool to foresee the regions where a robot can achieve its maximum speed with respect to a direction of interest. In order to better identify the correlation between the index and the real performance, the maximum value of the measured speed has been plotted in Figure 10 for each investigated point with respect to the value given by the KDI at the same point. Moreover, the linear regression between the KDI and the speed of each robot has been computed and plotted by means of a red line. In Figure 10, the proximity of the points on the scatter diagram to the red lines clearly shows a strong correlation with the KDI values of the robots' velocities; therefore, it proves the effectiveness of the novel proposed index.

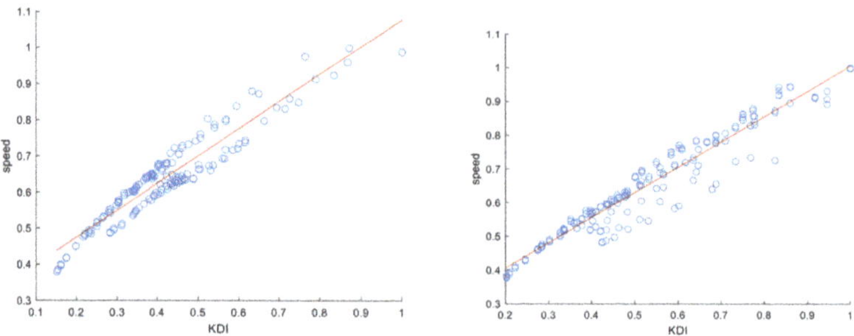

Figure 10. Scatter diagram of the KDI versus speeds for the SCARA (**left**) and articulated (**right**) robots.

6. Conclusions

A novel performance index called the KDI has been proposed. This index is not affected by the non-homogeneous Jacobian matrix and does not depend on the choice of the units of measurement and on the position of the absolute reference frame. The index was computed for a SCARA robot and an articulated robot to analyze the behavior of the manipulators for two main purposes: finding the direction of maximum velocity with respect to a point, and finding the area of maximum velocity with respect to a direction of interest in the workspace. Being a fully kinematic index (i.e., it does not consider the dynamics and the control of the robots), it gives a high-quality description of the robot's behavior, and the experimental tests demonstrate the effectiveness of the proposed index.

Future works can be addressed by following some steps: first, the use of the index for redundant manipulators will be investigated; second, the same reasoning will be followed to synthesize a dynamic performance index that is able to consider the direction of motion; afterwards, the use of these indexes for other issues, such as the optimal design of manipulators and/or robot trajectory planning, will be investigated.

Funding: This research received no external funding.

Conflicts of Interest: The author declares no conflict of interest.

References

1. Yoshikawa, T. Manipulability of Robotic Mechanisms. *Int. J. Robot. Res.* **1985**, *4*, 3–9. [CrossRef]
2. Salisbury, J.K.; Craig, J.J. Articulated hands: Force control and kinematic issues. *Int. J. Robot. Res.* **1982**, *1*, 4–17. [CrossRef]
3. Angeles, J.; López-Cajún, C.S. Kinematic Isotropy and the Conditioning Index of Serial Robotic Manipulators. *Int. J. Robot. Res.* **1992**, *11*, 560–571. [CrossRef]
4. Gosselin, C.; Angeles, J. A Global Performance Index for the Kinematic Optimization of Robotic Manipulators. *J. Mech. Des.* **1991**, *113*, 220–226. [CrossRef]
5. Gao, F.; Liu, X.; Gruver, W.A. Performance evaluation of two-degree-of-freedom planar parallel robots. *Mech. Mach. Theory* **1998**, *33*, 661–668. [CrossRef]
6. Merlet, J.P. Jacobian, Manipulability, Condition Number, and Accuracy of Parallel Robots. *J. Mech. Des.* **2005**, *128*, 199–206. [CrossRef]
7. Gosselin, C. The optimum design of robotic manipulators using dexterity indices. *Robot. Auton. Syst.* **1992**, *9*, 213–226. [CrossRef]
8. Kim, S.-G.; Ryu, J. New dimensionally homogeneous jacobian matrix formulation by three end-effector points for optimal design of parallel manipulators. *IEEE Trans. Robot. Autom.* **2003**, *19*, 731–737. [CrossRef]

9.   Kim, S.-G.; Ryu, J.-H. Force transmission analyses with dimensionally homogeneous Jacobian matrices for parallel manipulators. *KSME Int. J.* **2004**, *18*, 780–788. [CrossRef]

10.   Mansouri, I.; Ouali, M. A new homogeneous manipulability measure of robot manipulators, based on power concept. *Mechatronics* **2009**, *19*, 927–944. [CrossRef]

11.   Mansouri, I.; Ouali, M. The power manipulability – A new homogeneous performance index of robot manipulators. *Robot. Comput. Manuf.* **2011**, *27*, 434–449. [CrossRef]

12.   Cardou, P.; Bouchard, S.; Gosselin, C. Kinematic-Sensitivity Indicies for Dimensionally Nonhomogeneous Jacobian Matrices. *IEEE Trans. Robot.* **2010**, *26*, 166–173. [CrossRef]

13.   Wang, J.; Liu, X.; Wu, C. Optimal design of a new spatial 3-DOF parallel robot with respect to a frame-free index. *Sci. China Ser. E: Technol. Sci.* **2009**, *52*, 986–999. [CrossRef]

14.   Wang, J.; Wu, C.; Liu, X. Performance evaluation of parallel manipulators: Motion/force transmissibility and its index. *Mech. Mach. Theory* **2010**, *45*, 1462–1476. [CrossRef]

15.   Brinker, J.; Corves, B.; Takeda, Y. Kinematic performance evaluation of high-speed Delta parallel robots based on motion/force transmission indices. *Mech. Mach. Theory* **2018**, *125*, 111–125. [CrossRef]

16.   Wang, C.; Zhao, Y.; Dong, C.; Liu, Q.; Niu, W.; Liu, H.; Liu, H. Kinematic Performance Comparison of Two Parallel Kinematics Machines. In *Advances in Mechanism and Machine Science. IFToMM WC 2019. Mechanisms and Machine Science*; Uhl, T., Ed.; Springer: Cham, Switzerland, 2019; Volume 73.

17.   Boschetti, G.; Trevisani, A. Cable Robot Performance Evaluation by Wrench Exertion Capability. *Robotics* **2018**, *7*, 15. [CrossRef]

18.   La Mura, F.; Romanó, P.; Fiore, E.; Giberti, H. Workspace Limiting Strategy for 6 DOF Force Controlled PKMs Manipulating High Inertia Objects. *Robotics* **2018**, *7*, 10. [CrossRef]

19.   Wu, X. Performance Analysis and Optimum Design of a Redundant Planar Parallel Manipulator. *Symmetry* **2019**, *11*, 908. [CrossRef]

20.   Alvarado, R.R.; Castañeda, E.C. Optimum design of the reconfiguration system for a 6-degree-of-freedom parallel manipulator via motion/force transmission analysis. *J. Mech. Sci. Technol.* **2020**, *34*, 1339–1349. [CrossRef]

21.   Yu, G.; Wu, J.; Wang, L.; Gao, Y. Optimal Design of the Three-Degree-of-Freedom Parallel Manipulator in a Spray-Painting Equipment. *Robotics* **2019**, *38*, 1064–1081. [CrossRef]

22.   Wu, X.; Yan, R.; Xiang, Z.; Zheng, F.; Tan, R. Performance Analysis and Comparison of Three Planar Parallel Manipulators. *Power Transm.* **2019**, *79*, 270–279. [CrossRef]

23.   Baena, A.H.; Valdez, S.I.; Romero, F.D.J.T.; Montes, M.M. Comparison of Parallel Versions of SA and GA for Optimizing the Performance of a Robotic Manipulator. *Power Transm.* **2020**, *86*, 290–303. [CrossRef]

24.   Boschetti, G. A Picking Strategy for Circular Conveyor Tracking. *J. Intell. Robot. Syst.* **2016**, *81*, 241–255. [CrossRef]

25.   Bottin, M.; Rosati, G. Trajectory Optimization of a Redundant Serial Robot Using Cartesian via Points and Kinematic Decoupling. *Robot.* **2019**, *8*, 101. [CrossRef]

26.   Bottin, M.; Ceccarelli, M.; Morales-Cruz, C.; Rosati, G. Design and Operation Improvements for CADEL Cable-Driven Elbow Assisting Device. *Mech. Mach. Sci.* **2021**, *91*, 503–511.

27.   Bottin, M.; Rosati, G.; Boschetti, G. Working Cycle Sequence Optimization for Industrial Robots. *Mech. Mach. Sci.* **2021**, *91*, 228–236.

28.   Boschetti, G.; Rosa, R.; Trevisani, A. Optimal robot positioning using task-dependent and direction-selective performance indexes: General definitions and application to a parallel robot. *Robot. Comput. Manuf.* **2013**, *29*, 431–443. [CrossRef]

29.   Sharkawy, A.-N.; Papakonstantinou, C.; Papakostopoulos, V.; Moulianitis, V.; Aspragathos, N. Task Location for High Performance Human-Robot Collaboration. *J. Intell. Robot. Syst.* **2020**, 1–20. [CrossRef]

30.   Boschetti, G.; Trevisani, A. Direction Selective Performance Index for Parallel Manipulators. In Proceedings of the First International Conference on Multibody System Dynamics IMSD 2010, Lappeenranta, Finland, 25–27 May 2010.

 *applied sciences*

Article

Energy-Efficiency Improvement and Processing Performance Optimization of Forging Hydraulic Presses Based on an Energy-Saving Buffer System

Xiaopeng Yan and Baijin Chen *

State Key Laboratory of Materials Processing and Die & Mould Technology, School of Materials Science and Engineering, Huazhong University of Science and Technology, Wuhan 430074, China; yanxp@hust.edu.cn
* Correspondence: chenbaijin@hust.edu.cn

Received: 14 July 2020; Accepted: 29 August 2020; Published: 31 August 2020

Abstract: This paper proposes an energy-saving system based on a prefill system and a buffer system to improve the energy efficiency and the processing performance of hydraulic presses. Saving energy by integrating such systems into the cooling system of a hydraulic press has not been previously reported. A prefill system, powered by the power unit of the cooling system, is used to supply power simultaneously with the traditional power unit during the pressurization stage, thus reducing the usage of pumps and installed power of the hydraulic press. In contrast to the traditional prefill system, the proposed energy-saving system is controlled by a servo valve to adjust flow according to the load profile. In addition, a buffer system is employed to the cooling system to absorb the hydraulic shock generated at the unloading stage, store those shares of hydraulic energy as a recovery accumulator, and then release this energy to power the prefill system and the hydraulic actuator in the subsequent productive process. Finally, through a series of comparative experiments, it was preliminarily validated that the proposed system could reduce the installed power and pressure shock by up to 22.85% and 41%, respectively, increase energy efficiency by up to 26.71%, and provide the same processing characteristics and properties as the traditional hydraulic press.

Keywords: hydraulic press; energy saving; energy efficiency; installed power; processing performance

1. Introduction

Hydraulic presses are commonly used for forging, molding, blanking, punching, deep drawing, and other metal forming operations because of their high load capacity, high power-to-mass ratio, and large force/torque output capacity. However, they are also known for their high energy consumption, low energy efficiency, and poor processing characteristics. As a result of the significant difference between the installed power and the demanded power, 70% of the total energy consumption is attributed to power dissipation and actuation of the hydraulic system. Therefore, improving the energy efficiency of hydraulic presses is an urgent issue that manufacturing industries must resolve [1–5].

Energy dissipation generates from each part of the hydraulic press system when motion or power are transmitted, as shown in Figure 1. It is estimated that only 9.32% of the input energy is transmitted into the forming energy. To better study the energy-saving methods for hydraulic presses, it is necessary to be clear on the energy consumption characteristics of hydraulic system. For this purpose, Zhao [6] established the basic energy flow model of the hydraulic press system, revealed the energy dissipation mechanism of each component, and indicated that the imbalance between installed power and demanded power is the main cause of low energy efficiency. Installed power is designed to meet the maximum power requirements of PS. However, as the same power unit also serves other low-power operations, mismatch between installed power and demanded power occurs.

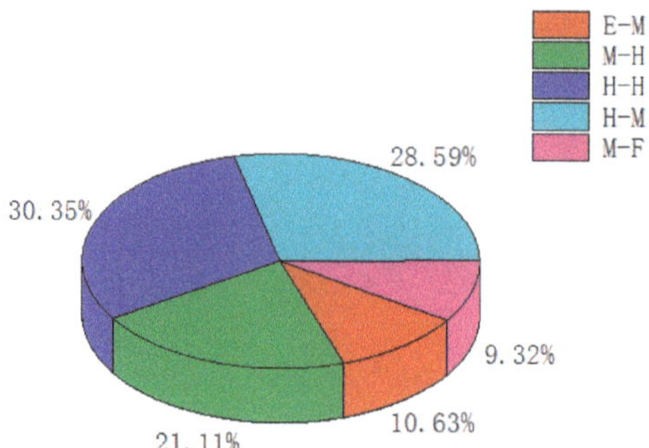

Figure 1. Energy dissipation of each energy conversion unit during a working cycle. (E-M, electrical–mechanical energy unit; M-H, mechanical–hydraulic energy unit; H-H, hydraulic–hydraulic energy unit; H-M, hydraulic–mechanical energy unit; M-F, mechanical-forming energy unit).

During the past decades, researchers are applying an increasing amount of effort to achieve the match between the installed power and the required power for hydraulic presses. As reported by Zhao [6] and Huang [7], energy-matching methods for hydraulic presses can be divided into two categories: energy-matching methods and energy-recovery methods.

A common approach based on an energy-matching mechanism is the volume control electrohydraulic system driven directly by various kinds of variable-speed motors and variable-displacement pumps. The control of pressure, flow, and direction of working liquid is achieved by changing the rotation speed or output displacement. Quan and Helduser [8] applied variable-speed motors and constant-displacement pumps in hydraulic drive units to match load variations. Su et al. [9] applied variable-frequency motors in the hydraulic press to reduce the mismatch between output power and load power. Camoirano and Dellepiane [10] employed variable-frequency drive (VFD) technology to achieve more efficient energy management as well as the precise control of torque and speed of AC motors. Wang et al. [11] adopted a calibration model with a genetic algorithm to adjust the variable-speed pump flow rate to their designated value and achieve an energy-saving ratio of at least 16.1%. Ge [12] adopted a variable-speed motor to drive a variable-displacement pump and employed a matching method based on segmented speed and continuous displacement control of the pump to reduce the throttle loss. The energy-saving ratio under partial load condition can be up to 33%. Since there is limited scope for further increasing the electrohydraulic unit's efficiency, researchers have also focused on the design of hydraulic control circuits to reduce the mismatch. Load sensing systems [13], hydraulic adaptive systems [14], fuzzy control systems [15], close-loop volume control systems [16], negative flow control systems [17], and secondary regulation systems are all useful methods to match the load by regulating operating parameters and system states. As these methods achieve matching by adjusting the output flow, speed, and pressure, the installed power remains unaltered.

Many papers on energy-recovery methods have been published in recent years. A lot of energy-recovery circuits have been applied in hydraulic press machines. Yan et al. [18] proposed a flywheel energy-saving system (FESS) that can store the redundant energy at no-load stages and low-load stages and then release the stored energy at high-load stages. Dai et al. [19] applied a hydraulic accumulator to a 20 MN fast forging hydraulic press to realize energy conversion by absorbing large flow–pressure pulses and hydraulic shock. The results show that the hydraulic accumulator has promising energy-saving effects. Triet and Ahn [20] utilized a hydraulic accumulator and a flywheel to realize energy recovery and presented a control strategy for this energy-saving method. Ven [21] developed a novel hydraulic accumulator that can keep the hydraulic system pressure constant by using a piston with an area that varies with stroke. Compared with conventional recovery accumulators, this new accumulator could significantly increase the energy storage density. Xia et al. [22] proposed an integrated drive and energy recuperation system for a hydraulic excavator, and the large gravitational potential energy of the boom was recovered by a three-chamber hydraulic cylinder. Lin et al. [23] combined the advantages of hydraulic accumulators and electric accumulators and presented a compound energy recovery system to improve the energy efficiency of hydraulic equipment. Through a series of experiments, they validated that the compound system could increase the energy efficiency by approximately 39%. Fu et al. [24] studied the energy-saving potential of the boom cylinder with an accumulator in the hybrid excavator system. The results of simulations and experiments showed that the closed hydraulic regeneration system had a high recovery efficiency. Examples of applying energy recovery systems also include generator–super capacitor energy recovery and the gas cylinder energy recovery system [25]. Figure 2 gives the energy density of different energy recovery circuits. However, Huang [26] pointed out that the low utilization efficiency of the recovery energy is also an important problem that cannot be ignored. In general terms, an energy recovery process is composed of two sub-processes: the recovery process and the reutilization process. The number of energy conversions increases with the integration of an energy recovery system, thereby increasing the system's structural complexity and causing low energy efficiency as well.

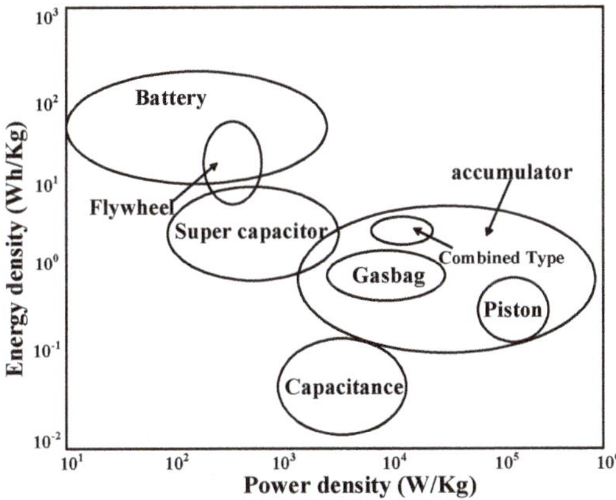

Figure 2. Power density and energy density of different energy-recovery systems.

In summary, energy efficiency and mismatch between the demanded power and the installed power can be significantly improved by using the aforementioned methods. However, most of those methods only consider energy efficiency and ignore the impact that they have on the hydraulic press itself. Therefore, most of the energy-saving presses present complicated structures and poor practicality. Furthermore, most of the existing studies in the literature only focus on the hydraulic power units and hydraulic control system; they ignore that the hydraulic press is a forming machine composed of various functional systems, such as the hydraulic actuating system, hydraulic cooling system, and other auxiliary systems. Each of these systems waste a large amount of energy during the forming process and has great potential to save energy.

Therefore, in contrast to existing energy-saving methods, the system in this paper focuses on the hydraulic cooling system as a novel starting point and proposes an energy-saving buffer system to improve the energy efficiency of a single press. A reduction of the installed power and improvement in energy efficiency can be achieved by integrating the prefill system into the hydraulic cooling system, whereas noise pollution, processing properties, and structural complexity can be improved by adding a buffer system. In addition, a servo valve is employed to adjust the supplied flow rate of the energy-saving buffer system according to the load profiles. Finally, the proposed energy-saving system was applied to a 13MN forging hydraulic press as a case study, and the results shows its significant economic and energy-saving potential.

2. Energy-Saving Method of the Hydraulic Press

2.1. Energy Characteristics of the Hydraulic Press

In Figure 3, a diagram showing the working cycle of a hydraulic press is presented. The operations performed by the hydraulic press include waiting within a working cycle (Stages 1 and 8, WT), fast falling (Stage 2, FF), pressing with slow falling (Stage 3, PS), pressure maintaining (Stage 4, PM), unloading (Stage 5, UL), fast returning (Stage 6, FR), and slow returning (Stage 7, SR). All these stages are also part of the forming process.

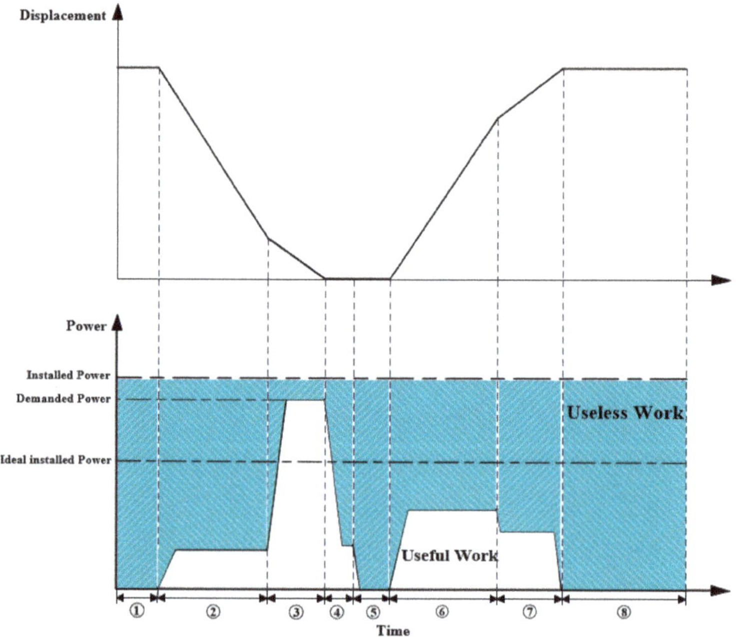

Figure 3. Hydraulic press working cycle diagram and comparison between installed and demanded power.

In traditional methods, installed power is designed to meet the maximum power requirements of Stage 3, which is much larger than the operating power of the other working stages. However, the duration of Stage 3 is only of a very small percentage of the total operating time, leading to high energy consumption and low energy efficiency. Furthermore, hydraulic presses operate performing periodic movements; the intermittent time between two adjacent cycles is almost equal to the duration of a cycle. Once hydraulic presses start running, they would not stop unless the hydraulic system reaches the unloaded state and the demanded power is zero; this leads to a large amount of power lost inadvertently. As shown in Figure 1, the white and blue areas represent useful work and useless work, respectively. Useless work is converted into heat or another useless form of energy during the operation of hydraulic presses; this is harmful to the processing properties.

When the hydraulic press is unloading, a large flow is unloaded during the processing cycle, leading to a considerable amount of energy being wasted by the traditional power unit. In addition, when the hydraulic cylinder rises rapidly with high working pressure, the high-pressure oil cannot be completely unloaded. In consequence, the residual high-pressure oil enters the hydraulic pipeline causing high vibration and huge noise pollution. The phenomena presented above constitute the main reasons for the low energy efficiency and poor processing performance in hydraulic presses.

Theoretically, if an external power system could supply energy to the hydraulic press simultaneously with the main power system during Stage 3, the installed power would be reduced and the energy efficiency would be increased.

2.2. Methodology

By analyzing in detail the reasons for the high installed power of hydraulic presses, this paper envisages the addition of an instantaneous power source during the PS stage (Stage 3). The power source can energize the hydraulic press simultaneously with the traditional power unit so as to reduce the power burden of the traditional power unit and thereby reduce the installed power of the hydraulic press. Based on this idea, an energy-saving method featuring a prefill system and a buffer system is proposed.

The proposed energy-saving system is developed by integrating a prefill system and a buffer system into the cooling system of the traditional hydraulic press, as shown in Figure 4. The pivot of the proposed system is the prefill system, which could supply high-pressure oil to the main cylinder similar to a motor-pump unit. In order to ensure the efficiency of the supplemental power and reduce pipeline losses, the prefill system is installed next to the hydraulic cylinder instead of tens of meters away from the hydraulic press, as in the traditional power unit. In addition, a servo valve is employed to adjust the output flow according to the load profiles, thereby achieving high-precision energy supplement. Based on this method, the installed power can be set at the ideal installed power as shown in Figure 1. During Stages 1, 2, 4, 5, 6, 7, and 8, the traditional power unit based on the ideal installed power can energize the press by itself, the prefill system is not used, and the servo valve outside the liquid-filling tank closes to reduce energy loss. When the hydraulic press operates at Stage 3, the power (or oil flow) required by the hydraulic press increases and exceeds the maximum value, it is difficult for the traditional hydraulic power unit alone to complete the pressurization process. At this time, the servo valve outside the liquid-filling tank opens with high accuracy, and the prefill system starts to supply power to the hydraulic press together with the traditional power unit. In this manner, power reliance on the traditional power unit is significantly reduced.

Furthermore, as depending on the power unit of the traditional cooling system, the addition of the prefill system does not require an increase in the number of motors and pumps to meet the flow and power requirements of Stage 3. Therefore, the usage of motor-pump units in the whole hydraulic system can be significantly reduced; the energy dissipation caused by the high installed power can be remarkably reduced in the low-load and no-load stage. Moreover, by reducing the number of pumps, the equipment footprint can be reduced, the structure of the equipment as a whole can be more compact, and the equipment utilization rate can be significantly improved.

When the hydraulic press operates at Stages 4 and 5, since the hydraulic press moves extremely quickly during the unloading stage, high-pressure oil cannot be completely unloaded. In consequence, the oil returning from the returning cylinder has high pressure and thus has a great impact on the prefill system and causes massive large noise pollution. To ensure the power supply of the prefill system successively improves the processing performance of the hydraulic press, the advantages of hydraulic accumulators are adopted in the method described in this paper by integrating a buffer system as an auxiliary facility. The buffer system, installed after the prefill system, can temporarily store the returning oil, absorb the high pressure in the oil to reduce the pressure shock in the system, and thus improve the processing performance of the hydraulic press. Furthermore, the high-pressure oil temporarily stored in the buffer system can also energize the hydraulic press together with the prefill system during Stage 3, allowing the energy-saving system to have a higher kinetic energy output and thus attain higher energy efficiency.

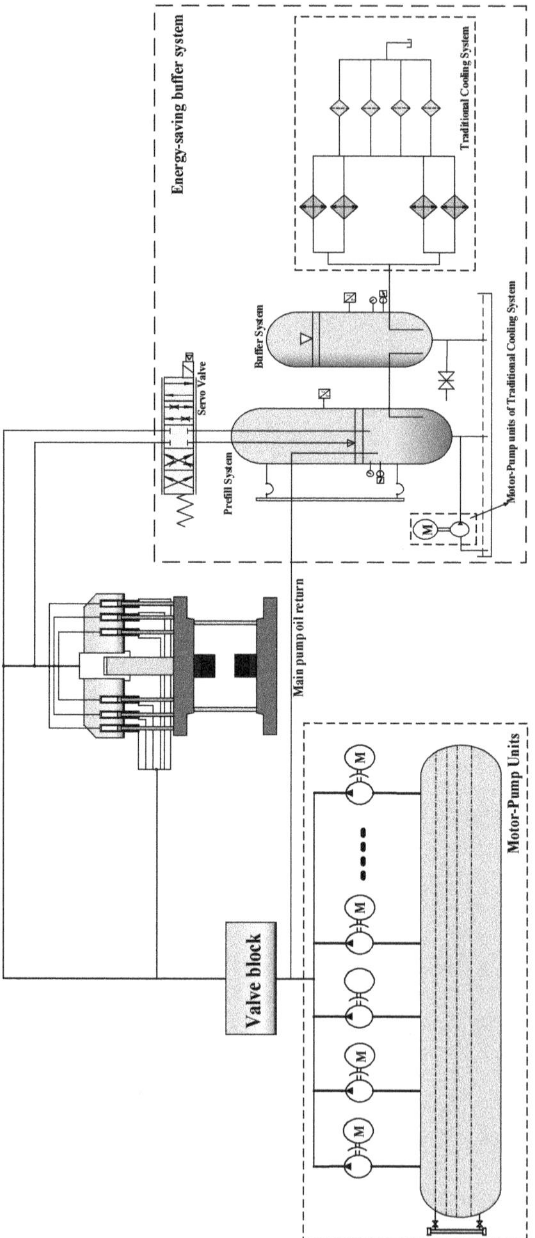

Figure 4. Diagram of the proposed energy-saving buffer system.

3. Energy Consumption Model and Theoretical Analysis

A hydraulic press is a closed energy conversion system composed of electric energy, mechanical energy, hydraulic energy, and forming energy, and the total energy consumption of a single hydraulic press is in the form of electrical energy.

The required electrical energy of a hydraulic press is determined by the flow rate q and the working pressure p of the actuator, which changes with the operation stage. Taking into account the energy conversion efficiency of the hydraulic system, the electrical energy demanded ($E_{electric-1}$) by the hydraulic machine to complete forming actions can be expressed as Equation (1):

$$E_{electric-1} = \sum_i \int_{T_0}^{T_0+T_1} p \cdot q \frac{1}{\eta_{Drive}} dt \tag{1}$$

where stage i includes the PS stage and the FF stage; η_{Drive} is the energy efficiency of motor-pump units; T_0 is the start time of stage i; and T_1 is the duration of stage i.

In terms of a hydraulic press with an energy-saving buffering system, Equation (1) can be divided into two parts,

$$\begin{aligned} E_{electric-1} &= E_1 + E_2 \\ &= \sum_i \left(\int_{T_0}^{T_0+T_1} p_{TMP} \cdot q_{TMP} \frac{1}{\eta_{Drive}} dt \right. \\ &\left. + \int_{T_0}^{T_0+T_1} p_{PFS} \cdot q_{PFS} dt \right) \end{aligned} \tag{2}$$

where E_1 and E_2 are the energy provided by the traditional motor pumps and the prefill system, respectively; p_{TMP} and q_{TMP} are the pressure and flow rate provided by the traditional motor pumps, respectively; and p_{PFS} and q_{PFS} are the pressure and flow rate provided by the prefill system, respectively.

Apart from the electrical energy demanded by the hydraulic main circuit during the FF and PS stages, other electrical energy ($E_{electric-2}$) demanded by the overflow circuit and the unloading circuit during the PM, PR, and WT stages can be expressed as Equation (3):

$$E_{electric-2} = \sum_e \left(\int_{Tm} P_1 dt + \int_{Tn} P_2 dt + \int_{Ti} P_3 dt \right) \tag{3}$$

where P_1, P_2, and P_3 are the motor power of the UL stage, the FR stage, and the WT stage, respectively; and Tm, Tn, and Ti are the duration of the UL, FR, and WT stages.

Therefore, total energy consumption in a working cycle ($E_{electric-Total}$) can be obtained:

$$E_{electric-Total} = E_1 + E_{electric-2}. \tag{4}$$

Since the hydraulic actuator does not output mechanical energy until the forming process completes, the forming energy ($E_{Forming}$) of one working cycle can be expressed as:

$$E_{Forming} = \sum_i \int_i F_{NP} \cdot \dot{h} dt \tag{5}$$

where h is the height of the piston rod; and F_{NP} is the forming pressure of the hydraulic press.

Therefore, the energy efficiency (η_{E-F}, electrical-forming energy) of the proposed hydraulic press can be expressed as Equation (6):

$$\eta_{E-F} = \frac{E_{Forming}}{E_{electric-Total}}. \tag{6}$$

As shown in the above analysis, the integration of the prefill system and the buffer system can significantly reduce the power burden of traditional motor pumps during the high-load stage, and a reduction of installed power can be realized. In addition, both overflow losses during the PM stage and throttling losses during the idle state of the press could be greatly reduced because of the reduction of installed power, thereby improving the energy efficiency of the hydraulic press. Detailed energy-efficiency improvement is discussed in the following section.

4. Case Study

4.1. Experimental Scheme

To assess the energy-saving efficiency and processing performance of the proposed system, the system was tested on a 13 MN hydraulic press machine. The experimental setup (see Figure 5) consisted of a liquid-providing tank (volume and flow rate of 23 m^3 and 5800 L/min, respectively, Huawei, Yancheng, China), a buffering tank (volume of 6 m^3, Huawei, Yancheng, China), a hydraulic power unit containing an asynchronous motor (Y315S-4; rated power and speed of 110 kW and 1480 rpm, respectively, YKP, Botou, China), an oil pump (A2F250R2P2; rated speed and flow rate of 1480 rpm and 360 L/min, respectively, Rexroth, Shanghai, China), coolers, and filters. In addition, a DT-8850 noise meter is employed to measure the noise generated during the machining process, and body vibration is measured by the technical operator at the processing site.

Figure 5. Photograph of the energy-saving buffering system.

To test the operating stability of the system and the dynamic responsiveness during the rehydration process, two common forging methods, fast forging and constant forging, were selected and performed both on the proposed hydraulic press and a traditional hydraulic press. The forging frequency of constant forging and fast forging was set at 20 times per minute and 90 times per minute, respectively. The maximum forming pressure was 35 MPa. The installed power of the proposed forging press was set at 810 kW, while the traditional hydraulic press was 1050 kW.

Table 1 details the experimental parameters of this experiment. The pressure and flow rate of the prefill system (PFS) and the traditional power units (TPU) were recorded by a pressure gauge and a flow sensor and then uploaded to a computer.

Table 1. Parameters of the upsetting process.

Process Type	Forging Frequency	Load Pressure	Main Pump Unit Power	Displacement
Regular forging	20 times per minute	35 MPa	810 kW	450 mm
Fast forging	90 times per minute	35 MPa	810 kW	200 mm

4.2. Processing Performance Analysis

The forging frequency was set to 90 times per minute and 20 times per minute, respectively. The nominal pressure was set to 35 MPa. The relationships between TPU pressure and prefill pressure are shown in Figures 6 and 7.

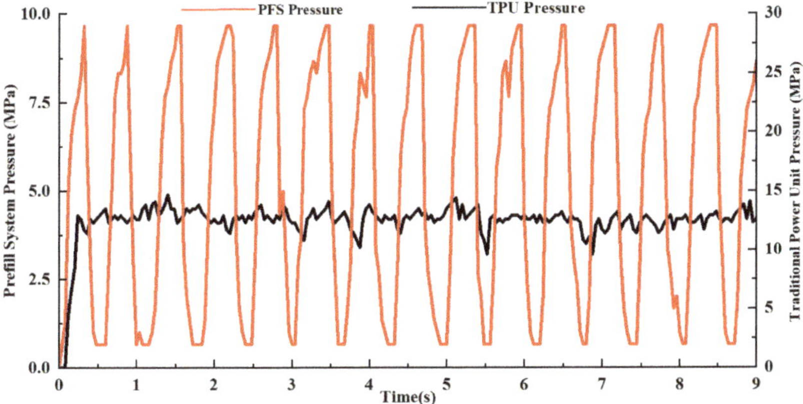

Figure 6. Output pressure of the prefill system and the main power unit during fast forging.

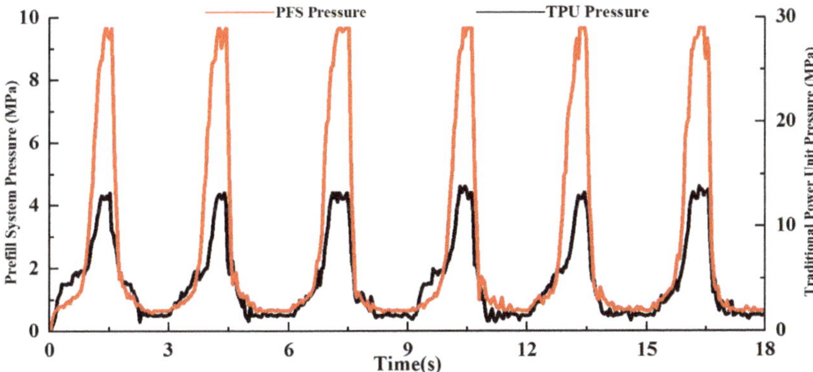

Figure 7. Output pressure of the prefill system and the main power unit during constant forging.

As it can be seen in Figures 6 and 7, the hydraulic press ran smoothly (without vibration) in the no-load and low-load stages. The TPU was used to supply energy to maintain the normal operation of the press, while the servo valve outside the liquid-providing tank was closed so that the output pressure and flow of the liquid-filling system was approximately zero. When the hammer of the hydraulic press contacted the forging workpiece, the load power rose and exceeded the nominal power supplied by the TPU; the servo valve outside the PFS opened with high accuracy, and the output power and flow of the PFS were used in conjunction with those of the TPU as the load pressure was reaching its maximum pressure. In the fast forging process, the servo valve remained open in order to ensure timely rehydration, since the timespan of one fast forging cycle was extremely short (see Figure 6).

When the press operated at the unloading stage, the traditional hydraulic system generated an impact pressure of up to 20 MPa in 0.04 s, causing a great impact on hydraulic pipes and beams. Many oil pipe ruptures and beam bending deformations are caused by this phenomenon, seriously affecting machining accuracy and the service life of the hydraulic press. Figures 8 and 9 show that the pressure-relief impact was well improved by integrating a buffer system. Since the PFS and the buffer system were installed close to the hydraulic cylinder, the high-pressure oil generated at the unloading stage directly entered the liquid-filling tank and the buffer tank without traveling a long distance along the hydraulic pipelines. Therefore, hydraulic impact and oil leakage could be reduced to a large extent, so as to improve the hydraulic working environment and reduce environmental pollution. According to experimental statistics, regardless of the forging method, adding the proposed system to a hydraulic press can reduce hydraulic impact by at least 9.2 MPa.

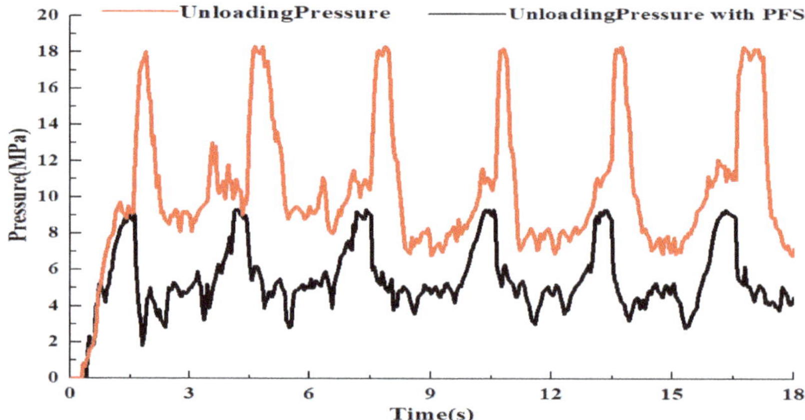

Figure 8. Unloading pressure of the hydraulic press with the energy-saving system and the traditional hydraulic press during fast forging.

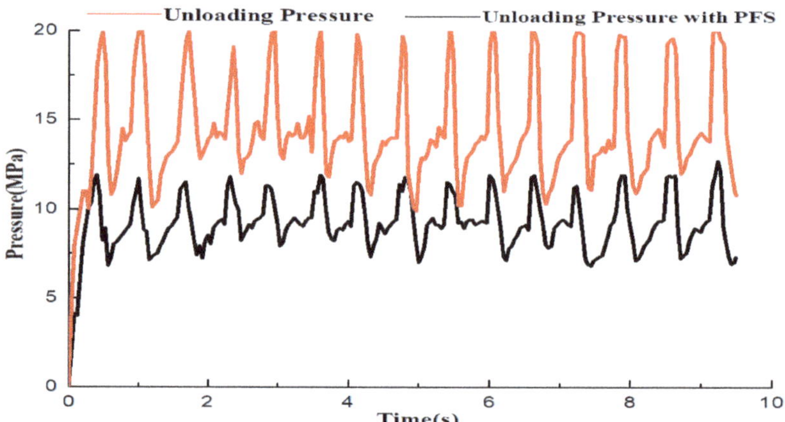

Figure 9. Unloading pressure of the hydraulic press with the energy-saving system and the traditional hydraulic press during fast forging.

Through several test experiments, the experimental data from Figures 6–9 are summarized in Tables 2 and 3, in which Tables 2 and 3 are the results of the experiments at the PS and unloading stage for different forging methods, respectively.

When the forging hydraulic press operates at the low-load or no-load stage, the energy supplied by the TPU is sufficient for the hydraulic press to operate normally, and the PFS does not work. As seen from Figures 6 and 7 and Table 3, when the hydraulic press operates at Stage 3, the load power rises and exceeds the maximum power the TPU could output, the servo valve outside the PFS opens, and then the PFS outputs high-pressure oil and energizes the main working cylinder together with the TPU; the maximum output pressure and flow rate of the TPU are 28.1 MPa and 2900 L/min, respectively, while the load pressure and flow rate of the press are 35 MPa and 4400 L/min. The difference between the load pressure and the output pressure of the TPU is the energy provided by the PFS, which are 5.4 MPa and 1480 L/min, respectively. The average pressure-saving rate for the two forging methods is

19.3% and 17.5%, respectively. Combined with the flow rate and the formula $P = p * q$, the maximum power saving rate of 20.1% could be obtained through calculations.

Table 2. Results of the experiments at the PS stage for different forging methods. PFS: prefill system, PS: pressing with slow falling, TPU: traditional power units.

Forging Method	Fast Forging	Constant Forging
Load pressure	35.0 MPa	35.0 MPa
Peak pressure of the TPU	28.1 MPa	28.0 MPa
Peak pressure of the PFS	5.4 MPa	4.9 MPa
Mean pressure of the TPU	27.1 MPa	26.9 MPa
Mean pressure of the PFS	4.9 MPa	4.5 MPa
Pressure saving rate	19.3%	17.5%
Peak flow rate of the TPU	2900 L/min	2870 L/min
Peak flow rate of the PFS	1480 L/min	1460 L/min
Power-saving rate	20.1%	19.6%

Table 3. Results of the experiments at the unloading stage for different forging methods.

Forging Method	Fast Forging	Constant Forging
Pipeline pressure	9.7 MPa	9.1 MPa
Traditional pipeline pressure	20.3 MPa	18.5 MPa
Buffering system pressure	6.3 MPa	5.2 MPa
Noise intensity	75 dB	72 dB
Traditional noise intensity	105 dB	100 dB
Oil temperature	35.7 °C	40.2 °C
Body vibration	Slight	Slight
Pressure impact reduction	10.6 MPa	9.2 MPa
Pressure decline rate	52.2%	49.73%
Noise decline rate	28.6%	28%

From Table 3, when the forging frequency was set at 90 times per minute (fast forging), the pipeline pressure dropped to 9.7 MPa, while the pressure shock and noise pollution decreased by 52.2% and 28.6%, respectively, compared with the traditional hydraulic press. When the forging frequency was set at 20 times per minute (constant forging), the pipeline pressure dropped to 9.2 MPa, the pressure shock and noise pollution decreased by 49.7% and 28%, respectively. In addition, although the cooling system was retrofitted, the system temperature remained within a low range, which was generally lower than that of the traditional hydraulic press.

4.3. Energy Consumption Analysis of the Two Presses

During the forging process, each motor pump in the drive unit changes the state of its access to the hydraulic system according to the energy demand of different operations, and one drive unit, which is useless in an operation, is set at an unloading state by switching the pressure-relief valve. Table 4 shows the compositions of the drive system of the two hydraulic presses and the number of motor pumps connected for each processing stage.

Table 4. Compositions of the drive system and the number of motor pumps used in different operations. WT: waiting within a working cycle, Stages 1 and 8, WT; FF: fast falling, Stage 2; UL: unloading, Stage 5; FR: fast returning, Stage 6; SR: slow returning, Stage 7.

	Total Number	WT	FF	PS	UL	FR	SR
Hydraulic press with PFS	7	0	3	7	1	7	2
Traditional hydraulic press	9	0	3	9	1	9	3

In this study, the energy consumption of a hydraulic press was scaled by measuring the active electricity consumption of motor pumps. A power meter and an data acquisition card were selected to measure and receive the voltage and current through the motors at any time, respectively. Then, the real-time measured data were transmitted to the PC after processing.

During the forming process, the energy efficiency of a hydraulic press can be expressed as follows,

$$\eta = \frac{E_{Forming}}{\sum_i E_{i-Input}} \cdot 100\% \tag{7}$$

where i is all the forming operations, including FF, PS, UL, FR, SR, and WT. $E_{i-Input}$ is the energy consumed by the hydraulic machine to complete the i-th operation, which can be obtained by Equation (8).

$$E_{i-Input} = \int_{t_{i-Start}}^{t_{i-End}} P_i(t)dt = \sum_{t_{i-Start}}^{t_{i-Start}} P_i \cdot \Delta t \tag{8}$$

where P_i is the active power of each operation, which can be obtained by the power meter, and $t_{i-Start}$ and $t_{i-Start}$ are the start time and the end time of the i-th operation, respectively.

According to Equation (8), the active electrical energy consumption of motors was obtained by experiments. The electricity consumption, useful energy, and energy efficiency of the two presses from experimental testing are shown in Tables 5 and 6.

Table 5. Energy dissipation of each operation in the hydraulic press with PFS (Constant Forging).

Operations (i)		FF	PS	UL	FR	SR	WT	Total
Energy (KJ)	Time (s)	0–0.5	0.5–1.4	1.2–1.3	1.3–1.9	1.9–2.2	2.2–3	3
Input energy ($E_{i-Input}$)		389.91	703.28	68.21	472.37	189.11	559.88	2382.76
Useful energy		165.41	692.87	9.28	439.78	68.19	0	1375.53
Energy efficiency		42.41%	98.42%	13.23%	93.00%	35.98%	0	57.72%
η					29.05%			

Table 6. Energy consumption of each operation of the traditional hydraulic press (Constant Forging).

Operations (i)		FF	PS	UL	FR	SR	WT	Total
Energy (KJ)	Time (s)	0–0.5	0.5–1.4	1.2–1.3	1.3–1.9	1.9–2.2	2.2–3	3
Input energy ($E_{i-Input}$)		492.28	901.89	99.38	604.22	297.09	832.80	3227.66
Energy efficiency		33.60%	72.82%	9.05%	74.00%	22.95%	0	37.04%
$\Delta E_{i-Input}$		+102.37	+198.61	+31.17	+121.85	+107.98	+242.92	+804.90
η					17.35%			

The traditional hydraulic press is equipped with 9 sets of motor pumps, of which the installed power is up to 1050 kW. According to Table 6, the total electric energy consumed to complete a working cycle is 3227.66 KJ, and only 31.01% of the input energy is converted into useful energy. Besides this, as a result of the pipe-valve energy loss, overflow energy loss, and other energy losses, only 17.35% of the input energy is used to form the workpiece. However, by integrating PFS into the press, the energy consumption of each stage significantly improved. The experimental data from Tables 5 and 6 compared with the traditional hydraulic press are summarized in Figure 10.

As shown above, since the installed power of the hydraulic press equipped with PFS was lower than that of the traditional press by 22.86%, the throttling loss and the no-load loss of the press were significantly reduced, and the average electrical energy consumption was minimized in each stage, as shown in Figure 10a. Total electrical energy consumption was reduced from 3227.66 KJ to 2382.76 KJ, so 804.90 KJ energy was saved per working cycle. During the PS stage, as the hammer contacted the forming workpiece, the load power rose and exceeded the nominal power supplied by the TPU, the servo valve outside the PFS opened, and the output power and flow of the PFS were used in conjunction with those of the TPU as the load pressure was reaching its maximum pressure. The difference between the input energy of the two presses is the energy provided by the PFS, which is about 198.61 KJ. After several experimental tests, compared with the traditional hydraulic press, the overall energy efficiency and the forming efficiency of the hydraulic press with PFS were improved by 26.71% and 11.70%, respectively.

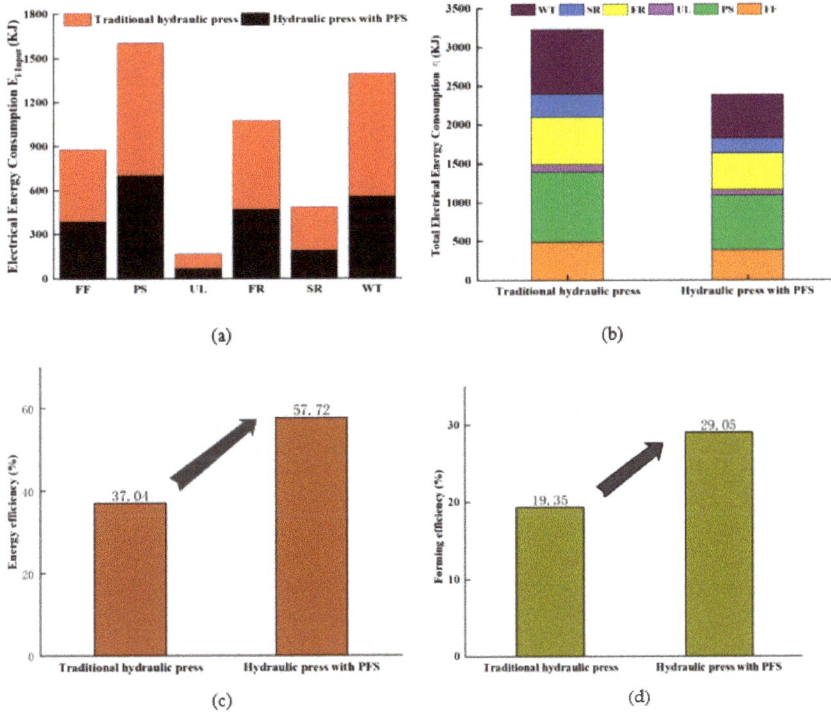

Figure 10. (**a**) Energy consumption under different processing stages; (**b**) Total energy consumption; (**c**) Total energy efficiency; (**d**) Forming efficiency.

5. Conclusions

Given the low energy efficiency and poor processing performance of the hydraulic press, this paper proposes a method featuring an energy-saving buffer system to reduce the installed power and improve processing performance. In this method, a liquid-filling system is used to supply power and an oil flow rate that the TPU cannot provide at the high-load stage. As it is not necessary to increase the number of motor pumps to meet the flow requirements, the motor-pump usage of the whole system is significantly reduced. In consequence, the installed power of the hydraulic press is reduced, and cost reduction and energy savings are achieved. Furthermore, to mitigate the high-pressure impact and noise pollution at the unloading stage, a buffer system is integrated into the system to absorb the high-pressure returning oil. In consequence, the pressure shock problem is addressed, and the service lifespan of the hydraulic press is increased. The proposed system was tested in a 13 MN hydraulic press in which an industrial press was used. Through a series of comparative experiments, it was preliminarily validated that the proposed system can reduce pump usage and pressure shock by up to 30% and 41%, respectively, increase energy efficiency by up to 26.71%, reduce noise pollution and installed power by 28% and 22.85%, respectively, and provide the same processing characteristics and properties as the traditional hydraulic press.

To better improve the energy efficiency of the hydraulic press, our future work will concentrate on the control strategies of the PFS to achieve higher precision output. Other research directions include the optimal design of the PFS and the buffering system for a given hydraulic press.

Author Contributions: Conceptualization, X.Y.; Methodology, X.Y. and B.C.; validation, X.Y. and B.C.; formal analysis, X.Y. and B.C.; data curation, X.Y.; writing original draft preparation, X.Y.; Funding, B.C.; writing-review and editing, B.C. All authors have read and agreed to the published version of the manuscript.

Funding: This work is financed by Huazhong University of Science and Technology and Jiangsu Huawei Machinery Manufacturing Co. Ltd.

Acknowledgments: The work is supported by Huazhong University of Science and Technology and Jiangsu Huawei Machinery Manufacturing Co. Ltd.

Conflicts of Interest: The authors declare no conflict of interest.

Abbreviations

TPU	traditional power units
PFS	prefill system
FF	fast falling
PS	pressure with slow speed
PM	pressure maintaining
UL	unloading
FR	fast returning
SR	slow returning
WT	waiting stage

References

1. Li, L.; Huang, H.H.; Liu, Z.F.; Li, X.Y.; Triebe, M.J.; Zhao, F. An energy-saving method to solve the mismatch between installed and demanded power in hydraulic press. *J. Clean. Prod.* **2016**, *139*, 636–645. [CrossRef]
2. Duflou, J.R.; Sutherland, J.W.; Dornfeld, D.; Herrmann, C.; Jeswiet, J.; Kara, S.; Hauschild, M.; Kellens, K. Towards energy and resource efficient manufacturing: A processes and systems approach. *CIRP Ann. Manuf. Technol.* **2012**, *61*, 587–609. [CrossRef]
3. Lin, T.; Chen, Q.; Ren, H.; Huang, W.; Chen, Q.; Fu, S. Review of boom potential energy regeneration technology for hydraulic construction machinery. *Renew. Sustain. Energy Rev.* **2017**, *79*, 358–371. [CrossRef]
4. Xu, Z.; Liu, Y.; Hua, L.; Zhao, X.; Guo, W. Energy analysis and optimization of main hydraulic system in 10,000 kN fine blanking press with simulation and experimental methods. *Energy Convers. Manag.* **2019**, *181*, 143–158. [CrossRef]
5. Mousa, E.; Kazemi, M.; Larsson, M.; Karlsson, G.; Persson, E. Potential for Developing Biocarbon Briquettes for Foundry Industry. *Appl. Sci.* **2019**, *9*, 5288. [CrossRef]
6. Zhao, K.; Liu, Z.F.; Yu, S.R.; Li, X.Y.; Huang, H.H.; Li, B.T. Analytical energy dissipation in large and medium-sized hydraulic press. *J. Clean. Prod.* **2015**, *103*, 908–915. [CrossRef]
7. Huang, H.H.; Zou, X.; Li, L.; Li, X.Y.; Liu, Z.F. Energy-Saving Design Method for Hydraulic Press Drive System with Multi Motor-Pumps. *Int. J. Precis. Eng. Manuf. Green Technol.* **2019**, *6*, 223–234. [CrossRef]
8. Quan, L.; Helduser, S. Energy Saving and High Dynamic Hydraulic Power Unit Based on Speed Variable Motor and Constant Hydraulic Pump. *China Mech. Eng.* **2003**, *14*, 606–609.
9. Su, C.L.; Chung, W.L.; Yu, K.T. An Energy-Savings Evaluation Method for Variable-Frequency-Drive Applications on Ship Central Cooling Systems. *IEEE Trans. Ind. Appl.* **2014**, *50*, 1286–1294. [CrossRef]
10. Camoirano, R.; Dellepiane, G. Variable frequency drives for MSF desalination plant and associated pumping stations. *Desalination* **2005**, *182*, 53–65. [CrossRef]
11. Wang, H.; Wang, H.Y.; Zhu, T. A new hydraulic regulation method on district heating system with distributed variable-speed pumps. *Energy Convers. Manag.* **2017**, *147*, 174–189. [CrossRef]
12. Ge, L.; Quan, L.; Zhang, X.G.; Zhao, B.; Yang, J. Efficiency improvement and evaluation of electric hydraulic excavator with speed and displacement variable pump. *Energy Convers. Manag.* **2017**, *150*, 62–71. [CrossRef]

13. Lovrec, D.; Kastrevc, M.; Ulaga, S. Electro-hydraulic load sensing with a speed-controlled hydraulic supply system on forming-machines. *Int. J. Adv. Manuf. Technol.* **2009**, *41*, 1066–1075. [CrossRef]

14. Feng, L.; Yan, H. Nonlinear Adaptive Robust Control of the Electro-Hydraulic Servo System. *Appl. Sci.* **2020**, *10*, 4494. [CrossRef]

15. Chen, B.J.; Huang, S.H.; Jin, L.; Gao, J.F. Control Strategy for Free Forging Hydraulic Press. *Chin. J. Mech. Eng.* **2008**, *44*, 304–307. [CrossRef]

16. Ferreira, J.A.; Sun, P.; Grácio, J.J. Close loop control of a hydraulic press for springback analysis. *J. Mater. Process. Technol.* **2006**, *177*, 377–381. [CrossRef]

17. Gao, F.; Pan, S.X. Experimental study on model of negative control of hydraulic system. *China Mech. Eng.* **2005**, *41*, 49–54. [CrossRef]

18. Yan, X.P.; Chen, B.J.; Zhang, D.W.; Wu, C.X.; Luo, W.X. An energy-saving method to reduce the installed power of hydraulic press machines. *J. Clean. Prod.* **2019**, *233*, 538–545. [CrossRef]

19. Dai, M.Q.; Zhao, S.D.; Yuan, X.M. The Application Study of Accumulator Used in Hydraulic System of 20MN Fast Forging Machine. *J. Appl. Mech.* **2011**, *80*, 870–874. [CrossRef]

20. Triet, H.H.; Ahn, K.K. Comparison and assessment of a hydraulic energy- saving system for hydrostatic drives. *Proc. Inst. Mech. Eng. Part I J. Syst. Control Eng.* **2011**, *225*, 21–34. [CrossRef]

21. Van de Ven, J.D. Constant pressure hydraulic energy storage through a variable area piston hydraulic accumulator. *Appl. Energy* **2013**, *105*, 262–270. [CrossRef]

22. Xia, L.P.; Quan, L.; Ge, L.; Hao, Y.X. Energy efficiency analysis of integrated drive and energy recuperation system for hydraulic excavator boom. *Energy Convers. Manag.* **2018**, *156*, 680–687. [CrossRef]

23. Lin, T.l.; Huang, W.P.; Ren, H.L.; Fu, S.J.; Liu, Q. New compound energy regeneration system and control strategy for hybrid hydraulic excavators. *Autom. Constr.* **2016**, *68*, 11–20. [CrossRef]

24. Fu, S.; Chen, H.; Ren, H.; Lin, T.; Miao, C.; Chen, Q. Potential Energy Recovery System for Electric Heavy Forklift Based on Double Hydraulic Motor-Generators. *Appl. Sci.* **2020**, *10*, 3996. [CrossRef]

25. Fan, Y.J.; Mu, A.L.; Ma, T. Design and control of a point absorber wave energy converter with an open loop hydraulic transmission. *Energy Convers. Manag.* **2016**, *121*, 13–21. [CrossRef]

26. Huang, N.; Yang, S.; Chen, Y. Development of machine tools and large tonnage hydraulic automatic hydraulic fine blanking press. *Mach. Tool Hydraul.* **2011**, *2011*, 4.

applied
sciences

MDPI

Article

Driving Force Distribution and Control for Maneuverability and Stability of a 6WD Skid-Steering EUGV with Independent Drive Motors

Hui Zhang [1,2,3,4], Huawei Liang [1,3,4,*], Xiang Tao [1,3,4], Yi Ding [1,3,4], Biao Yu [1,3,4] and Rengui Bai [1,2,3,4]

[1] Hefei Institutes of Physical Science, Chinese Academy of Sciences, Hefei 230031, China; gloryzh@mail.ustc.edu.cn (H.Z.); xtao@hfcas.ac.cn (X.T.); yding@hfcas.ac.cn (Y.D.); byu@hfcas.ac.cn (B.Y.); zaowuzhu@mail.ustc.edu.cn (R.B.)
[2] University of Science and Technology of China, Hefei 230026, China
[3] Anhui Engineering Laboratory for Intelligent Driving Technology and Application, Hefei 230031, China
[4] Innovation Research Institute of Robotics and Intelligent Manufacturing, Chinese Academy of Sciences, Hefei 230031, China
* Correspondence: hwliang@iim.ac.cn; Tel./Fax: +86-551-6539-3191

Featured Application: The paper proposes a hierarchical driving force distribution and control strategy for six-wheel drive (6WD) skid-steering electric unmanned ground vehicles (EUGVs) with independent drive motors to improve the vehicle maneuverability and stability. The proposed method can be applied to control the longitudinal velocity and yaw rate for distributed, independently driven electrical skid-steering vehicles, for instance, military vehicles, agricultural vehicles, and other special vehicles.

Abstract: In this paper, a hierarchical driving force distribution and control strategy for a six-wheel drive (6WD) skid-steering electric unmanned ground vehicle (EUGV) with independent drive motors is presented to improve the vehicle maneuverability and stability. The proposed hierarchical strategy is based on a nine-degrees-of-freedom (DOFs) dynamics model of 6WD skid-steering EUGV with a vehicle system dynamics model, wheel dynamics model, and tire model. In the proposed hierarchical strategy, the upper layer controller calculates the resultant driving force and yaw moment to control the vehicle motion states to track the desired ones by using the integral sliding mode control (ISMC) and proportion–integral–differential (PID) control methods. In the lower layer controllers, the driving force distribution method is adopted to allocate torques to the six motors. An objective function is proposed and composed of the longitudinal tire workload rates and weighting factors, considering the inequality constraints and equality constraints, which is solved by using the active set method. In order to evaluate the effectiveness of the proposed method, experiments with two types of scenarios were conducted. Comparative studies were also conducted with the other two methods used in the literature. The experimental results show that better performance can be achieved with the proposed control strategy in vehicle maneuverability and stability.

Keywords: six-wheel drive (6WD); skid steering; electric unmanned ground vehicle (EUGV); driving force distribution; vehicle motion control; maneuverability and stability

Citation: Zhang, H.; Liang, H.; Tao, X.; Ding, Y.; Yu, B.; Bai, R. Driving Force Distribution and Control for Maneuverability and Stability of a 6WD Skid-Steering EUGV with Independent Drive Motors. *Appl. Sci.* **2021**, *11*, 961. https://doi.org/10.3390/app11030961

Academic Editor: Alessandro Gasparetto
Received: 23 December 2020
Accepted: 18 January 2021
Published: 21 January 2021

Publisher's Note: MDPI stays neutral with regard to jurisdictional claims in published maps and institutional affiliations.

1. Introduction

In recent years, with the development and wide application of intelligent technology, sensor technology, and control technology, the unmanned ground vehicle (UGV) is being promoted quickly and widely for military and civil usage [1–6]. UGVs are employed to replace humans in various dangerous situations for executing many tasks, for instance, battle, research, rescue, mining, firefighting, and so on. In order to reduce fuel consumption, exhaust gas emissions, and noise, many types of vehicle power source have been widely studied in industry and academic fields, such as hybrid electric vehicles (HEVs), plug-in hybrid electric vehicles (PHEVs), and pure electric vehicles (EVs) [2,7–10]. Currently, the electric UGV (EUGV) is attracting great research focus by reason of its wide applications and the development of electric motor and control technology. The drive forms for EUGVs are mainly divided into centralized drive systems and distributed drive systems. Compared to the centralized drive system, the distributed drive system is composed of two or more independent drive wheels. With the development of independent drive technology, EUGVs with distributed drive systems have tremendous potential to improve vehicle maneuverability and lateral stability. That is because the independent drive wheels can be directly actuated and independently controlled; the six-wheel independent drive EUGV will be focused on in this paper.

In order to realize and improve the maneuverability and lateral stability of EVs with the distributed drive system, various methods have been studied and developed for four-wheel drive (4WD) EVs driven by independent motors, and research into 6WD and 8WD EVs has been completed based on those achievements. An EV with independent drive wheels is an over-actuated system, and each wheel can be controlled directly and individually. Therefore, these methods mainly concentrate on vehicle motion control and driving force distribution. In vehicle motion control research, the longitudinal and lateral motion controllers have attracted great research efforts. The algorithm utilized for motion controller design in prior literature includes a proportion–integral–differential (PID) controller [11,12], H-infinite controller [13], sliding mode controller (SMC) [14–16], linear quadratic regulator (LQR) [17], and model-predicted controller (MPC) [18,19]. The motion controller has been applied to calculate the vehicle resultant force for tracking the longitudinal motion and vehicle's resultant yaw moment for tracking the later motion, according to the vehicle's desired input and actual output. This vehicle resultant force and yaw moment are supported by each wheel with independent drive motors. Then, the driving force distribution is the next research point. In driving force distribution research, many distribution algorithms have been proposed for realizing the EV's turning stability, reducing the tire-road slip, or improving the performance in energy consumption [20–22]. Rule-based distribution and optimization-based distribution methods are mainly applied to allocate the torque of each wheel. Due to the lack of optimality and accuracy of the rule-based distribution algorithm [23], methods based on the optimization theory are widely used to obtain optimal control inputs. The optimization theory is the law that establishes an objective function and solves its minimum by taking into account system constraints [24]. Novellis et al. [25] investigated different objective functions and constraints on the basis of the motor performance, battery capacity, and tire-road adhesion for 4WD EVs. Zhao et al. [26] proposed a nonlinear control distribution scheme for 4WD EVs to improve vehicle stability with the constraints of the driven motors and tire-road adhesion. Saeid Khosravani et al. [27] proposed a novel optimization allocation strategy for a vehicle's longitudinal and lateral control and verified its effectiveness by simulation and experiments. Xiong et al. [28,29] investigated the directed yaw control method based on driver operation intention and designed allocation control algorithms with a dynamic efficiency matrix for the stability control of 4WD EVs. Not only have distribution algorithms for 4WD EVs been developed, but distribution algorithms for 6WD and 8WD EVs have also seen many research achievements. Kyongsu et al. [30] utilized the optimal driving force distribution for 6WD/6WS EVs, in which the objective function is based on the sum of each tire's workload rate while considering tire-road adhesion and friction circle. Kim [31] investigated an objective function consisting of the sum of six tire utilization rates and minimized its value using the Lagrange multiplier method for tire force distribution. Zhang [23,32] and Kim [33] researched the driving force allocation schemes for 8WD EVs to enhance stability and maneuverability.

Nevertheless, the aforementioned literature is focused on passenger EVs, EVs with Ackerman steering gears, and few papers have concentrated on skid-steering EUGVs. Due to the difference in vehicle dynamics models between differential-steering vehicles and Ackerman-steering vehicles, there are some problems that need to be solved: (1) For passenger EVs and Ackerman-steering EVs, the vehicle dynamics models are established by considering the wheel steering angle. However, each wheel steering angle of the skid-steering EUGV is zero in its dynamics model. This increases the understeer behavior performance and steering difficulty factor. (2) By reason of the steering difficulty, it is a challenge to design a driving force control and distribution algorithm for tracking the desired lateral yaw rate. To this end, a novel driving force distribution and control strategy with a hierarchical controller structure is proposed in this paper and was implemented on a skid-steering EUGV. The major contributions lie in the following aspects. A nine-degrees-of-freedom (DOFs) dynamics model of a 6WD skid-steering EUGV with a vehicle system dynamics model, wheel dynamics model, and tire model was established. A hierarchical controller that is composed of the upper layer controller and lower layer controller is proposed. The upper layer controller is devised to determine the resultant force and yaw moment to make sure the vehicle motion states track the desired values by using the integral sliding mode controller and PID controller. In the lower layer controller, a driving force distribution scheme is proposed based on the optimization allocation. The objective function is formulated based on the longitudinal tire workload rate and weighting factors. The weighting factors are established based on the ratio of the tire vertical force considering the vehicle motion conditions and vertical load change of each tire. Lastly, the two types of experimental scenarios with the proposed controller were tested to validate the effectiveness of the control algorithm for the skid-steering EUGV.

This paper is organized as follows. Section 2 introduces a nonlinear dynamics model of a skid-steering EUGV actuated by independent drive wheels. In Section 3, the hierarchical driving force distribution and control scheme for improving the maneuverability and stability of a 6WD skid-steering EUGV are described. Experiments with two cases of scenarios were conducted to verify the performance of the proposed controller, as described in Section 4. Finally, the conclusions are drawn in Section 5.

2. Vehicle and Wheel Dynamics Model

For the design of a vehicle driving force distribution control system, a 6WD skid-steering EUGV dynamics model with the wheel motor torque as the input and vehicle motion states—such as the lateral and longitudinal acceleration, longitudinal velocity, and yaw rate—as output is required. In this section, the establishment of a six-wheel vehicle dynamics model is described that includes three parts: a nonlinear vehicle system dynamics model, a wheel dynamics model, and a tire model.

2.1. Vehicle System Dynamics Model

In this paper, a 6WD skid-steering EUGV dynamics model was established as an object with nine degrees of freedom (DOFs). The full vehicle system dynamics model was comprised of three-DOF rotational and translational dynamic models of the body and six-wheel dynamic models. The vehicle system dynamics model considers longitudinal, lateral, and yaw dynamics and ignores roll and pitch dynamics, as shown in Figure 1. The nomenclature for the parameters, all the Equations, and all the symbols seen in Figure 1, can be found in Nomenclature. In Figure 1, ox represents the longitudinal direction of the coordinate system and oy represents the lateral direction of the coordinate system. According to the Newtonian and Euler Theorem, the longitudinal and lateral dynamic equations are

$$\begin{cases} m(\dot{v}_x - \omega v_y) = F_x \\ m(\dot{v}_y + \omega v_x) = F_y \end{cases},$$

(1)

where

$$\begin{cases} F_x = F_{x1} + F_{x2} + F_{x3} + F_{x4} + F_{x5} + F_{x6} \\ F_y = F_{y1} + F_{y2} + F_{y3} + F_{y4} + F_{y5} + F_{y6} \end{cases},$$

(2)

and the rotational dynamic equation is

$$I_z \dot{\omega} = M_{Fx} + M_{Fy},$$

(3)

where

$$\begin{cases} M_{Fx} = \frac{B}{2}\left(F_{x2} + F_{x4} + F_{x6} - F_{x1} - F_{x3} - F_{x5}\right) \\ M_{Fy} = (l_1 + a)\left(F_{y1} + F_{y2}\right) + a\left(F_{y3} + F_{y4}\right) - (l_2 - a)\left(F_{y5} + F_{y6}\right) \end{cases} , \tag{4}$$

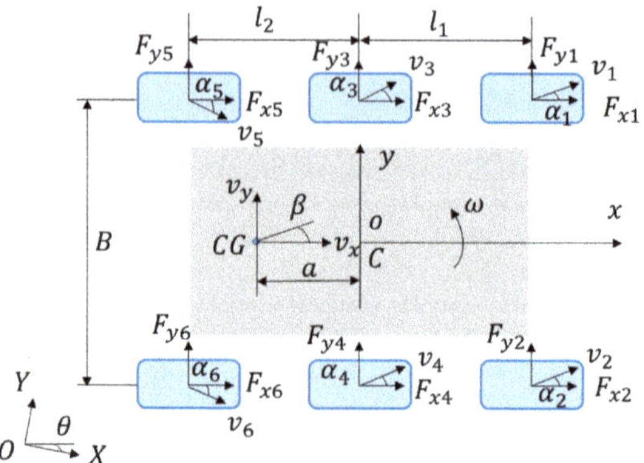

Figure 1. Schematic diagram of the vehicle dynamics model.

Thus, the wheel sideslip angles are calculated according to

$$\begin{cases} \alpha_1 = \arctan\dfrac{v_y + (l_1 + a)\omega}{v_x - \frac{B}{2}\omega} \\ \alpha_2 = \arctan\dfrac{v_y + (l_1 + a)\omega}{v_x + \frac{B}{2}\omega} \\ \alpha_3 = \arctan\dfrac{v_y + a\omega}{v_x - \frac{B}{2}\omega} \\ \alpha_4 = \arctan\dfrac{v_y + a\omega}{v_x + \frac{B}{2}\omega} \\ \alpha_5 = \arctan\dfrac{v_y - (l_2 - a)\omega}{v_x - \frac{B}{2}\omega} \\ \alpha_6 = \arctan\dfrac{v_y - (l_2 - a)\omega}{v_x + \frac{B}{2}\omega} \end{cases} , \tag{5}$$

The vertical forces of the six wheels are calculated according to

$$\begin{cases} F_{z1} = \dfrac{\left(l_2{}^2 - al_1 - al_2 + l_1 l_2\right)\left(\frac{1}{2}mg - \frac{H}{B}ma_y\right) - \frac{1}{2}(l_1 + l_2)ma_x H}{2(l_1{}^2 + l_1 l_2 + l_2{}^2)} \\ F_{z2} = \dfrac{\left(l_2{}^2 - al_1 - al_2 + l_1 l_2\right)\left(\frac{1}{2}mg + \frac{H}{B}ma_y\right) - \frac{1}{2}(l_1 + l_2)ma_x H}{2(l_1{}^2 + l_1 l_2 + l_2{}^2)} \\ F_{z3} = \dfrac{\left(l_1{}^2 + al_1 - al_2 + l_2{}^2\right)\left(\frac{1}{2}mg - \frac{H}{B}ma_y\right) + \frac{1}{2}(l_1 - l_2)ma_x H}{2(l_1{}^2 + l_1 l_2 + l_2{}^2)} \\ F_{z4} = \dfrac{\left(l_1{}^2 + al_1 - al_2 + l_2{}^2\right)\left(\frac{1}{2}mg + \frac{H}{B}ma_y\right) + \frac{1}{2}(l_1 - l_2)ma_x H}{2(l_1{}^2 + l_1 l_2 + l_2{}^2)} \\ F_{z5} = \dfrac{\left(l_1{}^2 + al_2 + l_1 l_2\right)\left(\frac{1}{2}mg - \frac{H}{B}ma_y\right) + \frac{1}{2}(l_1 + 2l_2)ma_x H}{2(l_1{}^2 + l_1 l_2 + l_2{}^2)} \\ F_{z6} = \dfrac{\left(l_1{}^2 + al_2 + l_1 l_2\right)\left(\frac{1}{2}mg + \frac{H}{B}ma_y\right) + \frac{1}{2}(l_1 + 2l_2)ma_x H}{2(l_1{}^2 + l_1 l_2 + l_2{}^2)} \end{cases} , \tag{6}$$

2.2. Wheel Dynamics Equation

A wheel dynamics model constructs the relationship between the driving or braking torque, wheel rotational acceleration, and longitudinal forces. Here, the rolling resistance was considered. The wheel dynamics equation is shown in Figure 2.

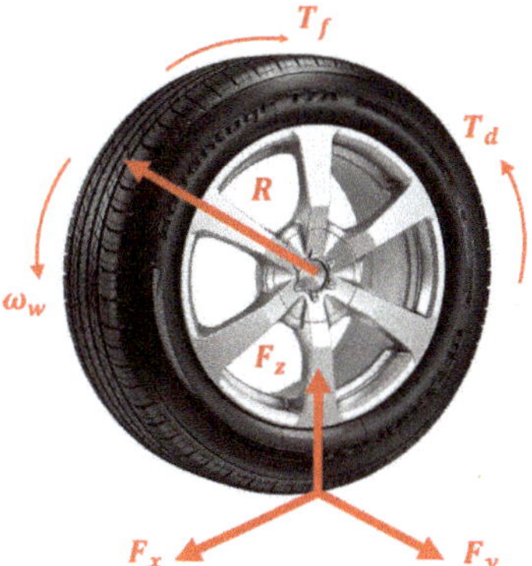

Figure 2. Schematic diagram of the wheel dynamics equation.

In Figure 2, T_d is the motor driving or braking torque ($T_d > 0$ for the driving torque and $T_d < 0$ for the braking torque), F_x is the tire longitudinal force, R is the tire effective radius, T_f is the rolling resistance torque of the tires, and its direction is opposite to T_d.

The wheel dynamics model is expressed by Equation (7) as follows:

$$J_w \dot{\omega}_w = T_d - T_f - F_x R, \tag{7}$$

where J_w is the wheel moment of inertia, and $\dot{\omega}_w$ is the derivative of the wheel rotation speed.

2.3. Tire Model

When a vehicle is driven on the road, the interaction forces between the vehicle and the road are determined by the interaction between the tires and the road. Therefore, tire forces are a key factor in the dynamic control of the vehicle, and tire force calculation models are necessary. It is well known that there are many tire models for calculating interaction forces, such as the Burckhardt tire model, Dugoff tire model, UniTire tire model, and Magic Formula tire model. Since the Magic Formula tire model has high accuracy in calculating longitudinal and lateral tire forces, especially when the tire has a large longitudinal slip rate and large lateral sideslip angle, it was chosen for calculating the tire forces in this paper.

The Magic Formula tire model is a semi-empirical tire model that is widely used in vehicle engineering. It uses a uniform formula to calculate the longitudinal and lateral forces of tires. The formula of the model is shown in Equation (8) [34].

$$\begin{cases} x = X + S_x \\ y(x) = D\sin\{C\arctan[Bx - E(Bx - \arctan(Bx))]\} \\ Y = y + S_y \end{cases}, \tag{8}$$

where X represents the longitudinal slip rate or lateral sideslip angle of the tire in a pure situation, and Y represents the longitudinal force or lateral force of the tire in a pure situation. In Equation (8), when X represents the longitudinal slip rate, Y represents the longitudinal force, and when X represents the lateral sideslip angle, Y represents the lateral force. S_x and S_y are correcting parameters. B, C, D, and E are the performance parameters determined from the testing data. Figure 3 shows the longitudinal force and lateral force in pure slip circumstances, varying with the different vertical loads of the tire. ($F_z = 1000$, 2000, 3000, 4000, and 5000 N). Figure 3a presents the longitudinal force of the tire in pure slip circumstances, according to varying values of the tire vertical load from 1000 to 5000 N and varying values of the tire longitudinal slip rate from -0.3 to 0.3. Figure 3b presents the lateral force of the tire in pure slip circumstances, according to varying values of the tire vertical load from 1000 to 5000 N and varying values of the tire lateral sideslip angle from $-30°$ to $30°$.

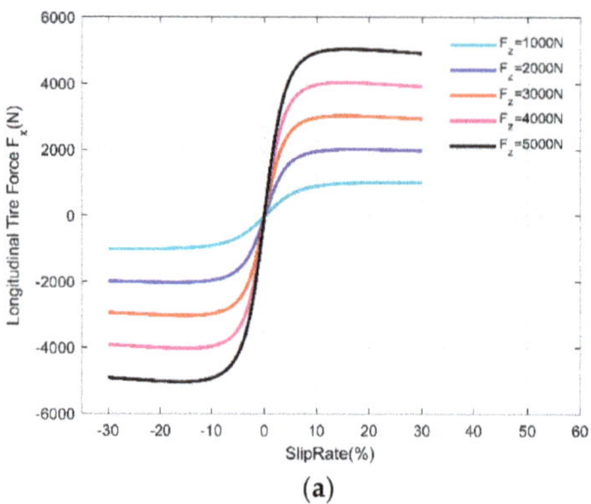 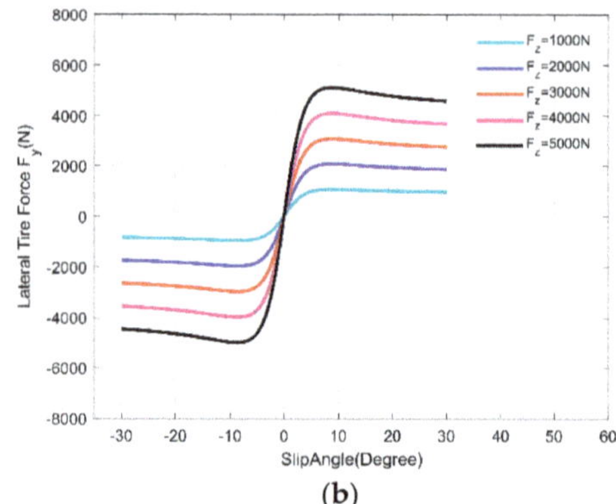

(a)   **(b)**

Figure 3. Longitudinal force and lateral force in pure slip circumstances. (**a**) Longitudinal force in pure slip circumstances, with slip rates from -0.3 to 0.3; (**b**) Lateral force in pure slip circumstances, with sideslip angles from $-30°$ to $30°$.

Furthermore, the tire works with longitudinal slip and lateral slip in many cases, which are called combined slip situations. In the case of combined slip situations, the calculation of the longitudinal and lateral forces of the tire can be performed as follows [34]:

$$\begin{cases} F_x^* = F_{x0} - \varepsilon\left(F_{x0} - F_{y0}\right)\left(\dfrac{\sigma_y^*}{\sigma_{total}^*}\right)^2 \\ F_y^* = F_{y0} - \varepsilon\left(F_{y0} - F_{x0}\right)\left(\dfrac{\sigma_x^*}{\sigma_{total}^*}\right)^2 \end{cases}, \tag{9}$$

where the total normalized slip $\sigma_{total}^* = \sqrt{\sigma_x^{*2} + \sigma_y^{*2}}$ and $\sigma_x^* = \dfrac{\sigma_x}{\sigma_{xmax}}$, $\sigma_y^* = \dfrac{\sigma_y}{\sigma_{ymax}}$. $\varepsilon = min\left(\sigma_{total}^*, 1\right)$. The theoretical slips are normalized by the peak slip values, σ_{xmax} and σ_{ymax}. σ_x and σ_y are calculated according to

$$\begin{cases} \sigma_x = -\dfrac{k}{1+k} \\ \sigma_y = \dfrac{\tan\alpha}{1+k} \end{cases}, \tag{10}$$

The equivalent longitudinal and lateral slips are calculated from the normalized total theoretical slip according to

$$\begin{cases} k' = -\dfrac{\sigma_{total}^*\sigma_{xmax}sign(\sigma_x)}{1+\sigma_{total}^*\sigma_{xmax}sign(\sigma_x)} \\ \alpha' = \tan^{-1}\left[\sigma_{total}^*\sigma_{ymax}sign\left(\sigma_y\right)\right] \end{cases}, \tag{11}$$

Using the equivalent longitudinal and lateral slips and ignoring the effect of the friction ratio, the longitudinal force and lateral force in Equation (12) are obtained according to

$$\begin{cases} F_{x0} = FX(Fz, k') \\ F_{y0} = FY(Fz, \alpha') \end{cases} \tag{12}$$

where FX and FY are the pure slip situation tire force functions according to Equation (8).

The longitudinal force and lateral force in the combined slip circumstances are shown in Figure 4; the tire vertical load is set to 2000 N. Figure 4a presents the longitudinal force of the tire when the tire vertical load is 2000 N, according to varying values of the tire longitudinal slip rate from 0 to 0.6 and varying values of the tire lateral sideslip angle from $0°$ to $30°$. Figure 4b presents the lateral force of tire when the tire vertical load is 2000 N, according to varying values of the tire longitudinal slip rate from 0 to 0.6 and varying values of the tire lateral sideslip angle from $0°$ to $30°$.

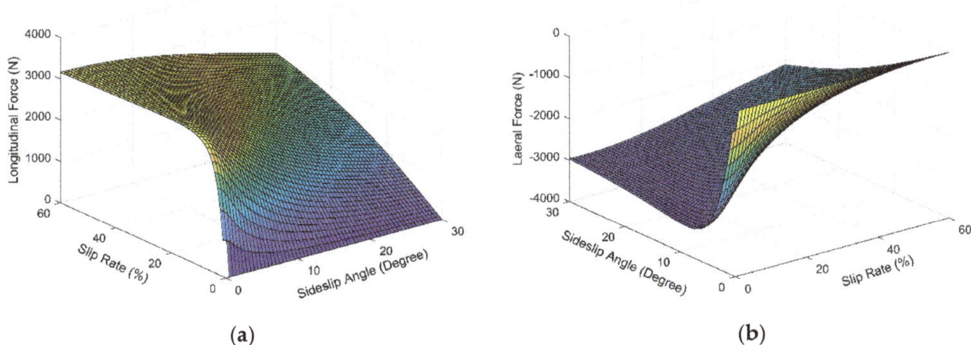

(a)   (b)

Figure 4. Longitudinal force and lateral force in combined slip circumstances; the slip rate is from 0 to 0.6, and the sideslip angle is from $0°$ to $30°$. (**a**) Longitudinal force; (**b**) Lateral force.

3. Driving Force Distribution and Control Strategy

3.1. Hierarchical Control Scheme

In this paper, vehicle motion control was implemented through a driving force distribution and control algorithm. There are two layers of processing in the hierarchical scheme: the upper layer controller and the lower layer controller, as shown in Figure 5. In the parsing layer, the desired state, including the desired longitudinal speed and the desired yaw rate, is obtained by parsing the inputs from the remote control. In the tire force estimation, the tire lateral forces are estimated by referring to [35]. In order to track the desired input state of the vehicle, an upper layer controller is proposed for calculating the desired resultant force and the desired resultant yaw moment. Since the desired resultant force and the desired resultant yaw moment are generated by the interaction between the six tires and the road, a lower layer controller was designed to distribute the driving force or torque of the six motors.

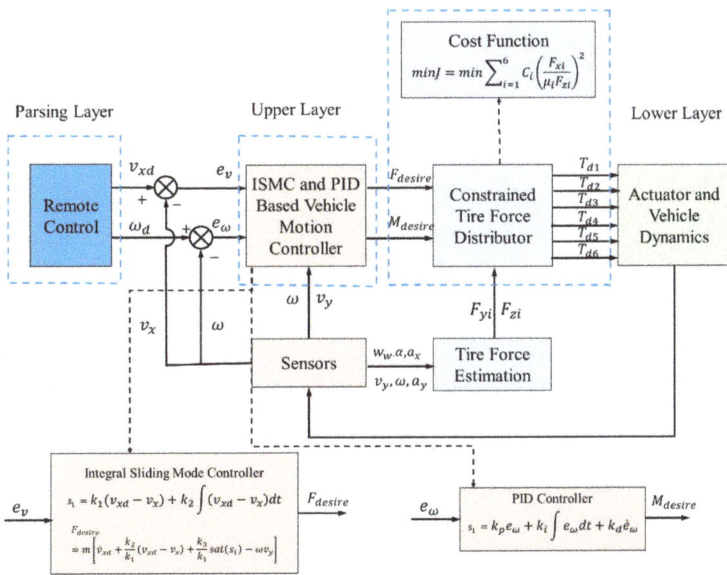

Figure 5. Schematic diagram of the controller.

3.2. Upper Layer Controller for Vehicle Motion Control

In order to eliminate the error between the desired state input of the vehicle and the feedback measurement state, the upper layer controller was designed to calculate the desired resultant force and the desired resultant yaw moment for tracking the desired inputs. According to Equations (1) and (3), the desired longitudinal velocity and the desired yaw rate can be independently controlled by the desired resultant force and the desired resultant yaw moment, respectively. Due to the external perturbations and nonlinear dynamics in the vehicle motion, the integral sliding mode control (ISMC) algorithm and PID can handle these perturbations and nonlinear dynamics with good robustness. Therefore, ISMC and PID were applied to solve the problem in the upper layer controller as described in the following sections of this paper.

3.2.1. Longitudinal Velocity Control

Based on Equation (1), the longitudinal velocity is determined by the longitudinal tire forces. Thus, the longitudinal velocity controller was designed to calculate the desired resultant force to track the desired longitudinal velocity. To decrease the tracking error, a sliding surface was defined by

$$s = k_1(v_{xd} - v_x) + k_2 \int (v_{xd} - v_x)dt, \tag{13}$$

Because of the chattering characteristics of the ISMC algorithm, it causes the tracking error to oscillate within the sliding surface. In order to reduce the chattering, an exponential reaching law with a saturation function is given:

$$\dot{s} = -k_3 sat(s) \tag{14}$$

where

$$sat(s) = \begin{cases} sgn(s), |s| \geq \phi \\ \frac{s}{\phi} \quad , |s| < \phi \end{cases} \tag{15}$$

where $k_1 > 0$ and ϕ is the sliding boundary layer thickness; $0 < \phi < 1$.

Combined with Equations (13)–(15), the following can be used to deduce the desired longitudinal force for tracking the desired longitudinal velocity:

$$\dot{v}_x = \dot{v}_{xd} + \frac{k_2}{k_1}(v_{xd} - v_x) + \frac{k_2}{k_1} sat(s)$$

$$F_{desire} = m\left[\dot{v}_{xd} + \frac{k_2}{k_1}(v_{xd} - v_x) + \frac{k_2}{k_1} sat(s) - \omega v_y\right], \tag{16}$$

3.2.2. Yaw Rate Control

The yaw rate controller was designed to calculate the desired resultant yaw moment to force the vehicle to follow the desired yaw rate using PID control as follows.

$$M_{desire} = I_z \dot{\omega}_d, \tag{17}$$

where

$$\dot{\omega}_d = [K_p(\omega_d - \omega) + K_i \int (\omega_d - \omega)dt + K_d \frac{d}{dt}(\omega_d - \omega)].$$

3.3. Lower Layer Controller for Driving Force Distribution

In the previous section, the ISMC and PID controller were used to calculate the desired resultant force and the desired resultant yaw moment. The vehicle has six motors driving each of the six wheels; that is to say that it is an over-actuation system. Thus, it is necessary to propose a practical driving force distribution algorithm. Since the motors can only provide longitudinal force to the wheels, the longitudinal force required for the wheels needs to be derived from Equation (18) as follows.

$$\begin{cases} F_{xdesire} = F_{x1d} + F_{x2d} + F_{x3d} + F_{x4d} + F_{x5d} + F_{x6d} \\ M_{Fxdesire} = \frac{B}{2}(F_{x2d} + F_{x4d} + F_{x6d} - F_{x1d} - F_{x3d} - F_{x5d}) \end{cases}, \tag{18}$$

where $F_{xid}(i = 1, 2, \cdots, 6)$ denote the desired longitudinal forces provided by the motors. According to the 6WD skid-steering EUGV dynamics model and upper layer controller, $F_{xdesire} = F_{desire}$ and $M_{Fxdesire} = M_{desire} - \sum M_{Fy}$. $\sum M_{Fy}$ can be calculated with Equation (4) based on the lateral force estimation of the tire. Therefore, the desired driving forces F_{xid} of the motors are obtained, and the desired driving torques T_{di} of the motors are calculated by

$$T_{di} = RF_{xid}, \tag{19}$$

Equation (18) describes the mathematical relationship between the driving forces of the six motors, but working out how to determine the output force of each motor is a problem. The driving force distribution algorithm is the solution to this problem. Now, the longitudinal tire workload rate η is defined as

$$\eta = \frac{F_x}{\mu F_z}, \tag{20}$$

where the tire longitudinal workload rate η influences the vehicle driving capability and dynamic stability. To achieve optimal maneuverability and stability control, an objective function is defined according to

$$minJ = min \sum_{i=1}^{6} C_i \eta_i^2 = min \sum_{i=1}^{6} C_i \left(\frac{F_{xi}}{\mu_i F_{zi}}\right)^2, \tag{21}$$

where $C_i(i = 1, 2, \cdots, 6)$ are weighting factors. In this paper, we propose that the weighting factors in the objective function can dynamically change according to the vehicle motion conditions and the vertical load of each tire. Since the vertical loads of the same side tires of the 6WD skid-steering EUGV are different, in order to dynamically adjust the longitudinal tire workload rates of the same-side wheels, the weighting factors are determined according to

$$\begin{cases} C_1 = C_2 = 1 \\ C_3 = \frac{F_{z3}}{F_{z1}} \\ C_4 = \frac{F_{z4}}{F_{z2}} \\ C_5 = \frac{F_{z5}}{F_{z1}} \\ C_6 = \frac{F_{z6}}{F_{z2}} \end{cases}, \tag{22}$$

There are several types of inequality constraints in the solving process for the objective function, including actuator constraints, road adhesion constraints, and tire friction circle constraints.

$$\begin{cases} \frac{T_{imin}}{R} \leq F_{xi} \leq \frac{T_{imax}}{R} \\ -\mu_i F_{zi} \leq F_{xi} \leq \mu_i F_{zi} \\ -\sqrt{(\mu_i F_{zi})^2 - (F_{yi})^2} \leq F_{xi} \leq \sqrt{(\mu_i F_{zi})^2 - (F_{yi})^2} \end{cases}, \tag{23}$$

Thus, a comprehensive cost function for the driving force distribution with equality and inequality constraints is established according to

$$\begin{cases} minJ = \frac{1}{2}x^T Wx \\ s.t. \begin{cases} b - Ax = 0 \\ max\left(\frac{T_{imin}}{R}, -\sqrt{(\mu_i F_{zi})^2 - (F_{yi})^2}\right) \leq F_{xid} \leq \\ min\left(\frac{T_{imax}}{R}, \sqrt{(\mu_i F_{zi})^2 - (F_{yi})^2}\right) \end{cases} \end{cases}, \tag{24}$$

where

$$x = [F_{x1d}, F_{x2d}, F_{x3d}, F_{x4d}, F_{x5d}, F_{x6d}]^T, \tag{25}$$

$$A = \begin{bmatrix} 1 & 1 & 1 & 1 & 1 & 1 \\ -\frac{B}{2} & \frac{B}{2} & -\frac{B}{2} & \frac{B}{2} & -\frac{B}{2} & \frac{B}{2} \end{bmatrix}, \tag{26}$$

$$W = \begin{bmatrix} C_1/(\mu_1 F_{z1})^2 & 0 & 0 & 0 & 0 & 0 \\ 0 & C_2/(\mu_2 F_{z2})^2 & 0 & 0 & 0 & 0 \\ 0 & 0 & C_3/(\mu_3 F_{z3})^2 & 0 & 0 & 0 \\ 0 & 0 & 0 & C_4/(\mu_4 F_{z4})^2 & 0 & 0 \\ 0 & 0 & 0 & 0 & C_5/(\mu_5 F_{z5})^2 & 0 \\ 0 & 0 & 0 & 0 & 0 & C_6/(\mu_6 F_{z6})^2 \end{bmatrix}, \tag{27}$$

$$b = [F_{xdesire}, M_{Fxdesire}]^T, \tag{28}$$

The optimal solution F_{xi} from solving the comprehensive cost function (24) is transformed into a problem of quadratic programming (QP) with equality constraints and inequality constraints, which can be solved by using the active set method.

4. Experimental Results

4.1. Experimental Platform and Projects

In order to verify the performance of the proposed distribution control method, experiments were implemented as described in this section. The experimental platform used for the validation is shown in Figure 6.

Figure 6. The experimental platform.

The studied vehicle can be regarded as a combination of six wheel corners, a body, and other components. Each wheel corner consists of an independent electrical drive motor system and a suspension system. Each wheel cannot steer, and the platform steers by skidding. Therefore, the experimental platform is a 6WD skid-steering EUGV. The tire model configured in the platform is 205/55 R16. The other components include a GPS/IMU, computer, and wireless receiver module. The model of the GPS/IMU is XW-GI7660. The platform parameters are listed in Table 1.

Table 1. Parameters of the experimental platform.

Symbol	Parameters	Value and Units
m	Vehicle total mass	2020 kg
I_z	Yaw moment of inertia of the vehicle	1897 N·m
a	Distance from center of gravity to the middle axle	0.2 m
B	Vehicle width	2.2 m
H	Height of center of gravity	0.68 m
l_1	Distance from the front axle to the middle axle	1.2 m
l_2	Distance from the middle axle to the rear axle	1.206 m
J_w	Rotation moment of the wheel	0.85 kg·m^2
R	Wheel effective radius	0.308 m

Figure 7 shows the details of the hardware of the platform. The desired control signals are sent by the operator via the operating remote control and received by the computer via the wireless receiver module. The motion status of the vehicle, including the longitudinal and lateral velocity, heading, yaw rate, and accelerations, was measured with the GPS/IMU system in the vehicle. The proposed distribution and control algorithm was employed in the computer, while the computer sent the control signals to the motor controller and received the feedback signals from the independent drive motors through the CAN bus.

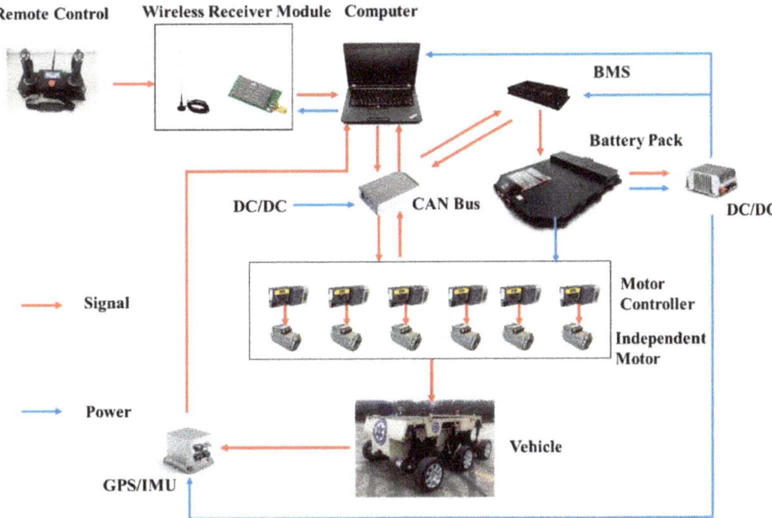

Figure 7. The hardware of the six-wheel-drive (6WD) skid-steering electric unmanned ground vehicle (EUGV) platform.

In order to demonstrate the excellent performance of the algorithm proposed in this paper, three control methods for experiments were used:

Method 1: Platform under the rule-based driving force even distribution method in [23], which is marked by Ref1.

Method 2: Platform under the optimal driving force vectoring distribution method in [15], which is marked by Ref2.

Method 3: Platform under the proposed distribution control.

The 6WD skid-steering EUGV platform requires straight-line and steering tracking capabilities. Thus, two types of testing scenarios were considered:

Scenario 1: Platform motion in straight conditions.

Scenario 2: Platform motion in curved conditions.

The figures were drawn by using MATLAB R2018b.

4.2. Experimental Results Discussion

4.2.1. Experiments in Straight Conditions

The desired longitudinal velocity of the vehicle was set to vary between 0 and 5 km/h. Figure 8 shows the desired speed at different times. At the beginning of the time, the desired speed was set to 1.8 km/h, and then, at 2 s, the desired speed was set to 5 km/h. Figure 8 compares the response of the actual speed resulting from the three control methods. From 13 to 22 s, the desired longitudinal velocity was set to 3 km/h. After that, from 22 to 24 s, the desired speed was set to 0; from 24 to 34 s, the desired speed was set to -5 km/h; and from 34 to 42 s, the desired speed was set to -3 km/h, and then, it was set to 0 until the end. Here, a negative desired speed means an opposite motion direction relative to the starting desired motion direction. The locally magnified graph in Figure 8 shows the response of the actual velocity when the desired velocity was increased at 13 s. As mentioned above, the longitudinal velocity tracking accuracy of the three control methods is approximate in straight conditions. Figure 9 shows the yaw rate tracking performance of the three control methods, which is similar to in the above case. Figure 9a presents the yaw rate tracking performance. The tracking errors of the three controllers are very small and nearly the same. Figure 9b–d present the torque on each wheel of the three controllers. The torques on each wheel with Ref2 and the proposed controller are nearly same, and are a little larger than those with Ref1.

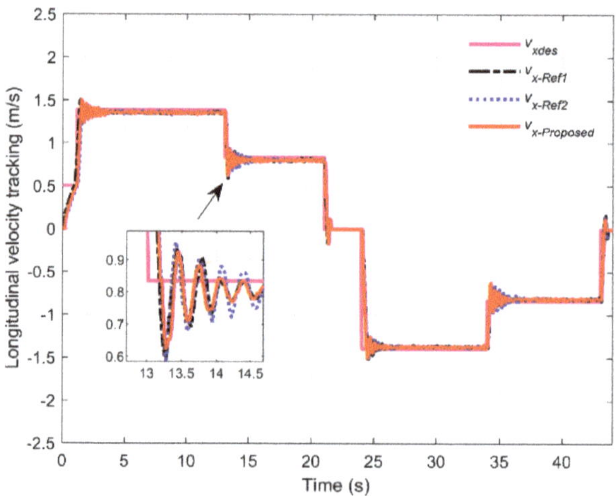

Figure 8. Longitudinal velocity tracking.

Figure 10a,b compare the responses to heading and heading error. However, while the yaw rate tracking accuracy is approximately the same for the three control methods in Figure 9, the heading and heading errors are different. Compared to at the starting point, from 0 to 25 s, the heading errors with the Ref1 and Ref2 controllers were almost same, and the maximums were about $-0.94°$, but the maximum of the heading error with the proposed controller was about $-0.8°$, which was the smallest among the three controllers. After 25 s, the heading errors with the Ref1 and the proposed controllers were almost same, and Ref2 had the largest among the three controllers. In other words, the proposed controller reduces the heading error in straight conditions. As a result, the proposed control method has better tracking performance when tracking straight conditions.

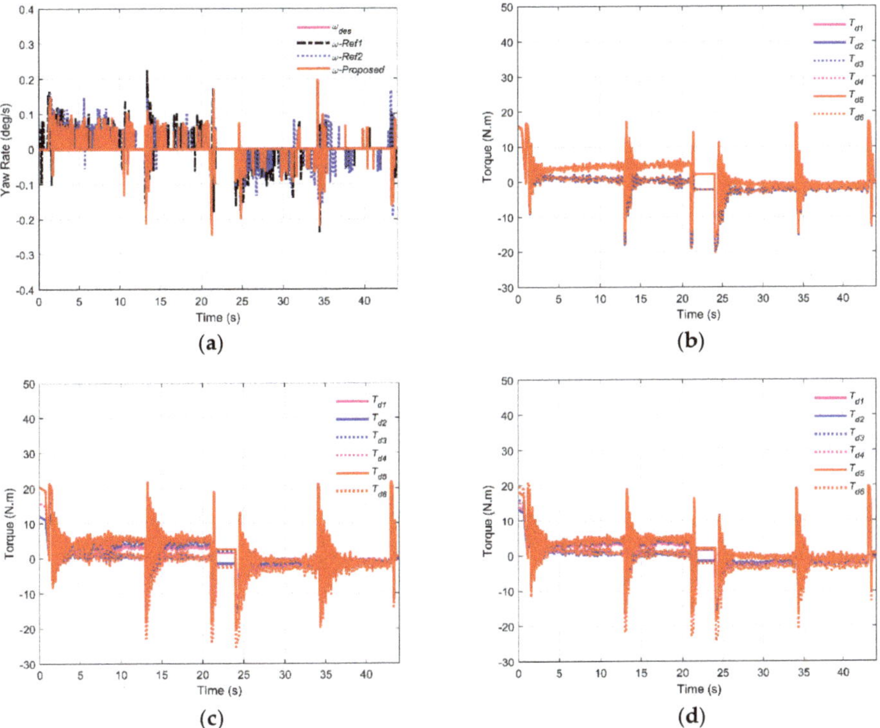

Figure 9. Comparison of yaw rate and the desired torque on each wheel with three controllers: (**a**) Comparison of yaw rate; (**b**) Torque on each wheel with Ref1 controller; (**c**) Torque on each wheel with Ref2 controller; (**d**) Torque on each wheel with the proposed controller.

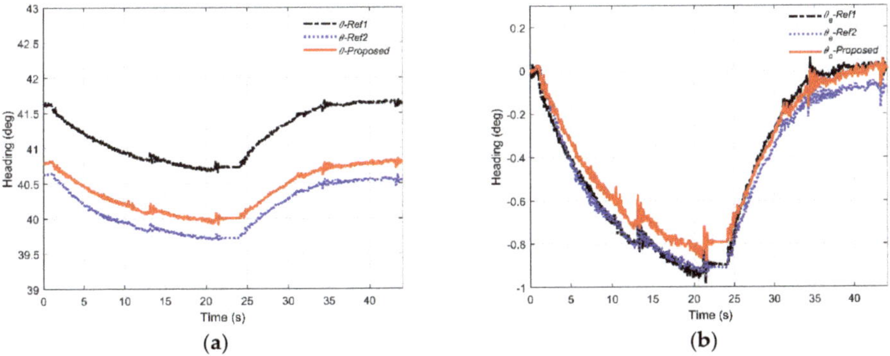

Figure 10. Comparison of heading and heading error with three controllers: (**a**) Comparison of heading; (**b**) Comparison of heading error.

Figure 11 shows the response of the sideslip angle of the vehicle. When the platform was moving in straight conditions, the value of the sideslip angle was small with the three controllers. According to Figure 8, when the desired speed underwent a step-jump change, the motor torque of each wheel, heading, and sideslip angle of the platform underwent a jump change, as shown in Figures 9, 10a and 11. To depict the performance more directly, the phase planes of the sideslip angle vs. yaw rate are shown in Figure 12. It can be seen that the movement of the boundary of the vehicle motion states with the proposed controller is smaller than that with the other two controllers. This means that the proposed controller has better stability performance.

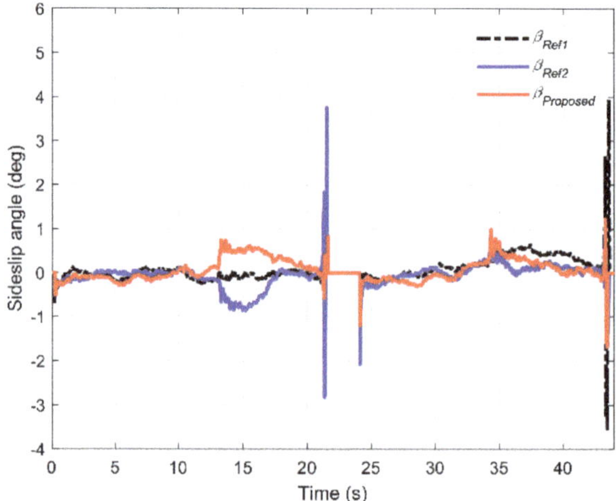

Figure 11. Comparison of sideslip angles.

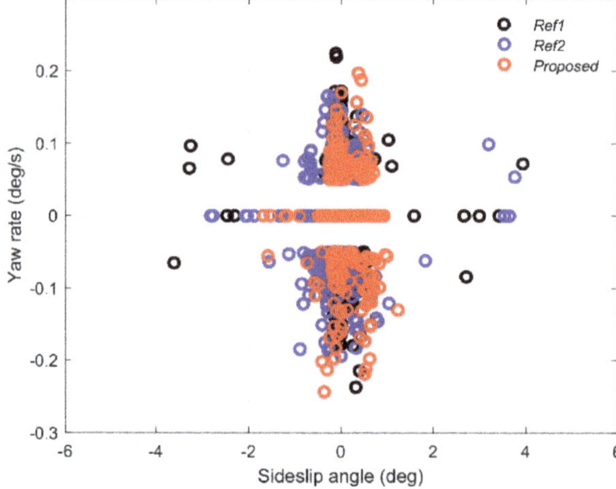

Figure 12. Comparison of phase planes.

4.2.2. Experiments in Curved Condition

In the curved condition experiments, the desired longitudinal velocity of the platform was set to 1.8 km/h for 0–1.5 s, 5 km/h for 1.5–13 s, and 0 for 13–13.5 s, as shown in Figure 13. The desired yaw rate was set to 2.87 degrees/s from 3 to 11 s and 0 in other periods. The longitudinal and lateral tracking in the curve condition more rigorously verified the performance of the driving force distribution control algorithm compared to in the straight conditions. Therefore, the purpose of these experiments was to verify the maneuverability and stability performance of the proposed driving force distribution algorithm.

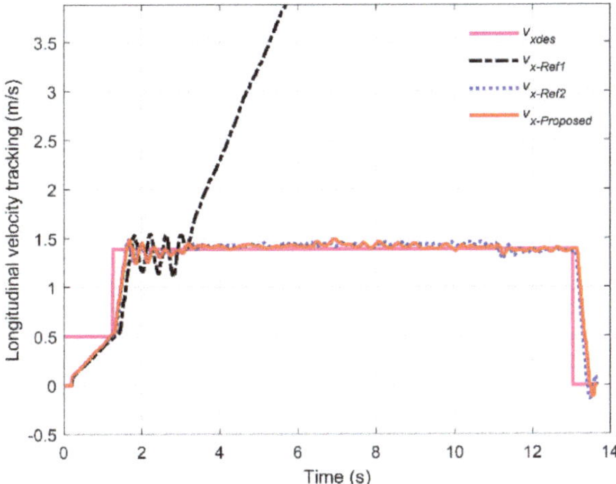

Figure 13. Longitudinal velocity tracking.

Figure 13 compared the longitudinal velocity tracking. The longitudinal velocity of the UGV platform with the Ref1 controller diverged when the desired yaw rate was not zero. At 3 s, the desired yaw rate was set to 2.87 deg/s. Then, the Ref1 controller distributed the torque on each wheel to control the platform to track the desired longitudinal velocity and yaw rate.

Since the torque on the same side of the vehicle was equal with the Ref1 controller, the vehicle's longitudinal velocity could not converge to the desired value, and the experiment was stopped at 5.8 s. Figure 14a,b present the longitudinal velocity tracking error and its percentage in stable tracking from 3 to 12 s with Ref2 and the proposed controller. There was little difference in terms of the longitudinal speed tracking accuracy between the two controllers. Figure 15 compared the yaw rate tracking. However, according to Figures 15 and 16, the yaw rate tracking error with the proposed controller was smaller than that with the Ref2 controller in stable tracking from 5 to 11 s. In order to observe the tracking error and accuracy more directly and clearly, the mean absolute error (MAE), root mean square error (RMSE), and standard deviation (SD) values of the errors with two controllers were calculated and are listed in Table 2 From Table 2, it can be observed that the MAEs of the longitudinal velocity and yaw rate tracking with the proposed controller were smaller than those with the Ref2 controller. According to the MAEs, compared to Ref2, the longitudinal velocity tracking error and yaw rate tracking error decreased by 33.9% and 54.5% with the proposed controller in the tracking stabilization stage, respectively. The RMSE of the yaw rate tracking with the proposed controller was smaller than that with the Ref2 controller, and the other two metrics including the RMSE of the longitudinal velocity tracking and SDs of the longitudinal and yaw rate tracking with the proposed controller were all smaller than those with the Ref2 controller. This indicates that the proposed controller has better performance in terms of longitudinal and lateral tracking accuracy.

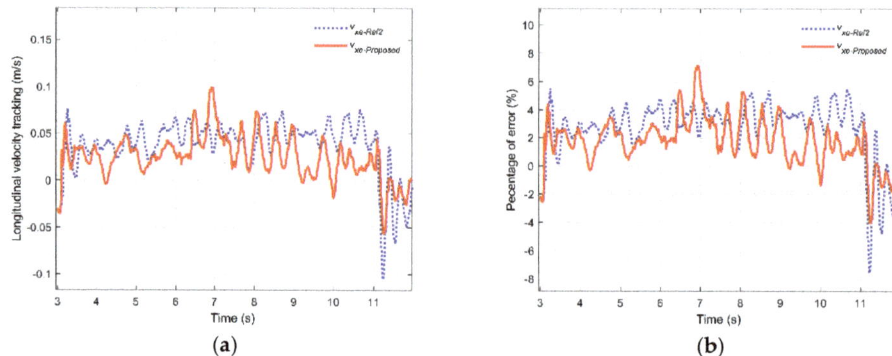

Figure 14. Comparison of longitudinal velocity tracking with two controllers: (**a**) Comparison of longitudinal velocity tracking error; (**b**) Comparison of percentage of longitudinal velocity tracking error.

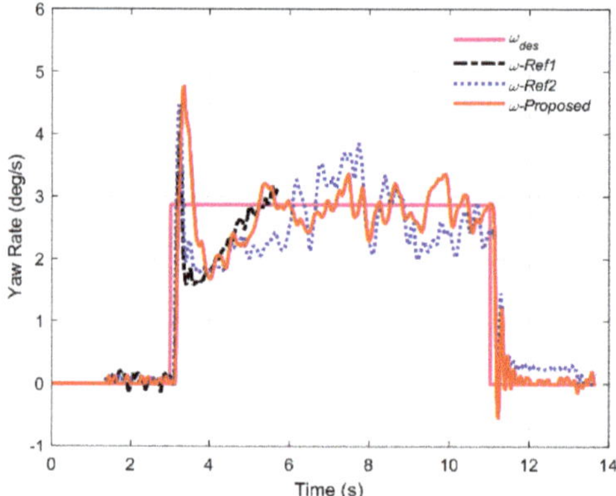

Figure 15. Yaw rate tracking.

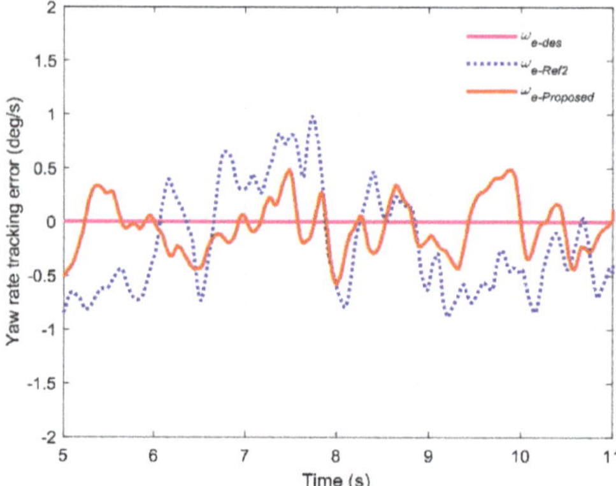

Figure 16. Yaw rate tracking error.

Table 2. Tracking error analysis.

Title 1	MAE	RMSE	SD
v_{xe}-Ref2	0.0433	0.0471	0.0294
v_{xe}-Proposed	0.0286	0.0341	0.0237
w_e-Ref2	0.4688	0.0418	0.4922
w_e-Proposed	0.2133	0.0417	0.2509

Figure 17 shows the desired torque of each motor with the three controllers. Figures 18 and 19 present the sideslip angles of the vehicle and the phase planes of the sideslip angle vs. yaw rate. It can be seen that the proposed controller has better stability performance than Ref2.

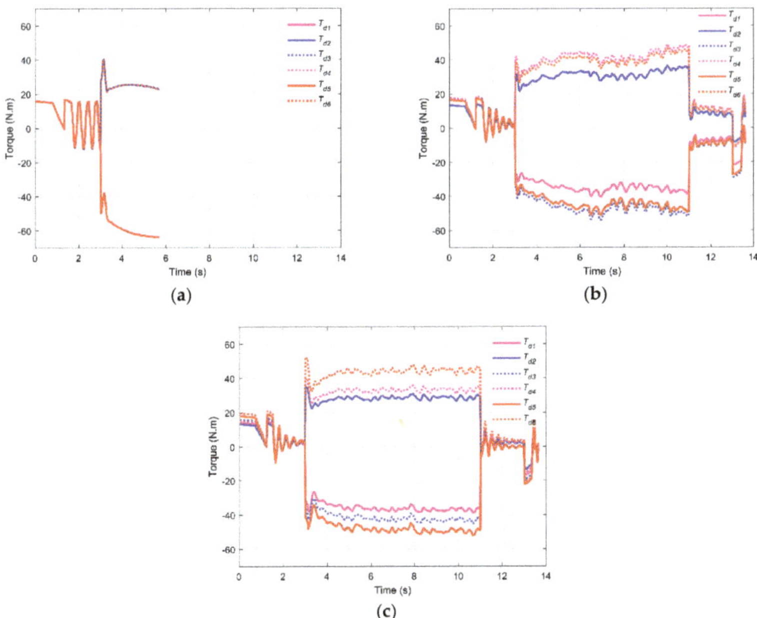

Figure 17. Torque on each wheel with three controllers: (**a**) Torque on each wheel with Ref1 controller; (**b**) Torque on each wheel with Ref2 controller; (**c**) Torque on each wheel with the proposed controller.

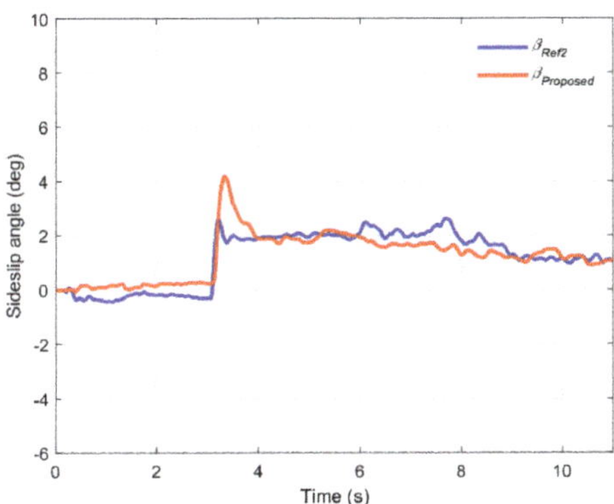

Figure 18. Comparison of sideslip angles.

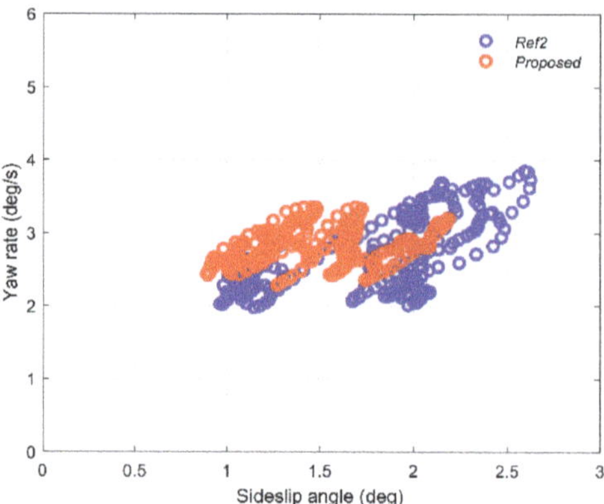

Figure 19. Comparison of phase planes.

5. Conclusions

In this paper, a driving force distribution and control strategy for a 6WD skid-steering EUGV actuated by independent drive wheels is proposed to improve the maneuverability and stability. The key conclusions include the following:

(1) The architecture of a 6WD skid-steering EUGV actuated by independent drive wheels was introduced. Focusing on longitudinal, lateral, and yaw motion, a nonlinear vehicle system dynamics model with nine DOFs was established. By using the Magic Formula tire model, the longitudinal and lateral tire forces in pure circumstances were introduced. In order to describe the actual forces of the tire, a dynamics model of the tire in the combined slip circumstances was developed.

(2) To improve the maneuverability and stability, a driving force distribution and control strategy was proposed. The control strategy was composed of hierarchical controllers, including an upper layer controller and lower layer controller. In the upper layer controller, the controlled resultant driving force and yaw moment were calculated to control the vehicle motion states to track the desired input by using the integral sliding mode control (ISMC) and proportion–integral–differential (PID) control method. In the lower layer controller, the controlled resultant driving force and yaw moment were realized by distributing the actuator motor's torque. An objective function consisting of the longitudinal tire workload rates and weight factors of the tire was established by considering the inequality constraints, including the actuator constraints, road adhesion constraints, and tire friction circle constraints. The weight factors were composed of the ratio of the tire vertical force considering the vehicle motion conditions and vertical load change of each tire. Based on the active set algorithm, the control distribution method was used to solve the objective function for optimal driving force allocation.

(3) Experimental tests were conducted to verify the effectiveness of the proposed control algorithm. Two types of testing scenarios were studied: motion in straight conditions and curved conditions. In addition, the rule-based driving force even allocation method and optimal driving force vectoring distribution method used in the related references were experimented with for comparison. The experimental results demonstrate that the proposed control strategy had excellent performance in vehicle maneuverability and stability. In the curve-tracking condition, compared to Ref2, the longitudinal velocity tracking error and yaw rate tracking error were decreased by 33.9% and 54.5% with the proposed controller in the tracking stabilization stage, respectively. In the future, we will test the experimental platform in different scenarios with different terrains or banks or other more complex situations to test and investigate the robustness of our proposed control method.

Author Contributions: Conceptualization, H.Z. and H.L.; investigation, H.Z.; methodology, H.Z.; software, H.Z. and X.T.; validation, H.Z., H.L., and X.T.; data curation, H.Z. and R.B.; writing—

original draft preparation, H.Z.; writing—review and editing, H.L. and B.Y.; supervision, H.L. and B.Y.; project administration, Y.D.; funding acquisition, H.L. All authors have read and agreed to the published version of the manuscript.

Funding: This work was supported by the National Key Research and Development Program of China (Nos. 2016YFD0701401, 2017YFD0700303 and 2018YFD0700602); Youth Innovation Promotion Association of the Chinese Academy of Sciences (Grant No. 2017488); Key Supported Project in the Thirteenth Five-year Plan of the Hefei Institutes of Physical Science, Chinese Academy of Sciences (Grant No.KP-2017-35, KP-2017-13 and KP-2019-16); Independent Research Project of the Research Institute of Robotics and Intelligent Manufacturing Innovation, Chinese Academy of Sciences (Grant No. C2018005); and Technological Innovation Project for New Energy and Intelligent Networked Automobile Industry of Anhui Province.

Data Availability Statement: The data used to support the findings of this study are available from the corresponding author upon request.

Conflicts of Interest: The authors declare that there is no conflict of interest.

Abbreviations

a	Distance from center of gravity to the middle axle
β	Vehicle sideslip angle
a_x	Longitudinal acceleration
a_{xdes}	Desired longitudinal acceleration
a_y	Lateral acceleration
B	Vehicle width
H	Height of center of gravity
F_{xi}	Longitudinal force of the ith tire ($i = 1,2,\cdots,6$)
F_{xid}	Desired longitudinal force of the ith tire ($i = 1,2,\cdots,6$)
F_{yi}	Lateral force of the ith tire ($i = 1,2,\cdots,6$)
F_{zi}	Vertical force of the ith tire ($i = 1,2,\cdots,6$)
F_x	Total longitudinal force at the center of gravity
F_y	Total lateral force at the center of gravity
g	Acceleration of gravity
I_z	Yaw moment of inertia of the vehicle
J_w	Rotation moment of the wheel
l_1	Distance from front axle to middle axle
l_2	Distance from middle axle to rear axle
m	Vehicle total mass
M_{Fx}	Resultant moment produced by longitudinal force of tire
M_{Fy}	Resultant moment produced by lateral force of tire
R	Wheel effective radius
T_i	Torque of the ith wheel ($i = 1,2,\cdots,6$)
T_{di}	Desired torque of the ith wheel ($i = 1,2,\cdots,6$)
v_x	Vehicle longitudinal velocity
v_{xdes}	Desired vehicle longitudinal velocity
v_y	Vehicle lateral velocity
ω	Vehicle yaw rate
ω_{des}	Desired vehicle yaw rate
v_{xi}	Longitudinal velocity of the ith wheel ($i = 1,2,\cdots,6$)
v_{yi}	Lateral velocity of the ith wheel ($i = 1,2,\cdots,6$)
ω_{wi}	Rotation velocity of the ith wheel ($i = 1,2,\cdots,6$)
α_i	Sideslip angle of the ith tire ($i = 1,2,\cdots,6$)
λ_i	Slip rate of the ith tire ($i = 1,2,\cdots,6$)
η_i	Longitudinal force workload rate of the ith tire ($i = 1,2,\cdots,6$)
μ_i	Coefficient of friction between road and the ith tire ($i = 1,2,\cdots,6$)

References

1. Ahmed, M.; El-Gindy, M.; Lang, H.; Omar, M. *Development of Active Rear Axles Steering Controller for 8X8 Combat Vehicle*; SAE Technical Paper; SAE International: Warrendale, PA, USA, 2020.
2. Shuai, Z.; Li, C.; Gai, J.; Han, Z.; Zeng, G.; Zhou, G. Coordinated motion and powertrain control of a series-parallel hybrid 8 x 8 vehicle with electric wheels. *Mech. Syst. Sig. Process.* **2019**, *120*, 560–583. [CrossRef]
3. D'Urso, P.; El-Gindy, M. Development of control strategies of a multi-wheeled combat vehicle. *Int. J. Autom. Control* **2018**, *12*, 325–360. [CrossRef]
4. Lam, A.Y.S.; Leung, Y.-W.; Chu, X. Autonomous-Vehicle Public Transportation System: Scheduling and Admission Control. *IEEE Trans. Intell. Transp. Syst.* **2016**, *17*, 1210–1226. [CrossRef]
5. Kayacan, E.; Ramon, H.; Saeys, W. Robust Trajectory Tracking Error Model-Based Predictive Control for Unmanned Ground Vehicles. *IEEE–ASME Trans. Mechatron.* **2016**, *21*, 806–814. [CrossRef]
6. Huang, W.; Wen, D.; Geng, J.; Zheng, N.-N. Task-Specific Performance Evaluation of UGVs: Case Studies at the IVFC. *IEEE Trans. Intell. Transp. Syst.* **2014**, *15*, 1969–1979. [CrossRef]
7. Wu, X.; Hu, X.; Moura, S.; Yin, X.; Pickert, V. Stochastic control of smart home energy management with plug-in electric vehicle battery energy storage and photovoltaic array. *J. Power Sources* **2016**, *333*, 203–212. [CrossRef]
8. Boulon, L.; Hissel, D.; Bouscayrol, A.; Pape, O.; Pera, M.-C. Simulation Model of a Military HEV with a Highly Redundant Architecture. *IEEE Trans. Veh. Technol.* **2010**, *59*, 2654–2663. [CrossRef]
9. Li, Z.; Zheng, L.; Gao, W.; Zhan, Z. Electromechanical Coupling Mechanism and Control Strategy for In-Wheel-Motor-Driven Electric Vehicles. *IEEE Trans. Ind. Electron.* **2019**, *66*, 4524–4533. [CrossRef]
10. De Carvalho Pinheiro, H.; Ferraris, A.; Galanzino, E.; Sisca, L.; Carello, M.; Airale, A.; Messana, A. *All-Wheel Drive Electric Vehicle Modeling and Performance Optimization*; SAE Technical Paper 2019-36-0197; SAE International: Warrendale, PA, USA, 2020. [CrossRef]
11. Asiabar, A.N.; Kazemi, R. A direct yaw moment controller for a four in-wheel motor drive electric vehicle using adaptive sliding mode control. *Proc. Inst. Mech. Eng. Part K J. Multi-Body Dyn.* **2019**, *233*, 549–567. [CrossRef]
12. Carvalho Pinheiro, H.; Messana, A.; Sisca, L.; Ferraris, A.; Airale, A.G.; Carello, M. Torque Vectoring in Electric Vehicles with In-wheel Motors. In *Advances in Mechanism and Machine Science*; Springer: Cham, Germany, 2019; Volume 73, pp. 3127–3136. [CrossRef]
13. Huang, X.Y.; Zhang, H.; Zhang, G.G.; Wang, J.M. Robust Weighted Gain-Scheduling H-infinity Vehicle Lateral Motion Control With Considerations of Steering System Backlash-Type Hysteresis. *IEEE Trans. Control Syst. Technol.* **2014**, *22*, 1740–1753. [CrossRef]
14. Zhao, H.Y.; Chen, W.X.; Zhao, J.Y.; Zhang, Y.L.; Chen, H. Modular Integrated Longitudinal, Lateral, and Vertical Vehicle Stability Control for Distributed Electric Vehicles. *IEEE Trans. Veh. Technol.* **2019**, *68*, 1327–1338. [CrossRef]
15. Song, Y.T.; Shu, H.Y.; Chen, X.B.; Luo, S. Direct-yaw-moment control of four-wheel-drive electrical vehicle based on lateral tyre-road forces and sideslip angle observer. *IET Intel. Transp. Syst.* **2019**, *13*, 303–312. [CrossRef]
16. Kang, J.; Yoo, J.; Yi, K. Driving Control Algorithm for Maneuverability, Lateral Stability, and Rollover Prevention of 4WD Electric Vehicles with Independently Driven Front and Rear Wheels. *IEEE Trans. Veh. Technol.* **2011**, *60*, 2987–3001. [CrossRef]
17. Xiong, L.; Yu, Z.; Wang, Y.; Yang, C.; Meng, Y. Vehicle dynamics control of four in-wheel motor drive electric vehicle using gain scheduling based on tyre cornering stiffness estimation. *Veh. Syst. Dyn.* **2012**, *50*, 831–846. [CrossRef]
18. Zhou, H.L.; Jia, F.J.; Jing, H.H.; Liu, Z.Y.; Guvenc, L. Coordinated Longitudinal and Lateral Motion Control for Four Wheel Independent Motor-Drive Electric Vehicle. *IEEE Trans. Veh. Technol.* **2018**, *67*, 3782–3790. [CrossRef]
19. Nahidi, A.; Kasaiezadeh, A.; Khosravani, S.; Khajepour, A.; Chen, S.-K.; Litkouhi, B. Modular integrated longitudinal and lateral vehicle stability control for electric vehicles. *Mechatronics* **2017**, *44*, 60–70. [CrossRef]
20. Zhai, L.; Sun, T.M.; Wang, J. Electronic Stability Control Based on Motor Driving and Braking Torque Distribution for a Four In-Wheel Motor Drive Electric Vehicle. *IEEE Trans. Veh. Technol.* **2016**, *65*, 4726–4739. [CrossRef]
21. Wang, Y.F.; Fujimoto, H.; Hara, S. Torque Distribution-Based Range Extension Control System for Longitudinal Motion of Electric Vehicles by LTI Modeling with Generalized Frequency Variable. *IEEE-ASME Trans. Mechatron.* **2016**, *21*, 443–452. [CrossRef]
22. Yuan, X.B.; Wang, J.B. Torque Distribution Strategy for a Front- and Rear-Wheel-Driven Electric Vehicle. *IEEE Trans. Veh. Technol.* **2012**, *61*, 3365–3374. [CrossRef]
23. Liu, M.; Huang, J.; Cao, M. Handling Stability Improvement for a Four-Axle Hybrid Electric Ground Vehicle Driven by In-Wheel Motors. *IEEE Access* **2018**, *6*, 2668–2682. [CrossRef]
24. Boyd, S.; Boyd, S.P.; Vandenberghe, L. *Convex Optimization*; Cambridge University Press: Cambridge, UK, 2004.
25. De Novellis, L.; Sorniotti, A.; Gruber, P. Wheel Torque Distribution Criteria for Electric Vehicles with Torque-Vectoring Differentials. *IEEE Trans. Veh. Technol.* **2014**, *63*, 1593–1602. [CrossRef]
26. Zhao, H.Y.; Gao, B.Z.; Ren, B.T.; Chen, H.; Deng, W.W. Model predictive control allocation for stability improvement of four-wheel drive electric vehicles in critical driving condition. *IET Control Theory Appl.* **2015**, *9*, 2688–2696. [CrossRef]
27. Khosravani, S.; Jalali, M.; Khajepour, A.; Kasaiezadeh, A.; Chen, S.K.; Litkouhi, B. Application of Lexicographic Optimization Method to Integrated Vehicle Control Systems. *IEEE Trans. Ind. Electron.* **2018**, *65*, 9677–9686. [CrossRef]
28. Leng, B.; Xiong, L.; Yu, Z.P.; Zou, T. Allocation control algorithms design and comparison based on distributed drive electric vehicles. *Int. J. Automot. Technol.* **2018**, *19*, 55–62. [CrossRef]

29. Xiong, L.; Teng, G.W.; Yu, Z.P.; Zhang, W.X.; Feng, Y. Novel Stability Control Strategy for Distributed Drive Electric Vehicle Based on Driver Operation Intention. *Int. J. Automot. Technol.* **2016**, *17*, 651–663. [CrossRef]
30. Kim, W.; Yi, K.; Lee, J. An optimal traction, braking, and steering coordination strategy for stability and manoeuvrability of a six-wheel drive and six-wheel steer vehicle. *Proc. Inst. Mech. Eng. Part D J. Automob. Eng.* **2012**, *226*, 3–22. [CrossRef]
31. Kim, C.; Ashfaq, A.M.; Kim, S.; Back, S.; Kim, Y.; Hwang, S.; Jang, J.; Han, C. Motion Control of a 6WD/6WS Wheeled Platform with In-wheel Motors to Improve Its Maneuverability. *Int. J. Control Autom. Syst.* **2015**, *13*, 434–442. [CrossRef]
32. Zhao, Y.; Zhang, C.N. Electronic Stability Control for Improving Stability for an Eight In-Wheel Motor-Independent Drive Electric Vehicle. *Shock Vib.* **2019**, 1–21. [CrossRef]
33. Kim, W.; Yi, K.; Lee, J. Drive control algorithm for an independent 8 in-wheel motor drive vehicle. *J. Mech. Sci. Technol.* **2011**, *25*, 1573–1581. [CrossRef]
34. Pacejka, H. *Tire and Vehicle Dynamics*; Elsevier: Amsterdam, The Netherlands, 2005.
35. Cordeiro, R.A.; Victorino, A.C.; Azinheira, J.R.; Ferreira, P.A.V.; de Paiva, E.C.; Bueno, S.S. Estimation of Vertical, Lateral, and Longitudinal Tire Forces in Four-Wheel Vehicles Using a Delayed Interconnected Cascade-Observer Structure. *IEEE/ASME Trans. Mechatron.* **2019**, *24*, 561–571. [CrossRef]

applied sciences

Article

An Estimator for the Kinematic Behaviour of a Mobile Robot Subject to Large Lateral Slip

Andrea Botta *, Paride Cavallone , Luigi Tagliavini, Luca Carbonari, Carmen Visconte and Giuseppe Quaglia

Department of Mechanical and Aerospace Engineering, Politecnico di Torino, 10129 Torino, Italy; paride.cavallone@polito.it (P.C.); luigi.tagliavini@polito.it (L.T.); luca.carbonari@polito.it (L.C.); carmen.visconte@polito.it (C.V.); giuseppe.quaglia@polito.it (G.Q.)
* Correspondence: andrea.botta@polito.it

Abstract: In this paper, the effects of wheel slip compensation in trajectory planning for mobile tractor-trailer robot applications are investigated. Firstly, a kinematic model of the proposed robot architecture is marked out, then an experimental campaign is done to identify if it is possible to kinematically compensate trajectories that otherwise would be subject to large lateral slip. Due to the close connection to the experimental data, the results shown are valid only for Epi.q, the prototype that is the main object of this manuscript. Nonetheless, the base concept can be usefully applied to any mobile robot subject to large lateral slip.

Keywords: mobile robots; tractor-trailer; wheel slip compensation; kinematics

Citation: Botta, A.; Cavallone, P.; Tagliavini, L.; Carbonari, L.; Visconte, C.; Quaglia, G. An Estimator for the Kinematic Behaviour of a Mobile Robot Subject to Large Lateral Slip. *Appl. Sci.* **2021**, *11*, 1594. https://doi.org/10.3390/app11041594

Received: 23 January 2021
Accepted: 5 February 2021
Published: 10 February 2021

Publisher's Note: MDPI stays neutral with regard to jurisdictional claims in published maps and institutional affiliations.

1. Introduction

In recent years, autonomous navigation applied to mobile robots is a rising trend. Within this field, trajectory planning and control is one of the most studied themes. Along with the growth of the mobile robotic fields of application, new and better performance is required from robots. In particular, off-road capabilities, quick manoeuvring at high speed, and particular locomotion architectures require a deep understanding of how the robot dynamically behaves to develop proper trajectory planning and control.

Lateral and longitudinal wheel slips are two of the most common phenomena to be faced in order to compensate undesired robot behaviour. Many of the existing works in mobile robotics overcome these issues assuming pure rolling conditions of the wheels, neglecting, therefore, wheel slip [1–5]. These issues are instead largely studied in the automotive field due to similar interest in handling control [6,7]. Wheel slip is almost impossible to be measured directly; therefore, in most of the cases, it has to be estimated. Thus, many slip estimation approaches have been proposed [8], such as sliding mode control [9,10], motor current sensing control [11], Kalman [12,13] or particle filter [14,15] based estimations, IMU-based slip estimation [16], and vision-based estimation [17]. All these methods have advantages and limitations reported in depth in the literature. Generally speaking, the best-performing solutions require a structured environment (i.e., vision-based systems) or intensive computation load in order to model, estimate, and control the robot correctly. Therefore, it is not always possible to meet the requirements of the various solutions, either because it is not possible to setup in advance the work environment, because the control unit is not capable to manage the slip compensation complexity, or maybe because it is already overburdened by other heavy tasks. For this reason, this paper proposes a very simple approach based on adjusting kinematics parameters to compensate the large lateral wheel slip. A vaguely similar approach was adopted in a couple of cases among tracked robots in order to correlate the complex tracks interaction with the soil with a simpler wheel–ground contact [18,19].

The idea behind this approach is to characterise the mobile robot through an experimental campaign with the aim of finding a simple correlation between the ideal and

167

expected kinematic trajectory and the measured one. Then, by means of the experimental correlation between the desired and observed behaviour, it is possible to adjust some robot control reference input signals with the aim of obtaining a trajectory closer to the kinematic one by compensating undesired handling behaviours due to dynamic phenomena. The experimental correlation between key parameters, such as the yaw rate and path curvature radius, should be simple enough to be easily and quickly computed by any robot control unit. This approach is strongly related to the robot (and the environment) that has been characterised. However, if a good identification is done, it is possible to obtain a very simple correlation between parameters that can be easily used to compute the required corrections.

This paper briefly introduces a kinematic model of the mobile robot that is then used to fit the experimental results. Post-processing the data, a correlation between the theoretical and experimental results is found and used to compensate the divergence between the two.

2. Kinematic Model

This section summarises the theoretical background of the study. Initially, the robot is briefly presented, then the kinematic model is derived.

2.1. Robot Architecture

This study focus on Epi.q (Figure 1), the last model of a family of modular surveillance UGV [20]. It is composed of two practically identical modules linked together by a 2 DOF (relative yaw and relative roll rotations) joint. Its most notable feature is the architecture of its hybrid legged-wheeled locomotion unit (Figure 2): the unique design is an underactuated epicyclic gearing system driven by a single electric motor coupled to the sun gear; depending on the forces acting on the wheels and the carrier, the motor imposes a rotation motion to the wheels, to the carrier or both of them [21]. Hence, this locomotion unit enables to overcome little obstacles using the carrier as a legged rotating unit. While this locomotion unit enables interesting behaviour, during more conventional navigation, all wheels of the same unit are constrained to spin at the same speed due to the nature of the gearing system. This inevitably leads to a very evident phenomenon of large lateral wheel slip during curved trajectories, even at low speed.

This drawback can be addressed in several ways to obtain the desired trajectories, compensating the divergences from the kinematic model. The authors have proposed an initial attempt to dynamically model the behaviour of a robot with a similar architecture [22]. In this study, however, a different approach, purely based on kinematic models, is taken.

Figure 1. Epi.q, modular surveillance UGV.

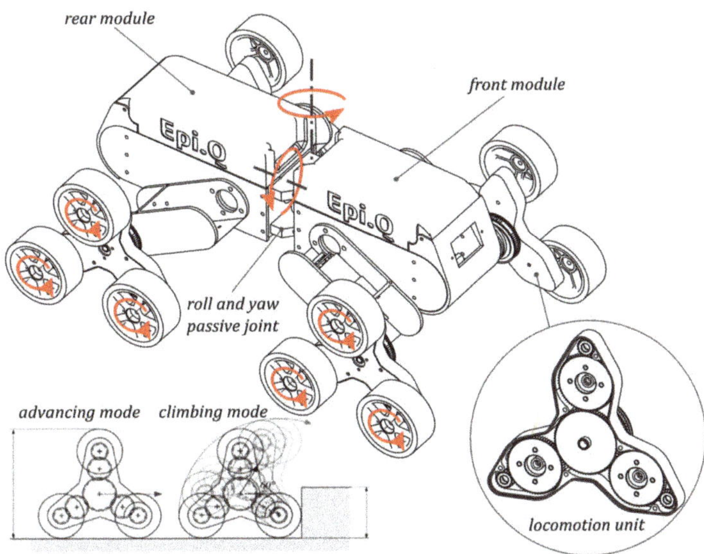

Figure 2. Schematic representation of Epi.q and its locomotion unit.

2.2. Kinematic Model

The kinematic model is based on the following hypothesis:

- The model is purely kinematic.
- The model considers only planar motion on a flat surface; out-of-plane motions are neglected.
- The locomotion units are simplified considering a single equivalent wheel per side with the axis passing through the centre of mass of the module.

Figure 3 represents the kinematic module, where (x_n, y_n, φ_n) is the pose of the n-th module ($n = 1$ for the front module, $n = 2$ for the back one), δ is the relative yaw rotation between the front and rear module, i is the track of the module, $a = 0.132\ m$ and $b = 0.139\ m$ are the distances between the central joint and the front and rear modules respectively, and v_n is the longitudinal speed of the n-th module in its reference frame.

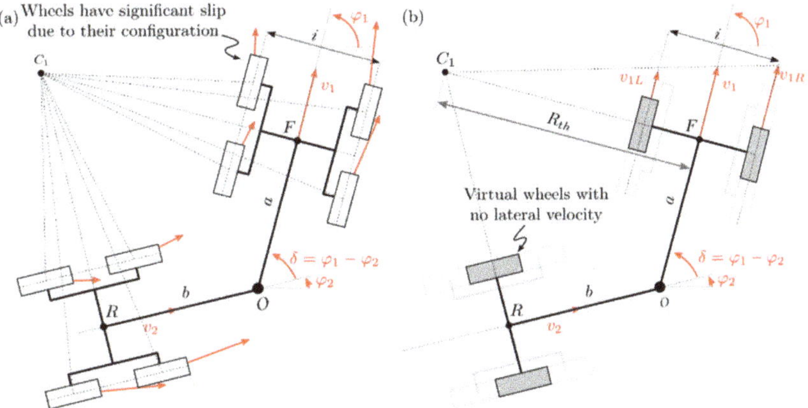

Figure 3. (**a**) Epi.q kinematic model. (**b**) The kinematic model with virtual non-slipping wheels.

The geometric relations between the modules impose the following constraints:

$$x_2 = x_1 - b \cos \varphi_2 - a \cos \varphi_1 \tag{1}$$

$$y_2 = y_1 - b \sin \varphi_2 - a \sin \varphi_1 \tag{2}$$

$$\delta = \varphi_1 - \varphi_2 \tag{3}$$

Moreover, two additional holonomic constraints can be defined:

$$\dot{x}_1 \sin \varphi_1 - \dot{y}_1 \cos \varphi_1 = 0 \tag{4}$$

$$\dot{x}_2 \sin \varphi_2 - \dot{y}_2 \cos \varphi_2 = 0 \tag{5}$$

The velocity in the world reference frame of each module is defined as:

$$\dot{x}_1 = v_1 \cos \varphi_1 \tag{6}$$

$$\dot{y}_1 = v_1 \sin \varphi_1 \tag{7}$$

$$\dot{x}_2 = v_2 \cos \varphi_2 \tag{8}$$

$$\dot{y}_2 = v_2 \sin \varphi_2 \tag{9}$$

Hence, the kinematic system can be defined as:

$$\dot{q} = \begin{bmatrix} \dot{x}_1 \\ \dot{y}_1 \\ \dot{\varphi}_1 \\ \dot{\delta} \end{bmatrix} = \mathbf{A} \begin{bmatrix} v_1 \\ \dot{\varphi}_1 \end{bmatrix} \tag{10}$$

The relation of $\dot{\delta}$ as a function of v_1 and $\dot{\varphi}_1$ can be obtained differentiating Equations (1)–(3) in time and substituting the results in Equation (5). With some manipulation, it is possible to get:

$$\dot{\delta} = \left(\frac{a}{b} \cos \varphi_1 + 1 \right) \dot{\varphi}_1 - \frac{1}{b} \sin \delta v_1 \tag{11}$$

Thus, the kinematic model can be explicitly written as:

$$\dot{q} = \begin{bmatrix} \dot{x}_1 \\ \dot{y}_1 \\ \dot{\varphi}_1 \\ \dot{\delta} \end{bmatrix} = \begin{bmatrix} \cos \varphi_1 & 0 \\ \sin \varphi_1 & 0 \\ 0 & 1 \\ -\frac{1}{b} \sin \delta & \frac{a}{b} \cos \varphi_1 + 1 \end{bmatrix} \begin{bmatrix} v_1 \\ \dot{\varphi}_1 \end{bmatrix} \tag{12}$$

Considering that each module behaves as a skid steering robot, it is possible to rewrite the latter relations in an alternative form considering the two following relations:

$$v_1 = \frac{1}{2} \left(v_{1R} + v_{1L} \right) \tag{13}$$

$$\dot{\varphi}_1 = \frac{1}{i} \left(v_{1R} - v_{1L} \right) \tag{14}$$

Therefore:

$$\dot{q} = \begin{bmatrix} \dot{x}_1 \\ \dot{y}_1 \\ \dot{\varphi}_1 \\ \dot{\delta} \end{bmatrix} = \begin{bmatrix} \frac{1}{2} \cos \varphi_1 & \frac{1}{2} \cos \varphi_1 \\ \frac{1}{2} \sin \varphi_1 & \frac{1}{2} \sin \varphi_1 \\ \frac{1}{i} & -\frac{1}{i} \\ \frac{a \cos \delta + b - \frac{i}{2} \sin \delta}{bi} & -\frac{a \cos \delta + b + \frac{i}{2} \sin \delta}{bi} \end{bmatrix} \begin{bmatrix} v_{1R} \\ v_{1L} \end{bmatrix} \tag{15}$$

$v_{1R}v_{1L}$ are easily related to the motor angular speeds ω_{1R} and ω_{1L} by knowing the wheel radius $r_{wheel} = 32$mm and the transmission ratio of the gearing system $\tau_{advancing} = 4.1$.

3. Experimental Identification of the Estimator

3.1. Experimental Setup

This section delineates the main points of the experimental setup required to study the behaviour of the robot while performing curve trajectories. As already introduced, Epi.q was the robot used in this study. Its front module motors were velocity controlled to achieve the desired curved trajectory, while, for these tests, the rear motors were disabled; therefore the robot behaved similarly to a tractor-trailer system instead of two independent active modules linked together. Relevant on-board measurements, such as motor angular speed, were transmitted in real-time to a PC, where they were logged. The visual-based tracking system, described in depth in [23] and summarised in the following section, was used to track the actual trajectory of the robot. Several tests at different motor speeds were performed and then used to compare the actual trajectory with the theoretical one.

3.2. Tracking Method

As can be seen in Figure 4, two ArUco markers, a set of open source square fiducial markers, designed for the fast processing of high quality images [24–26], were fixed to the robot (one per module), while one was fixed to the ground at an arbitrary location. The camera recording the tests was free to move in the environment; its position only affects the accuracy of the measure. The test recordings were then post-processed by a custom marker-tracking algorithm that first searched for a set of markers and then estimated the homogeneous transformation $\mathbf{T}_n^C$, a representation of the pose of the nth marker in the camera reference frame. The poses (x_1, y_1, φ_1) and (x_2, y_2, φ_2) of the two moving markers were then estimated with respect to the marker fixed to the ground, representing the world reference frame, by applying a concatenation of homogeneous transformations such as $\mathbf{T}_n^0 = \mathbf{T}_n^C \mathbf{T}_C^0 = \mathbf{T}_n^C \mathbf{T}_0^{C-1}$. With this approach, the camera can move freely, and the tracking results could still cohere to the desired reference frame. As anticipated, the camera position only influenced the accuracy of the measure, since the system used just a single camera; therefore, the picture depth estimation was noisy.

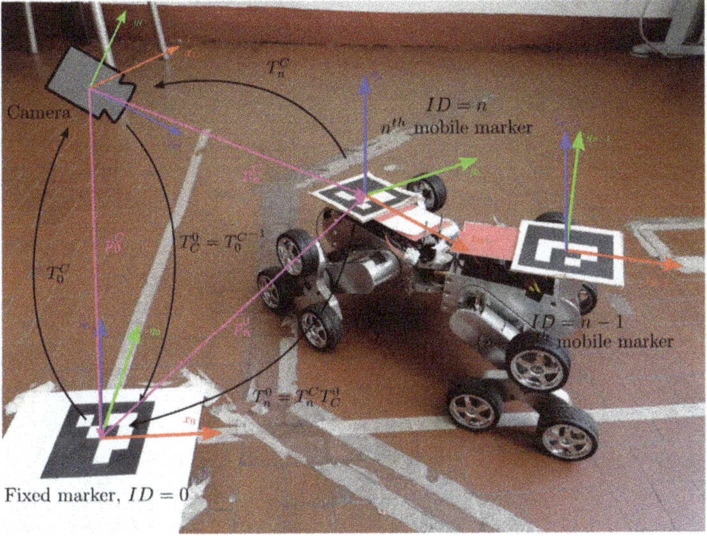

Figure 4. Experimental setup with a schematic representation of the visual-based tracking system.

3.3. Results

Table 1 collects the test performed with the robot together with the kinematic inputs and the main parameter used as comparisons, such as the curve radiuses and the estimated track between the wheels. The theoretical radius R_{th} was computed using the kinematic relation $R_{th} = \frac{v_1}{\dot{\varphi}_1}$, which is the expected curve radius if the mobile robot behaviour is purely kinematic. The radius $\hat{R}$ is the radius of the trajectory curve estimated from the experimental data collected by the vision-based tracking system. As expected, the estimated radius was always larger than the theoretical one, reflecting the under-steering behaviour due to the large lateral wheel slip. The estimated track $\hat{i}$, instead, was obtained fitting the tracked trajectory with the kinematic model described in the previous section, where the same kinematic input v_{1R} and v_{1L} that have been logged experimentally have been used as input to the model. By recalling the model described before, this trend can be interpreted as that the virtual axle used in the model is wider than the actual track in order to mathematically represent the under-steering behaviour of the robot. $\hat{R}$ and $\hat{i}$ are two different way to represent the robot under-steering tendency with a basic difference: $\hat{R}$ is purely based on the fitting of the measured trajectory with a circle, while $\hat{i}$ is obtained fitting the whole kinematic model shown before using the measured kinematic input in order to match as close as possible the robot behaviour.

Table 1. List of tests and their parameters.

Test	Longitudinal Speed v_1 [m/s]	Yaw Rate $\dot{\varphi}_1$ [rad/s]	Theoretical Radius R_{th} [m]	Estimated Radius $\hat{R}$ [m]	Estimated Track $\hat{i}$ [m]
1	0.412	0.961	0.429	0.452	0.303
2	0.393	0.915	0.430	0.458	0.296
3	0.310	0.616	0.504	0.621	0.319
4	0.319	0.596	0.535	0.598	0.314
5	0.384	0.940	0.408	0.427	0.333
6	0.405	1.001	0.405	0.525	0.336
7	0.299	0.305	0.983	1.225	0.333
8	0.285	0.311	0.916	0.965	0.337

Figure 5 depicts a comparison between the various trajectories taken into account for each test. Figure 5a shows a comparison of the measured trajectory and the resulting curves of the two kinematic models when the measured input data are used. The difference between the two kinematic models and the actual measured data can be interpreted as the lack of dynamic phenomena modelling, such as robot inertia, wheel longitudinal, and lateral slip. By making a comparison between the two kinematic models instead, the results described before are again evident: the kinematic model with the estimated track length $\hat{i}$ ("Fit Kin.") has a more under-steering behaviour compared to the kinematic model with the actual track length ("Th. Kin."). Since the fitted kinematic model is a better approximation of the actual robot trajectory, it could be said that a kinematic model with virtual wheels with a slightly increased track length $\hat{i}$ could partially compensate unmodelled dynamic behaviours better than the kinematic model with the actual track length.

Figure 5b shows a comparison between two circles and the steady-state part of the measured trajectory: the actual trajectory is not an exact circle, but the two circles approximate it, one better than the other. Again, it appears from the results that the kinematic circle of radius R_{th} is smaller and a worse approximation of the actual data than the circle with the estimated radius $\hat{R}$. This figure depicts more clearly that, if a purely steady-state comparison is made, the perfectly circular kinematic trajectories could be a good approximation of the real one.

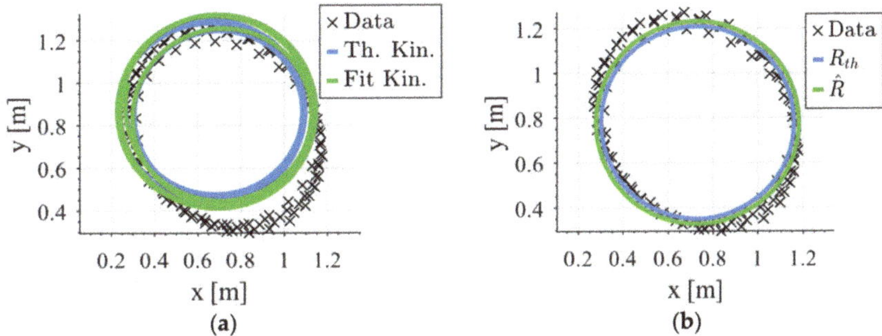

Figure 5. Test 1: trajectories comparison. (**a**) Comparison between kinematic models and measured data. The theoretical kinematic curve ("Th. Kin.") uses the actual robot parameters, while the other trajectory ("Fit Kin.") use the best fitting track length $\hat{i}$. (**b**) Comparison between measured data, steady-state kinematic trajectory with radius R_{th} and best fitting circle with radius $\hat{R}$.

Figure 6 highlights the latter statements even better; the figure shows the instantaneous curvature radius of the two kinematic models over time, and how the two curves reach a steady-state very close to the value of the radius of the circle described before. The theoretical kinematic model tends toward $R_{th} = \frac{v_1}{\dot{\varphi}_1}$, while the kinematic model with the corrected track length $\hat{i}$ tends to the value of $\hat{R}$, the radius of the best-fitting circle.

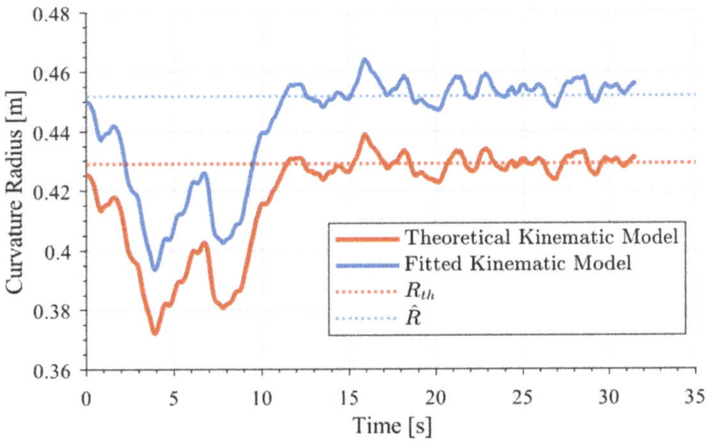

Figure 6. Test 1: Trajectory radius comparison.

As described before, the virtual track length that better fits the experimental trajectories $\hat{i}$ is always larger than the robot actual track length $i = 0.286$ m to mathematically describe the under-steering behaviour of the robot that, otherwise, the theoretical kinematic model could not represent. Figure 7, in particular, illustrates that there is a strong correlation between $\hat{i}$ and the front module yaw rate $\dot{\varphi}_1$: the quicker the robot turns, the higher the fitted virtual length is, or, from another point of view, the quicker the robot turns, the more evident the dynamic effects on the handling, such as under-steer and wheel slip, become more relevant.

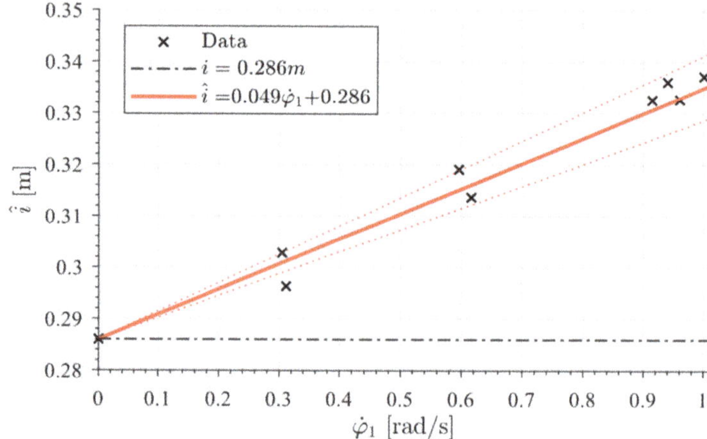

Figure 7. Test 1: Estimate track length $\hat{i}$ as a function of the yaw rate $\dot{\varphi}_1$, experimental data and linear trend.

Figure 7 also shows that the data trend could be approximated by the linear function:

$$\hat{i} = p_1 \dot{\varphi}_1 + p_2 \tag{16}$$

where $p_1 = 0.049\,\mathrm{m\,s}$ and $p_0 = 0.286\,\mathrm{m}$ are the coefficients of the linear function.

This very simple relation could be used to predict and, therefore, partially compensate the unmodelled dynamic effects of the kinematic model. It is important to state that this particular relation holds only for working scenarios identical, or at least very similar, to the experimental setup. Unfortunately, this simple approach has as a downside the limitation of being valid only in tested and specific scenarios. Nevertheless, a robot working in a well-known and uniform environment could benefit from this approach.

The latter equation can be rewritten in order to explicitly show the relation between $\hat{i}$ and the radius R by recalling that $R = v_1 / \dot{\varphi}_1$:

$$\hat{i} = p_1 \frac{v_1}{R} + p_0 \tag{17}$$

Due to its simplicity, the relation can also be easily rewritten as:

$$R = \frac{p_1 v_1}{\hat{i} - p_0} \tag{18}$$

Hence, depending on the case and the known and unknown parameters, it is possible to compute, using a simple linear equation, the correction parameters that could compensate for undesired behaviour in a robot.

4. Conclusions

In this paper, a simple and computationally efficient solution is proposed to predict, estimate, and compensate complex dynamic behaviour, such as significant lateral slip, which highly influences the mobile robot handling employing an experimental kinematic characterisation. Compared to the commonly used approaches to similar issues, the results shown herein are highly dependent on a particular robot and a particular environment. The fitting method behaves sufficiently well at lower speeds and yaw rates, but when velocities increase, or more generally, when the dynamic phenomena are more relevant, the robot handling behaviour diverges from a purely kinematic one in a manner that is not easily compensated by the proposed fitting method. However, its simplicity makes its application feasible to any kind of kinematic architecture of mobile platforms where the wheels transversal slip is significant. Moreover, its computational efficiency allows its implementation at low-power, low-performance, or overburdened control units. As a future work, the validity of the proposed method can be tested on other machines of the class of eight-wheeled mobile articulated robots developed at Politecnico di Torino (such as the Agri.q rover for precision agriculture, or even the Rese.q snake rescue robot). Prior to a proper identification of the virtual track, in fact, the methodology can be used both to estimate the actual trajectories of the robot modules and, above all, to enhance the precision of a robot's motion planning with a consequent improvement of the robot's control within its workspace.

Author Contributions: Conceptualisation, G.Q.; methodology, A.B., P.C., L.T., L.C., C.V. and G.Q.; formal analysis, A.B., P.C., L.T., L.C., C.V. and G.Q.; data curation, A.B.; writing—original draft preparation, A.B.; writing—review and editing, A.B., P.C., L.T., L.C., C.V. and G.Q.; visualisation, A.B.; supervision, G.Q.; project administration, G.Q.; funding acquisition, G.Q. All authors have read and agreed to the published version of the manuscript.

Funding: This research received no external funding.

Institutional Review Board Statement: Not applicable.

Informed Consent Statement: Not applicable.

Data Availability Statement: Some or all data generated or used during the study are available from the corresponding author by request.

Conflicts of Interest: The authors declare no conflict of interest.

References

1. Shih, C.-L.; Lin, L.-C. Trajectory Planning and Tracking Control of a Differential-Drive Mobile Robot in a Picture Drawing Application. *Robotics* **2017**, *6*, 17. [CrossRef]
2. Begnini, M.; Bertol, D.W.; Martins, N.A. A Robust Adaptive Fuzzy Variable Structure Tracking Control for the Wheeled Mobile Robot: Simulation and Experimental Results. *Control Eng. Pract.* **2017**, *64*, 27–43. [CrossRef]
3. Miah, S.; Shaik, F.; Chaoui, H. Universal Dynamic Tracking Control Law for Mobile Robot Trajectory Tracking. In Proceedings of the 2017 IEEE International Conference on Industrial Technology (ICIT), Toronto, ON, Canada, 22–25 March 2017; pp. 896–901.
4. Matraji, I.; Al-Durra, A.; Haryono, A.; Al-Wahedi, K.; Abou-Khousa, M. Trajectory Tracking Control of Skid-Steered Mobile Robot Based on Adaptive Second Order Sliding Mode Control. *Control Eng. Pract.* **2018**, *72*, 167–176. [CrossRef]
5. Wang, C.; Liu, X.; Yang, X.; Hu, F.; Jiang, A.; Yang, C. Trajectory Tracking of an Omni-Directional Wheeled Mobile Robot Using a Model Predictive Control Strategy. *Appl. Sci.* **2018**, *8*, 231. [CrossRef]
6. Singh, K.B.; Arat, M.A.; Taheri, S. Literature Review and Fundamental Approaches for Vehicle and Tire State Estimation. *Veh. Syst. Dyn.* **2019**, *57*, 1643–1665. [CrossRef]
7. Piyabongkarn, D.; Rajamani, R.; Grogg, J.A.; Lew, J.Y. Development and Experimental Evaluation of a Slip Angle Estimator for Vehicle Stability Control. *IEEE Trans. Control Syst. Technol.* **2009**, *17*, 78–88. [CrossRef]
8. Chindamo, D.; Lenzo, B.; Gadola, M. On the Vehicle Sideslip Angle Estimation: A Literature Review of Methods, Models, and Innovations. *Appl. Sci.* **2018**, *8*, 355. [CrossRef]
9. Alipour, K.; Robat, A.B.; Tarvirdizadeh, B. Dynamics Modeling and Sliding Mode Control of Tractor-Trailer Wheeled Mobile Robots Subject to Wheels Slip. *Mech. Mach. Theory* **2019**, *138*, 16–37. [CrossRef]
10. Tian, Y.; Sarkar, N. Control of a Mobile Robot Subject to Wheel Slip. *J. Intell. Robot. Syst.* **2014**, *74*, 915–929. [CrossRef]
11. Kim, D.; Yoon, H.; Kim, K.; Sreejith, M.S.; Lee, J. Using Current Sensing Method and Fuzzy PID Controller for Slip Phenomena Estimation and Compensation of Mobile Robot. In Proceedings of the 2017 14th International Conference on Ubiquitous Robots and Ambient Intelligence (URAI), Jeju, Korea, 28 June–1 July 2017; pp. 397–401.
12. Liu, F.; Li, X.; Yuan, S.; Lan, W. Slip-Aware Motion Estimation for Off-Road Mobile Robots via Multi-Innovation Unscented Kalman Filter. *IEEE Access* **2020**, *8*, 43482–43496. [CrossRef]
13. Cui, M.; Liu, H.; Liu, W.; Qin, Y. An Adaptive Unscented Kalman Filter-Based Controller for Simultaneous Obstacle Avoidance and Tracking of Wheeled Mobile Robots with Unknown Slipping Parameters. *J. Intell. Robot. Syst.* **2018**, *92*, 489–504. [CrossRef]
14. Toledo-Moreo, R.; Colodro-Conde, C.; Toledo-Moreo, J. A Multiple-Model Particle Filter Fusion Algorithm for GNSS/DR Slide Error Detection and Compensation. *Appl. Sci.* **2018**, *8*, 445. [CrossRef]
15. Zhou, H.; Yao, Z.; Fan, C.; Wang, S.; Lu, M. Rao-Blackwellised Particle Filtering for Low-Cost Encoder/INS/GNSS Integrated Vehicle Navigation with Wheel Slipping. *Iet RadarSonar Amp Navig.* **2019**, *13*, 1890–1898. [CrossRef]

16. Yi, J.; Zhang, J.; Song, D.; Jayasuriya, S. IMU-based localization and slip estimation for skid-steered mobile robots. In Proceedings of the 2007 IEEE/RSJ International Conference on Intelligent Robots and Systems, San Diego, CA, USA, 29 October–2 November 2007; pp. 2845–2850.
17. Reina, G.; Ishigami, G.; Nagatani, K.; Yoshida, K. Vision-Based Estimation of Slip Angle for Mobile Robots and Planetary Rovers. In Proceedings of the 2008 IEEE International Conference on Robotics and Automation, Pasadena, CA, USA, 19–23 May 2008; pp. 486–491.
18. Martínez, J.L.; Mandow, A.; Morales, J.; Pedraza, S.; García-Cerezo, A. Approximating Kinematics for Tracked Mobile Robots. *Int. J. Robot. Res.* **2005**, *24*, 867–878. [CrossRef]
19. Gonzalez, R.; Iagnemma, K. Slippage Estimation and Compensation for Planetary Exploration Rovers. State of the Art and Future Challenges. *J. Field Robot.* **2018**, *35*, 564–577. [CrossRef]
20. Quaglia, G.; Nisi, M. *Design and Construction of a New Version of the Epi.q UGV for Monitoring and Surveillance Tasks*; American Society of Mechanical Engineers Digital Collection: Houston, TX, USA, 2016.
21. Bruzzone, L.; Quaglia, G. Locomotion Systems for Ground Mobile Robots in Unstructured Environments. *Mech. Sci.* **2012**, *3*, 49–62. [CrossRef]
22. Botta, A.; Cavallone, P.; Carbonari, L.; Tagliavini, L.; Quaglia, G. Modelling and Experimental Validation of Articulated Mobile Robots with Hybrid Locomotion System. In Proceedings of the Advances in Italian Mechanism Science; Niola, V., Gasparetto, A., Eds.; Springer International Publishing: Cham, Switzerland, 2021; pp. 758–767.
23. Botta, A.; Quaglia, G. Performance Analysis of Low-Cost Tracking System for Mobile Robots. *Machines* **2020**, *8*, 29. [CrossRef]
24. Garrido-Jurado, S.; Muñoz-Salinas, R.; Madrid-Cuevas, F.J.; Marín-Jiménez, M.J. Automatic Generation and Detection of Highly Reliable Fiducial Markers under Occlusion. *Pattern Recognit.* **2014**, *47*, 2280–2292. [CrossRef]
25. Garrido-Jurado, S.; Muñoz-Salinas, R.; Madrid-Cuevas, F.J.; Medina-Carnicer, R. Generation of Fiducial Marker Dictionaries Using Mixed Integer Linear Programming. *Pattern Recognit.* **2016**, *51*, 481–491. [CrossRef]
26. Romero-Ramirez, F.J.; Muñoz-Salinas, R.; Medina-Carnicer, R. Speeded up Detection of Squared Fiducial Markers. *Image Vis. Comput.* **2018**, *76*, 38–47. [CrossRef]

 *applied sciences*

Article

Anti-Disturbance Control for Quadrotor UAV Manipulator Attitude System Based on Fuzzy Adaptive Saturation Super-Twisting Sliding Mode Observer

Ran Jiao [1,*], Wusheng Chou [1,2], Yongfeng Rong [1] and Mingjie Dong [3]

[1]   School of Mechanical Engineering and Automation, Beihang University, Beijing 100191, China; wschou@buaa.edu.cn (W.C.); ryf_2018@buaa.edu.cn (Y.R.)

[2]   The State Key Laboratory of Virtual Reality Technology and Systems, Beihang University, Beijing 100191, China

[3]   School of Mechanical Engineering and Applied Electronics Technology, Beijing University of Technology, Beijing 100124, China; dongmj@bjut.edu.cn

*   Correspondence: jiaoran@buaa.edu.cn

Received: 24 April 2020; Accepted: 20 May 2020; Published: 27 May 2020

Abstract: Aerial operation with unmanned aerial vehicle (UAV) manipulator is a promising field for future applications. However, the quadrotor UAV manipulator usually suffers from several disturbances, such as external wind and model uncertainties, when conducting aerial tasks, which will seriously influence the stability of the whole system. In this paper, we address the problem of high-precision attitude control for quadrotor manipulator which is equipped with a 2-degree-of-freedom (DOF) robotic arm under disturbances. We propose a new sliding-mode extended state observer (SMESO) to estimate the lumped disturbance and build a backstepping attitude controller to attenuate its influence. First, we use the saturation function to replace discontinuous sign function of traditional SMESO to alleviate the estimation chattering problem. Second, by innovatively introducing super-twisting algorithm and fuzzy logic rules used for adaptively updating the observer switching gains, the fuzzy adaptive saturation super-twisting extended state observer (FASTESO) is constructed. Finally, in order to further reduce the impact of sensor noise, we invite a tracking differentiator (TD) incorporated into FASTESO. The proposed control approach is validated with effectiveness in several simulations and experiments in which we try to fly UAV under varied external disturbances.

Keywords: super-twisting; sliding mode extended state observer; saturation function; fuzzy logic; attenuate disturbance

1. Introduction

Unmanned aerial vehicles (UAVs) have become a popular and active research topic among scholars worldwide [1,2]. It could work at locations where entry is difficult and humans cannot access [3]. Recently, they are not only used in traditional scenes, such as monitoring, aerial photography, surveying, and patrol, but are also employed in new application scenarios requiring physically interacting with external environment. Then, to conduct physical interaction, aerial robots are either equipped with a rigid tool [4] or an n-degree-of-freedom (DoF) robotic arm [5–8], which are called UAV manipulators. As for the UAV equipped with a robotic arm, several disturbances cannot be avoided, such as wind gust and model uncertainties, during task execution. Therefore, to meet the aerial operation task requirements with high reliability, disturbance rejection control problem should be investigated.

Many controllers have been proposed for disturbance rejection. The Robust Model Predictive Control (MPC) method is used for multirotor UAV to provide robust performance under an unknown but bounded disturbance [9]. Meanwhile, the adaptive control structure is built to obtain coefficients of system with uncertainty online. A controller based on Model Reference Adaptive Control is proposed in [10], which performs better on disturbance rejection compared with non-adaptive controllers. However, as what we know, one of the biggest drawbacks of adaptive controllers is that they perform little robust under the bursting phenomenon.

Adopting the disturbance and uncertainty estimation and attenuation (DUEA) strategy, such as Disturbance observer (DOB) [11], unknown input observer (UIO) [12], and extended state observer (ESO) [13], would be a potential solution for disturbance rejection. Meanwhile, there are several types of DOBs applied to UAVs. Work [14] proposes an inner-loop control structure to recover the dynamics of a multirotor combined with additional objects similar to the bare multirotor using DOB-based method. A linear dual DOB is built in [15] to reject modeling error and external disturbance when designing the control system. The authors of [16] propose a DOB-based tracking flight control method for a quadrotor to reject disturbance, which is supposed to be composed of some harmonic elements, and applies it to flight control of the UAV Quanser Qball 2. A disturbance observer with finite time convergence (FTDO), which could conduct online estimation of the unknown uncertainties and disturbances, is incorporated into an hierarchical controller to solve the problem of path tracking of a small coaxial rotor-type UAV [17]. The authors of [18] introduce a robust DOB for an aircraft to compensate the uncertain rotational dynamics into a nominal plant and proposes a nonlinear feedback controller implemented for desired tracking performance.

As for ESO, it was first proposed by Han [19] and has been introduced in many control methods [20]. Traditionally, ESO methods mainly focus on coping with disturbances which slowly change [21]. Nevertheless, it is evident that the disturbance caused by the wind gust sometimes performs drastic change so that it can not be estimated by traditional ESO thoroughly. Thus, an enhanced ESO that could quickly estimate the disturbance is necessary in this field. A high order sliding mode observer is built to estimate unmodeled dynamics and external disturbances for an aerial vehicle to track the trajectory [22]. Moreover, a sliding mode observer is proposed for equivalent-input-disturbance approach to control the under-actuated subsystem of a quadrotor UAV [23]. However, chattering phenomenon is normal in traditional sliding mode observer (SMO) because of the discontinuous sign function. In order to reduce the chattering problem, the super-twisting algorithm is introduced in a ESO to mitigate estimation chattering [24], but there is still a little chattering. In order to solve it, the authors of [25] proposed a new SMO control strategy to reduce the estimated speed chattering of the motor, in which the switching function is replaced by a sigmoid function. The sigmoid function has also been used in SMO for UAV in [26]; however, it is conducted without experiment.

The main contributions of this study are given as follows.

(1) In order to alleviate the estimation chattering problem of traditional SMESO, a new observer named fuzzy adaptive saturation super-twisting extended state observer (FASTESO) is proposed, in which a saturation function is invited to replace the discontinued sign function, meanwhile a super-twisting algorithm is introduced to prevent excessively high observer gain and TD [19] and is incorporated for avoiding directly using acceleration information, which is full of noise.

(2) To stay robustness under disturbances with unknown bounds, fuzzy logic rules are introduced as an adaptive algorithm to adaptively adjust the observer switching gains. Furthermore, it also contributes to the chattering attenuation under high switching gain with low estimation value and observer performance improvement under low switching gain with high estimation value.

(3) The proposed method is verified with effectiveness on a quadrotor UAV manipulator prototype in several simulations and experiments.

The rest of this article is organized as follows. Section 2 introduces some preliminaries of this work. Section 3 describes the kinematic and dynamic models of the quadrotor UAV manipulator. The construction of proposed FASTESO and attitude controller are given in Section 4. Additionally, several simulations are conducted in Section 5. Moreover, Section 6 shows the experiment. Finally, Section 7 presents the conclusion.

2. Preliminaries

In this section, some mathematical preliminaries are provided for understanding the whole paper more easily.

2.1. Notation

The 2-norm of a vector or a matrix is provided by $\|\cdot\|$. $\lambda_{\max}(Z)$ and $\lambda_{\min}(Z)$ represent the maximal and minimum eigenvalue of the matrix Z, respectively. Moreover, the operator $S(\cdot)$ denotes a vector $\kappa = \begin{bmatrix} \kappa_1 & \kappa_2 & \kappa_3 \end{bmatrix}^T$ to a skew symmetric matrix as

$$S(\kappa) = \begin{bmatrix} 0 & -\kappa_3 & \kappa_2 \\ \kappa_3 & 0 & -\kappa_1 \\ -\kappa_2 & \kappa_1 & 0 \end{bmatrix} \tag{1}$$

The sign function is given as

$$sign(\kappa) = \begin{cases} \frac{\kappa}{|\kappa|}, & |\kappa| \neq 0 \\ 0, & |\kappa| = 0 \end{cases} \tag{2}$$

2.2. Quaternion Operations

The unit quaternion $q = \begin{bmatrix} q_0 & q_v \end{bmatrix}^T \in R^4$, $\|q\| = 1$, is used to represent the rotation of the quadrotor in this paper. Several corresponding operations are given as follows.

The quaternion multiplication:

$$q \otimes \sigma = \begin{bmatrix} q_0\sigma_0 - q_v^T\sigma_v \\ q_0\sigma_v + \sigma_0 q_v - S(\sigma_v)q_v \end{bmatrix} \tag{3}$$

The relationship between rotation matrix C_A^B and unit quaternion q is represented by:

$$C_A^B = (q_0^2 - q_v^T q_v)I_3 + 2q_v q_v^T + 2q_0 S(q_v) \tag{4}$$

Take the time derivative of Equation (4), we could obtain

$$\dot{C}_A^B = -S(\omega)\, C_A^B \tag{5}$$

The derivative of a quaternion and the quaternion error q_e are provided, respectively, as

$$\dot{q} = \begin{bmatrix} \dot{q}_0 \\ \dot{q}_v \end{bmatrix} = \frac{1}{2}q \otimes \begin{bmatrix} 0 \\ \omega \end{bmatrix} = \frac{1}{2}\begin{bmatrix} -q_v^T \\ S(q_v) + q_0 I_3 \end{bmatrix}\omega \tag{6}$$

$$q_e = q_d^* \otimes q \tag{7}$$

where q_d represents the desired quaternion whose conjugate is $q_d^* = \begin{bmatrix} q_{d0} & -q_{dv} \end{bmatrix}^T$. Moreover, ω denotes the angular rate of the system.

$$\omega_e = \omega - C_d^b \omega_d \tag{8}$$

$$\dot{q}_e = \frac{1}{2} q_e \otimes \begin{bmatrix} 0 \\ \omega_e \end{bmatrix} = \frac{1}{2} \begin{bmatrix} -q_{ev}^T \\ S(q_{ev}) + q_{e0} I_3 \end{bmatrix} \omega_e \tag{9}$$

3. Models of UAV

In this section, we present a mathematical description of the quadrotor UAV manipulator system model. The abstract graph is shown in Figure 1, in which the robotic arm is fixed at the geometric center of the UAV.

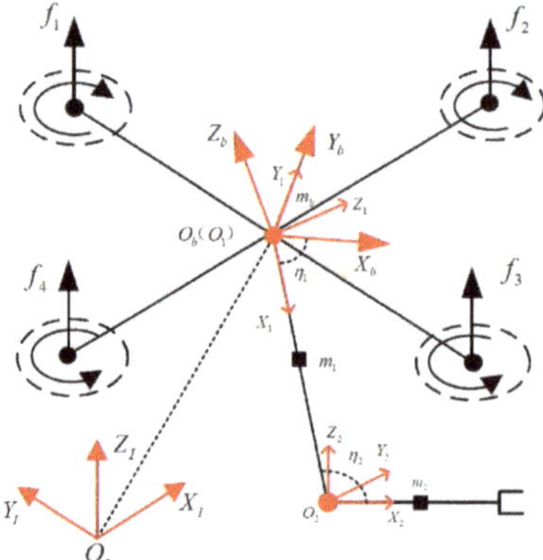

Figure 1. Quadrotor and robotic arm system with coordinate reference frames.

3.1. Kinematic Model

Figure 1 shows several coordinates in UAV that we consider.

O_I: the world-fixed inertial reference frame. O_b: the body-fixed reference frame centered in the quadrotor Center of Gravity (CoG). O_i: the coordinate of robotic arm i, where $(i = 1, 2)$ denotes the link number.

Moreover, several parameters definition are given as follows.

$p^I = [x, y, z]^T$: the absolute position of O_b with respect to O_I. $\Psi = [\varphi, \theta, \psi]^T$: Euler angles used for representing the UAV attitude angle. $\eta = [\eta_1, \eta_2]^T$: the joint angles of the robotic arm relative to the zero position. $H = [(p^I)^T, \Psi^T, \eta^T]^T \in R^{8\times1}$: a vector used for including all the generalized variables. $\dot{p}^I$: the absolute linear velocity of UAV described in O_I. $\dot{p}^b$: the linear velocity of UAV with respect to O_b. $\omega^I = [\omega_x^I, \omega_y^I, \omega_z^I]^T$: UAV's absolute rotational velocity described in O_I. $\omega^b = [\omega_x^b, \omega_y^b, \omega_z^b]^T$: the rotational velocity of UAV relative to O_b. Then we could get:

$$\begin{cases} \dot{p}^I = R_b \dot{p}^b \\ \omega^I = R_t \dot{\Psi} \\ \omega^b = (R_b)^T \omega^I = (R_b)^T R_t \dot{\Psi} = R_r \dot{\Psi} \end{cases} \tag{10}$$

where $R_t \in R^{3\times3}$ represents the transformation matrix for converting Euler angle rates $\dot{\Psi}$ into ω^I. $R_b \in R^{3\times3}$ denotes the rotation matrix representing the orientation of O_b relative to O_I, and $R_r = (R_b)^T R_t \in R^{3\times3}$ is used to map the derivative of Ψ into UAV angular rate expressed in O_b.

Where

$$\begin{cases} R_b = \begin{bmatrix} c\theta c\psi & s\varphi s\theta c\psi - c\varphi s\psi & c\varphi\, s\theta\, c\psi + s\varphi s\psi \\ c\theta s\psi & s\varphi s\theta s\psi + c\varphi c\psi & c\varphi\, s\theta s\psi - s\varphi c\psi \\ -s\theta & s\varphi c\theta & c\varphi\, c\theta \end{bmatrix} \\[4pt] R_r = \begin{bmatrix} 1 & 0 & -s\theta \\ 0 & c\varphi & s\varphi c\theta \\ 0 & -s\varphi & c\varphi c\theta \end{bmatrix} \end{cases} \tag{11}$$

$c(\cdot)$ and $s(\cdot)$ denote $cos(\cdot)$ and $sin(\cdot)$, respectively.

Moreover, the position and angular rate of the frame O_i, which is fixed to the robotic arm, with respect to O_I are provided by

$$\begin{cases} p_i^I = p^I + R_b p_i^b \\ \omega_i^I = \omega^I + R_b \omega_i^b \end{cases} \tag{12}$$

where $i = (x, y, z)$. The vectors p_i^b, ω_i^b, which are expressed in O_b, represent the position of O_i and the angular rate of the i-th robotic arm frame with respect to O_b, respectively. Additionally, more relationships are provided,

$$\begin{cases} \dot{p}_i^b = J_{pi}\dot{\eta} \\ \omega_i^b = J_{ri}\dot{\eta} \end{cases} \tag{13}$$

where $J_{pi} \in R^{2\times2}$ and $J_{ri} \in R^{2\times2}$ are Jacobian matrices representing the translational and angular velocities of each robotic arm link to the $\dot{\eta}$, respectively. According to Equations (12) and (13), we could get the translational and angular velocity of O_i with respect to O_I as

$$\begin{cases} \dot{p}_i^I = \dot{p} + S(\omega^I)R_b p_i^b + R_b J_{pi}\dot{\eta} \\ \omega_i^I = \omega^I + R_b J_{ri}\dot{\eta} \end{cases} \tag{14}$$

where $S(\cdot)$ represents the skew-symmetric matrix.

3.2. Dynamic Model

In order to obtain the dynamic model of the quadrotor UAV manipulator system, the Euler-Lagrange equation is introduced.

$$\frac{d}{dt}\frac{\partial L}{\partial H_k} - \frac{\partial L}{\partial H_k} = u_k + d_k, L = K - U \tag{15}$$

where $k = (1, ..., 8)$. L: the Lagrangian with kinetic energy K and potential energy U of the integrated system. u_k: the generalized driving force. d_k: external disturbance applied to the system. Additionally, K could be given in detail as

$$\begin{cases} K = K_b + \sum\limits_{i=1}^{2} K_i \\ K_b = \frac{1}{2}m_b(\dot{p}^I)^T \dot{p}^I + \frac{1}{2}(\omega^I)^T R_b I_b R_b^T \omega^I \\ K_i = \frac{1}{2}m_i(\dot{p}_i^I)^T \dot{p}_i^I + \frac{1}{2}(\omega_i^I)^T R_b R_i^b I_i R_i^b R_b^T \omega_i^I \end{cases} \tag{16}$$

where m_b is the quadrotor base mass, m_i is the i-th robotic arm mass ($i = (1,2)$), I is the inertia matrix, and R_i^b the rotation matrix between the frame fixed to the center of mass of the i-th link and O_b. Then, the total potential is provided as

$$\begin{cases} U = U_b + \sum\limits_{i=1}^{2} U_i \\ U_b = m_b g e_3^T p^I \\ U_i = m_i g e_3^T p_i^I \end{cases} \tag{17}$$

where e_3 denotes unit vector $[0,0,1]^T$ and g represents the gravity constant. By substituting Equations (16) and (17) into Equation (15), the dynamic model of the quadrotor UAV manipulator system could be obtained as

$$M(\mathbf{H})\ddot{\mathbf{H}} + C(\mathbf{H},\dot{\mathbf{H}}) + G(\mathbf{H}) = u + d \tag{18}$$

Moreover, the total kinetic energy can be expressed:

$$K = \frac{1}{2}\dot{\mathbf{H}}^T M(\mathbf{H})\dot{\mathbf{H}} \tag{19}$$

where

$$M(\mathbf{H}) = \begin{bmatrix} M_{11} & M_{12} & M_{13} \\ M_{21} & M_{22} & M_{23} \\ M_{31} & M_{32} & M_{33} \end{bmatrix} \tag{20}$$

Details of the inertia matrix $M(\mathbf{H})$ and Coriolis matrix $C(\mathbf{H},\dot{\mathbf{H}})$ can be found in [27]. The $G(\mathbf{H})$ could be obtained via partial derivative Equation (21):

$$G(\mathbf{H}) = \frac{\partial U}{\partial \mathbf{H}} \tag{21}$$

Express Equation (18) in detail as

$$\begin{bmatrix} M_{11} & M_{12} & M_{13} \\ M_{21} & M_{22} & M_{23} \\ M_{31} & M_{32} & M_{33} \end{bmatrix}\begin{bmatrix} \ddot{p}^I \\ \ddot{\Psi} \\ \ddot{\eta} \end{bmatrix} + \begin{bmatrix} G_1 \\ G_2 \\ G_3 \end{bmatrix} + \begin{bmatrix} C_{11} & C_{12} & C_{13} \\ C_{21} & C_{22} & C_{23} \\ C_{31} & C_{32} & C_{33} \end{bmatrix}\begin{bmatrix} \dot{p}^I \\ \dot{\Psi} \\ \dot{\eta} \end{bmatrix} = \begin{bmatrix} u_t \\ u_r \\ u_l \end{bmatrix} + \begin{bmatrix} d_t \\ d_r \\ d_l \end{bmatrix} \tag{22}$$

where u_t, u_r, u_l represent the generalized control inputs corresponding to p^I, Ψ, η. u_l is not used, as we do not consider a dynamic robotic arm in this work. Vector $d = [d_t, d_l, d_r]^T$ are the lumped external disturbance. As for the quadrotor attitude loop subsystem, we would transform u_r to the quadrotor inputs $f_v = [\omega_1^2, \omega_2^2, \omega_3^2, \omega_4^2]^T$, which is a positive correlation vector with forces produced from the quadrotor propellers. Moreover, ω_i represents rotor rate of quadrotor propeller ($i = 1,2,3,4$). Then, convert the generalized control inputs to the propeller rate:

$$u_r = R_r{}^T \Xi f_v \tag{23}$$

$$\Xi = \begin{bmatrix} \frac{\sqrt{2}l}{2}\Lambda_T & \frac{\sqrt{2}l}{2}\Lambda_T & -\frac{\sqrt{2}l}{2}\Lambda_T & -\frac{\sqrt{2}l}{2}\Lambda_T \\ \frac{\sqrt{2}l}{2}\Lambda_T & -\frac{\sqrt{2}l}{2}\Lambda_T & -\frac{\sqrt{2}l}{2}\Lambda_T & \frac{\sqrt{2}l}{2}\Lambda_T \\ -l\Lambda_C & l\Lambda_C & -l\Lambda_C & l\Lambda_C \end{bmatrix} \tag{24}$$

where Λ_T and Λ_C are the thrust and drag coefficients, respectively. Additionally, l denotes the distance from each motor to the quadrotor CoG.

According to Equation (22), we could obtain in detail the dynamics of the attitude loop subsystem which is the base of next section:

$$M_{21}\ddot{p}^I + M_{22}\ddot{\Psi} + M_{23}\ddot{\eta} + C_{21}\dot{p}^I + C_{22}\dot{\Psi} + C_{23}\dot{\eta} + G_2 = u_r + d_r \tag{25}$$

4. Method

Based on the UAV model, the proposed control strategy is described in detail in this section. As shown in Figure 2, quadrotor attitude control problem could be briefly divided into two components: FASTESO, used for estimating and compensating the disturbance, plays a role of the feedforward loop, and Backstepping controller, built for regulating the orientation to track the desired attitude timely, plays a role of the feedback loop. Meanwhile, TD, saturation function, and fuzzy logic methods are incorporated into the whole control method.

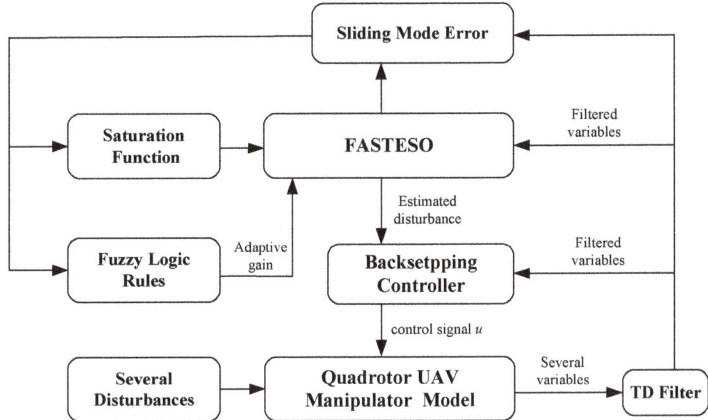

Figure 2. Whole control structure of the proposed method.

4.1. FASTESO

UAVs with robotic arms have more model uncertainty than traditional UAVs. Additionally, UAVs usually face some other disturbances such as external wind when flying outside. Moreover, the control performance of closed loop system will be largely determined by the observation performance. Therefore, in this part, in order to enhance the performance of feedback controller, a new SMO named FASTESO is built to estimate the lumped disturbance exerted on quadrotor UAV manipulator in finite time.

As for the traditional SMO, the sign function is generally adopted as the control function, and owing to switch time and space lag the SMO performs serious chattering problem [28,29]. In order to alleviate the chattering phenomenon, the sign function is replaced by a saturation function in this work. Moreover, the chattering phenomenon would also be invited by large switching gain because when the control signal crosses the sliding surface, larger switching gain would produce faster and bigger switching control parts and perform terrible chattering phenomenon. Additionally, the invited saturation function has a slower performance than the sign function on arithmetical speed, so there would be a time delay relatively. To further settle the problems of chattering and time delay mentioned before, fuzzy logic rules are employed in FASTESO to adjust the switching gains according to the states of the sliding surface.

4.1.1. Construction of Traditional Super-Twisting Extended State Observer (STESO)

In this part, according to our previous work [30], a traditional SMESO named STESO for quadrotor UAV manipulator is constructed. In this work, the situation, in which the robotic arm keeps constant while the UAV faces external disturbances during the flight, is mainly considered: ($\dot{\eta} = \ddot{\eta} = 0$). That is a general scenario that the quadrotor UAV manipulator often encounters when conducting aerial tasks. Nevertheless, usually there is unwanted vibration of the equipped robotic arm caused by the high-speed motors and propellers. To consider all mentioned disturbances, the model (25) could be reconstructed as

$$(M_{21} + \Delta M_{21})\ddot{p}^I + (M_{22} + \Delta M_{22})\dddot{\Psi} + C_{21}\dot{p}^I + (M_{23} + \Delta M_{23})\Delta\ddot{\eta} + C_{22}\dot{\Psi} + C_{23}\Delta\dot{\eta} + G_2 = u_r + d_r \quad (26)$$

where ΔM_{21}, ΔM_{22}, ΔM_{23} represent model uncertainties. $\Delta\dot{\eta}$ and $\Delta\ddot{\eta}$, considered as a portion of model uncertainty, denote the residuals from the temporary variation of the robotic arm parameters $\dot{\eta}, \ddot{\eta}$.

In order to build the observer, reconstruct the attitude dynamic Equation (26) as

$$M_{22}\dddot{\Psi} = u_r + d_r - M_{21}\ddot{p}^I - \Delta M_{21}\ddot{p}^I - \Delta M_{22}\dddot{\Psi} - (M_{23} + \Delta M_{23})\Delta\ddot{\eta} - C_{23}\Delta\dot{\eta} - C_{21}\dot{p}^I - C_{22}\dot{\Psi} - G_2 \quad (27)$$

Introduce the lumped disturbance as

$$d_r{}^* = d_r - \Delta M_{21}\ddot{p}^I - \Delta M_{22}\ddot{\Psi} - (M_{23} + \Delta M_{23})\Delta\ddot{\eta} - C_{23}\Delta\dot{\eta} \tag{28}$$

Combining Equations (27) and (28), we get

$$M_{22}\ddot{\Psi} = u_r + d_r{}^* - M_{21}\ddot{p}^I - C_{21}\dot{p}^I - C_{22}\dot{\Psi} - G_2 \tag{29}$$

where the variable $\ddot{p}^I$ is measured directly from the sensors or velocity differential is too noisy to use, so TD is introduced here to estimate the system acceleration for noise alleviation. Additionally, $\dot{p}^I$ and $\dot{\Psi}$ measured from the sensors are also a little noisy, in that case, TD would be used to reduce noise, too. Moreover, the items M_{21}, C_{21} and G_2 could be obtained in advance according to the presented UAV model. As for dynamics model (29), considering the feedback linearization method, we could reformulate the control input as

$$u_r = u_r^* + M_{21}\ddot{p}^I + C_{21}\dot{p}^I + C_{22}\dot{\Psi} + G_2 \tag{30}$$

Combining Equations (29) and (30), we get

$$M_{22}\ddot{\Psi} = u_r^* + d_r{}^* \tag{31}$$

When building the STESO, it is supposed that every channel is independent from each other. Therefore, only one channel will be presented here and the other two are completely identical. As for model Equation (31), the one-dimensional model for STESO design is provided as

$$J_i\ddot{\Psi}_i = u_{ri}^* + d_{ri}{}^* \tag{32}$$

Introduce a new extended state vector $\zeta = [\zeta_1 \quad \zeta_2]^T$, where $\zeta_1 = J_i\dot{\Psi}_i$ and $\zeta_2 = d_{ri}{}^*$, $(i = \varphi, \theta, \psi)$. Reconstruct the model (32) as

$$\begin{cases} \dot{\zeta}_1 = u_{ri}^* + \zeta_2 \\ \dot{\zeta}_2 = \chi \end{cases} \tag{33}$$

where χ denotes the derivative of $d_{ri}{}^*$, supposed by $|\chi| < f^+$. It means that the differentiation of lumped disturbance is bounded.

Then, according to the work in [30], build STESO for the observable system (33) as

$$\begin{cases} \dot{z}_1 = z_2 + u_{ri}^* + \alpha_1|e_1|^{\frac{1}{2}}sign(e_1) \\ \dot{z}_2 = \alpha_2 sign(e_1) \end{cases} \tag{34}$$

where z_1 and z_2 are estimations of ζ_1 and ζ_2, respectively. $e_1 = \zeta_1 - \hat{\zeta}_1$ and $e_2 = \zeta_2 - \hat{\zeta}_2$ denote estimation errors. We could obtain that the system estimation errors e_1 and e_2 would converge to zero within finite time under suitable observer gains α_1, α_2. Moreover, the details of the convergence analysis and parameter selection rules could be found in [30].

4.1.2. Saturation Function

To mitigate the chattering problem, we replace the sign function, usually expressing the discontinuous control in the observer, with a saturation function. The traditional STESO Equation (34) could be rewritten as

$$\begin{cases} \dot{z}_1 = z_2 + u_{ri}^* + \alpha_1|e_1|^{\frac{1}{2}}sign(e_1) \\ \dot{z}_2 = k_f * \alpha_2 sat(e_1) \end{cases} \tag{35}$$

where k_f is the adaptive proportional factor of switching gain generated from fuzzy logic rules which will be introduced in next part. $sat()$ represents the saturation function whose curve is shown in Figure 3, where particularly e is the difference between the real system and the estimated one. k_e is the output of the saturation function. We could adjust the sliding mode effect by regulating values of e and k_e. Obviously the chattering would be reduced via increasing e and decreasing k_e. If e is big enough and k_e is small enough, the high-frequency chattering could be avoided. However, meanwhile the robustness of the sliding mode method would also be reduced. Therefore the tuning of these parameters is a tradeoff between the chattering alleviation performance and the system robustness. Therefore, e and k_e should be considered overall according to real applications.

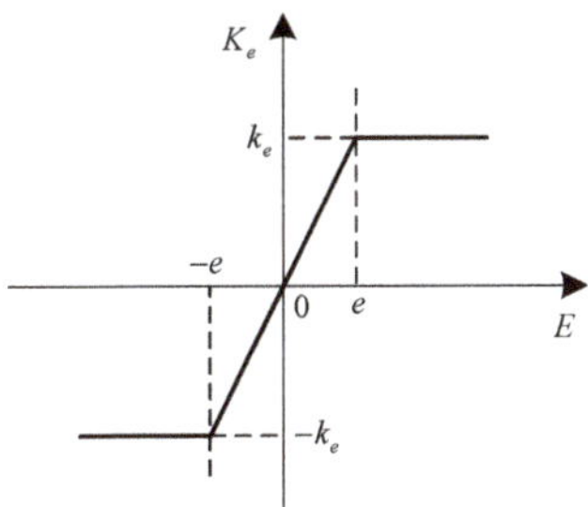

Figure 3. Sketch of the saturation function.

4.1.3. Adaptive Switching Gains with Fuzzy Logic Rules

The fuzzy logic controller was proposed many years ago [31]. In this part, the fuzzy rules are designed according to the fuzzy control theory for effectively optimizing the switching gain based on the states of the sliding-mode surface.

Normally, for purpose of enhancing the ability of observer on anti-disturbance and undertaking the generation of sliding mode, the sliding mode switching gain is often chosen to be too large, but this would further enhance the chattering noise in the disturbance estimation. Therefore, it is feasible to introduce the intelligent method to effectively estimate the switching gain according to the sliding mode arrival condition to alleviate the system chattering problem. For example, as the system state is far from the sliding surface, it means that the absolute value of difference $|e|$ is relatively large, the proportional factor k_f should be enlarged to drive the system state back. Similarly, as the system state is close to the sliding surface, proportional factor k_f should be smaller.

The proposed fuzzy logic system in this work consists of one input variable and one output variable which would be quantized first. The input and output variables are divided into four fuzzy subsets. Let $|e| = \left\{ \begin{array}{cccc} ZR & PS & PM & PB \end{array} \right\}$ denote the input variable and $k_f = \left\{ \begin{array}{cccc} ZR & PS & PM & PB \end{array} \right\}$ represent the output variable, respectively. Each rank is depicted by a membership function of trigonometric function, as shown in Figures 4 and 5, where the fuzzy language is defined as ZR(zero), PS(positive small), PM(positive middle), and PB(positive big).

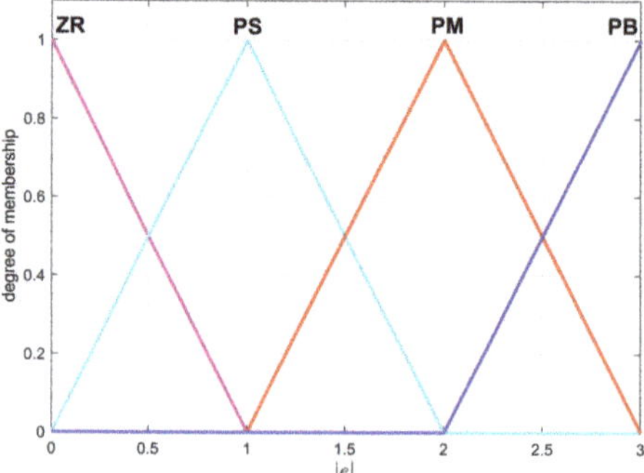

Figure 4. Fuzzy input membership functions.

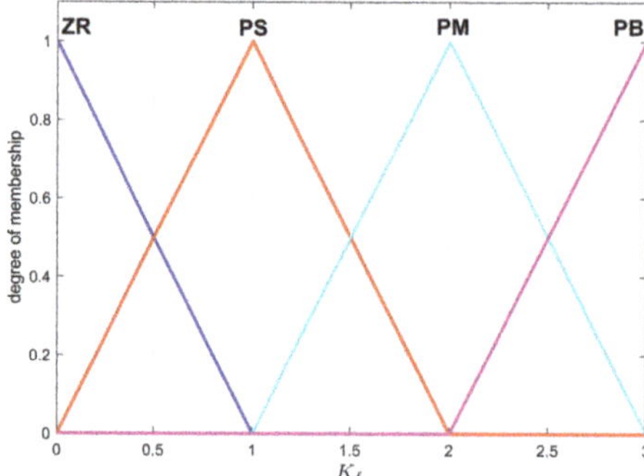

Figure 5. Fuzzy output membership functions.

Then define the fuzzy rules as

Rule 1: If $|e|$ is PB, then k_f is PB.

Rule 2: If $|e|$ is PM, then k_f is PM.

Rule 3: If $|e|$ is PS, then k_f is PS.

Rule 4: If $|e|$ is ZR, then k_f is ZR.

Finally, as for the fuzzy outputs, according to the work in [32], the center of gravity algorithm is adopted as the defuzzification method to convert the fuzzy subset duty cycle changes to real numbers, which is given as

$$k_f = \frac{\int_0^3 k_f \cdot \mu_{out} \cdot dk_f}{\int_0^3 \mu_{out} \cdot dk_f} \tag{36}$$

where μ_{out} represents the resulting output membership function. It takes every rule into account performing the union of the resulting output membership function $\mu_{out,i}$ of each *Rule* i $(i = 1, ..., 4)$, which means the maximum operation between them.

4.2. Attitude Controller

In order to achieve high-precision attitude stabilization in the presence of external wind and model uncertainties, a backstepping controller is built here combined with FASTESO for UAV attitude system. The main objective of this part is to guarantee that the state of attitude q and angular rate w converge to the reference values q_d and w_d in real time. As this part is identical to our previous work [30] except the UAV model, which would not impact the controller design. Therefore we are not going to describe it in detail.

According to the authors of [30], the control signal vector u_r is designed as

$$u_r = -M_{22}(S(\omega_e)C_d^b\omega_d - C_d^b\dot{\omega}_d) + M_{21}\ddot{p}^I + M_{23}\ddot{\eta} + C_{21}\dot{p}^I + C_{22}\dot{\Psi}$$
$$+C_{23}\dot{\eta} + G_2 - M_{22}K_{b1}\dot{q}_{ev} - q_{ev} - K_{b2}\tilde{\omega}_e - \hat{d}_r \tag{37}$$

where $\hat{d}_r$ is the estimated result of disturbance from FASTESO. Additionally, like what we did at FASTESO, TD will also be adopted here to process several variables in Equation (37) to reduce noise. Moreover, K_{b1}, K_{b2} are controller gains to be designed, and ω_e, $\tilde{\omega}_e$, and q_{ev} are defined in Equation (38), details of which can be found in [30].

$$\begin{cases} \omega_e = \omega - C_d^b\omega_d \\ \tilde{\omega}_e = \omega_e + K_{b1}q_{ev} \\ q_e = \begin{bmatrix} q_0 & q_{ev} \end{bmatrix}^T \\ q_e = q_d^* \otimes q \end{cases} \tag{38}$$

5. Simulation

In this section, the performance of proposed FASTESO-based control method is validated on the PX4/Gazebo platform [33], which provides simulations of physics close to the real world. A manually designed scalable disturbance, which includes a sudden change in some special points, is defined in Equation (39) and adopted in the simulations. Moreover, its amplitude could be adjusted by a. The main nominal coefficients of the quadrotor base are generated by the online toolbox of Quan and Dai [34] given in Table 1.

$$d_e = a * (0.25\sqrt{(|0.02\sin(0.1\pi t)|)} + 0.015\sin(0.3\pi t - 5) - 0.02) \tag{39}$$

Table 1. Quadrotor base coefficients used in simulation.

Coefficients	Description	Value
m_b	Mass of the quadrotor base	1.8 kg
J_{bx}	Pitch inertia	1.319×10^{-2} kg $\cdot$ m^2
J_{by}	Roll inertia	1.319×10^{-2} kg $\cdot$ m^2
J_{bz}	Yaw inertia	2.283×10^{-2} kg $\cdot$ m^2
l	Quadrotor base size	0.205 m

5.1. CASE 1 (Observers Performance Comparison)

In this section, several comparison simulations on observer performance of disturbance estimation are conducted.

5.1.1. Domestic Observers Comparison

To verify the effectiveness of the proposed fuzzy logic rules for adaptively adjusting the observer gain to cope with disturbances with different amplitudes, comparative simulations are conducted for a quadrotor UAV manipulator performing a hovering flight with the manually designed scalable disturbance (39) between FASTESO and ASTESO combined with a backstepping controller (BAC). The ASTESO is defined as a simplified version of FASTESO, which is not equipped with adaptive gain under fuzzy logic rules. Their coefficients are given as FASTESO: $\alpha_1 = diag(0.4,0.4,0.5)$; $\alpha_2 = diag(0.7,0.7,0.9)$; $e = 0.0001$; $k_e = 0.05$. ASTESO: $\alpha_1 = diag(0.4,0.4,0.5)$; $\alpha_2 = diag(0.7,0.7,0.9)$. BAC: $K_{b1} = diag(5,5,8)$; $K_{b2} = diag(3,3,4)$. The simulations are divided into two groups classified by different scalars of disturbances, $a = 10, 20$. The corresponding results could be found in Figures 6–9.

As shown in Figure 6, both FASTESO and ASTESO have good estimation performance under scalar $a = 10$. The resultant observer gain of FASTESO is described in Figure 7. Nevertheless, the ASTESO starts to break down when the disturbance is larger, which could be found in Figure 8. Under $a = 20$, the ASTESO fails to follow the larger disturbance, but the proposed FASTESO still performs well. This is because the fixed parameter of ASTESO is not big enough to estimate the disturbance with the amplitude in Figure 8. As shown in Figure 9, the adaptive gain of FASTESO also shows why it is effective under disturbances with different amplitudes. To summarize, this part shows that the proposed FASTESO could estimate a wider range of disturbance thanks to the fuzzy logic rules for adaptive observer gain.

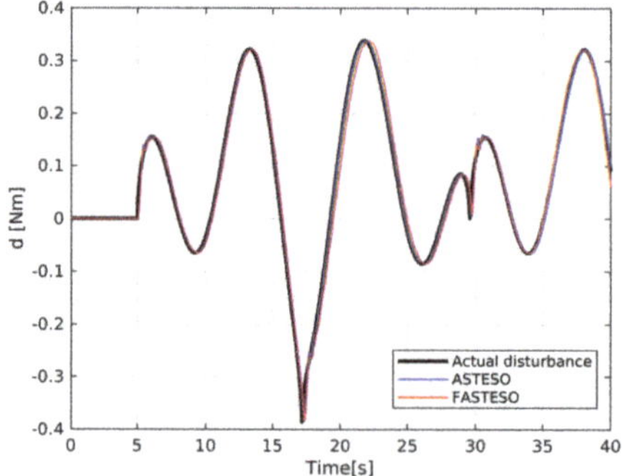

Figure 6. Estimation results from ASTESO and fuzzy adaptive saturation super-twisting extended state observer (FASTESO) with scalar $a = 10$.

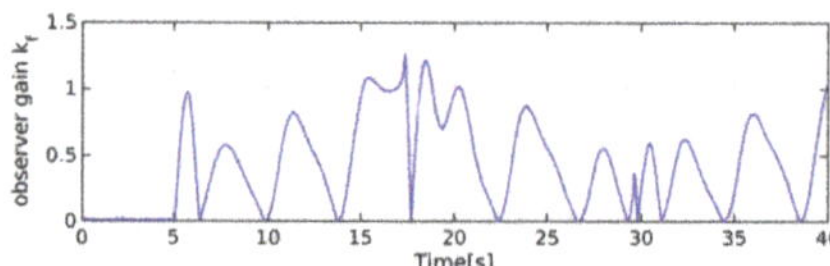

Figure 7. Observer gain of FASTESO during adaptation process with scalar $a = 10$.

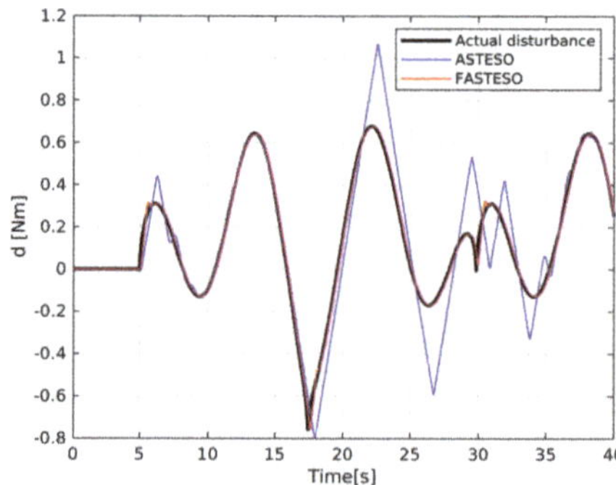

Figure 8. Estimation results from ASTESO and FASTESO with scalar $a = 20$.

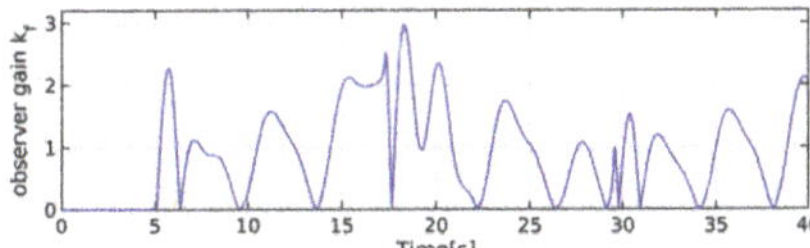

Figure 9. Observer gain of FASTESO during adaptation process with scalar $a = 20$.

5.1.2. Various Observers Comparison

To demonstrate the effectiveness of the introduction of the saturation function into SMO, a simulation similar with last part is conducted under FASTESO with an applied disturbance under $a = 5$ shown in Figure 10. Meanwhile the 2nd-order linear ESO (ESO2) and STESO are also invited for comparison. The parameters of the mentioned observers are listed. FASTESO:$\alpha_1 = diag(0.4, 0.4, 0.5)$; $\alpha_2 = diag(0.7, 0.7, 0.9)$; $e = 0.0001$; $k_e = 0.05$. ESO2:$k_1 = diag(4.5, 4.5, 6.5)$; $k_2 = diag(22, 22, 28)$; $k_3 = diag(5, 5, 8)$; STESO:$\alpha_1 = diag(0.2, 0.2, 0.3)$; $\alpha_2 = diag(0.3, 0.3, 0.4)$; BAC: $K_{b1} = diag(5, 5, 8)$; $K_{b2} = diag(3, 3, 4)$.

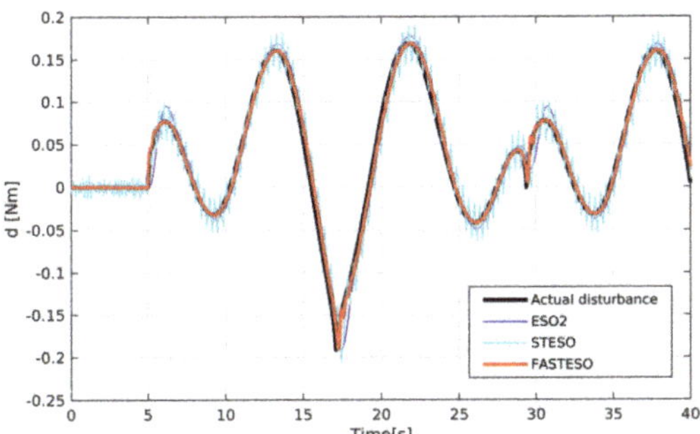

Figure 10. Estimation results from ESO2, STESO, and FASTESO with scalar $a = 5$.

From the response results in Figure 10, we could find that FASTESO has a perfect performance on estimation even at the special point with a sudden change. Although the curve under ESO2 could usually follow the actual value, it performs an overestimation near the peak. As for STESO, it could follow the fast variation of disturbance; however, it has a terrible chattering problem which is also the main reason for the introduction of saturation function in this work. To summarize, the FASTESO performs well at all range of disturbance ($a = 5, 10, 20$). Meanwhile, it follows better than other traditional observers under varied disturbance.

5.2. CASE 2 (Composite Comparison)

To further illustrate the effectiveness of the whole proposed control strategy, we would show all-sided simulation results from Section 5.1.2. Moreover, a single backstepping controller without any observer is also conducted on quadrotor UAV manipulator system for comparison. Three-axis components of disturbance torque $d_r = \begin{bmatrix} d_{r\varphi} & d_{r\theta} & d_{r\psi} \end{bmatrix}^T$ are equal and shown in Figure 10, which are applied on the UAV simultaneously. The result curves of both attitude and angular rate are shown in Figures 11–16.

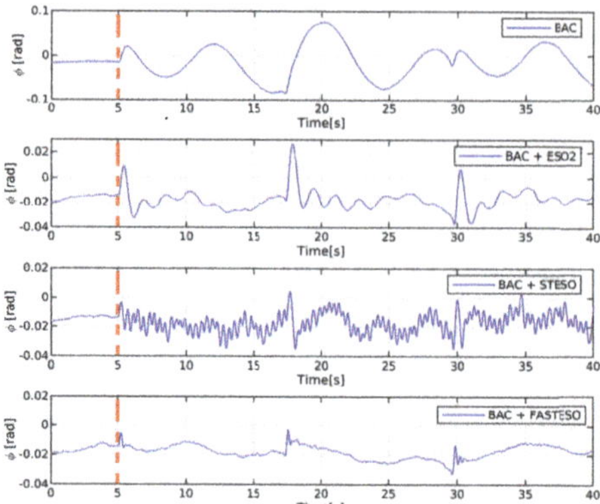

Figure 11. Results of roll angle with unknown external torque disturbance under backstepping controller combined with nothing, ESO2, STESO and FASTESO.

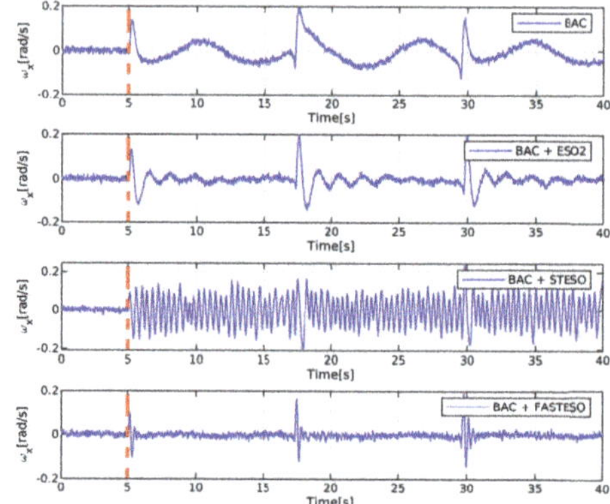

Figure 12. Results of angular rate ω_x with unknown external torque disturbance under backstepping controller combined with nothing, ESO2, STESO and FASTESO.

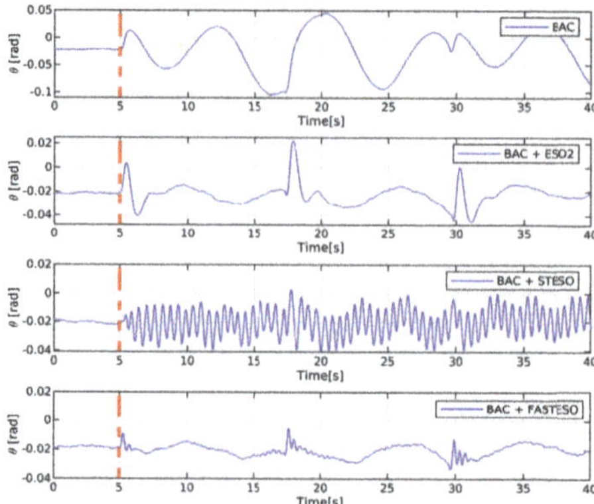

Figure 13. Results of pitch angle with unknown external torque disturbance under backstepping controller combined with nothing, ESO2, STESO and FASTESO.

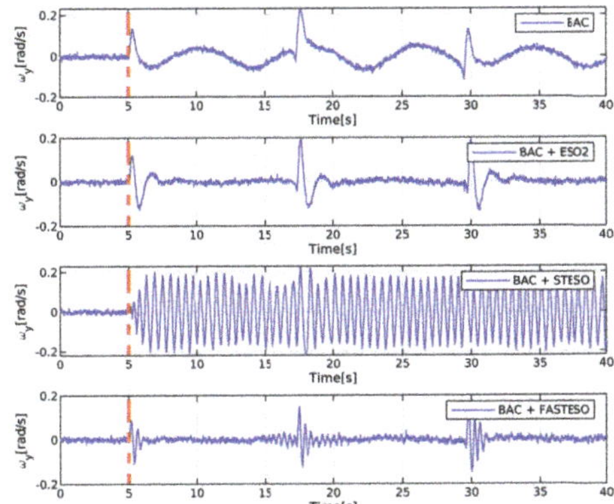

Figure 14. Results of angular rate ω_y with unknown external torque disturbance under backstepping controller combined with nothing, ESO2, STESO and FASTESO.

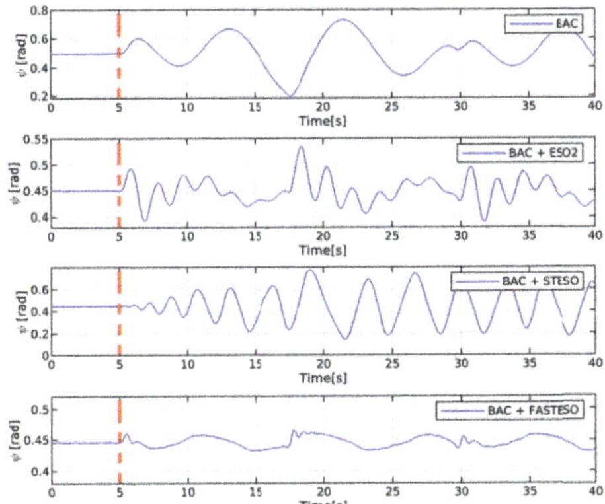

Figure 15. Results of yaw angle with unknown external torque disturbance under backstepping controller combined with nothing, ESO2, STESO and FASTESO.

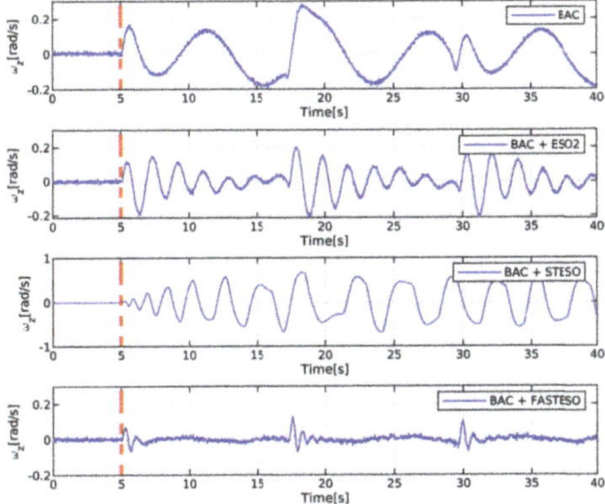

Figure 16. Results of angular rate ω_z with unknown external torque disturbance under backstepping controller combined with nothing, ESO2, STESO and FASTESO.

We could easily find that, without an observer, the single BAC has biggest attitude offset and angular rate fluctuation, which proves its bad robustness under disturbances. Specifically, the attitude curves even follows the applied torque disturbance well and angular rate has a dramatic change at special points. As for BAC+ESO2, although the estimation value follows the disturbance well to some extent, it has a performance of drastic change on both attitude and angular rate at peak point owing to the over estimation near the peak. As for BAC+STESO, in spite of the good performance of STESO on following the disturbance, it has a terrible chattering phenomenon and leads to a mess on UAV system. Compared to other situations, the UAV under BAC+FASTESO has the best performance on both attitude offset and angular rate fluctuation with varied torque disturbance.

6. Experiment

In this section, to verify the effectiveness of the whole proposed method, the quadrotor UAV manipulator is assigned to conduct a hovering task beside an electrical fan, which could generate several level gear wind as torque disturbance acting on the UAV system. BAC+FASTESO is adopted for UAV disturbance rejection test, meanwhile a single BAC without any observer is also conducted for comparison. The experimental scene is shown in Figure 17. The whole proposed control scheme is built in Pixhawk [35]. The parameters are chosen as follows. For BAC+FASTESO: $\alpha_1 = diag(0.2, 0.2, 0.3)$; $\alpha_2 = diag(0.3, 0.3, 0.4)$ $e = 0.001$; $k_e = 0.1$; $K_{b1} = diag(3, 3, 5)$; $K_{b2} = diag(2, 2, 2.5)$. For BAC: $K_{b1} = diag(6, 6, 9)$; $K_{b2} = diag(3.5, 3.5, 5)$.

The fan is placed to mainly apply the external torque in roll channel. The results of UAV hovering under first and second gear wind are shown in Figures 18–23. We found that the attitude of hovering UAV changes following the variation of wind gear. Moreover, the attitude vibration offset is reduced by help of FASTESO compared to the situation with single BAC. Meanwhile, the angular rate fluctuation is also reduced. They intuitively verify the effectiveness of FASTESO. Additionally, Figures 22 and 23 show the control output generated only from BAC in the control situations single BAC and BAC+FASTESO, respectively. We could find that the control signal is around zero with FASTESO; however, the control signal is impacted so much owing to external disturbance in situation without any observer. Moreover, the controller gain in BAC+FASTESO could keep smaller to make the whole system stable thanks to the disturbance estimation and compensation from FASTESO, which also contributes to the better performance of BAC+FASTESO. Meanwhile, the estimation of the lumped disturbance including external wind and model uncertainties is shown in Figure 20. Although we do not know the ground truth of the torque disturbance generated from the fan and model uncertainties to verify if the estimation result in Figure 20 is right or not, according to Figures 20 and 22, we could see that their results are almost the same, which also demonstrates the effectiveness of FASTESO to an extent.

Figure 17. Experimental scene.

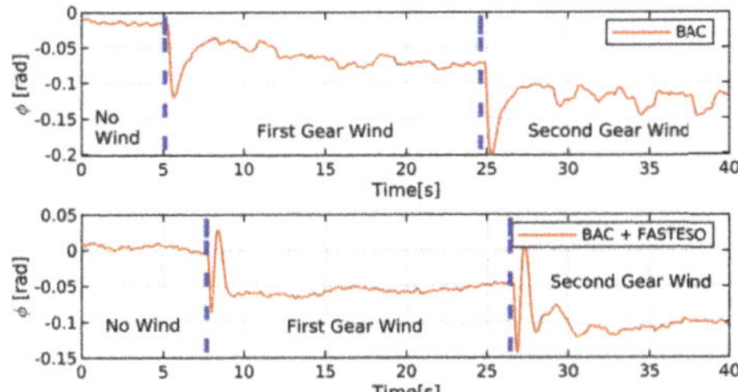

Figure 18. Experimental results of attitude with BAC and BAC+FASTESO under external wind.

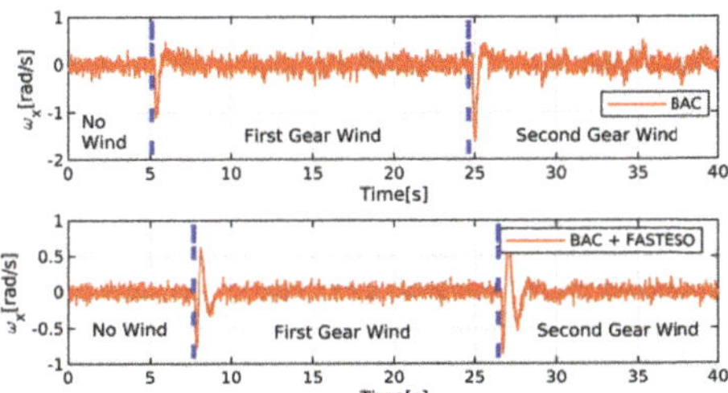

Figure 19. Experimental results of angular rate with BAC and BAC+FASTESO under external wind.

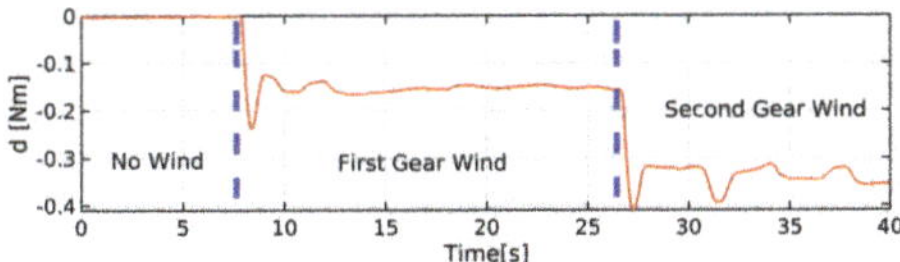

Figure 20. Disturbance estimation result from FASTESO.

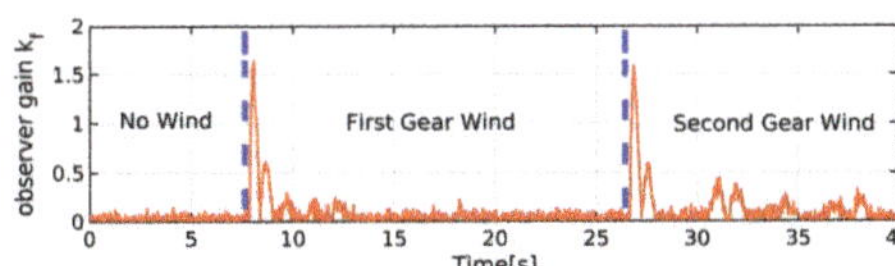

Figure 21. Observer gain of FASTESO during adaptation process.

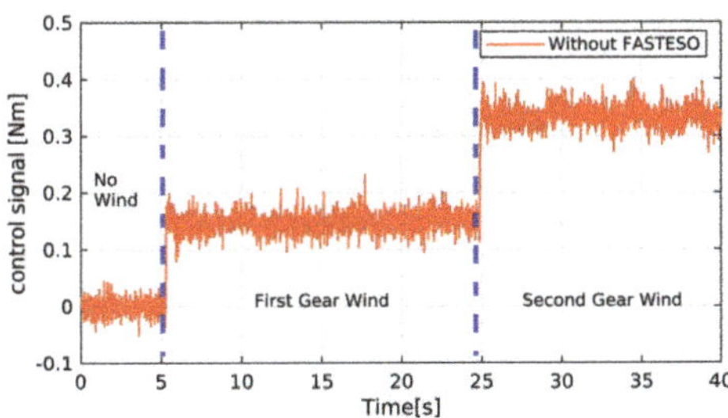

Figure 22. Control output signal only from BAC without FASTESO.

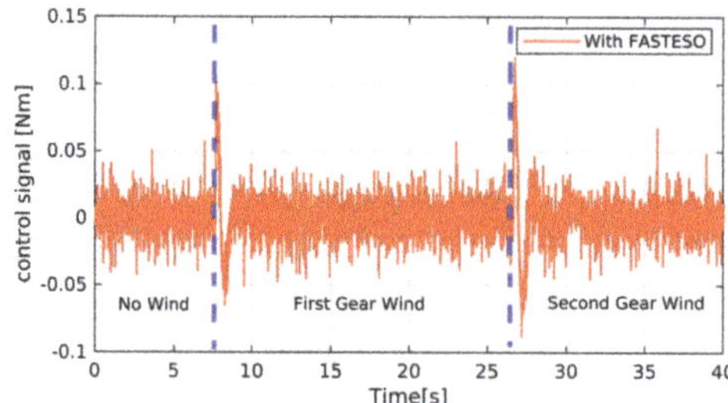

Figure 23. Control output signal only from BAC with FASTESO.

7. Conclusions

In this study, a new observer named FASTESO is proposed, which is incorporated with the saturation function, TD and fuzzy logic rules. Combined with backstepping controller, the whole control scheme for the quadrotor UAV manipulator system is finally built for rejection of lumped disturbance including external wind and model uncertainties. The effectiveness of the whole control structure is veritied in both several simulations and experiments.

In the future, we would pay more attention to the estimation separation of the model uncertainties from the external disturbance.

Author Contributions: Conceptualization, R.J.; writing–original draft preparation, R.J.; software, R.J. and Y.R.; validation, R.J. and Y.R.; methodology, R.J. and M.D.; writing—review and editing, R.J., W.C. and M.D.; supervision, W.C. All authors have read and agreed to the published version of the manuscript.

Funding: This work was supported by National Key R&D Program of China under grant number 2017YFB1302503 and 2019YFB1310802.

Conflicts of Interest: The authors declare no conflicts of interest.

Abbreviations

Additional abbreviation illustration:

FASTESO   Fuzzy adaptive saturation super-twisting extended state observer

References

1. Premachandra, C.; Ueda, D.; Kato, K. Speed-Up Automatic Quadcopter Position Detection by Sensing Propeller Rotation. *IEEE Sens. J.* **2018**, *19*, 2758–2766. [CrossRef]
2. Nakajima, K.; Premachandra, C.; Kato, K. 3D environment mapping and self-position estimation by a small flying robot mounted with a movable ultrasonic range sensor. *J. Electr. Syst. Inf. Technol.* **2017**, *4*, 289–298. [CrossRef]
3. Premachandra, C.; Otsuka, M.; Gohara, R.; Ninomiya, T.; Kato, K. A study on development of a hybrid aerial/terrestrial robot system for avoiding ground obstacles by flight. *IEEE/CAA J. Autom. Sin.* **2018**, *6*, 327–336. [CrossRef]
4. Ryll, M.; Muscio, G.; Pierri, F.; Cataldi, E.; Antonelli, G.; Caccavale, F.; Bicego, D.; Franchi, A. 6D interaction control with aerial robots: The flying end-effector paradigm. *Int. J. Robot. Res.* **2019**, *38*, 1045–1062. [CrossRef]
5. Kim, S.; Seo, H.; Shin, J.; Kim, H.J. Cooperative aerial manipulation using multirotors with multi-dof robotic arms. *IEEE/ASME Trans. Mechatronics* **2018**, *23*, 702–713. [CrossRef]
6. Jiao, R.; Dong, M.; Chou, W.; Yu, H.; Yu, H. Autonomous Aerial Manipulation Using a Hexacopter Equipped with a Robotic Arm. In Proceedings of the 2018 IEEE International Conference on Robotics and Biomimetics (ROBIO), Kuala Lumpur, Malaysia, 12–15 December 2018; pp. 1502–1507.
7. Ruggiero, F.; Trujillo, M.A.; Cano, R.; Ascorbe, H.; Viguria, A.; Peréz, C.; Lippiello, V.; Ollero, A.; Siciliano, B. A multilayer control for multirotor UAVs equipped with a servo robot arm. In Proceedings of the 2015

IEEE international conference on robotics and automation (ICRA), Seattle, WA, USA, 26–30 May 2015; pp. 4014–4020.

8. Jiao, R.; Chou, W.; Ding, R.; Dong, M. Adaptive robust control of quadrotor with a 2-degree-of-freedom robotic arm. *Adv. Mech. Eng.* **2018**, *10*, 1687814018778639. [CrossRef]

9. Alexis, K.; Papachristos, C.; Siegwart, R.; Tzes, A. Robust model predictive flight control of unmanned rotorcrafts. *J. Intell. Robot. Syst.* **2016**, *81*, 443–469. [CrossRef]

10. Achtelik, M.; Bierling, T.; Wang, J.; Höcht, L.; Holzapfel, F. Adaptive control of a quadcopter in the presence of large/complete parameter uncertainties. In Proceedings of the Infotech@ Aerospace 2011, St. Louis, MO, USA, 29–31 March 2011; p. 1485.

11. Chen, W.H.; Yang, J.; Guo, L.; Li, S. Disturbance-observer-based control and related methods—An overview. *IEEE Trans. Ind. Electron.* **2015**, *63*, 1083–1095. [CrossRef]

12. Cristofaro, A.; Johansen, T.A. Fault tolerant control allocation using unknown input observers. *Automatica* **2014**, *50*, 1891–1897. [CrossRef]

13. Pu, Z.; Yuan, R.; Yi, J.; Tan, X. A class of adaptive extended state observers for nonlinear disturbed systems. *IEEE Trans. Ind. Electron.* **2015**, *62*, 5858–5869. [CrossRef]

14. Kim, S.; Choi, S.; Kim, H.; Shin, J.; Shim, H.; Kim, H.J. Robust control of an equipment-added multirotor using disturbance observer. *IEEE Trans. Control Syst. Technol.* **2017**, *26*, 1524–1531. [CrossRef]

15. Wang, C.; Song, B.; Huang, P.; Tang, C. Trajectory tracking control for quadrotor robot subject to payload variation and wind gust disturbance. *J. Intell. Robot. Syst.* **2016**, *83*, 315–333. [CrossRef]

16. Chen, M.; Xiong, S.; Wu, Q. Tracking flight control of quadrotor based on disturbance observer. *IEEE Trans. Syst. Man, Cybern. Syst.* **2019**. [CrossRef]

17. Mokhtari, M.R.; Cherki, B.; Braham, A.C. Disturbance observer based hierarchical control of coaxial-rotor UAV. *ISA Trans.* **2017**, *67*, 466–475. [CrossRef]

18. Wang, L.; Su, J. Robust disturbance rejection control for attitude tracking of an aircraft. *IEEE Trans. Control Syst. Technol.* **2015**, *23*, 2361–2368. [CrossRef]

19. Han, J. From PID to active disturbance rejection control. *IEEE Trans. Ind. Electron.* **2009**, *56*, 900–906. [CrossRef]

20. Liu, J.; Vazquez, S.; Wu, L.; Marquez, A.; Gao, H.; Franquelo, L.G. Extended state observer-based sliding-mode control for three-phase power converters. *IEEE Trans. Ind. Electron.* **2016**, *64*, 22–31. [CrossRef]

21. Madoński, R.; Herman, P. Survey on methods of increasing the efficiency of extended state disturbance observers. *ISA Trans.* **2015**, *56*, 18–27. [CrossRef]

22. Muñoz, F.; Bonilla, M.; Espinoza, E.S.; González, I.; Salazar, S.; Lozano, R. Robust trajectory tracking for unmanned aircraft systems using high order sliding mode controllers-observers. In Proceedings of the 2017 International Conference on Unmanned Aircraft Systems (ICUAS), Miami, FL, USA, 13–16 June 2017; pp. 346–352.

23. Cai, W.; She, J.; Wu, M.; Ohyama, Y. Disturbance suppression for quadrotor UAV using sliding-mode-observer-based equivalent-input-disturbance approach. *ISA Trans.* **2019**, *92*, 286–297. [CrossRef]

24. Shi, D.; Wu, Z.; Chou, W. Super-twisting extended state observer and sliding mode controller for quadrotor UAV attitude system in presence of wind gust and actuator faults. *Electronics* **2018**, *7*, 128. [CrossRef]

25. Kim, H.; Son, J.; Lee, J. A high-speed sliding-mode observer for the sensorless speed control of a PMSM. *IEEE Trans. Ind. Electron.* **2010**, *58*, 4069–4077.

26. Rong, Y.; Jiao, R.; Kang, S.; Chou, W. Sigmoid Super-Twisting Extended State Observer and Sliding Mode Controller for Quadrotor UAV Attitude System With Unknown Disturbance. In Proceedings of the 2019 IEEE International Conference on Robotics and Biomimetics (ROBIO), Dali, China, 6–8 December 2019; pp. 2647–2653.

27. Lippiello, V.; Ruggiero, F. Cartesian impedance control of a UAV with a robotic arm. *IFAC Proc. Vol.* **2012**, *45*, 704–709. [CrossRef]

28. Zhang, Z.; Xu, H.; Xu, L.; Heilman, L.E. Sensorless direct field-oriented control of three-phase induction motors based on "Sliding Mode" for washing-machine drive applications. *IEEE Trans. Ind. Appl.* **2006**, *42*, 694–701. [CrossRef]

29. Kang, K.L.; Kim, J.M.; Hwang, K.B.; Kim, K.H. Sensorless control of PMSM in high speed range with iterative sliding mode observer. In Proceedings of the Nineteenth Annual IEEE Applied Power Electronics Conference and Exposition, 2004 (APEC'04), Anaheim, CA, USA, 22–26 February 2004; Volume 2, pp. 1111–1116.

30. Jiao, R.; Wang, Z.; Chu, R.; Dong, M.; Rong, Y.; Chou, W. An Intuitional End-to-End Human-UAV Interaction System for Field Exploration. *Front. Neurorobot.* **2019**, *13*, 117. [CrossRef] [PubMed]
31. Lee, C.C. Fuzzy logic in control systems: Fuzzy logic controller. I. *IEEE Trans. Syst. Man, Cybern.* **1990**, *20*, 404–418. [CrossRef]
32. Palm, R. Robust control by fuzzy sliding mode. *Automatica* **1994**, *30*, 1429–1437. [CrossRef]
33. Meier, L.; Honegger, D.; Pollefeys, M. PX4: A node-based multithreaded open source robotics framework for deeply embedded platforms. In Proceedings of the 2015 IEEE international conference on robotics and automation (ICRA), Seattle, WA, USA, 26–30 May 2015; pp. 6235–6240.
34. Quan, Q. Flight Performance Evaluation of UAVs. 2018. Available online: http://flyeval.com/ (accessed on 15 April 2020).
35. Meier, L.; Tanskanen, P.; Heng, L.; Lee, G.H.; Fraundorfer, F.; Pollefeys, M. PIXHAWK: A micro aerial vehicle design for autonomous flight using onboard computer vision. *Auton. Robot.* **2012**, *33*, 21–39. [CrossRef]

applied sciences

Article

Towards the Development of an Automatic UAV-Based Indoor Environmental Monitoring System: Distributed Off-Board Control System for a Micro Aerial Vehicle

Jorge Solis [1,*], Christoffer Karlsson [1], Simon Johansson [1] and Kristoffer Richardsson [2]

[1]   Department for Engineering and Physics, Karlstad University, 651 88 Karlstad, Sweden; christoffer@ljung.io (C.K.); simon_9@live.se (S.J.)
[2]   Bitcraze AB, 212 22 Malmö, Sweden; kristoffer@bitcraze.io
*   Correspondence: solis@ieee.org; Tel.: +46-54-700-1953

Abstract: This research aims to develop an automatic unmanned aerial vehicle (UAV)-based indoor environmental monitoring system for the acquisition of data at a very fine scale to detect rapid changes in environmental features of plants growing in greenhouses. Due to the complexity of the proposed research, in this paper we proposed an off-board distributed control system based on visual input for a micro aerial vehicle (MAV) able to hover, navigate, and fly to a desired target location without considerably affecting the effective flight time. Based on the experimental results, the MAV was able to land on the desired location within a radius of about 10 cm from the center point of the landing pad, with a reduction in the effective flight time of about 28%.

Keywords: micro aerial vehicles; visual-based control; Kalman filter

Citation: Solis, J.; Karlsson, C.; Johansson, S.; Richardsson, K. Towards the Development of an Automatic UAV-Based Indoor Environmental Monitoring System: Distributed Off-Board Control System for a Micro Aerial Vehicle. *Appl. Sci.* **2021**, *11*, 2347. https://doi.org/10.3390/app11052347

Academic Editors: Juan-Carlos Cano and Alessandro Gasparetto

Received: 6 November 2020
Accepted: 2 March 2021
Published: 6 March 2021

Publisher's Note: MDPI stays neutral with regard to jurisdictional claims in published maps and institutional affiliations.

1. Introduction

During the last few decades, the forest industry has progressed from manual to machine monitoring and harvesting to reduce hazards to the worker and increase productivity. Thanks to the recent advances in robot technology, remote control, vision, etc., autonomous machines now are being employed in agriculture, construction, medicine, and manufacturing. As a consequence, harvesters nowadays are controlled with a control area network (CAN)-based distributed control system and information system, with GPS localization to log data [1]. However, forestry is still a demanding area for robot technology, where reliability, precision, and adaptability to the dynamic changes are required to develop more enhanced autonomous or tele-operated operations. So far, mobile robots have been introduced as fundamental data-gathering tools for taking and processing high-resolution samples [1]. However, it is still hard to penetrate the forest autonomously due to the rough terrain and wheel slip. More recently, the introduction of unmanned aerial vehicles (UAVs) has provided a more efficient means of gathering environmental data where the operator can choose the spatial and time resolution of the data to be acquired while covering large areas [2]. Even though different kinds of UAVs have been proposed for automatic environmental monitoring, there are still technical challenges in terms of perception, control, and locomotion capabilities [3].

In particular, the automatic environmental monitoring in indoor farming (e.g., in greenhouse agriculture) is still limited [4]. By means of micro-aerial vehicles (MAV), it may be possible to measure climate features, monitor the plant growth, etc. However, there is still a range of limitations (e.g., power computation, autonomy, payload capacity, battery, etc.). Therefore, in this paper we proposed a vision-based control system to achieve autonomous navigation and landing in indoor environments without requiring advanced calibration procedures or additional external sensors.

Our main contribution is to propose a distributed off-board method capable of detecting a predefined fiducial marker and inferring information from a single defined marker in

197

order to allow the MAV to localize itself relative to the target location even in the existence of disturbances in indoor environments (e.g., occlusion). The main reason why we have selected a distributed off-board method was due to the computation power and battery constrains of the MAV. Moreover, we have analyzed and compared different experimental methods for tuning the control parameters in order to select the most proper method that not only reduce the time required for tuning when the payload is changed (by simulating different payloads). Finally, the selected control parameters were also analyzed in order to verify the effect on the effective flight time, which has scarcely been investigated in the literature.

Related Work

Falanga et al. [5] introduced an on-board sensing approach for a quadcopter system capable of achieving autonomous navigation and landing on a movable target location. Lange et al. [6] proposed a similar approach with an on-board computation based on optical flow and image-based detection where the target landing location was denoted by means of placing several concentric white rings on a black background to enhance the landing marker contrast. Moreover, Foster et al. [7] introduced a simultaneous localization and mapping (SLAM) method based on a monocular camera to build a map of an environment for achieving autonomous hovering for a micro aerial vehicle. Based on the monocular SLAM method, the micro aerial vehicle was capable of stabilizing while navigating by using only the on-board embedded sensors. However, the proposed method required an external sensor (in this case, a RGB-D sensor was used) for the pose estimation computation, and therefore a calibration procedure is required to achieve the autonomous navigation.

On the other hand, several researchers have proposed different methods for the navigation control of a quadcopter other than the conventional proportional-integral and derivative (PID) controller (which it is one of the most widely used controller in industry). For example, Reizenstein proposed in [8] a method where a linear-quadratic (LQ)-controller is used together with a PID controller for the autonomous navigation of a quadcopter. However, if different kinds of environmental sensors have to be embedded onboard a micro aerial vehicle, the model-based control has to be re-designed as the model needs to be coherent with the changes in the system specifications. Therefore, the control design process may be even more complex in order to compensate for the model discrepancies, and therefore it may require more time to find a suitable control strategy. On the other hand, the tuning of the PID controllers may be a tedious time-consuming process. However, there are several tuning methods that have been introduced in industry to obtain a fast and acceptable performance.

As indicated above, most of the proposed approaches depend on still external sensors, advanced calibration procedures, as well as complex control strategies without analyzing the effect on the effective flight time due to the use of additional on-board sensors (e.g., camera, environmental sensors, etc.). Moreover, the most suitable PID tuning method for a micro aerial vehicle while changing the specifications of the system (due to the possibility of exchanging different on-board environmental sensors) is an issue that has been scarcely analyzed.

2. Materials and Methods

2.1. Hardware

In this research, as a first approach the micro aerial vehicle Crazyflie 2.0 was selected as a development platform, as it is an open-source and open-hardware system [9]. The Crazyflie 2.0 (Figure 1a) features an expansion port where one can connect expansion boards. In particular, the Flow deck V2 adds a micro aerial vehicle the ability to discern when it is moving in any direction above the ground by means of an optical flow sensor which compensates for the inertial measurement unit (IMU) data errors [9].

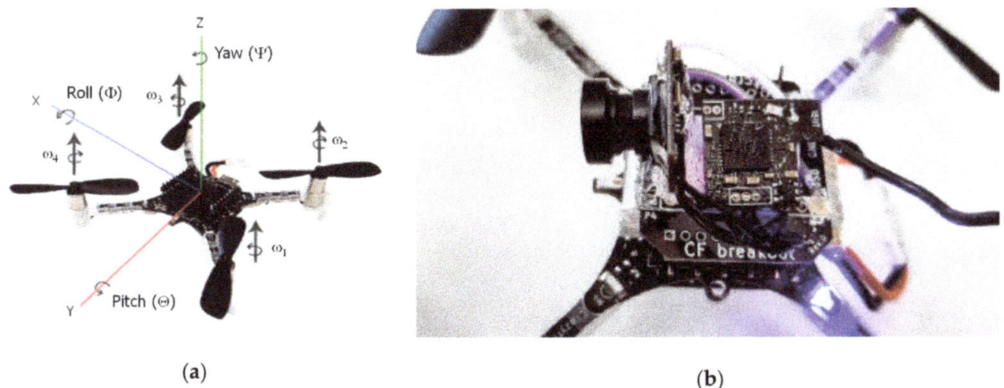

(**a**)　　　　　　　　　　　　　　　　(**b**)

Figure 1. (**a**) Crazyflie 2.0 commercialized by Bitcraze AB; (**b**) the camera and transmitter module was mounted on top of the Crazyflie 2.0.

In order to be able to localize the marker and determine the relative pose of the quadcopter, a vision sensor was needed. Therefore, the Eachine module was selected, which it is sold as a spare part for their M80S RC Drone [10]. The camera and transmitter module were soldered onto a Breakout deck for easy connection and disconnection (Figure 1b). The broadcasted video signal was picked up by an Eachine 5.8 GHz OTG USB video class (UVC) receiver [10].

2.2. Camera Pose Estimation and Target Detection

In this research, in order perform the camera pose estimation and target detection, we selected the ArUco library proposed Muñoz and Garrido [11]. The library contains a set of ArUco markers which are represented by binary square fiducial markers with its own unique identifier that can be used for camera estimation (see Section 3.2).

When the four corners of the ArUco marker are detected from the camera's point of view, the camera pose can be estimated. As the marker is planar, the transformation can be carried out by means of a homography function. The function returns the transformation as two vectors: one for the translation and the other one for the relative rotation. After the calculation of those vectors, the rotation matrix, R(Ψ, Θ, Φ), can be computed as shown in Equation (1), which indicates the rotation between the two reference frames. Rotations may also occur about an arbitrary axis—i.e., an axis other than the unit vectors representing the reference frame. In such cases, we need a way of determining the final orientation and the unique axis about which the point vector is rotated (often referred to as orientation kinematics). A rotation of a generic vector p about an arbitrary axis u through an angle ϕ can be written as Equation (2), where $\hat{u}$ is the skew-symmetric matrix representation of the vector u and is defined as Equation (3). Equation (2) is called Rodrigues' rotation formula and has proved to be especially useful for our purpose, with the main reason being that by knowing the numerical values of the elements contained within the rotation matrix, it is possible to compute $\hat{u}$ and ϕ using Equations (4) and (5).

$$R(\Psi, \Theta, \Phi) = \begin{bmatrix} cos\Theta cos\Psi & sin\Phi sin\Theta cos\Psi - cos\Phi sin\Psi & cos\Phi sin\Theta cos\Psi + sin\Phi sin\Psi \\ cos\Theta sin\Psi & sin\Phi sin\Theta sin\Psi + cos\Phi cos\Psi & cos\Phi sin\Theta sin\Psi - sin\Phi cos\Psi \\ -sin\Theta & sin\Phi cos\Theta & cos\Phi cos\Theta \end{bmatrix} \tag{1}$$

$$Rot(u, \phi) = I cos\phi + uu^{T}(1 - cos\phi) + \hat{u} sin\phi, \tag{2}$$

$$\hat{u} = \begin{bmatrix} 0 & -u_3 & u_2 \\ u_3 & 0 & -u_1 \\ -u_2 & u_1 & 0 \end{bmatrix} \tag{3}$$

$$\hat{u} = \frac{1}{2sin\phi}\left(R - R^T\right) \tag{4}$$

$$cos\phi = \frac{\tau - 1}{2}, \tag{5}$$

where τ is the sum of the scalar values in the main diagonal of the rotation matrix, R: $\tau = r_{11} + r_{22} + r_{33}$.

In order to detect the ArUco markers, at first, in order to find the candidate corners of the markers, a local adaptive thresholding and extraction of contours is carried out. The adaptive thresholding is conducted by applying a window sliding ($3 \times 3, 5 \times 5, 11 \times 11$, etc.) and finding optimum greyscale window. Values that are below the calculated value will be black and those above are white. After obtaining the thresholded image, if the result is not similar to a square, then the candidate corner is not selected. After selecting the corners, the white and black elements in the inner binary matrix are extracted in order to validate them. This process is conducted using the homography matrix and Otsu's method [11]. The resultant binary bits are then compared against all the available ArUco markers, so the respective marker is detected.

Through the ArUco library, camera pose estimation and marker detection can be achieved. However, due to the possibility of camera occlusion, data losses, etc., the relative pose of the Crazyflie 2.0 cannot be conducted. In order to overcome such issues, we have proposed to implement a Kalman filter [12]. The Kalman filter is an optimal estimator based on a recursive algorithm for estimating the track of an object by means of the given measurements to keep the accuracy of the state estimation aiming to predict future states. In our case, we have proposed to use the linear Kalman filter, which minimizes the mean-square error of the state. If we assume a current state x_k (including both the position and velocity), as shown in Equation (6), the filtering algorithm will then assume a correlation between the elements contained in x_k, captured by a covariance matrix, P_k.

Based on this, the state estimation of $\hat{x}_k$ will be computed by means of the previously estimated state, $\hat{x}_{k-1}$, the covariance matrix, and the current measurement from time step k (by assuming that the error between the measured and estimated state are Gaussian distributed). The state is then updated based on the given measurements and the previously estimated state.

$$x_k = (p,\ v). \tag{6}$$

Equation (7) indicates the state transition matrix, F_k. If we assume an acceleration, α, and the estimation shown in Equation (8), we can compute the state estimation as shown in Equation (9), where B_k is the control matrix and u_k is the control vector used as an input to the system. By defining the covariance Q, it is possible to take into account the noise in every prediction step, as shown in Equation (10).

$$F_k = \begin{bmatrix} 1 & \Delta t \\ 0 & 1 \end{bmatrix}, \tag{7}$$

$$\hat{x}_k = \left\{ \begin{array}{c} p_k = p_{k-1} + \Delta t v_{k-1} + \frac{1}{2}\alpha\Delta t^2 \\ v_k = v_{k-1} + \alpha\Delta t \end{array} \right., \tag{8}$$

$$\hat{x}_k = F_k\hat{x}_{k-1} + B_k u_k, \tag{9}$$

$$P_k = F_k P_{k-1} F_k^T + Q_k. \tag{10}$$

The given new measurement data are then stored in the vector, z. As the data may include noise, denoted as R, the vector of the logged data can be defined as Equation (11), where H is a general matrix that represents the measurement matrix. By defining the state estimation as Equation (12), where K is known as the Kalman gain, the final equation to update state can be defined as Equations (13) and (14), where K' is determined as Equation (15).

$$z_k = Hx_k + R_k, \tag{11}$$

$$\hat{x}_{k+1} = Fx_k + Ky, \tag{12}$$

$$\hat{x}'_k = \hat{x}_k + K'(z_k - H_k\hat{x}_k), \tag{13}$$

$$P_k = P_k - K'H_kP_k, \tag{14}$$

$$K' = P_kH_k^T\left(H_kP_kH_k^T + R_k\right)^{-1}. \tag{15}$$

In cases where the desired marker is not identified after some time, the velocity for each state will be exponentially decreased such that for each time step, $k_n \leq n_{max}$, where k_n denotes a time step with an undetected marker, the updated estimation has been computed as Equation (16), where α is the diminishing factor.

$$\hat{x}'_{k_n}(v_k) = \hat{x}'_{k_n}(v_{k-1})\alpha. \tag{16}$$

2.3. Control Architecture

The physical representation of the proposed system is shown in Figure 2a. As can be observed, a PC is used to process the grabbed images from the camera mounted on-board of the MAV though the video transmitter (Eachine TX801 5.8 GHz) and the command signals to the MAV through a bi-directional USB radio communication (Crazyradio PA 2.4 GHz).

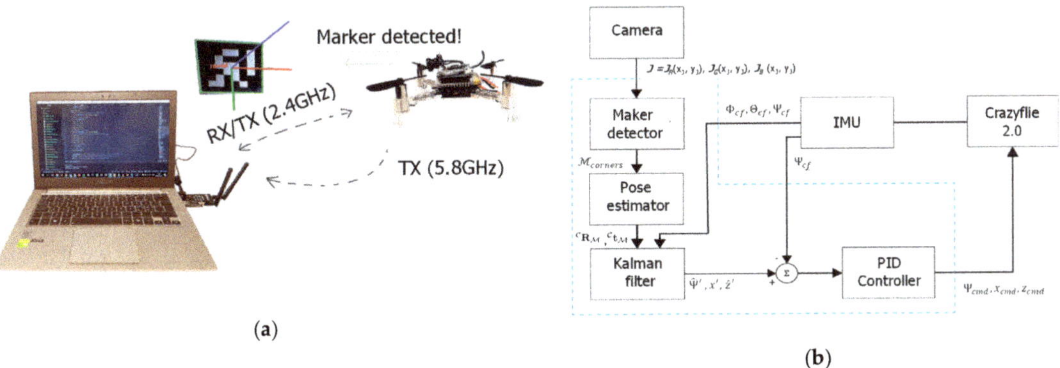

(a)

(b)

Figure 2. (a) Overview of the proposed off-board vision-based control system; (b) schematic diagram for the vision-based control of the MAV.

After computing in the PC the camera posture estimation and identifying the ArUco marker, the relative MAV posture is then given as new measurement data to the linear Kalman filter algorithm. As shown in Figure 2, the estimated pose of the MAV is then used as a reference signal to the compute the error signal. In order to reduce the error, the PID controllers the compute the commanded control signal in order to align the MAV pose towards the target location. As a first approach, we only considered the yaw rotation (Ψ-angle) and the x and z translations.

The on-board controller consists of two PID controllers in a cascade. A rule of thumb for cascaded PID controllers is that in order to have a stable and robust control scheme, the inner loop should operate at a higher frequency than the outer loop [13]. If the outer loop runs at a lower frequency than the inner loop, synchronization problems may occur, which entails in that the steady state value of the outer loop is reached before the inner loop has computed its output response, causing instability [13]. In this case, the on-board attitude and attitude rate controllers operate at 250 and 500 Hz, respectively, and the off-board controllers operate at about 10 Hz.

The set of off-board PID controllers has three different command variables: the relative yaw angle and translation in x- and z direction. In order to determine the best suitable experimental tuning method for the PID controllers, three different methods were tested: Ziegler–Nichols, Lambda, and Approximate M-constrained Integral Gain Optimisation (AMIGO).

The Ziegler–Nichols method was published in 1942 [14] and was developed by performing a large number of simulations on pneumatic analogue machines [15]. The parameters are determined from a step response analysis according to the Table 1. The AMIGO method was introduced in 2004. It has similarities to the Ziegler–Nichols method, but is instead based on simulations performed with computers [16]. The parameters are determined from a step response experiment according to Table 2. The lambda method differs from the methods above by a parameter that the user can adjust themselves. The method is widely used in the paper industry and was developed in the 1960s. The original method only dealt with the PI controller, but it is possible to derive formulas for PID controllers as well. The lambda method does not provide parameters for integrating processes [15]. For the PID controller in parallel, the parameters are taken from a step response according to Table 3.

Table 1. Summary of values for k, k_i, and k_d according to the Ziegler–Nichols method.

K	k_i	k_d
$\dfrac{1.2 \cdot T_{100\%}}{K_p \cdot L}$	$\dfrac{1.2 \cdot T_{100\%}}{2 \cdot K_p \cdot L^2}$	$\dfrac{1.2 \cdot T_{100\%}}{2 \cdot K_p}$

Table 2. Summary of values for k, k_i, and k_d according to the approximated *M*-constrained integral gain optimization (AMIGO) method.

k	k_i	k_d
$\dfrac{1}{K_p}\left(0.2 + 0.45 \cdot \dfrac{T_{63\%}}{L}\right)$	$\dfrac{\dfrac{1}{K_p}\left(0.2 + 0.45 \cdot \dfrac{T_{63\%}}{L}\right)}{\dfrac{0.4 \cdot L + 0.8 \cdot T_{63\%}}{L + 0.1 \cdot T_{63\%}} \cdot L}$	$\dfrac{\left(0.2 + 0.45 \cdot \dfrac{T_{63\%}}{L}\right)}{K_p} \cdot \dfrac{0.5 \cdot L \cdot T_{63\%}}{0.3 \cdot L + T_{63\%}}$

Table 3. Summary of values for k, k_i, and k_d according to the lambda method.

k	k_i	k_d
$\dfrac{1}{K_p} \cdot \dfrac{\dfrac{L}{2} + T_{63\%}}{\dfrac{L}{2} + T_{cl}}$	$\dfrac{1}{K_p \cdot \left(\dfrac{L}{2} + T_{cl}\right)}$	$\dfrac{1}{K_p} \cdot \dfrac{\dfrac{L}{2} + T_{63\%}}{\dfrac{L}{2} + T_{cl}} \cdot \dfrac{T_{63\%} \cdot L}{L + 2 \cdot T_{63\%}}$

As is shown in Tables 1–3, the times calculated are $T_{63\%}$, $T_{100\%}$, and L. L is the dead time of the system and this is calculated by determining the point of intersection of the previously calculated maximum derivative and the time axis. $T_{100\%}$ is obtained by calculating the intersection point of the maximum derivative with the line Ys, then obtained a time that is $T_{100\%} + L$. To get $T_{100\%}$, L is then subtracted. $T_{63\%}$ is obtained by examining when the first measurement data point exceeds the value of 0.63Ys. The time for that measurement data point is then time $T_{63\%} + L$ and, the same as for $T_{100\%}$, the dead time L must be subtracted away. In the lambda method, the factor $T_{cl} = T_{63\%}$ will be used to obtain an aggressive regulation similar to the results of Ziegler–Nichols and AMIGO.

To test the step response of the Crazyflie 2.0 in indoor conditions, different additional weights (0, 4.7, 6.1, and 10.8 g) were considered. To calculate which parameters would be used for PID according to the Ziegler–Nichols, lambda, and AMIGO methods, the same program was used in Matlab. Four tests for each case were performed and they resulted in the mean values of the PID parameters according to Tables 4–7.

Table 4. Proportional-Integral and derivative (PID) parameters according to the average values from the step responses without any additional load.

	k	k_i	k_d
Lambda	1.1064	1.5796	0.0961
AMIGO	3.5127	7.6882	0.2779
Ziegler-Nichols	6.7968	21.5534	0.5548

Table 5. PID parameters according to the average values from the step responses with an additional load of 4.7 g.

	k	k_i	k_d
Lambda	0.7208	0.4440	0.1597
AMIGO	1.9126	1.5705	0.3935
Ziegler-Nichols	2.6293	3.1875	0.5573

Table 6. PID parameters according to the average values from the step responses with an additional load of 6.1 g.

	k	k_i	k_d
Lambda	0.3448	0.4902	0.3690
AMIGO	0.7161	0.2540	0.6309
Ziegler-Nichols	0.5812	0.1372	0.6230

Table 7. PID parameters according to the average values from the step responses with an additional load of 10.8 g.

	k	k_i	k_d
Lambda	0.4016	0.4748	0.8708
AMIGO	1.0585	0.2012	1.8373
Ziegler-Nichols	0.2500	0.0290	0.5398

As is shown in the experimental results, with an additional load of 0, 4.7, and 6.1 g in Tables 8–10, respectively, the AMIGO method in each case has a noticeably shorter settling time than the Ziegler–Nichols method (as is shown in Table 8, the lambda method is much slower than the others and is therefore excluded when placing an additional load of 4.7 and 6.1 g). The AMIGO method is considered the better of them even though in some cases it has a slower rise time and higher overshoot than the Ziegler–Nichols method, but these two parameters do not differ much from the different methods. As an example, we can observe the experimental results in Figures 3 and 4 for both a case without any additional weight and a case with a load of 4.7 g, respectively.

Table 8. The rising time, maximum overshoot and settling time for the step response without any additional load.

	Rising Time (s)	Maximum Overshoot (cm)	Settling Time (s)
AMIGO	0.25	8	1.5
Ziegler-Nichols	0.25	24	2.5
Lambda	1.0	12	9

Table 9. The rising time, maximum overshoot, and settling time for the step response with an additional load of 4.7 g.

	Rising Time (s)	Maximum Overshoot (cm)	Settling Time (s)
AMIGO	0.90	19	4
Ziegler-Nichols	0.55	14	8

Table 10. The rising time, maximum overshoot, and settling time for the step response with an additional load of 6.1 g.

	Rising Time (s)	Maximum Overshoot (cm)	Settling Time (s)
AMIGO	0.90	12.1	4.5
Ziegler-Nichols	0.5	11.9	7.5

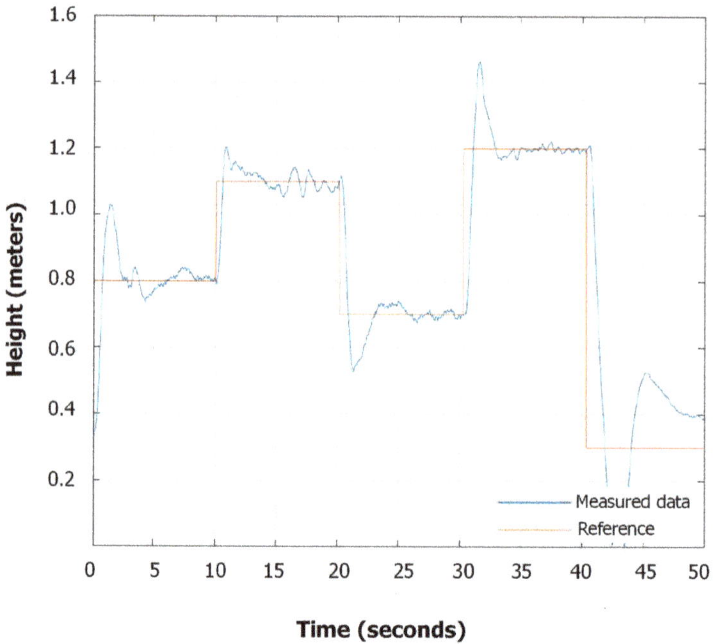

Figure 3. Experimental results when following the predefined trajectory without any additional load according to the AMIGO method.

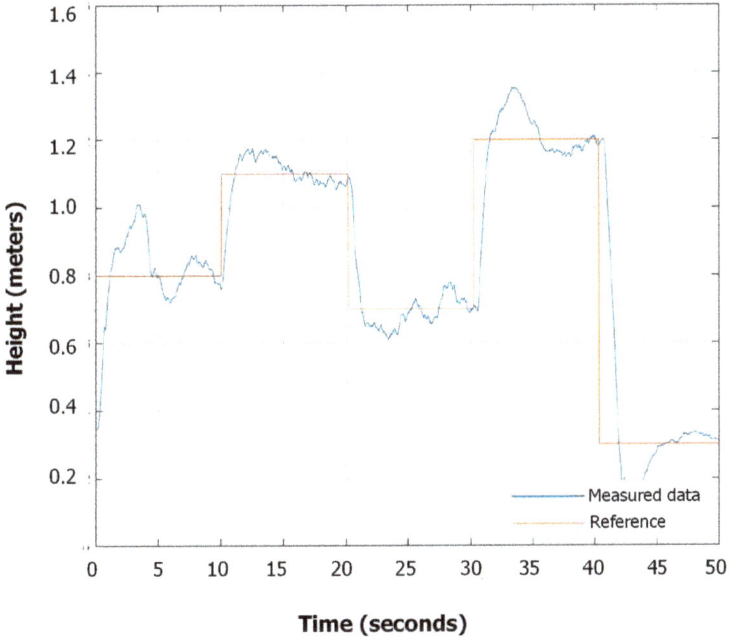

Figure 4. Experimental results when following the predefined trajectory with an additional load of 4.7 g according to the AMIGO method.

Therefore, based on the preliminary experimental results presented above, the AMIGO method was chosen as the primary tuning method as it has been proven to be a reliable method for tuning the PID controllers of the Crazyflie 2.0.

3. Experimental Results

3.1. Pose Estimation and Fiducial Marker Detection

In order to verify the effectiveness of the optimized pose estimation, we pre-defined a motion sequence of the ArUco marker in indoor conditions. In particular, the ArUco marker was rotated at different angles in front of the Crazyflie 2.0 whilst the MAV was programmed to always have its y-axis pointing collinear with the z-axis of the reference frame of the marker, radiating out from its center. As it is shown in Figure 5a, at time $t \approx 5.5$ s, the proposed method achieved a good estimation of the movement of the marker even though the loss of measurement data. Additionally, at the time $t \approx 12.5$ s, the detection of the marker is lost for approximately one second and the proposed method still predicted the movement of the marker.

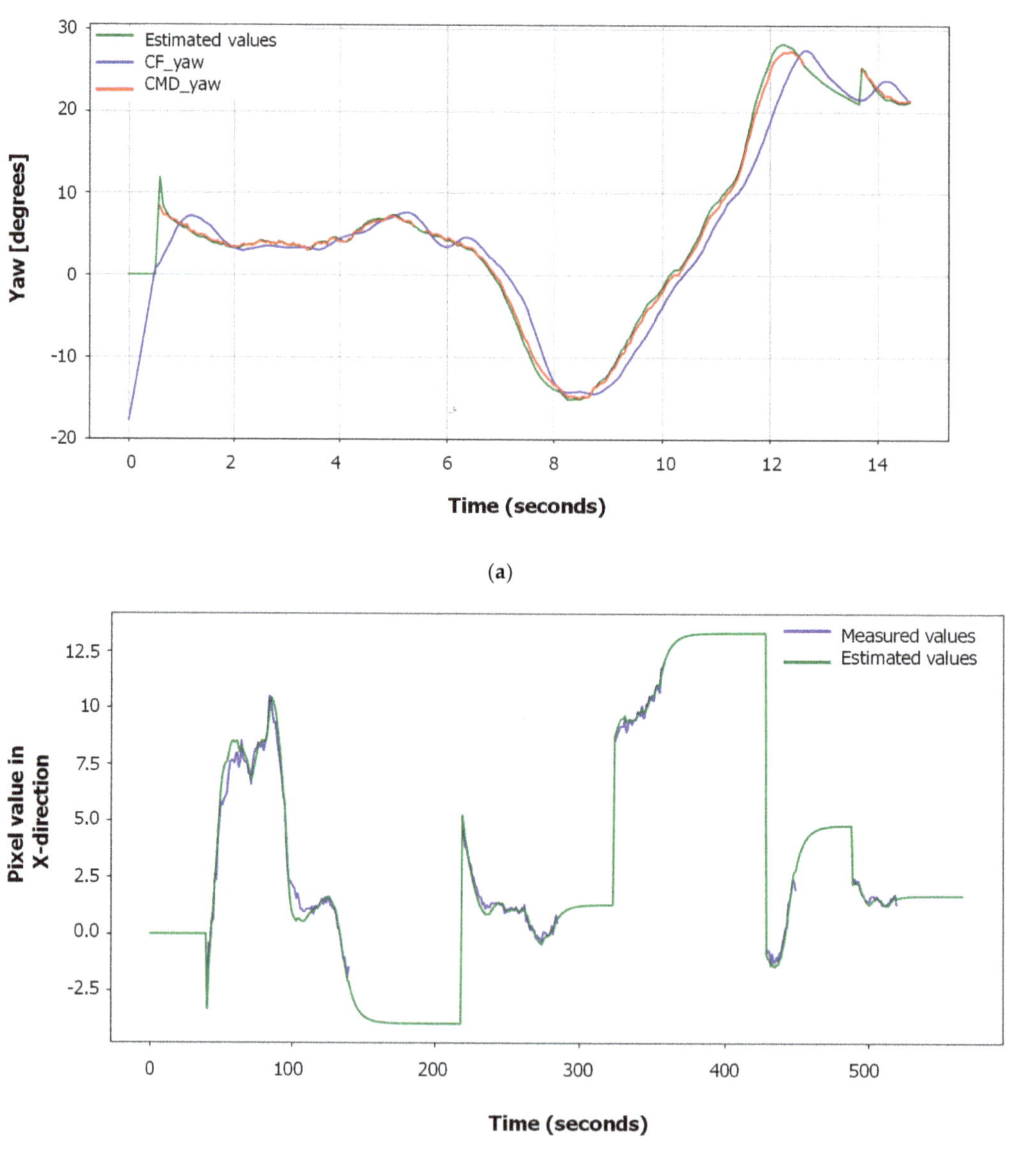

(a)

(b)

Figure 5. Experimental results: (**a**) estimation along the yaw; (**b**) estimation of marker center along the x-direction.

Similarly, in Figure 5b the Kalman filter was used in an experiment for tracking the center point of the ArUco marker. In this experiment, the camera view was obstructed at different points in time and moved to a new location. The filter was limited to output only an estimate for a maximum of 25 frames in a row when the marker was not detected and data were sampled only when the marker was detected or estimated by the Kalman filter; hence, the peaks in a green line. Similar to the plot in Figure 5a, the factor $\alpha = 0{:}85$ was set for diminishing the velocity $\hat{v}_k = \alpha \hat{v}_{k-1}$ for each iteration where an estimation was made in the previous time step for a maximum n_{max} number of frames.

In order to verify the effectiveness of the ArUco marker detection, we examined at which distance, x, the marker can be recognized when placed directly in front of the camera. The experiment was conducted by placing the marker at various distances, $x \in [10, 220]$ centimeters away from the camera. In order to keep the results of the experiments consistent, the marker used is characterized by a 200 x 200 mm square with a 6×6 internal matrix and has the identifier ID = 24. As can be observed in Figure 6, the pose estimation algorithm is able to estimate the distance between the camera and the marker with only about 2 cm of deviation from the actual value (in this case, the distance from the MAV is about 1 m). The error in the estimated distance increases with the distance with a maximum deviation of about 14 cm (in this case, the distance from the MAV is around 2 m). From the experimental results, we could confirm a detection success ratio for the ArUco marker of 92.8% (39 out of 42).

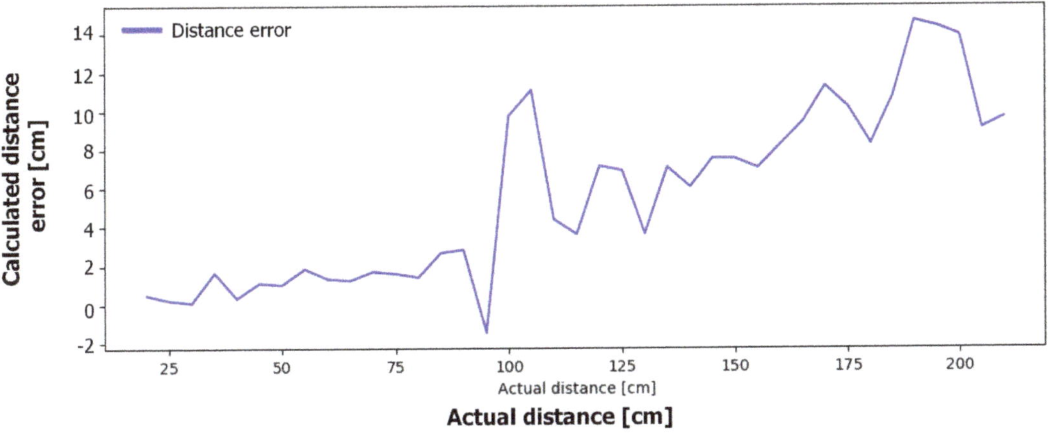

Figure 6. Actual distance between the camera and the marker versus the distance calculated by the marker detection and pose estimation function.

3.2. PID Controller: Yaw and x-Direction

After tuning the two PID controllers for the yaw angle and translation in x-direction by means of the AMIGO method (with some further fine-tuning), we determined the gains as $K_p = 1.0$, $K_i = 0.10$, $K_d = 0.045$ and $K_p = 5.10$, $K_i = 5.40$, $K_d = 0.62$. respectively. Based on that, an experiment was conducted where the quadcopter was placed in a position approximately 30 cm away from the marker in x-direction (x_{cmd}) with a 35° relative angle (Ψ_{cmd}). The experimental results are shown in Figure 7. Based on the experimental results, the performance is considered to be quite accurate with a deviation around the set point of only about ±5 degrees. The x-direction controller shows significant improvement over using a simple P-controller but still displays a minor oscillation around the set point. However, the results are very much sufficient for the scope of this project as it shows that the MAV is able to orient and position itself relative to the target marker with only a small amount of oscillations and overshoot in indoor conditions.

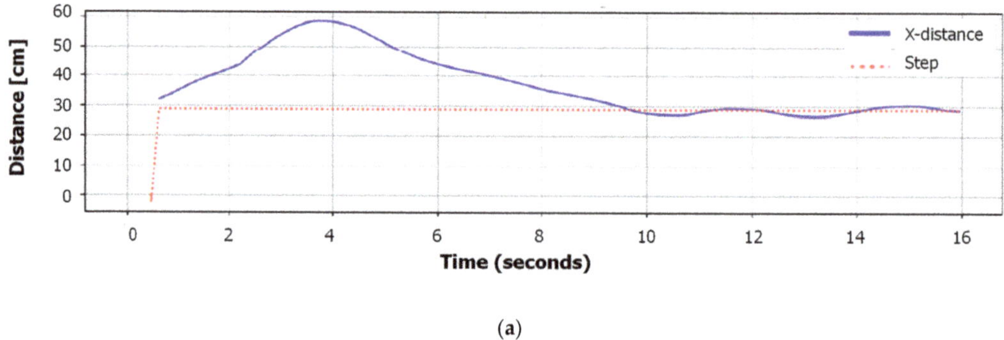

(**a**)

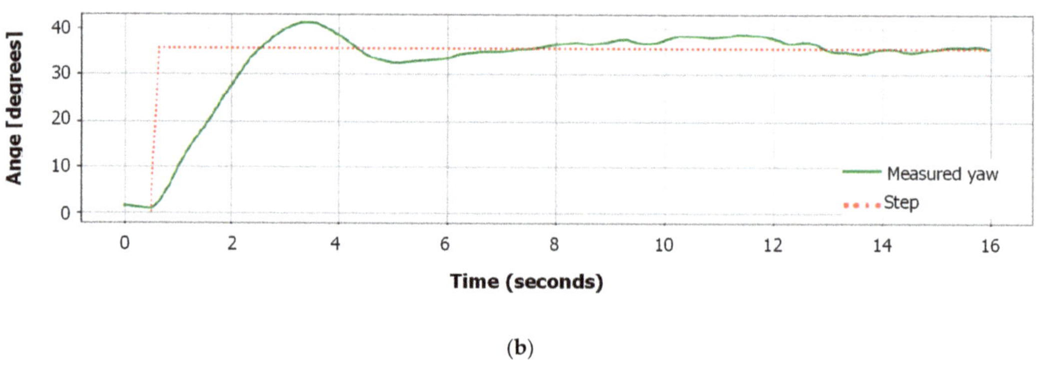

(**b**)

Figure 7. Experimental results: (**a**) x-direction controller; (**b**) yaw angle controller.

3.3. Target Landing

In order to determine the performance for targeted landing, an experiment was conducted indoors where the marker was placed along a certain distance away from the MAV and with a constant angle, α, as shown in Figure 8a. Both the target location and the MAV were placed on the same initial height and a signal was transmitted to the MAV to take off and hover above ground at a height of 30 cm. By using the measurement pad (Figure 8b), we could compute the exact landing place of the MAV and compare it with the desired one. For this purpose, we performed 10 trials while keeping the same posture in each trial. The experimental results are shown in Table 11 and Figure 9. From the results, we can conclude that there are some inconsistencies in terms of both precision and accuracy. The MAV is able to land in relatively close proximity to the center of the measurement pad with a standard deviation of about $\sigma \approx 5.32$ cm in both x- and y-direction.

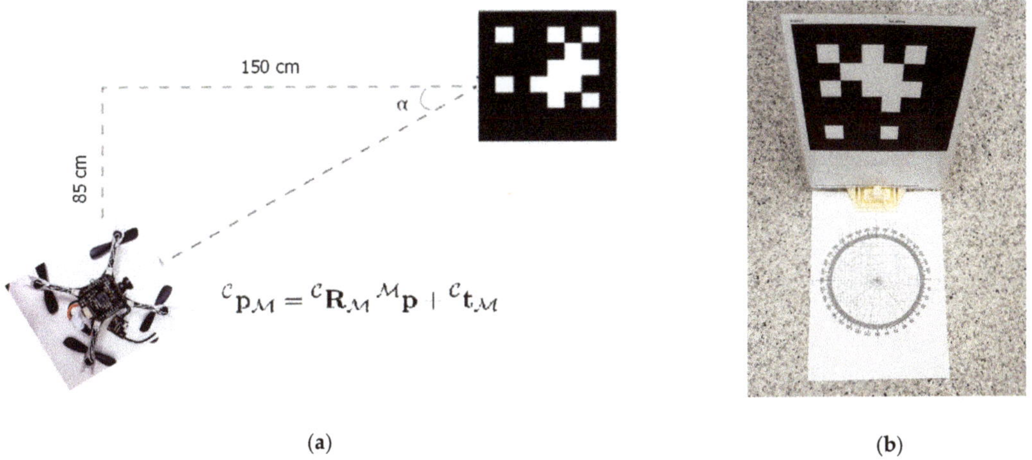

Figure 8. (**a**) The micro aerial vehicle (MAV) is located 1.5 m away from the target location and a rotation of 30 degrees; (**b**) image of the selected ArUco marker and measurement pad for target landing.

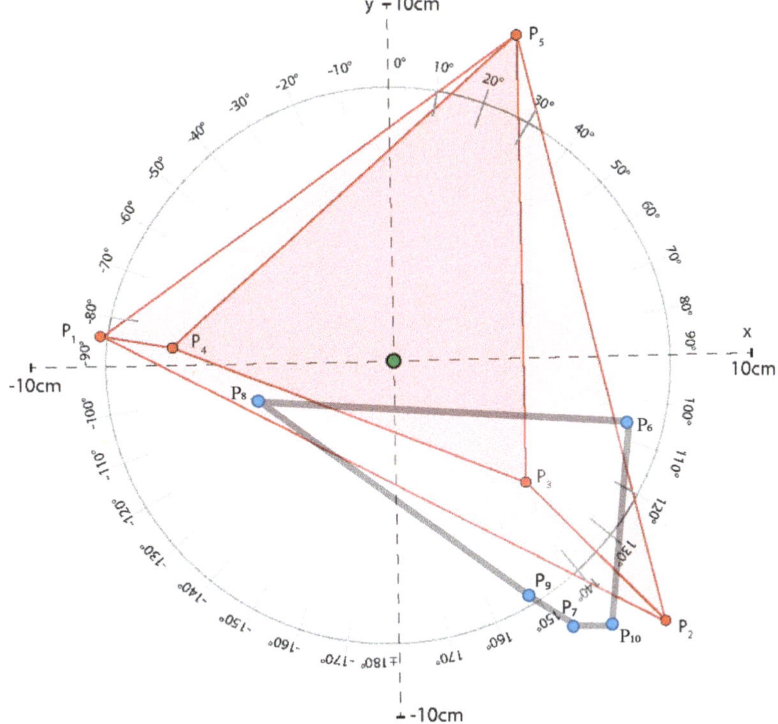

Figure 9. Each point, P_i, on the diagram corresponds to the landing point for each attempt.

Table 11. Results of the targeted landing experiment. The mean value μ and the standard deviation σ are indicated for each respective parameter.

x [cm] ($\mu = 1.88$, $\sigma = 5.65$)	y [cm] ($\mu = -2.27$, $\sigma = 4.99$)	Ψ [degrees] ($\mu = 0.085$, $\sigma = 4.06$)
−7.7	1.0	0.12
7.3	−7.1	−10.67
3.5	−3.4	5.19
−6.1	0.5	0.38
3.6	9.2	1.44
7.0	−2.5	1.75
5.1	−6.2	0.34
−4.3	−1.7	1.1
4.9	−5.2	0.14
5.5	−7.3	1.06

3.4. Battery Characterization

The Crazyflie 2.0 has a flight time of around 7 min [9] without any additional equipment. At first, the flight time was measured first without the camera and video transmission module mounted on the vehicle and then the flight time was measured with the module mounted and turned on, thus contributing to the effective flight time both by its net weight and by the power required for indoor operation. The results from measuring the battery level with and without the camera module activated can be seen in Figure 10. Based on these results, it can hover in place for roughly 6.7 min until the battery voltage level drops below 3.0 Volts.

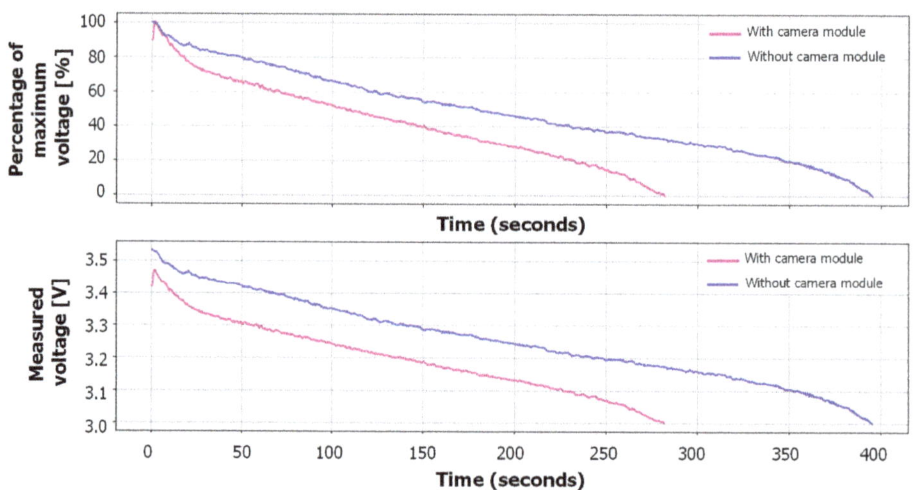

Figure 10. Changes in the effective flight time of the MAV with and without the camera module.

In Figure 11, a similar experiment was conducted but this time with the camera and transmission module mounted on-board the Crazyflie 2.0 and with the power to the module turned on. The intention of this experiment was to assess how the battery life is influenced by only its contribution by weight, excluding the power consumption for the operation of the module. However, at this point the battery had reached too many charge cycles to be able to maintain a similar operational time as for the previous experiments, making the battery discharge more rapidly and thus decreasing the effective flight time substantially.

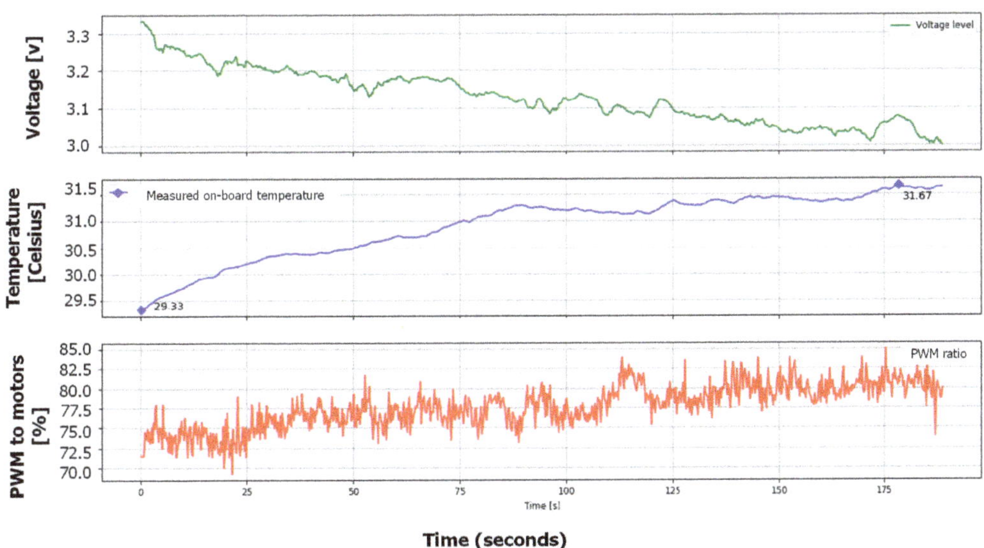

Time (seconds)

Figure 11. Sensor readings during hover with camera and video transmission module on-board (camera and transmitter (TX) module turned off).

Therefore, we cannot compare the results between the different measurements directly, but we can observe how other parameters influence the flight time. As the battery level decrease, the duty cycle of the PWM signal given as input to the metal–oxide–semiconductor field-effect (MOSFET) transistors which drive the motors must increase in order to keep the vehicle in equilibrium at the specified height, because the thrust generated by each motor must add up to support the total weight of the platform.

The plot in Figure 11 shows how the duty cycle of the PWM signal increases as the battery voltage decreases in order to keep the quadcopter stable and at a constant height. This means that since the thrust output from each motor will determine the position (in this case, the height) of the quadcopter, by increasing the cycle frequency the quadcopter is able to remain in its original pose, but this will also entail in an increased power consumption. We can also see from the graph that the readings from the on-board temperature sensor increase with the time of operation. This is interesting because it allows us to reason about the losses in the system that appear as heat losses. The internal resistance in the battery will cause the temperature in the battery to increase over time during operation and will cause the voltage to drop; however, since the sensor measures the ambient temperature, we cannot conclude that the increase in temperature is subject to only the heat generated by the battery, as it may also originate from, e.g., friction or other losses from the brushless motors which generate heat.

4. Conclusions

In this paper, in order to design and implement an off-board distributed control system based on visual input for a micro aerial vehicle (MAV) able to hover, navigate. and fly to a desired target location while taking into account the effective fly time due to the battery constrains, the authors have proposed camera pose estimation and marker detection for a MAV based on the ArUco library and the optimization of the relative MAV pose estimation based on the linear Kalman filter. From the experimental results, the MAV was able to land on the desired location within a radius of about 10 cm from the center point of the landing pad, with a reduction in the effective flight time of about 28%

As future work, a feed-forward term based on deep learning algorithms (e.g., LSTM) will be integrated to further improve the accuracy of the pose estimation and landing. In addition, sensors will be placed onboard the MAV to measure the indoor environmental conditions.

Author Contributions: Conceptualization, J.S.; supervision, J.S. and K.R.; writing—review and editing, J.S.; writing—original draft preparation, C.K., S.J. All authors have read and agreed to the published version of the manuscript.

Funding: This research received no external funding.

Conflicts of Interest: The authors declare no conflict of interest.

References

1. Viljamaa, P.; Koivo, H.N. Adaptive feed control of a forest harvester. In *International Conference in Mechatronics*; Wiley: Hoboken, NJ, USA, 2003; pp. 235–240.
2. Dunbabin, M.; Roberts, J.; Usher, K.; Corke, P. A new robot for environmental monitoring on the great barrier reef. In Proceedings of the Australian Conference on Robotics and Automation, Canberra, Australia, 6– 8 December 2004.
3. Hodgkinson, B. Environmental Monitoring with K-means Error Reduction Using UAVs Controlled by a Fluid Based Scheme. In Proceedings of the IROS 2012 Workshop on Robotics for Environmental Monitoring, Algarve, Portugal, 7 October 2012.
4. Roldán, J.J.; Joossen, G.; Sanz, D.; Del Cerro, J.; Barrientos, A. Mini-UAV Based Sensory System for Measuring Environmental Variables in Greenhouses. *Sensors* **2015**, *15*, 3334–3350. [CrossRef] [PubMed]
5. Falanga, D.; Zanchettin, A.; Simovic, A.; Delmerico, J.; Scaramuzza, D. Vision-based autonomous quadrotor landing on a moving platform. In Proceedings of the 15th International Symposium on Safety, Security, and Rescue Robotics, Shanghai, China, 11–13 October 2017; pp. 200–207.
6. Lange, S.; Sunderhauf, N.; Protzel, P. A vision based onboard approach for landing and position control of an autonomous multirotor UAV in GPS-denied environments. In Proceedings of the International Conference on Advanced Robotics, Munich, Germany, 22–26 June 2009; pp. 1–6.
7. Forster, C.; Pizzoli, M.; Scaramuzza, D. SVO: Fast semi-direct monocular visual odometry. In Proceedings of the IEEE International Conference on Robotics and Automation, Hong Kong, China, 31 May–7 June 2014; pp. 15–22.
8. Reizenstein, A. Position and Trajectory Control of a Quadcopter Using PID and LQ Controllers. Master's Thesis, Linköping University, Linköping, Sweden, 2017.
9. Bitcraze. Available online: https://store.bitcraze.io/products (accessed on 6 February 2020).
10. Eachine—FPV VTX Camera. Available online: https://www.eachine.com/FPV-SYSTEMS-c-153.html (accessed on 9 February 2020).
11. Ramirez, F.R.; Salinas, R.M.; Carnicer, R.M. Speeded Up Detection of Squared Fiducial Markers. *Image Vision Comput.* **2018**, *76*, 38–47. [CrossRef]
12. Welch, G.; Bishop, G. *TR 95-041: An Introduction to the Kalman Filter*; University of North Carolina: Chapel Hill, NC, USA, 2006; pp. 1–16.
13. Karagiannis, I. Design of Gyro Based Roll Stabilization Controller for a Concept Amphibious Commuter Vehicle. Master Thesis, Royal Institute of Technology, Stockholm, Sweden, 2015.
14. Ziegler, J.G.; Nichols, N.B. Optimum settings for automatic controllers. *Trans. ASME* **1942**, *64*, 759–768. [CrossRef]
15. Garpinger, O.; Hägglund, T. A Software Tool for Robust PID Design. *IFAC Proc. Vol.* **2008**, *41*, 6416–6421. [CrossRef]
16. Åström, K.J.; Hägglund, T. *Advanced PID Control. eng*; ISA—The Instrumentation, Systems, and Automation Society: Durham, NC, USA, 2006.

 *applied sciences*

Article

Motion Planning and Coordinated Control of Underwater Vehicle-Manipulator Systems with Inertial Delay Control and Fuzzy Compensator

Han Han [1], Yanhui Wei [1,*], Xiufen Ye [1] and Wenzhi Liu [2]

[1] College of Automation, Harbin Engineering University, Harbin 150001, China; hanhanheu@163.com (H.H.); yexiufen@hrbeu.edu.cn (X.Y.)

[2] College of Information and Communication Engineering, Harbin Engineering University, Harbin 150001, China; liuwenzhi@hrbeu.edu.cn

* Correspondence: wyheauini@hrbeu.edu.cn; Tel.: +86-150-0464-3531

Received: 28 April 2020; Accepted: 29 May 2020 ; Published: 5 June 2020

Abstract: This paper presents new motion planning and robust coordinated control schemes for trajectory tracking of the underwater vehicle-manipulator system (UVMS) subjected to model uncertainties, time-varying external disturbances, payload and sensory noises. A redundancy resolution technique with a new secondary task and nonlinear function is proposed to generate trajectories for the vehicle and manipulator. In this way, the vehicle attitude and manipulator position are aligned in such a way that the interactive forces are reduced. To resist sensory measurement noises, an extended Kalman filter (EKF) is utilized to estimate the UVMS states. Using these estimates, a tracking controller based on feedback Linearization with both the joint-space and task-space tracking errors is proposed. Moreover, the inertial delay control (IDC) is incorporated in the proposed control scheme to estimate the lumped uncertainties and disturbances. In addition, a fuzzy compensator based on these estimates via IDC is introduced for reducing the undesired effects of perturbations. Trajectory tracking tasks on a five-degrees-of-freedom (5-DOF) underwater vehicle equipped with a 3-DOF manipulator are numerically simulated. The comparative results demonstrate the performance of the proposed controller in terms of tracking errors, energy consumption and robustness against uncertainties and disturbances.

Keywords: underwater vehicle-manipulator system; motion planning; coordinated motion control; inertial delay control; fuzzy compensator; extended Kalman filter; feedback linearization

---

1. Introduction

With increasing interest in the field of marine research, autonomous underwater vehicle manipulator systems (UVMSs) [1] have rapidly developed into important devices for exploring the ocean, completing underwater tasks, underwater sampling and so on. It is a challenging problem to accurately control the UVMS in an energy-efficient manner due to the kinematic redundancy and underwater environment with hydrodynamic uncertainties, unknown external disturbances (such as ocean currents) and inaccurate sensor information. For solving these problems, inverse kinematics and robust coordinated control techniques have been developed for the UVMS.

For the inverse kinematics of the UVMS, the solution can be obtained through mapping the end-effector's velocities to the velocities of the vehicle and manipulator. As the UVMS has redundant degrees of freedom (DOFs), there are various combinations of vehicle and manipulator velocities without affecting the end-effector velocities. A common solution is to adopt the pseudo-inverse Jacobian matrix of the UVMS or its weighted form [2]. However, this method is not desirable for redundant exploration to avoid joint limits, improve system manipulability or save energy. Therefore, the task-priority redundancy resolution technique [3] was proposed in such a way that the fulfillment of the primary task has a higher priority than that of a secondary task. Generally, the secondary task is to optimize the performance index through assigning additional motion in the null space of the primary task. Sarkar and Podder [4] solved the inverse kinematics of the UVMS on the acceleration level to minimize the total hydrodynamic drag; however, the performance index of this method requires dynamic equations which can not be modeled exactly. Han et al. [5] proposed a new performance index designed to minimize restoring moments without using dynamic equations. However, this method was implemented for a specific configuration of the UVMS.

The task-priority strategy can be extensible to chain multiple tasks which have a lower order of priority (Siciliano and Slotine [6]). Antonelli et al. [7] used a fuzzy inference system (FIS) to handle multiple secondary tasks, such as reduction of fuel consumption and improvement of system manipulability. In such a way, a secondary task can be activated by FIS when the corresponding variable is without the safe range. Wang et al. [8] used a fuzzy logic algorithm to decide the priorities of secondary objectives, such as manipulator singularity avoidance and attitude optimization of the UVMS. The experimental validation of three difference kinematic control schemes was presented in [9]. In [10], a multitask kinematic control of the underwater biomimetic vehicle-manipulator system (UBVMS) was designed. A unifying framework for the kinematic control of UVMSs was proposed in [11]. A very recent work dedicated to motion planning for the UVMS was presented in [12].

To achieve trajectory tracking, it is very important to design a coordinated motion controller for the UVMS. The simple control methods (e.g., proportional-integral-derivative (PID) control) are not suitable for the UVMS due to the inherent nonlinear and coupled dynamics of the system [13–15]. Schjlberg and Fossen [16] proposed a control strategy in terms of feedback linearization. Sarkar and Podder [4] utilized a computed torque controller (CTC) for trajectory tracking of the UVMS. Taira et al. [17] proposed a model-based motion control for the UVMS, which can be applicable to three types of servo systems; i.e., a voltage-controlled, a torque-controlled and a velocity-controlled servo system. Korkmaz et al. [18] presented a trajectory tracking control for an underactuated underwater vehicle manipulator system (U-UVMS) based on the inverse dynamics. However, these model-based controllers are poor in terms of robustness against model uncertainties. In [19], a fuzzy logic control method was designed for a hybrid-driven UVMS to grasp marine products on the seabed. A model reference adaptive control approach for an UVMS was proposed in [20]. Antonelli et al. [21] proposed an adaptive controller based on virtual decomposition; however, a regressor matrix corresponding to parameter vector is required in this method. An indirect adaptive controller based on the extended Kalman filter (EKF) was proposed in [22]; meanwhile the performance would be degraded due to the estimated error via EKF. To eliminate the bias from the EKF estimation, Dai et al. [23] introduced a H∞ control in the indirect adaptive controller to achieve robust performance. However, this method results in a residual error when the bounds of the disturbance cannot be known prior. To reduce or omit the estimation error of the uncertainties and disturbances, a fuzzy compensator based on estimations was utilized [24]. To handle state and input constraints of the UVMS, a robust predictive control (RMPC) [24], a nonlinear MPC (NMPC) [25], a tube-based robust MPC [26], a fast MPC (FMPC) [27] and a fast tube MPC (FTMPC) [28] were used for the UVMS trajectory tracking, but these MPC approaches do not permit the self-motion utilized to perform energy efficient trajectory tracking.

For achieving precise and robust performance, the control design should be enhanced by the estimations of the system's uncertainties and disturbances. The popular estimation techniques include time delay control (TDC) [29,30], the extended state observer (ESO) [31], the disturbance observer [32], the nonlinear disturbance observer (NDO) [33,34] the uncertainty and disturbance estimator (UDE) [35,36] (redefined as inertial delay control (IDC) [37]) and so on. Among them, because IDC is simple in design and easy to complete, it is widely used to estimate the effect of the lumped uncertainties and disturbances. Generally, the IDC is applied to the sliding mode control (SMC) for ensuring precise and robust performance. The combined method does not require the bounds of uncertainties and disturbances, and it does not use the discontinuous function in the control law.

However, the above-mentioned methods are based on joint-space variables, which may not be suitable for a variety of underwater tasks with high-precision end-effector position requirements [38]. These task space control schemes can easily adapt to the online modification of the end-effector's motion [39,40]. However, the task-space controllers also have disadvantages. (a) The kinematic redundancy of the UVMS cannot be exploited. (b) The output of the task-space controller should be mapped into the joint space so as to be realized by thrusters and actuators. Li et al. [34] proposed a hybrid strategy-based coordinated controller for the UVMS. The hybrid strategy is to transform the joint-space controller (to exploit the system's redundancy) to the task-space controller (to ensure high-accuracy tracking performance).

Inspired by the above studies, new motion planning and coordinated control schemes of the UVMS are proposed in this paper. The contribution of this work is that the proposed scheme can ensure precise, energy-efficient and robust performance in the presence of model uncertainties, external disturbances, payload and sensory noises. First, a new redundancy resolution technique is proposed, where a new secondary task with a nonlinear function is inserted for generating energy-saving trajectories for the vehicle and manipulator. Second, an EKF estimation system is employed for resisting sensory noises. Third, a coordinated motion control with joint-space errors, end-effector errors, IDC and a fuzzy compensator is proposed as a robust tracking controller against uncertainties and disturbances. Last, the effectiveness of the proposed scheme is verified through numerical simulations.

The rest of the paper is organized as follows. Section 2 is concerned with the kinematic and dynamic modeling of the UVMS. In Section 3, an improved redundancy resolution technique is presented. The proposed control scheme is proposed in Section 4. Numerical simulations and the detailed performance analysis are presented in Section 5. Section 6 holds the conclusions.

2. Modeling

The UVMS investigated in this paper is composed of an underwater vehicle with a 3 DOFs manipulator. The coordinate system of the UVMS is shown in Figure 1. In the body-fixed frame $\{B\}$, we define that the vectors of vehicle's linear and angular velocities are v_1 and v_2, where $v_1 = [u, v, w]^T$, $v_2 = [p, q, r]^T$ and $v = [v_1^T, v_2^T]^T$. The vector of joint positions is assumed to be $q = \left[q_1, q_2 \cdots q_n\right]^T$, where n is the number of manipulator's joints. The position and orientation vector of the UVMS relative to the body-fixed frame is assumed to be $\zeta = [v_1^T, v_2^T, \dot{q}^T]^T$. In the inertial frame $\{I\}$, the vectors of end-effector's position and orientation are defined as η_{E1} and η_{E2}, and assume $x_E = [\eta_{E1}^T, \eta_{E2}^T]^T$.

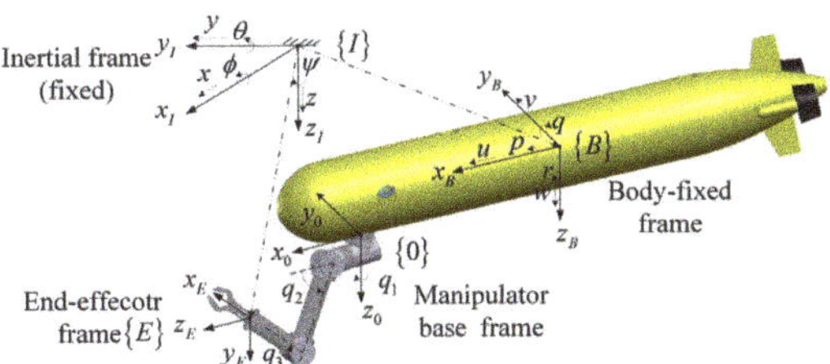

Figure 1. Coordinate systems of the underwater vehicle manipulator system (UVMS).

The kinematic model of the UVMS [21] can be obtained as shown in (1), where the velocities of the UVMS in the body-fixed frame (ζ) are mapped into end-effector velocities ($\dot{x}_E$) via $J(R_B^I, q)$.

$$\dot{x}_E = \begin{bmatrix} R_B^I & S(R_B^I p_0^B + R_0^I p_E^0) R_B^I & J_{pos,q}^I \\ 0_{3\times3} & R_B^I & J_{ori,q}^I \end{bmatrix} \left[v_1^T, v_2^T, \dot{q}^T\right]^T = J(R_B^I, q)\zeta \tag{1}$$

where $J(R_B^I, q) \in \mathbb{R}^{6\times(6+n)}$, $R_0^I = R_B^I R_0^B$, $J_{pos,q}^I = R_0^I J_{pos,q}^0$ and $J_{ori,q}^I = R_0^I J_{ori,q}^0$. R_0^B is the rotational matrix describing the transformation from the manipulator's base frame to the body-fixed frame, and p_E^0 is the position vector from manipulator's base to the center of the body-fixed frame. p_E^0 is the position vector from end-effector to the manipulator's base. $J_{pos,q}^0$ is the manipulator's linear Jacobian matrix, and $J_{ori,q}^0$ is the manipulator's angular Jacobian matrix. $S(\cdot)$ is the cross-product operator.

The vectors of vehicle's position and attitude relative to the inertial frame are defined as η_1 and η_2, where $\eta_1 = [x, y, z]^T$, $\eta_2 = [\phi, \theta, \psi]^T$ and $\eta = [\eta_1^T, \eta_2^T]^T$. The velocity vector of the UVMS defined in the body-fixed frame (ζ) can be obtained by (2).

$$\zeta = \begin{bmatrix} R_I^B & 0_{3\times3} & 0_{3\times3} \\ 0_{3\times3} & J_v & 0_{3\times3} \\ 0_{3\times3} & 0_{3\times3} & I_{3\times3} \end{bmatrix} \begin{bmatrix} \dot{\eta}_1 \\ \dot{\eta}_2 \\ \dot{q} \end{bmatrix} = J_\xi(\eta_2)\dot{\xi} \tag{2}$$

where $v_1 = R_I^B\dot{\eta}_1$, $v_2 = J_v\dot{\eta}_2$ and $\xi = [\eta_1^T, \eta_2^T, q^T]^T$. R_I^B is the linear rotational matrix describing the transformation from the inertial frame to the body-fixed frame, and J_v is the angular rotational matrix. The values of R_I^B and J_v can be referred to the literature [41]. $J_\xi(\eta_2)$ is the Jacobian matrix which relates the vehicle velocities with respect to the inertial frame and the body-fixed frame.

Dynamic Modeling

The nonlinear dynamic equations of the UVMS expressed in the body-fixed frame $\{B\}$ can be established as [16,21]:

$$M(q,\eta)\dot{\zeta}+C(q,\zeta)\zeta+D(q,\zeta)\zeta+g(q,R_I^B)=\tau+\tau_{dis} \tag{3}$$

where

$$M(q,\eta) = \begin{bmatrix} M_v + H_1(q) & H_2(q) \\ H_2^T(q) & M_m(q) \end{bmatrix}, \quad C(q,\zeta) = \begin{bmatrix} C_v(v) + C_1(q,\dot{q},v) & C_2(q,\dot{q}) \\ C_3(q,\dot{q},v) & C_m(q,\dot{q}) \end{bmatrix}$$

$$D(q,\zeta)=\begin{bmatrix} D_v(v)+D_1(q,\dot{q},v) & D_2(q,\dot{q},v) \\ D_3(q,\dot{q},v) & D_m(q)+D_4(q,\dot{q},v) \end{bmatrix}, \quad g(q,R_I^B) = \begin{bmatrix} g_v(\eta) + g_1(q) \\ g_m(q) \end{bmatrix}, \quad \tau = [\tau_v^T, \tau_m^T]^T$$

where $M(q,\eta) \in \mathbb{R}^{(6+n)\times(6+n)}$ is the inertia matrix including added mass terms, and $H_1(q)$ and $H_2(q)$ are matrices of the inertia effects due to the manipulator. $C(q,\zeta) \in \mathbb{R}^{(6+n)}$ is the Coriolis and centripetal matrix, and $C_i(q,\dot{q},v)(i = 1,3)/C_2(q,\dot{q})$ is the matrix of Coriolis and centripetal forces due to the coupling effects/due to the manipulator. $D(q,\zeta)\zeta \in \mathbb{R}^{(6+n)}$ is the vector of dissipative effects, and $D_i(q,\dot{q},v)$ $(i = 1\cdots4)$ is the matrix of drag effects due to the coupling effects. $g(q,R_B^I) \in \mathbb{R}^{(6+n)}$ is the vector of gravity and buoyancy effects, $g_v(\eta)$ is the restoring vector of the vehicle, $g_m(q)$ is the restoring vector of the manipulator and $g_1(q)$ is the restoring vector due to the manipulator. $\tau \in \mathbb{R}^{(6+n)}$ is the vector of generalized forces. τ_{dis} is the vector of disturbances. Generally, in a deep water environment, τ_{dis} comes from ocean currents, payload, etc. In particular, time-varying ocean currents increase the uncertainty of the UVMS hydrodynamic forces, making accurate control of the UVMS difficult.

As for the underwater manipulator, it is assumed that its links are composed of cylindrical elements. The hydrodynamic effects on cylinders can be referred to [16]. For a cylinder, the inertial matrix of added mass and added moment is a diagonal matrix, while the off-diagonal elements are neglected. The drag force can be expressed by a nonlinear function related to the velocity vector of the center of mass of the link. Generally, when calculating the hydrodynamic forces, the linear skin-friction force, quadratic drag force and lift force are considered. The third-order and higher order terms of the drag forces are neglected. In addition, based on the assumption that velocity of the ocean current is constant, the diffraction forces can be neglected.

In the real system, the above model parameters are usually difficult to accurately measure or estimate, especially the hydrodynamic forces acting on the UVMS. Thus, it is advisable to divide the model parameters into two parts: the normal value part and the bias part. The normal value is denoted as $(\cdot)^*$, which can be obtained through using strip theory, pool experiment analysis or CFD computation. The bias term is denoted as $\Delta(\cdot)$, which describes the difference between the real value and the nominal value. Then, (4) can be obtained. For control design, the normal values are available, while the bias parts are considered as model parameter uncertainties.

$$M = M^* + \Delta M, \quad C = C^* + \Delta C, \quad D = D^* + \Delta D \tag{4}$$

Considering that the vehicle is driven by thrusters and the manipulator is driven by motors, the generalized force vector τ is related to the vector of thruster forces and actuator torques F_{td} through (5).

$$\tau = \begin{bmatrix} B_v & 0_{6 \times n} \\ 0_{n \times p_v} & I_n \end{bmatrix} F_{td} = BF_{td} \tag{5}$$

where $F_{td} = [T^T, \tau_m^T]^T \in \mathbb{R}^{p_v + n}$. $T \in \mathbb{R}^{p_v}$ represents the vector of thruster forces, and $\tau_m \in \mathbb{R}^n$ represents the vector of actuator torques. $B_v \in \mathbb{R}^{6 \times p_v}$ is the thruster configuration matrix, and $B \in \mathbb{R}^{(6+n) \times (6+n)}$ is the thruster-actuator configuration matrix. It is known that for an under-actuated underwater vehicle, $p_v < 6$. Generally, for a manipulator, n joint motors are all available.

3. Proposed Redundancy Resolution

This section proposes a new redundancy resolution technique to generate energy-efficient trajectories for the vehicle and the manipulator. It is known that infinite solutions of the UVMS inverse kinematics can be obtained by inverting the mapping (1). The solution using the pseudo inverse of the Jacobian matrix is expressed as [2]

$$\zeta = J^+(R_B^I, q)\dot{x}_E \tag{6}$$

where $\dot{x}_E$ is the end-effector velocity vector. $J^+(R_B^I, q)$ is the pseudo inverse of the Jacobian matrix and $J^+(R_B^I, q) = J^T(R_B^I, q)(J(R_B^I, q)J^T(R_B^I, q))^{-1}$.

However, this solution does not exploit the redundant DOFs of the system, and it is not suitable from the perspective of energy consumption. Therefore, a new task-priority redundancy resolution technique is proposed in this section. In the proposed technique, the primary task is to map the end-effector variables into the joint-space variables, and two secondary tasks are provided to explore the kinematic redundancy for energy savings, joint limit avoidance and small roll and pitch angles kept for the vehicle, as shown in (7).

$$\zeta_d = J_W^+(\dot{x}_{Ed} - K_f e_E) + (I - J_W^+ J_W)[J_s^+(\eta, q)(\dot{\eta}_{sd} - K_s e_s) - \alpha K_\zeta J_\xi(\eta_2)\dot{\xi}] \tag{7}$$

where $J_W^+ = W^{-1}J^T(JW^{-1}J^T)^{-1}$ is the weighted pseudo-inverse Jacobian matrix. J is considered as the primary task Jacobian matrix and $W \in \mathbb{R}^{(6+n) \times (6+n)}$ is the motion distribution matrix with elements belonging to $[0, 1]$. When the diagonal elements of the former three rows of W are close to 1, the diagonal elements of the later n rows of W will be close to 0. This results in greater movement of the vehicle and less movement of the manipulator. Otherwise, when the diagonal elements of the former three rows of W are close to 0, the diagonal elements of the later n rows of W will be close to 1. This results in less movement of the vehicle and greater movement of the manipulator. The diagonal elements of the middle three rows of W correspond to the movement of the vehicle's attitude. The larger they are, the greater the movement of the vehicle's attitude. The off-diagonal elements of W describe the degrees of the coupling effects between the DOFs of the UVMS, which can refer to our previous work [15]. The closer the off-diagonal element of W to 1, the greater the corresponding coupling motion. $J_s(\eta, q)$ and $J_\zeta(\eta_2)$ are the secondary task Jacobian matrices. It can be recognized that the secondary tasks are fulfilled in the null space, which will not affect the motion of the primary task. $\dot{x}_{Ed}$ is the primary-task vector and $e_E = x_E - x_{Ed}$ is the error of the primary task. $\dot{\eta}_{sd}$ is a secondary-task vector to achieve system coordination between its rotational subsystem and translational subsystem, including the vehicle attitudes and joint angles, and $e_s = \dot{\eta}_s - \dot{\eta}_{sd}$ is its error. K_f and K_s are positive definite matrices. The other secondary task vector is the velocity vector of the UVMS $\dot{\zeta} = [\dot{\eta}_1^T, \dot{\eta}_2^T, \dot{q}^T]^T$, which contributes to the system's self-motion utilized for reducing energy requirements. K_ζ is a diagonal matrix whose elements belong to $[0, \infty)$. The larger the diagonal element of K_ζ, the greater the corresponding coupling motion. For instance, if the diagonal element of the secondary row of K_ζ is larger, the vehicle will have a larger yaw angle according to the coupling effects. Similarly, if the diagonal element of the third row of K_ζ is larger, the vehicle will have a larger pitch angle. α is a coefficient belonging to $[0, 1]$, which is used to adjust the values of K_ζ.

To effectively utilize the self-motion during the entire UVMS motion, α is defined as a nonlinear function related to time $t \geq 0$, as given in (8).

$$\alpha = \begin{cases} -0.5(1 - e^{-\lambda(t-t_s)}) + 0.5 & t > t_s \\ 0.5(1 - e^{-\lambda(t-t_s)}) + 0.5 & t \leq t_s \end{cases} \tag{8}$$

where t_s relatives to the time at which the system enters deceleration phase. λ is the coefficient and $\lambda > 0$.

The curve of the nonlinear function is shown in Figure 2. It can be recognized that the smaller the value of λ, the smoother the variations of the nonlinear function. Therefore, it is better to choose a small value of λ to ensure smooth movement of the UVMS.

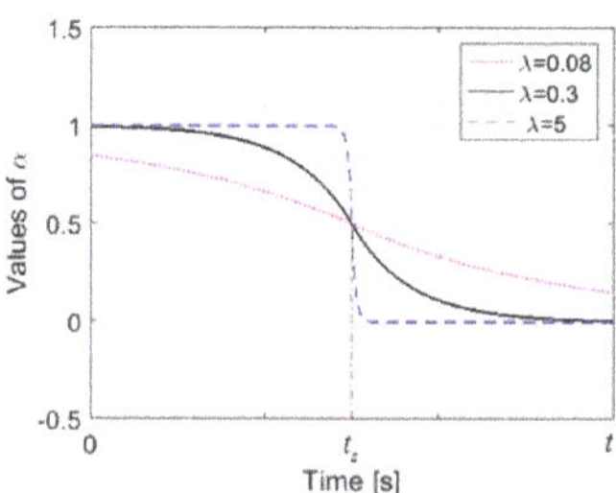

Figure 2. Nonlinear function of α.

4. Control Design

The purpose of control design is to obtain the values of thruster forces and actuator torques in order to drive the UVMS to the desired trajectory. In addition, the robustness of the designed controller is important in the presence of model parameter uncertainties, time-varying external disturbances, payload variations and sensory noises. In this section, for precise and robust control of the UVMS, a new coordinated motion controller including inertial delay control (IDC) and a fuzzy compensator is proposed. Besides, the proposed controller uses the estimated UVMS states via an EKF.

4.1. Design of an EKF

Due to the presence of sensory measurement noises, the vehicle and manipulator positions measured by sensors are not inaccurate. Therefore, it is necessary to utilize a nonlinear filter to estimate the system's states. As the extended Kalman filter (EKF) is simple and easy to complete and has low computational complexity, the EKF is used in this study. It is necessary to obtain a linear model during the KF design process. The dynamic equations of the UVMS can be linearized by ignoring the higher order terms in the expended Taylor series. The state vector of the system, e.g., position/attitude and velocity vectors, is defined as $X = [\xi^T, \zeta^T]^T$. The measurement model can be expressed as $Z = h(X) = \xi$.

Based on (2) and (3), the time derivative of the system state vector X can be obtained as

$$\dot{X} = f(X, t) = \begin{bmatrix} J_\xi^{-1}(\eta_2)\dot{\zeta} \\ M^{-1}(\tau - \tau_d - b) \end{bmatrix} \tag{9}$$

where $f(X, t)$ is considered as the estimated model of the system.

With the additive Gaussian white noise, the predicted system state vector $\hat{X}_{k+1}^-$ at $t = t_{k+1}$ is given in (10), and the predicted measurement state vector Z_{k+1}^- at $t = t_{k+1}$ is shown in (11). Then the covariance matrix of the predicted state vector can be obtained as shown in (12).

$$\hat{X}_{k+1}^- = \hat{X}_k + f(\hat{X}_k, t) + Q_k \tag{10}$$

$$Z_{k+1}^- = h(\hat{X}_{k+1}^-) + \Gamma_k \tag{11}$$

$$P_{k+1}^- = \phi_k P_k \phi_k^T + Q_k \tag{12}$$

where $\hat{X}_k$ is the predicted state vector at t_k. Q_k is the vector of system noises. Γ_k is the vector of measurement noises. P_k is the covariance matrix of the estimated system states.

Therefore, the estimated system states can be obtained through the following correction step:

$$K_{k+1} = P_{k+1}^- H_{k+1}^T (H_{k+1} P_{k+1}^- H_{k+1}^T + \Gamma_{k+1})^{-1} \tag{13}$$

$$\hat{X}_{k+1} = \hat{X}_{k+1}^- + K_{k+1}(Z_{k+1} - Z_{k+1}^-) \tag{14}$$

$$P_{k+1} = (I - K_{k+1} H_{k+1}) P_{k+1}^- \tag{15}$$

where $\phi_k = \left.\dfrac{\partial f(X,t)}{\partial X^T}\right|_{X=\hat{X}_k}$ and $H_{k+1} = \left.\dfrac{\partial h(X)}{\partial X^T}\right|_{X=\hat{X}_{k+1}^-}$.

4.2. Design of a Tracking Controller

The control objective is to make sure that the tracking errors quickly converge to zero under the conditions of model parameter uncertainties, time-varying external disturbances and payload. First, the tracking errors of the system are defined. The vector of end-effector tracking errors is shown in (16), and the vectors of tracking errors in the joint space are shown in (17)–(19).

$$e_E = \hat{x}_E - x_{Ed} \tag{16}$$

$$e = \begin{bmatrix} R_I^B(\hat{\eta}_1 - \eta_{1d}) \\ \eta \varepsilon_d - \eta \varepsilon_d + S(\varepsilon)\varepsilon_d \\ \hat{q} - q_d \end{bmatrix} \tag{17}$$

$$\dot{e} = \hat{\zeta} - \zeta_d \tag{18}$$

$$\ddot{e} = \dot{\hat{\zeta}} - \dot{\zeta}_d \tag{19}$$

where the superscript $(\hat{\cdot})$ denotes the corresponding estimated values via EKF, and the subscript $(\cdot)_d$ denotes the corresponding desired values. $[\varepsilon, \eta]$ and $[\varepsilon_d, \eta_d]$ are the quaternions of $\hat{\eta}_2$ and η_{2d}.

Then, the tracking controller based on feedback linearization is given as

$$u = -M^*(k_1 e + k_2 \dot{e} - \dot{\zeta}_d) + C^* \hat{\zeta} + D^* \hat{\zeta} + g - J^T k_E e_E - \hat{\delta} \tag{20}$$

where $J = J(R_B^I, q)$ is the Jacobian matrix of the system, as given in (1). k_1, k_2 and k_E are the positive symmetric matrices. $\hat{\delta}$ is the estimated vector of the lumped uncertainties and disturbances, which is described in the next subsection.

4.3. Inertial Delay Control (IDC)

Due to the underwater circumstances, the dynamic equations of the UVMS include unknown external disturbances and an amount of parameter uncertainties caused by identification errors. These lumped uncertainties and disturbances can be expressed as (21) with reference to (3) and (4).

$$\delta = -(\Delta M \hat{\dot{\zeta}} + \Delta C \hat{\zeta} + \Delta D \hat{\zeta}) + \tau_{dis} \tag{21}$$

where $\hat{\dot{\zeta}}$ and $\hat{\zeta}$ denote the estimates of the system states via EKF.

Then, based on (3), (4) and (21), the acceleration vector of the system can be obtained as

$$\dot{\zeta} = -(M^*)^{-1}(C^*\hat{\zeta} + D^*\hat{\zeta} + g) + (M^*)^{-1}(u + \delta) \tag{22}$$

Substituting the proposed control law (20) in (22), dynamical equation of the tracking errors is

$$\ddot{e} + k_2\dot{e} + k_1 e + (M^*)^{-1}J^T k_E e_E = (M^*)^{-1}e_\delta \tag{23}$$

where $e_\delta = \delta - \hat{\delta}$ is the estimated error vector.

It is assumed that a slow-varying signal can be approximated and estimated by a filter with appropriate bandwidth [37]. Based on this assumption, the uncertainty and disturbance estimator (UDE) is proposed for estimating slow-varying uncertainties [35,37]. Then, the estimations of lumped uncertainties and disturbances $\hat{\delta}$ can be given as

$$\hat{\delta} = G_f(s)\delta \tag{24}$$

where $G_f(s)$ is a strictly proper low-pass filter possessing a uniform steady-state gain and a sufficiently large bandwidth. Based on (24), it is found that by passing the lumped uncertainties and disturbances δ through a inertial filter $G_f(s)$, the estimation vector $\hat{\delta}$ can be obtained. The UDE method is redefined as inertial delay control (IDC) [37], because it is analogous to the time delay control (TDC) which delays the plant signals in time to obtain the estimates.

Based on (23) and (24), we can obtain

$$\hat{\delta} = G_f(s)[M^*(\ddot{e} + k_2\dot{e} + k_1 e) + J^T k_E e_E + \hat{\delta}] \tag{25}$$

A choice of $G_f(s)$ with first order is given by

$$G_f(s) = \frac{I}{I + Ts} \tag{26}$$

where T is a diagonal matrix with small positive constant. I is the identity matrix.

Then (25) can be rewritten as

$$T\hat{\dot{\delta}} + \hat{\delta} = M^*(\ddot{e} + k_2\dot{e} + k_1 e) + J^T k_E e_E + \hat{\delta} \tag{27}$$

Therefore, the estimates of the lumped uncertainties and disturbances can be obtained as

$$\hat{\delta} = T^{-1}M^*(\dot{e} + k_2 e + k_1 \int_0^t e \, dt) + T^{-1}J^T k_E \int_0^t e_E \, dt \tag{28}$$

From (25) and (26), the equation of estimated errors can be written as

$$\dot{e}_\delta = -T^{-1}e_\delta + \dot{\delta} \tag{29}$$

If the lumped uncertainties and disturbances δ are slowly varying, then $\dot{\delta}$ is small and $\dot{\delta} \approx 0$. Therefore, the estimated errors (e_δ) go to zero asymptotically. If $\dot{\delta}$ is not small, but $\ddot{\delta}$ is small, e_δ is ultimately bounded and the estimated accuracy can be improved by estimating δ and $\dot{\delta}$.

4.4. Fuzzy Compensator

Based on the estimates via IDC, the fuzzy compensator is given as

$$u_{fuzzy} = \rho\hat{\delta} + \epsilon \tag{30}$$

where $\rho = \mathrm{diag}(\rho_1, \rho_2 \cdots \rho_{6+n})$ is the parameter of the fuzzy compensator, and ϵ is a constant vector.

The fuzzy compensator is a multiple-inputs-single-output fuzzy logic controller (FLC) with the joint-space system errors e_i and e_j as two input variables and ρ_i after defuzzification and denormalization as an output variable. Denote the system error vector e (as given in (17)) as $e = [e_1, e_2 \cdots e_i \cdots e_{6+n}]$. The main advantage of this fuzzy compensator is that the required fuzzy rules take the dynamic coupling between the vehicle and the manipulator [15,16] into account. It is known that the roll, pitch and yaw motions of the vehicle are coupled with its surge, sway and heave motions. As the roll and pitch angles should be kept small for properly working of the bottom sensors, it is assumed that the surge and sway motions are mostly affected by the yaw angle. Note that the pitch and heave motions are interactive, and the manipulator's joints 2 and 3 are interactive. The position of manipulator's joint 1 is mostly affected by the sway motion. Based on these analysis, the fuzzy rules are given in Table 1. Table 2 shows the relationships between an output and two input variables.

Table 1. Rule base for ρ_i.

ρ_i \ $|e_j|$ / $|e_i|$	ZE	PS	PM	PB
ZE	ZE	ZE	PS	PM
PS	PS	PS	PM	PB
PM	PM	PM	PB	PB
PB	PB	PB	PB	PB

The following symbols are used in Table 1: ZE (zeros), PS (positive small), PM (positive medium), PB (positive big). Figure 3a,b shows the member functions of the normalized input and output variables respectively. After the fuzzification stage, the Mamdani inference method is used for fuzzy implication, and then the centroid method is used for defuzzification. Finally, based on denormalization the actual output variables can be obtained.

Table 2. Relationships between two input variables and an output variable.

inputs	$|e_i|$ / $|e_j|$	$|e_1|$ / $|e_6|$	$|e_2|$ / $|e_6|$	$|e_3|$ / $|e_5|$	$|e_5|$ / $|e_3|$	$|e_6|$ / $|e_2|$	$|e_7|$ / $|e_2|$	$|e_8|$ / $|e_9|$	$|e_9|$ / $|e_8|$
outputs	ρ_i	ρ_1	ρ_2	ρ_3	ρ_5	ρ_6	ρ_7	ρ_8	ρ_9

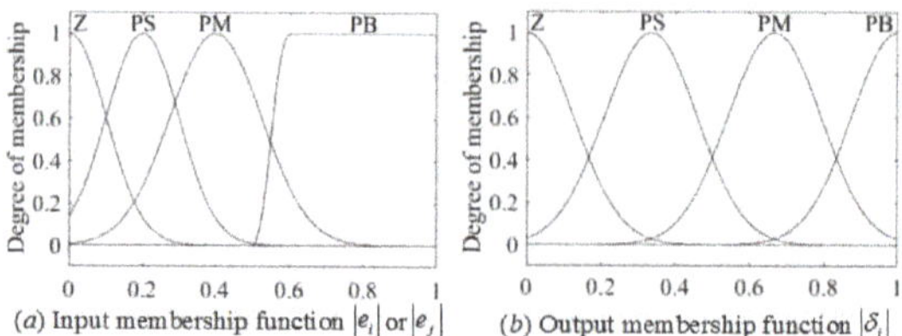

Figure 3. Input and output membership functions.

Incorporating the fuzzy compensator, the proposed coordinated motion controller is given as

$$u_c = u - u_{fuzzy} \tag{31}$$

Then we can obtain the vector of thruster forces and actuator torques F_{td}, as shown in (32).

$$F_{td} = B^+ u_c = \begin{bmatrix} B_v^+ & 0_{p_v \times n} \\ 0_{n \times 6} & I_n \end{bmatrix} u_c \tag{32}$$

where B_v^+ is the pseudo inverse of B_v and B^+ is the pseudo inverse of B.

Therefore, the generalized force vector τ can be obtained based on (5). For an under-actuated UVMS, $\tau = u$ except that the elements of τ corresponding to the underacted motions are zeros.

The proposed control system is schematically represented by a block diagram in Figure 4. The controller block includes five sub blocks to calculate a control vector; i.e., the tracking controller, IDC, fuzzy compensator, B^+ and B blocks. In addition to system dynamics, the tracking controller also requires tracking errors of end-effector positions and joint-space states. The end-effector position tracking errors are calculated according to the desired end-effector positions derived from the trajectory planning block and the estimated end-effector positions obtained from the forward kinematics block using the estimated joint-space states. The estimates of joint-space states are obtained from the EKF block. The proposed redundancy resolution block generates the required joint-space trajectories for the desired tasks. The IDC block estimates the lumped uncertainties and disturbances of the system. The fuzzy compensator reduces the influences of perturbation on the UVMS.

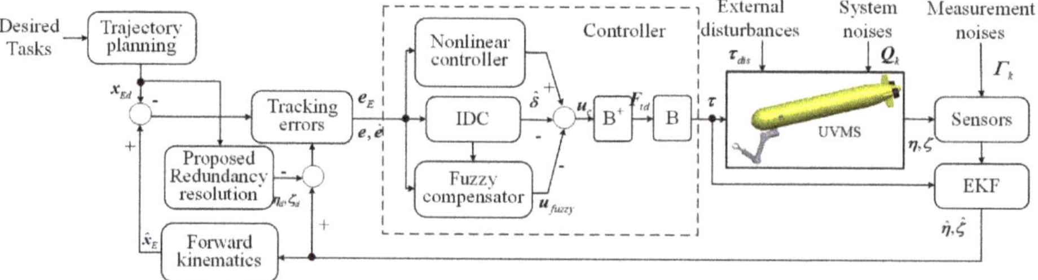

Figure 4. Block diagram of the proposed controller.

4.5. Stability Analysis

We define a Lyapunov function which is positive definite as:

$$V(e, \dot{e}, e_E, e_\delta) = \frac{1}{2}e^T k_1 k_2 e + \frac{1}{2}\dot{e}^T k_2 \dot{e} + \frac{1}{2}e_\delta{}^T e_\delta + \frac{1}{2}e_E^T k_2 (M^*)^{-1} k_E e_E \tag{33}$$

Differentiating $V(x_1, x_2, e_E)$ yields

$$\dot{V}(e,\dot{e},e_E,e_\delta)=e^T k_1 k_2 \dot{e}+\dot{e}^T k_2 \ddot{e}+e_\delta{}^T \dot{e}_\delta+e_E^T k_2 (M^*)^{-1} k_E \dot{e}_E \tag{34}$$

where M^* is assumed to be constant.

Substituting the proposed control law (31) in (22), and taking into account (29), (34) can be rewritten as

$$\begin{aligned}
\dot{V}&=e^T k_1 k_2 \dot{e}+e_E^T k_2 (M^*)^{-1} k_E \dot{e}_E+e_\delta{}^T(-T^{-1}e_\delta+\dot{\delta})\\
&\quad+\dot{e}^T k_2(-k_2\dot{e}-k_1 e-(M^*)^{-1}J^T k_E e_E+(M^*)^{-1}(e_\delta-\rho\hat{\delta}-\epsilon))\\
&=-\dot{e}^T\|k_2\|^2\dot{e}-e_\delta{}^T T^{-1}e_\delta+e_\delta{}^T\dot{\delta}+\dot{e}^T k_2(M^*)^{-1}((1+\rho)e_\delta-\rho\delta-\epsilon)\\
&=-\dot{e}^T k_2[k_2\dot{e}-(M^*)^{-1}((1+\rho)e_\delta-\rho\delta-\epsilon)]-e_\delta(T^{-1}e_\delta-\dot{\delta})
\end{aligned} \tag{35}$$

By choosing large enough values of k_2, small values of ρ and small enough values of T such that

$$k_2\dot{e}\geq (M^*)^{-1}((1+\rho)e_\delta-\rho\delta-\epsilon),\ \|\rho\delta\|\leq\sigma,\ T^{-1}e_\delta\geq\dot{\delta}, \tag{36}$$

$\dot{V}(e,\dot{e},e_E,e_\delta)$ is negative semi-definite, where $\sigma\rightarrow 0$ is a vector with smaller positive values. Consequently, the tracking errors and estimated errors of the system all converge to zero asymptotically; i.e.,

$$\lim_{t\rightarrow\infty}e\rightarrow 0,\ \lim_{t\rightarrow\infty}\dot{e}\rightarrow 0,\ \lim_{t\rightarrow\infty}e_E\rightarrow 0,\ \lim_{t\rightarrow\infty}e_\delta\rightarrow 0 \tag{37}$$

Therefore, the closed-loop system is asymptotically stable in the entire state space.

5. Simulation Studies

To verify the performance of the proposed technique, numerical simulations were performed on a UVMS with a torpedo-type AUV and a 3-DOF underwater manipulator [15] shown in Figure 1. The AUV is driven by five thrusters in total and its thruster configuration is shown in Figure 5. The thruster configuration matrix B_v and its pseudo inverse B_v^+ are shown in the Appendix A. The parameters for the AUV and manipulator are given in the Appendix A and Tables 3 and 4. From Table 4, it can be seen that the whole system is neutrally buoyant, while the manipulator's links have negative buoyancy. Thus, the UVMS used for numerical simulations in this paper can approximate to the real system.

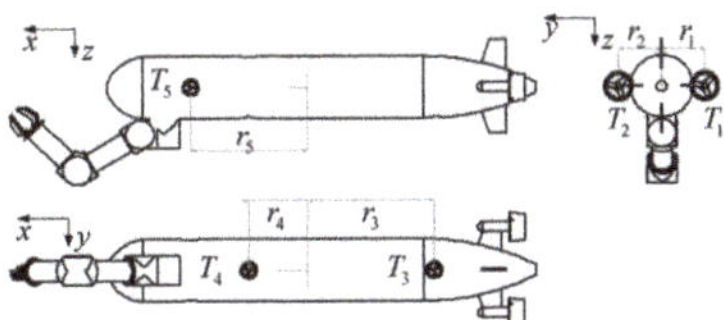

Figure 5. Thruster distribution of the AUV.

Table 3. D-H parameters of the manipulator.

Joint (k)	Offset (α_{k-1})	Length (a_{k-1})	Distance (d_k)	Angle (θ_k)
1	$0°$	0	0	q_1
2	$90°$	l_{q1}	0	q_2
3	$0°$	l_{q2}	0	q_3
4	$0°$	l_{q3}	0	$q_4=0°$

Table 4. The list of UVMS parameters.

	Length (m)	Diameter (m)	Mass (kg)	Buoyancy (N)	Weight (N)
AUV	1.78	0.26	78.2	771.59	767.14
Link1	0.1	0.06	1.28	11.43	12.54
Link2	0.3	0.0425	1.92	17.2	18.87
Link3	0.3	0.0425	1.92	17.2	18.87

5.1. Simulation Conditions

In the simulations, the UVMS's end-effector is commanded to follow a spatial circle with diameter 2.24 m and a straight line of length 7.0 m. The simulation time of the circular trajectory is 50 s, where the initial 10 s is used for acceleration, 30 s is used to follow the circle and the final 10 s is used for deceleration. The simulation time of the straight-line trajectory is 30 s, where in the initial 10 s the acceleration is a half-period sine function, and then it maintains zeros, and in the final 10 s it is a half-period negative sine function. The UVMS's end-effector maintains orientation during the two trajectory tracking tasks. The initial desired positions and orientations are the same as the initial actual positions and orientations. The initial desired and actual velocities and accelerations are zeros. The average speeds of the two trajectories both are 0.23 m/s. The sampling time for the simulation is $t = 20$ ms.

In this case, the model uncertainties, external disturbances, payload and sensory noises in position and orientation measurements are introduced for simulating the real working environment. To reflect the uncertainties, it is assumed that the modeling inaccuracy for each parameter is 10%. The vector of time-varying ocean currents in the inertial frame is assumed to be governed by (38). It is supposed that the end-effector of the manipulator is attached with a payload of 1 kg (in water). The following sensory noises are introduced: gaussian noise of 0.01 m mean and 0.01 m standard deviation for the vehicle position measurements; $0.5°$ mean and $0.5°$ standard deviation for the vehicle attitude measurements; and $0.05°$ mean and $0.05°$ standard deviation for the manipulator's joint position measurements. In addition, the thruster dynamic characteristics are inserted into the simulation. Suppose that the thruster response delay time is 50 ms, and its efficiency is 95%.

$$v_c = \left[0.15 + 0.1\cos(0.3t),\ 0.05\cos(0.1\pi t),\ 0.1\cos(0.2t),\ 0,\ 0,\ 0\right]^T \text{m/s} \tag{38}$$

To implement solution (7), the primary task vector is $x_{Ed} = [\eta_{E1d}^T, \eta_{E2d}^T]^T$. $k_f = \text{diag}(4,4,4,6,6,6)$ and $k_s = 25I$ for the two trajectories. Other parameters are shown in Table 5. A secondary task is designed to align the vehicle orientation and the joint position with the primary task in terms of reducing the coupling effects, as shown in (39).

$$\dot{\eta}_{sd} = \begin{bmatrix} \phi \\ \dot{\theta} \\ \dot{\psi} \\ \dot{q}_1 \\ \dot{q}_2 \end{bmatrix} = \begin{bmatrix} 0 & 0 & 0 \\ 0 & 0 & -\alpha_1 \\ 0 & \alpha_2 & 0 \\ 0 & -\alpha_3 & 0 \\ 0 & 0 & -\alpha_4 \end{bmatrix} \dot{\eta}_{E1d} \tag{39}$$

where $\dot{\eta}_{E1d}$ is the linear part of x_{Ed}.

Table 5. Parameters for the proposed redundancy resolution technique.

Trajectory	α_1	α_2	α_3	α_4	λ	t_s (s)	K_ζ
Straight line	−0.02	0.13	0	0.05	0.3	15	$\text{diag}(0,2,0.2,0,1,1,1,1,1)$
Circle	−0.06	0.2	−0.2	0.15	0.3	45	$\text{diag}(0.1,0.3,0.1,0,1,1,1,1,1)$

The distribution matrix is defined as

$$W^{-1} = \begin{bmatrix} 0.02I_{3\times3} & 0_{3\times3} & 0_{3\times3} \\ W_1 & 0.4I_{3\times3} & 0_{3\times3} \\ W_2 & 0_{3\times3} & 0.98I_{3\times3} \end{bmatrix}, \quad W_1 = \begin{bmatrix} 0 & 0 & 0 \\ 0 & 0 & 0.2 \\ 0 & 0.35 & 0 \end{bmatrix}, \quad W_2 = \begin{bmatrix} 0 & 0.25 & 0 \\ 0 & 0 & 0.5 \\ 0 & 0 & 0 \end{bmatrix} \tag{40}$$

To illustrate the effectiveness of the proposed redundancy resolution technique in terms of energy savings, the comparative redundancy resolution technique is given in (41).

$$\zeta_d = J_W^+(\dot{x}_{Ed} - K_f e_E) + (I - J_W^+ J_W)[J_s^+(\eta, q)(\dot{\eta}_{sd} - K_s e_s)] \tag{41}$$

where the difference between (7) and (41) is that the secondary task vector $\dot{\zeta}$ is not included in the compared technique.

The proposed control scheme is compared with the H∞-EKF method [23] which is given by

$$u_c = M(q)(\ddot{q}_d - V\dot{e} - Pe) + H(q, \dot{q}) + M(q)(-R^{-1}B^T Pe) + \tau_c \tag{42}$$

where V and P are the derivative and the proportional gain matrices. q and $\dot{q}$ are the estimated vectors from EKF. τ_c is the disturbance estimation from EKF as well. R is the given positive definite matrix.

For simple representation, the proposed redundancy resolution is termed case 1 (c1), and the comparative redundancy resolution is termed case 2 (c2). Hence, the proposed control scheme based on the proposed redundancy resolution is termed proposed control$_{c1}$; the H∞-EKF method based on the proposed redundancy resolution is termed H∞-EKF$_{c1}$; and the proposed control based on the comparative redundancy resolution is termed proposed control$_{c2}$.

5.2. Results and Discussion

The results of numerical simulations are shown in Figures 6–13. Figures 6–8 present the desired and actual spatial trajectories and their tracking errors. From these results it is observed that the proposed controller drives the UVMS to track the desired spatial linear and circular trajectories quite satisfactorily in both the proposed redundancy resolution technique (c1) and the comparative redundancy resolution technique (c2). Moreover, the proposed control scheme outperforms the H∞-EKF method, and has smaller tracking errors in both positions and orientations under the conditions of model uncertainties, time-varying ocean currents, payload and sensory noises. Even though the H∞-EKF method adopted a H∞ robust controller to compensate the estimated bias from the EKF, the residual tracking errors can not be fully eliminated, as shown in Figure 6b,c and Figure 6e,f. The proposed controller performs better than the H∞-EKF method in terms of robustness, which is dedicated to the IDC and fuzzy compensator for reducing the perturbation effects.

Figure 9 plots the norm of the vector F_{td} (i.e., thruster forces and actuator torques) and energy consumption of the UVMS. It can be noted that the comparative redundancy resolution technique (c2) is consuming more energy in generating trajectories for the vehicle and manipulator during both the linear and circular trajectories tracking. However in the proposed redundancy resolution technique (c1), the UVMS states are adjusted by self-motion to minimize interaction effects between the vehicle and the manipulator. This is because of the introduction of the secondary task vector $\dot{\zeta}$ and the nonlinear function.

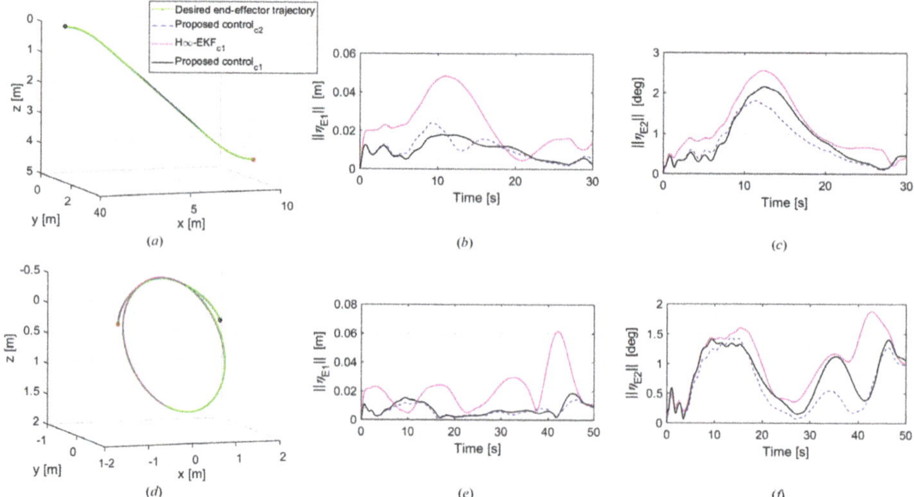

Figure 6. Spatial linear and circular trajectories and their tracking errors. (*a*) Desired linear trajectory and tracking control results; (*b,c*) position tracking errors in positions and orientations; (*d*) desired circular trajectory and tracking control results; (*e,f*) tracking errors in positions and orientations.

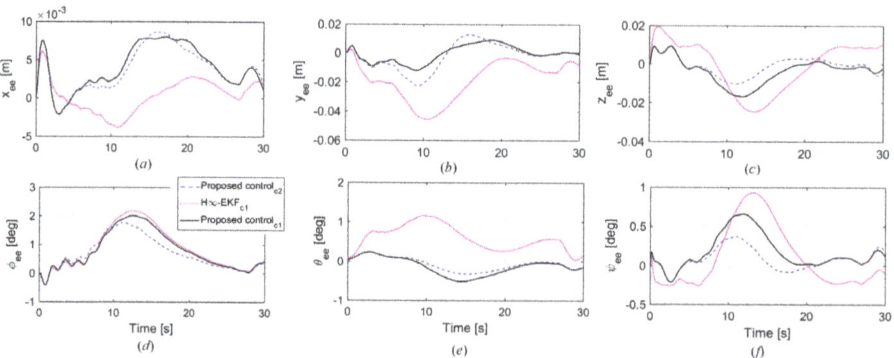

Figure 7. End-effector errors when tracking the linear trajectory. (*a*) x_{ee} error, (*b*) y_{ee} error, (*c*) z_{ee} error, (*d*) tracking errors in the end-effector roll direction, (*e*) tracking errors in the end-effector pitch direction, (*f*) tracking errors in the end-effector yaw direction.

For better understanding, the generated trajectories for the vehicle positions/attitudes and manipulator positions are presented in Figures 10 and 11. It can be seen from the results that the generated trajectories have larger differences on vehicle attitudes and joint angles than vehicle positions. This is because the adjustment of the vehicle position has little effect on reducing the interactive forces between the vehicle and the manipulator without affecting the primary task. Consequently, the energy consumption can be reduced by changing the vehicle attitude and joint angles. In addition, it is observed from Figures 10 and 11 that the small roll and pitch angles of the vehicle are kept in the proposed control scheme, which contributes to properly working of the vehicle's onboard sensors.

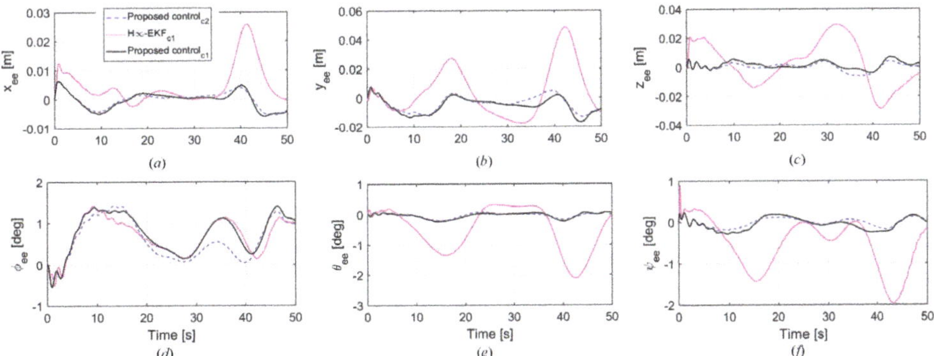

Figure 8. End-effector errors when tracking the circular trajectory. (*a*) x_{ee} error, (*b*) y_{ee} error, (*c*) z_{ee} error, (*d*) tracking errors in the end-effector roll direction, (*e*) tracking errors in the end-effector pitch direction, (*f*) tracking errors in the end-effector yaw direction.

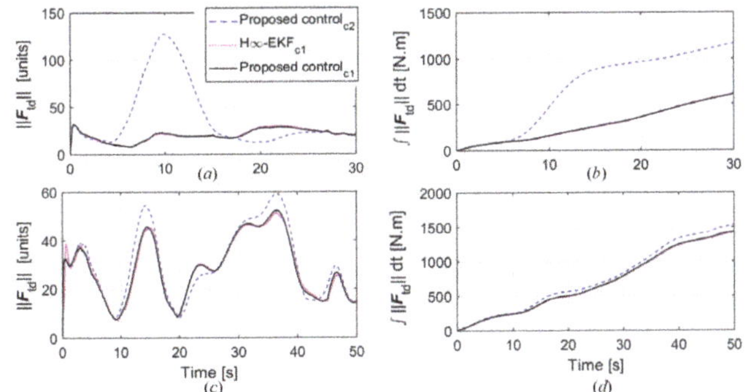

Figure 9. Time histories of the norm of the vector F_{td} and UVMS energy consumption. (*a*,*b*) Results from the linear trajectory tracking, (*c*,*d*) results from the circular trajectory tracking, (F_{td}; i.e., the vector of thruster forces and actuator torques).

Figures 12 and 13 show the required thruster forces for the vehicle and actuator torques for the manipulator during the linear and circular trajectory tracking. It is observed that the thruster forces for the two trajectories are less in the proposed redundancy resolution technique (c1), which results in the reduced energy consumption. In addition, the thruster forces and actuator torques for both trajectories in the proposed control$_{c1}$ are within their constraints (±60 N for the thrusters and ±3 N·m for the actuators).

The quantitative indexes of the time integral of tracking errors and energy consumption are listed in Table 6. From these indices, it is indicated that the tracking error in the proposed control$_{c1}$ is smaller than that in the H∞-EKF$_{c1}$ method, and the energy consumption in the proposed control$_{c1}$ is less than that in the proposed control$_{c2}$. Overall, the proposed control scheme based on the proposed redundancy resolution technique (c1) ensures the precise and robust performance with a reduced energy requirement under the conditions of model parameter uncertainties, time-varying ocean currents, payload and sensory noises.

In the simulations, we have taken the model parameter uncertainties, time-varying external disturbances, payload and sensory noises into consideration. However, in a practical case, these lumped uncertainties and disturbances may be more complicated, and hence can not be simulated. Even though the results from computer simulations are promising, it is necessary to validate the effectiveness of the proposed control scheme and the proposed redundancy resolution technique through experiments in a water pool or at sea. This is our future work.

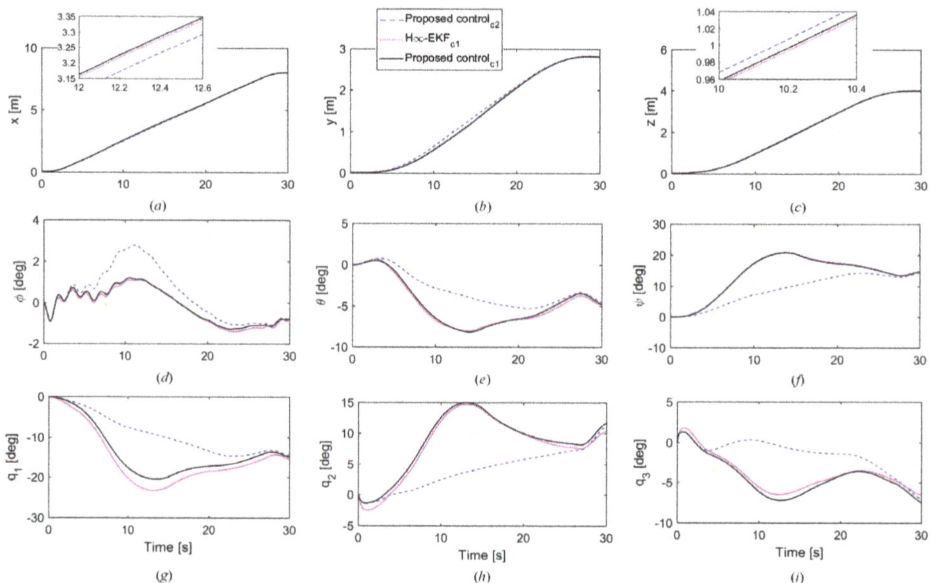

Figure 10. Joint-space positions for the straight line trajectory. (*a*) X position, (*b*) y position, (*c*) z position, (*d*) roll angle, (*e*) pitch angle, (*f*) yaw angle, (*g*) joint 1 angle, (*h*) joint 2 angle, (*i*) joint 3 angle.

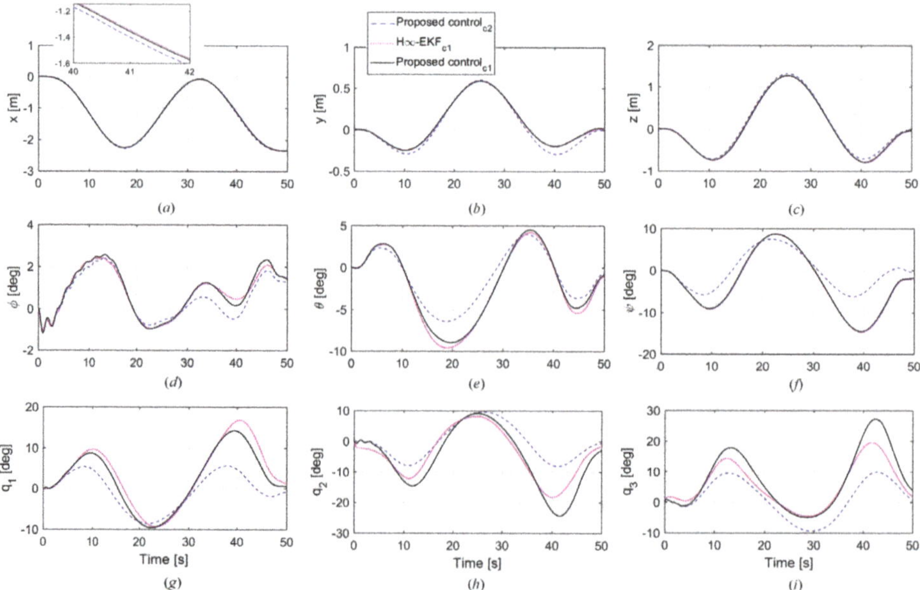

Figure 11. Joint-space positions for the circular trajectory. (*a*) X position, (*b*) y position, (*c*) z position, (*d*) roll angle, (*e*) pitch angle, (*f*) yaw angle, (*g*) joint 1 angle, (*h*) joint 2 angle, (*i*) joint 3 angle.

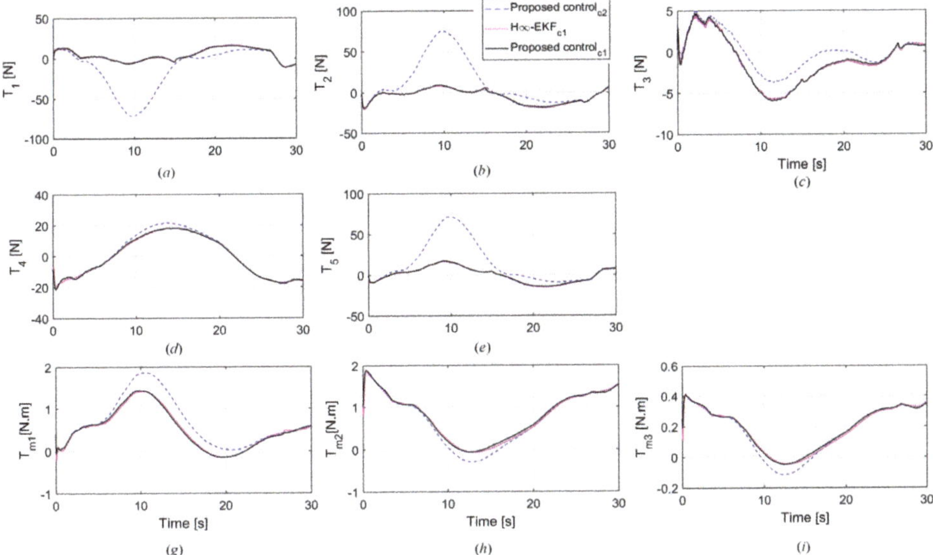

Figure 12. Thruster forces and actuator torques for the line trajectory tracking. (*a–e*) Thruster forces T_1, T_2, T_3, T_4 and T_5; (*g–i*) actuator torques T_{m1}, T_{m2} and T_{m3}.

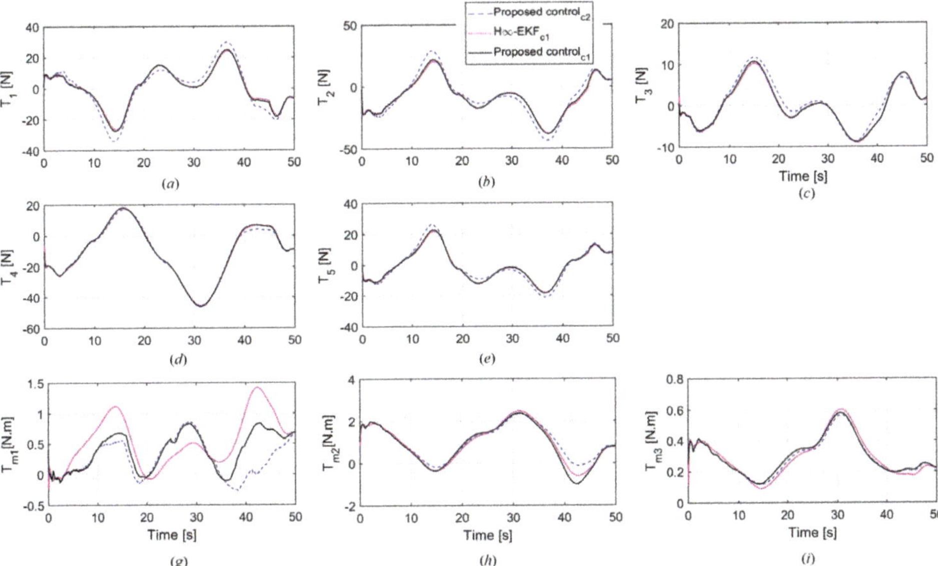

Figure 13. Thruster forces and actuator torques for the circular trajectory tracking. (*a–e*) Thruster forces T_1, T_2, T_3, T_4 and T_5; (*g–i*) actuator torques T_{m1}, T_{m2} and T_{m3}.

Table 6. Performance analysis of the UVMS for the linear and circular trajectories tracking.

Control Schemes	Linear Trajectory			Circular Trajectory		
	ISE_p	ISE_o	$\int \|\|F_{td}\|\|dt$	ISE_p	ISE_0	$\int \|\|F_{td}\|\|dt$
Proposed control$_{c2}$	0.3083	22.5597	1161.6	0.3318	32.2834	1532.6
H∞ − EKF$_{c1}$	0.6869	34.5853	607.2053	1.0775	63.1945	1420.4
Proposed control$_{c1}$	0.2905	26.1319	603.8734	0.3708	40.1401	1428.3

Note : $e_E = [(\tilde{\eta}_{E1})^T, (\tilde{\eta}_{E2})^T]^T$, $\text{ISE}_p = \int \|\|\tilde{\eta}_{E1}\|\|dt$, $\text{ISE}_o = \int \|\|\tilde{\eta}_{E2}\|\|dt$.

6. Conclusions

This paper presents a motion planning and coordinated control scheme for the trajectory tracking of the UVMS. A new secondary task with a nonlinear coefficient for redundancy resolution of the UVMS is proposed. In this way, the interactive effects between the vehicle and the manipulator can be minimized and the energy consumption of the UVMS is reduced. Simulation results show that the energy consumption based on the proposed redundancy resolution technique (proposed control$_{c1}$) is reduced by 48% (in the linear trajectory tracking) and 6% (in the circular trajectory tracking), compared with the comparative redundancy resolution technique (proposed control$_{c2}$). The proposed redundancy resolution technique is simple in design and easy to implement. Furthermore, a control scheme including a fuzzy compensator and a tracking control with joint-space errors, end-effector errors and inertial delay control (IDC) is proposed. The proposed control scheme ensures precise and robust tracking performance in the presence of model uncertainties, time-varying ocean currents, payload and sensory noises. Simulation results show that the position and orientation tracking precisions based on the proposed control$_{c1}$ are reduced by 57.7% and 24.4% (in the linear trajectory tracking) and 65.6% and 36.5% (in the circular trajectory tracking), compared with the H$_\infty$−EKF$_{c1}$ method [23]. Even though the effectiveness of the proposed redundancy resolution technique and coordinated motion control scheme were validated through numerical simulations, experiments should be carried out on the real UVMS to further enhance the computer simulation results, which will be done in the future.

Author Contributions: Conceptualization, H.H., Y.W. and X.Y.; data curation, H.H.; formal analysis, H.H.; funding acquisition, Y.W. and X.Y.; investigation, H.H.; methodology, H.H.; project administration, Y.W. and X.Y.; resources, Y.W. and W.L.; software, H.H.; supervision, Y.W., X.Y. and W.L.; validation, H.H., Y.W., X.Y. and W.L.; visualization, H.H.; writing—original draft, H.H.; writing—review editing, Y.W. and X.Y. All authors have read and agreed to the published version of the manuscript.

Funding: This research was supported by the National Natural Science Foundation of China under grant number 41876100, in part by the National key research and development program of China under grant number 2017YFC0306001, in part by the Heilongjiang Natural Science Foundation under grant number E2017024 and in part by the Key Program for International S and T Cooperation Projects of China under grant number 2014DFR10010.

Acknowledgments: The authors would like to thank Shuai Li, Fan Zhang and Shitong Du, who provided insight and expertise that greatly assisted the writing of this manuscript.

Conflicts of Interest: The authors declare no conflict of interest. The funders had no role in the design of the study; in the collection, analyses or interpretation of data; in the writing of the manuscript, or in the decision to publish the results.

Appendix A. Simulation Data of the AUV

For the torpedo-type AUV, its center of buoyancy (CB) is $(0, 0, 0.02)$ m, its center of gravity (CG) is $(0, 0, 0)$ m and its moment of inertia is $(0.69, 16.82, 16.82)$ kg $\cdot$ m^2. The AUV is equipped with an underwater manipulator, and the manipulator's base position is $p_0^B = (0.6, 0, 0.2)^T$ m. Some model parameters of the AUV given in (3) are composed of rigid-body terms and hydrodynamic terms [42], as shown in the following equations.

$$M_v = M_{\mathrm{RB}} + M_{\mathrm{A}}, \quad C_v(v_r) = C_{\mathrm{RB}}(v_r) + C_{\mathrm{A}}(v_r), \quad D_v(v_r) = D_{\mathrm{NL}}\mathrm{diag}\left(|v_r|\right) + D_{\mathrm{L}}\mathrm{diag}(|v_r|)$$

where M_{RB} and C_{RB} are the rigid-body terms which represent the inertia matrix and the Coriolis and centripetal matrix. M_{A}, C_{A}, D_{NL} and D_{L} are the matrices related to the hydrodynamic forces. M_{A} and C_{A} are the added mass matrix and the added Coriolis and centripetal matrix. D_{NL} and D_{L} are the quadratic damping matrix and the lift matrix. M_{A}, C_{A}, D_{NL} and D_{L} are given in the following equations [15].

$$M_{\mathrm{A}} = - \begin{bmatrix} X_{\dot{u}} & 0 & 0 & 0 & 0 & 0 \\ 0 & Y_{\dot{v}} & 0 & 0 & 0 & Y_{\dot{r}} \\ 0 & 0 & Z_{\dot{w}} & 0 & Z_{\dot{q}} & 0 \\ 0 & 0 & 0 & K_{\dot{p}} & 0 & 0 \\ 0 & 0 & M_{\dot{w}} & 0 & M_{\dot{q}} & 0 \\ 0 & N_{\dot{v}} & 0 & 0 & 0 & N_{\dot{r}} \end{bmatrix} = \begin{bmatrix} A_{11} & A_{12} \\ A_{21} & A_{22} \end{bmatrix}, \, C_{\mathrm{A}} = \begin{bmatrix} 0_{3\times3} & -S(A_{11}v_1 + A_{12}v_2) \\ -S(A_{11}v_1 + A_{12}v_2) & -S(A_{21}v_1 + A_{22}v_2) \end{bmatrix}$$

$$D_{NL} = - \begin{bmatrix} X_{u|u|} & 0 & 0 & 0 & 0 & 0 \\ 0 & Y_{v|v|} & 0 & 0 & 0 & Y_{r|r|} \\ 0 & 0 & Z_{w|w|} & 0 & Z_{q|q|} & 0 \\ 0 & 0 & 0 & K_{p|p|} & 0 & 0 \\ 0 & 0 & M_{w|w|} & 0 & M_{q|q|} & 0 \\ 0 & N_{v|v|} & 0 & 0 & 0 & N_{r|r|} \end{bmatrix}, D_{\mathrm{L}} = - \begin{bmatrix} 0 & 0 & 0 & 0 & 0 & 0 \\ 0 & Y_{uv} & 0 & 0 & 0 & Y_{ur} \\ 0 & 0 & Z_{uw} & 0 & Z_{uq} & 0 \\ 0 & 0 & 0 & 0 & 0 & 0 \\ 0 & 0 & M_{uw} & 0 & M_{uq} & 0 \\ 0 & N_{uv} & 0 & 0 & 0 & N_{ur} \end{bmatrix}$$

To obtain the above hydrodynamic coefficients, the strip theory is utilized for numerical calculation, where the fluid density is assumed to be 1030 kg/m^3, the linear-skin coefficient is assumed to be 0.4 and the drag coefficient is assumed to be 1. Moreover, some of the obtained coefficients are adjusted based on comparisons with data of the REMUS AUV according to dynamic similarity [15]. Then, the adjusted hydrodynamic coefficients are shown in Table A1.

Table A1. The list of AUV coefficients.

Added Mass Coefficients					
Force	**Value**	**Units**	**Moment**	**Value**	**Units**
$X_{\dot{u}}$	−2.33	Kg	$K_{\dot{p}}$	−0.3	Kg·m^2/rad
$Y_{\dot{v}}$	−90.8	Kg	$M_{\dot{q}}$	−20.3	Kg·m^2/rad
$Y_{\dot{r}}$	4.53	Kg·m	$M_{\dot{w}}$	−4.53	Kg·m
$Z_{\dot{w}}$	−90.8	Kg	$N_{\dot{r}}$	−20.3	Kg·m^2/rad
$Z_{\dot{q}}$	−4.53	Kg·m	$N_{\dot{v}}$	4.53	Kg·m
Drag Coefficients					
Force	**Value**	**Units**	**Moment**	**Value**	**Units**
$X_{u|u|}$	−2.96	Kg/m	$K_{p|p|}$	-0.558	Kg·m^2/rad^2
$Y_{v|v|}$	−2346	Kg/m	$M_{q|q|}$	−807	Kg·m^2/rad^2
$Y_{r|r|}$	0.759	Kg·m/rad^2	$M_{w|w|}$	8.76	Kg
$Z_{w|w|}$	−242	Kg/m	$N_{r|r|}$	−404	Kg·m^2/rad^2
$Z_{q|q|}$	−0.759	Kg·m/rad^2	$N_{v|v|}$	−8.76	Kg
Lift Coefficients					
Force	**Value**	**Units**	**Moment**	**Value**	**Units**
Y_{uv}	−56.5	Kg/m	M_{uq}	−8.9	Kg·m/rad
Y_{ur}	11.8	Kg/rad	M_{uw}	−24.9	Kg
Z_{uw}	−56.5	Kg/m	N_{ur}	−8.9	Kg·m/rad
Z_{uq}	−11.8	Kg/rad	N_{uv}	24.9	Kg

For the AUV, the thruster configuration matrix and its pseudo inverse are given as

$$
\boldsymbol{B}_v = \begin{bmatrix} 1 & 1 & 0 & 0 & 0 \\ 0 & 0 & 0 & 0 & 1 \\ 0 & 0 & 1 & 1 & 0 \\ 0 & 0 & 0 & 0 & 0 \\ 0 & 0 & r_3 & -r_4 & 0 \\ r_1 & -r_2 & 0 & 0 & r_5 \end{bmatrix}, \boldsymbol{B}_v^+ = \begin{bmatrix} \frac{r_2}{r_1+r_2} & -\frac{r_5}{r_1+r_2} & 0 & 0 & 0 & \frac{1}{r_1+r_2} \\ \frac{r_1}{r_1+r_2} & \frac{r_5}{r_1+r_2} & 0 & 0 & 0 & -\frac{1}{r_1+r_2} \\ 0 & 0 & \frac{r_4}{r_3+r_4} & 0 & \frac{1}{r_3+r_4} & 0 \\ 0 & 0 & \frac{r_3}{r_3+r_4} & 0 & -\frac{1}{r_3+r_4} & 0 \\ 0 & 1 & 0 & 0 & 0 & 0 \end{bmatrix}
$$

where $r_1 = 0.18$ m, $r_2 = 0.18$ m, $r_3 = 0.525$ m, $r_4 = 0.245$ m, $r_5 = 0.485$ m.

References

1. Podder, T.K.; Sarkar, N. Unified Dynamics-based Motion Planning Algorithm for Autonomous Underwater Vehicle-Manipulator Systems (UVMS). *Robotica* **2004**, *22*, 117–128. [CrossRef]
2. Antonelli, G.; Chiaverini, S. Fuzzy redundancy resolution and motion coordination for underwater vehicle-manipulator systems. *IEEE Trans. Fuzzy Syst.* **2003**, *11*, 109–120. [CrossRef]
3. Antonelli, G.; Chiaverini, S. Task-priority redundancy resolution for underwater vehicle-manipulator systems. In Proceedings of the IEEE International Conference on Robotics and Automation, Leuven, Belgium, 20 May 1998; pp. 768–773.
4. Sarkar, N.; Podder, T.K. Coordinated motion planning and control of autonomous underwater vehicle-manipulator systems subject to drag optimization. *IEEE J. Ocean. Eng.* **2001**, *26*, 228–239. [CrossRef]
5. Han, J.; Park, J.; Chung, W.K. Robust coordinated motion control of an underwater vehicle-manipulator system with minimizing restoring moments. *Ocean Eng.* **2011**, *38*, 1197–1206. [CrossRef]
6. Siciliano, B.; Slotine, J.-J.E. A general framework for managing multiple tasks in highly redundant robotic systems. In Proceedings of the Fifth International Conference on Advanced Robotics, Pisa, Italy, 19–22 June 1991; pp. 1211–1216.
7. Antonelli, G.; Chiaverini, S. A fuzzy approach to redundancy resolution for underwater vehicle-manipulator systems. *Control Eng. Pract.* **2003**, *11*, 445–452. [CrossRef]
8. Wang, Y.; Jiang, S.; Yan, F.; Gu, L.; Chen, B. A new redundancy resolution for underwater vehicle—Manipulator system considering payload. *Int. J. Adv. Robot. Syst.* **2017**, *14*. [CrossRef]
9. Haugalokken, B.O.A.; Joergensen, E.K.; Schjolberg, I. Experimental validation of end-effector stabilization for underwater vehicle-manipulator systems in subsea operations. *Robot. Auton. Syst.* **2018**, *109*, 1–12. [CrossRef]
10. Tang, C.; Wang, Y.; Wang, S.; Wang, R.; Tan, M. Floating Autonomous Manipulation of the Underwater Biomimetic Vehicle-Manipulator System: Methodology and Verification. *IEEE Trans. Ind. Electron.* **2018**, *65*, 4861–4870. [CrossRef]
11. Simetti, E.; Casalino, G.; Wanderlingh, F.; Aicardi, M. Task priority control of underwater intervention systems: Theory and applications. *Ocean Eng.* **2018**, *164*, 40–54. [CrossRef]
12. Youakim, D.; Ridao, P. Motion planning survey for autonomous mobile manipulators underwater manipulator case study. *Robot. Auton. Syst.* **2018**, *107*, 20–44. [CrossRef]
13. Mcmillan, S.; Orin, D.E.; Mcghee, R.B. Efficient dynamic simulation of an underwater vehicle with a robotic manipulator. *IEEE Trans. Syst. Man Cybern.* **1995**, *25*, 1194–1206. [CrossRef]
14. Dannigan, M.W.; Russell, G. Evaluation and reduction of the dynamic coupling between a manipulator and an underwater vehicle. *IEEE J. Ocean. Eng.* **1998**, *23*, 260–273. [CrossRef]
15. Han, H.; Wei, Y.; Ye, X.; Liu, W. Modelling and Fuzzy Decoupling Control of an Underwater Vehicle-Manipulator System. *IEEE Access* **2020**, *8*, 18962–18983. [CrossRef]
16. Schjolberg, I.; Fossen, T.I. Modelling and Control of Underwater Vehicle-Manipulator Systems. In Proceedings of the 3rd Conference on Marine Craft Maneuvering and Control, Southampton, UK, 7–9 September 1994; pp. 45–57.
17. Taira, Y.; Sagara, S.; Oya, M. Model-based motion control for underwater vehicle-manipulator systems with one of the three types of servo subsystems. *Artif. Life Robot* **2019**, *6*, 1–16. [CrossRef]

18. Korkmaz, O.; Ider, S.K.; Ozgoren, M.K. Trajectory Tracking Control of an Underactuated Underwater Vehicle Redundant Manipulator System. *Asian J. Control* **2016**, *18*, 1593–1607. [CrossRef]
19. Cai, M.; Wang, Y.; Wang, S.; Wang, R.; Ren, Y.; Tan, M. Grasping Marine Products with Hybrid-Driven Underwater Vehicle-Manipulator System. *IEEE Trans. Autom. Sci. Eng.* **2020**. [CrossRef]
20. Santhakumar, M.; Kim, J. Robust Adaptive Tracking Control of Autonomous Underwater Vehicle-Manipulator Systems. *J. Dyn. Syst. Meas. Control* **2014**, *136*, 054502. [CrossRef]
21. Antonelli, G.; Caccavale, F.; Chiaverini, S. Adaptive Tracking Control of Underwater Vehicle-Manipulator Systems Based on the Virtual Decomposition Approach. *IEEE Trans. Robot. Autom.* **2004**, *20*, 594–602. [CrossRef]
22. Mohan, S.; Kim, J. Indirect adaptive control of an autonomous underwater vehicle-manipulator system for underwater manipulation tasks. *Ocean Eng.* **2012**, *54*, 233–243. [CrossRef]
23. Dai, Y.; Yu, S. Design of an indirect adaptive controller for the trajectory tracking of UVMS. *Ocean Eng.* **2008**, *151*, 234–245. [CrossRef]
24. Esfahani, H.N. Robust Model Predictive Control for Autonomous Underwater Vehicle–Manipulator System with Fuzzy Compensator. *Pol. Marit. Res.* **2019**, *26*, 104–114. [CrossRef]
25. Cai, M.; Wang, Y.; Wang, S.; Wang, R.; Cheng, L.; Tan, M. Prediction-Based Seabed Terrain Following Control for an Underwater Vehicle-Manipulator System. *IEEE Trans. Syst. Man Cybern. Syst.* **2019**. [CrossRef]
26. Nikou, A.; Verginis, C.K.; Dimarogonas, D.V. A Tube-based MPC Scheme for Interaction Control of Underwater Vehicle Manipulator Systems. In Proceedings of the 2018 IEEE/OES Autonomous Underwater Vehicle Workshop (AUV), Porto, Portugal, 6–9 November 2018; pp. 1–8.
27. Dai, Y.; Yu, S.; Yan, Y. An Adaptive EKF-FMPC for the Trajectory Tracking of UVMS. *IEEE J. Ocean. Eng.* **2019**, 1–15. [CrossRef]
28. Dai, Y.; Yu, S.; Yan, Y.; Yu, X. An EKF-Based Fast Tube MPC Scheme for Moving Target Tracking of a Redundant Underwater Vehicle-Manipulator System. *IEEE ASME Trans. Mechatron.* **2019**, *24*, 2803–2814. [CrossRef]
29. Esfahani, H.N.; Azimirad, V.; Danesh, M. A Time Delay Controller included terminal sliding mode and fuzzy gain tuning for Underwater Vehicle-Manipulator Systems. *Ocean Eng.* **2015**, *107*, 97–107. [CrossRef]
30. Hosseinnajad, A.; Loueipour, M. Time Delay Controller Design for Dynamic Positioning of ROVs based on Position and Acceleration Measurements. In Proceedings of the 6th International Conference on Control, Instrumentation and Automation, Sanandaj, Iran, 30–31 October 2019.
31. Cui, R.; Chen, L.; Yang, C.; Chen, M. Extended State Observer-Based Integral Sliding Mode Control for an Underwater Robot With Unknown Disturbances and Uncertain Nonlinearities. *IEEE Trans. Ind. Electron.* **2017**, *64*, 6785–6795. [CrossRef]
32. Chen, W.H. Disturbance Observer Based Control for Nonlinear Systems. *IEEE ASME Trans. Mechatron.* **2005**, *9*, 706–710. [CrossRef]
33. Chen, W.; Wei, Y.; Zeng, J.; Han, H.; Jia, X. Adaptive Terminal Sliding Mode NDO-Based Control of Underactuated AUV in Vertical Plane. *Discret. Dyn. Nat. Soc.* **2016**. [CrossRef]
34. Li, J.; Huang, H.; Wan, L.; Zhou, Z.; Xu, Y. Hybrid Strategy-based Coordinate Controller for an Underwater Vehicle Manipulator System Using Nonlinear Disturbance Observer. *Robotica* **2019**, *37*, 1710–1731. [CrossRef]
35. Londhe, P.S.; Dhadekar, D.D.; Patre, B.M.; Waghmare, L.M. Uncertainty and disturbance estimator based sliding mode control of an autonomous underwater vehicle. *Intern. J. Dyn. Control* **2017**, *5*, 1122–1138. [CrossRef]
36. Han, H.; Wei, Y.; Guan, L.; Ye, X.; Wang, A. Trajectory Tracking Control of Underwater Vehicle-Manipulator Systems Using Uncertainty and Disturbance Estimator. In Proceedings of the OCEANS 2018 MTS/IEEE, Charleston, SC, USA, 22–25 October 2018.
37. Suryawanshi, P.V.; Shendge, P.D.; Phadke, S.B. Robust sliding mode control for a class of nonlinear systems using inertial delay control. *Nonlinear Dyn.* **2014**, *78*, 1921–1932. [CrossRef]
38. Mohan, S.; Kim, J. Coordinated motion control in task space of an autonomous underwater vehicle–manipulator system. *Ocean Eng.* **2015**, *104*, 155–167. [CrossRef]
39. Londhe, P.S.; Santhakumar, M.; Patre, B.M.; Waghmare, L.M. Task Space Control of an Autonomous Underwater Vehicle Manipulator System by Robust Single-Input Fuzzy Logic Control Scheme. *IEEE J. Ocean. Eng.* **2016**, *42*, 13–28. [CrossRef]

40. Londhe, P.S.; Mohan, S.; Patre, B.M.; Waghmare, L.M. Robust task-space control of an autonomous underwater vehicle- manipulator system by PID-like fuzzy control scheme with disturbance estimator. *Ocean Eng.* **2017**, *139*, 1–13. [CrossRef]

41. Antonelli, G. *Underwater Robots: Motion and Force Control of Vehicle-Manipulator Systems*, 3rd ed.; Springer: Berlin/Heidelberg, Germany, 2013; pp. 52–55.

42. Fossen, T.I. *Handbook of Marine Craft Hydrodynamics and Motion Control*; John Wiley and Sons: Hoboken, NJ, USA, 2011; pp. 128–131.

 *applied sciences*

Article

Local CPG Self Growing Network Model with Multiple Physical Properties

Ming Liu [1], Mantian Li [1], Fusheng Zha [1,2,*], Pengfei Wang [1], Wei Guo [1] and Lining Sun [1,*]

1   State Key Laboratory of Robotics and System, Harbin Institute of Technology, Harbin 150001, China; 14B308017@hit.edu.cn (M.L.); limt@hit.edu.cn (M.L.); wangpengfei1007@163.com (P.W.); wguo01@hit.edu.cn (W.G.)
2   Shenzhen Academy of Aerospace Technology, Shenzhen 518057, China
*   Correspondence: zhafusheng@hit.edu.cn (F.Z.); lnsun@hit.edu.cn (L.S.); Tel.: +86-451-8641-4462 (F.Z. & L.S.)

Received: 23 June 2020; Accepted: 5 August 2020; Published: 8 August 2020

Abstract: Compared with traditional control methods, the advantage of CPG (Central Pattern Generator) network control is that it can significantly reduce the size of the control variable without losing the complexity of its motion mode output. Therefore, it has been widely used in the motion control of robots. To date, the research into CPG network has been polarized: one direction has focused on the function of CPG control rather than biological rationality, which leads to the poor functional adaptability of the control network and means that the control network can only be used under fixed conditions and cannot adapt to new control requirements. This is because, when there are new control requirements, it is difficult to develop a control network with poor biological rationality into a new, qualified network based on previous research; instead, it must be explored again from the basic link. The other direction has focused on the rationality of biology instead of the function of CPG control, which means that the form of the control network is only similar to a real neural network, without practical use. In this paper, we propose some physical characteristics (including axon resistance, capacitance, length and diameter, etc.) that can determine the corresponding parameters of the control model to combine the growth process and the function of the CPG control network. Universal gravitation is used to achieve the targeted guidance of axon growth, Brownian random motion is used to simulate the random turning of axon self-growth, and the signal of a single neuron is established by the Rall Cable Model that simplifies the axon membrane potential distribution. The transfer model, which makes the key parameters of the CPG control network—the delay time constant and the connection weight between the synapses—correspond to the axon length and axon diameter in the growth model and the growth and development of the neuron processes and control functions are combined. By coordinating the growth and development process and control function of neurons, we aim to realize the control function of the CPG network as much as possible under the conditions of biological reality. In this way, the complexity of the control model we develop will be close to that of a biological neural network, and the control network will have more control functions. Finally, the effectiveness of the established CPG self-growth control network is verified through the experiments of the simulation prototype and experimental prototype.

Keywords: CPG; self-growing network; quadruped robot; trot gait

---

1. Introduction

Compared with traditional control methods, Central Pattern Generator (CPG) network control has been widely used for the motion control of bionic robots because it can significantly reduce the dimension of control variables without losing the complexity of its output motion mode; the method also has some other advantages [1–3]. However, the study of CPG control network to date has been polarized: one direction has focused on the control function [4–6] and ignored the biology rationality.

In this direction, a control network is designed only for certain fixed situations, which leads to the poor functional adaptability of the control network. Once there are new control requirements, it is difficult for the control network to be developed into a new, qualified one according to the previous research; instead, it has to be researched again from the basic link. Especially when applied to the motion control of a bionic robot, the above-mentioned CPG control network will have the following disadvantages: with excessive prior knowledge and excessive factors given by humans in each link of the control system, which includes the design of the CPG network, the setting method of network parameters and the establishment of the neuron model in the control network, the control function of the CPG network has lower complexity, worse adaptability and worse inheritance capabilities in future research.

Another direction has focused on biological rationality, but not the function of CPG control, which means that the control network is only similar to a real neural network in form but of no practical use [7–9]. Dehmamy et al. [10] modeled axon growth with driven diffusion. Their simulation showed that with background potential derived from the concentration guiders, even coupled random walkers can generate realistic paths for axons. Marinov et al. [11] proposed an ad-hoc growth model built upon the diffusion principle to reproduce the shape of Micro-Tissue Engineered Neural Networks (Micro-TENNs). The proposed model imposed various rules for individual neuronal growth to a 3D diffusion equation. Fard et al. [12] proposed a generative growth model to investigate growth rules for axonal branching patterns in cat area 17. The model achieves better statistical accuracy both locally and globally than the commonly used Galton–Watson model. Gafarov et al. [13] proposed a closed-loop growth model in which neural activity affects neurite outgrowth in an interactive way. Experiments showed that the model can be potentially used to form large-scale neural networks by self-organization. However, it is very difficult to study how the scattered neurons in different parts of the growth region form the whole network through interactive behaviors with biological methods, because the growth environment of cells is complex, and so it is difficult to precisely control their growth conditions to verify the specific growth mechanism. Therefore, the study of the growth process of a neural network by numerical simulation is very effective. In a simulation environment, researchers can set strict growth conditions and analyze the growth result of the network in a variety of ways and then compare it with a real biological neural network to verify the validity of the designed model [9,14–16].

However, the numerical simulation of the growth of a biological neural network has often been unable to simulate the control function of the network, and it can only compare the simulation model with the real biological network in terms of morphology and statistics. At the same time, in the control based on a neural network, it is precisely the complexity of the formation process of the biological neural network and the simplifications and assumptions made for the simulation model that make the network control model unrealistic and not universally applicable [17]. Therefore, it is necessary to combine the growth process with the control function in the control of bionic robots based on the CPG network to allow the two to develop in a coordinated way in the long term.

Therefore, we propose some physical characteristics (including axon resistance, capacitance, length and diameter, etc.) that can determine the corresponding parameters of the control model to combine the growth process and the function of the CPG control network. By coordinating the two, we aim to realize the control function of the CPG network under the condition of biological reality as far as possible. In this way, the complexity of the developed control model will be closer to that of a biological neural network, and the control network will have more control functions.

2. Establishment of Self-Growing Model of Neurons

In the formation process of a neural network, the position of each neuron is fixed, axons and dendrites grow from the neuron, and the axon of the former neuron and the dendrite of the latter neuron form a synaptic contact and finally realize the connection of the whole network. Studies have shown that, with the continuous elongation and swerving of the axon, the growth of the tail end of the neuron will not change the shape of the other part [9]. Therefore, the process of growing and eventually forming synaptic connections with another neuron in the neuron model can be simplified into the following model:

$$\vec{F} = \sum_{j=1}^{k} \vec{F}_j = \sum_{j=1}^{k} gm\frac{M_j}{d_j^2}\cdot\vec{e}_j \tag{1}$$

where k is the number of neurons in the gravitational field, d_j is the distance from the growth cone to the jth neuron and $\vec{e}_j$ is the direction vector of gravitation on the growth cone generated by the jth neuron.

In Equation (1), the growth cone can be regarded as a particle with a mass of m. Regard the neuron as a sphere with a mass M concentrated in the center; the distribution range of dendrites is understood as a sphere with radius r because the dendrites are numerous and dense. The guiding effect of the chemical molecules on the growth cone can be equivalent to the gravitational effect on particle m in the gravitational field generated by mass M, and the motion trajectory of the particle in the gravitational field can be regarded as the axon. The position of growth cone particle in the gravitational field is $P_i(x_i, y_i, z_i)$ at a certain time, with a growth rate of $V_i(v_{xi}, v_{yi}, v_{zi})$, and the gravitation on the growth cone can be determined from Newton's law of gravitation. Thus, the acceleration is

$$\vec{a} = \sum_{j=1}^{k} \vec{a}_j = \sum_{j=1}^{k} g\frac{M_j}{d_j^2}\cdot\vec{e}_j = (a_{xi}, a_{yi}, a_{zi}) \tag{2}$$

After a short period of time Δt (Δt is defined as 0.01 s in this paper), the displacement of the growth cone is

$$\begin{cases} \Delta x_i = v_{xi}\cdot\Delta t + \frac{1}{2}a_{xi}\cdot\Delta t^2 \\ \Delta y_i = v_{yi}\cdot\Delta t + \frac{1}{2}a_{yi}\cdot\Delta t^2 \\ \Delta z_i = v_{zi}\cdot\Delta t + \frac{1}{2}a_{zi}\cdot\Delta t^2 \end{cases} \tag{3}$$

Thus, the new position $P_{i+1}(x_{i+1}, y_{i+1}, z_{i+1})$ of the growth cone is

$$\begin{cases} x_{i+1} = x_i + \Delta x_i \\ y_{i+1} = y_i + \Delta y_i \\ z_{i+1} = z_i + \Delta z_i \end{cases} \tag{4}$$

and the new growth rate $V_{i+1}(v_{xi+1}, v_{yi+1}, v_{zi+1})$ is

$$\begin{cases} v_{xi+1} = v_{xi} + a_{xi}\cdot\Delta t \\ v_{yi+1} = v_{yi} + a_{yi}\cdot\Delta t \\ v_{zi+1} = v_{zi} + a_{zi}\cdot\Delta t \end{cases} \tag{5}$$

$P_{i+1}(x_{i+1}, y_{i+1}, z_{i+1})$ and $V_{i+1}(v_{xi+1}, v_{yi+1}, v_{zi+1})$ can be used for the next iteration. In this way, the continuous growth process of axons is discretized into a programmable iterative process.

Although it is reasonable to use the gravitational field model to describe the chemotaxis of growth cones, the growth of the model's axon is quite different from that in the actual situation. Experiments show that the overall deflection of the axon does not become more obvious as the concentration gradient increases gradually, because the axon grows faster on the side of the inverse gradient than the side along the gradient (growth rate regulation), which counteracts the chemotaxis. The results show that when the concentration gradient is large, the chemotactic deflection plays a leading role [8]. In addition, the guiding effect of chemical factors on the growth cone is either attraction or repulsion, but we only consider the case of attraction when modeling, because the mathematical model of the repulsion case is the same as that of the attraction case except with the opposite direction of force.

In fact, the growth direction of the growth cone depends on many intracellular and extracellular signals, which may lead to large fluctuations in the growth direction [9]. For example, the noise gradient in the chemoattractant orientation perception stage mentioned in [18,19] is one of the factors making the deflection direction of the growth cone uncertain when proceeding. In order to describe the uncertain component of the deflection direction of the growth cone, many studies on the mathematical modeling of the axon growing process introduce a random motion component based on the deterministic motion generated by gravity [20,21], as shown in Figure 1.

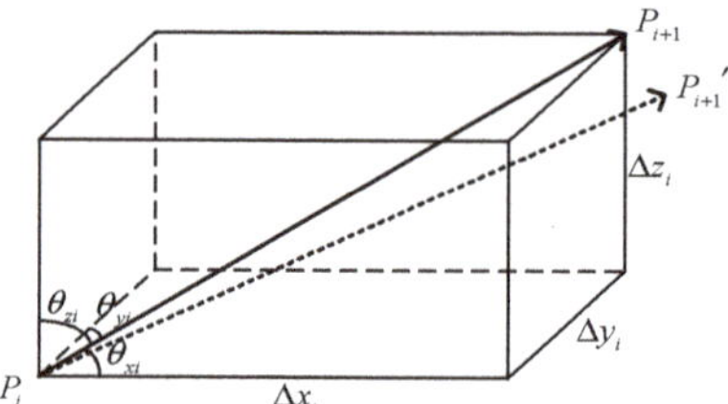

Figure 1. Diagram of a single growth iteration of an axon.

In Figure 1, P_i is the position of the growth cone after the i-1th growing process. P_{i+1} is the position of the growth cone after the ith growing process without the effect of a random motion component. Δx_i, Δy_i and Δz_i are the projected length of the ith growth length of the growth cone on the x, y, and z axes, and P_{i+1}' is the real position of the growth cone after the ith growing process under the effect of random motion component.

As can be seen from Figure 1, we can calculate the angle between the ith growth segment and the coordinate axis, $(\theta_{xi}, \theta_{yi}, \theta_{zi})$ using the following formula:

$$\begin{cases} \theta_{xi} = arctan \dfrac{\sqrt{\Delta y_i^2 + \Delta z_i^2}}{\Delta x_i} \\[2mm] \theta_{yi} = arc\,tan \dfrac{\sqrt{\Delta x_i^2 + \Delta z_i^2}}{\Delta y_i} \\[2mm] \theta_{zi} = arc\,tan \dfrac{\sqrt{\Delta y_i^2 + \Delta x_i^2}}{\Delta z_i} \end{cases} \tag{6}$$

Furthermore, we add a random motion component to this growing process:

$$\begin{cases} \theta_{xi}' = \theta_{xi} + \Delta\theta_x \\ \theta_{yi}' = \theta_{yi} + \Delta\theta_y \\ \theta_{zi}' = \theta_{zi} + \Delta\theta_z \end{cases} \tag{7}$$

where $\Delta\theta_x$, $\Delta\theta_y$ and $\Delta\theta_z$ meet uniform distribution in the interval $(-c\delta, c\delta)$. $\delta > 0$, which ensures that the random motions lead to the largest deflection angle, $c \in (0, 1)$ implies the intensity of random motion, and $(\theta_{xi}', \theta_{yi}', \theta_{zi}')$ is the real deflection angle of this growing process considering the random motion. Then, we can calculate the real displacement of the growth cone in this process, $(\Delta x_i', \Delta y_i', \Delta z_i')$, using the following formula:

$$\begin{cases} l_i = \sqrt{\Delta x_i^2 + \Delta y_i^2 + \Delta z_i^2} \\ \Delta x_i' = l_i \cdot \cos\theta_{xi}' \\ \Delta y_i' = l_i \cdot \cos\theta_{yi}' \\ \Delta z_i' = l_i \cdot \cos\theta_{zi}' \end{cases} \tag{8}$$

where l_i is the length of the axon segment of the ith growing process; the random motion only changes the deflection of each growing process, but not the step size. Finally, we calculate the real position of the growth cone after the ith growing process, $P_{i+1}'(x_{i+1}', y_{i+1}', z_{i+1}')$, using the following formula:

$$\begin{cases} x_{i+1}' = x_i + \Delta x_i' \\ y_{i+1}' = y_i + \Delta y_i' \\ z_{i+1}' = z_i + \Delta z_i' \end{cases} \tag{9}$$

Then, we use $P'_{i+1}(x'_{i+1}, y'_{i+1}, z'_{i+1})$ and $V_{i+1}(v_{xi+1}, v_{yi+1}, v_{zi+1})$ calculated from Equations (2)–(5) for the next iteration.

Three intensities of random motion values are selected as $c = 0.05$, $c = 0.15$ and $c = 0.56$, and 20 of the corresponding 200 axons are selected to draw the growing diagram of the growth cone under the effect of random motion of different intensities, as shown in Figure 2.

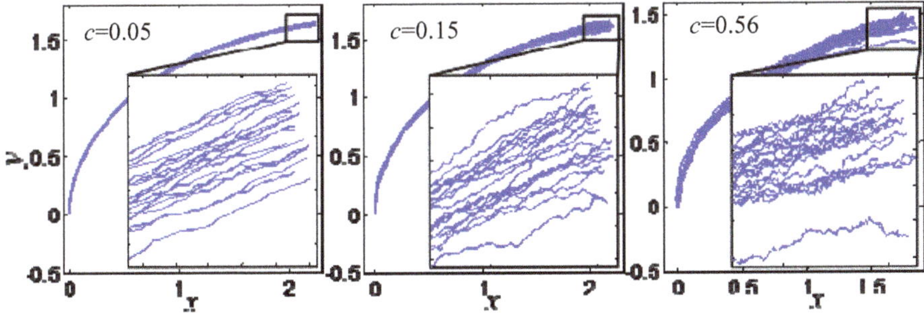

Figure 2. Growth of growth cone under different intensities of random motion.

According to the growth details that can be seen in Figure 2, the growth of the growth cone is relatively smooth around $c = 0.05$ with small random fluctuations. The random fluctuations increases gradually around $c = 0.15$, but the deflection angle of the growth cone does not change extremely, and the growth of the growth cone is still in order. However, when $c = 0.56$, the trajectory of the growth cone changes abruptly and the partial shape of the growth cone is obviously out-of-order. Therefore, the randomness of the growing process of the growth cone increases gradually with the increase of the effect intensity of random motion.

3. Establishment of Neuronal Axonal Signal Transmission Model

In order to describe the signal transmission process of the neurons listed above, a simplified model of neuron signal transmission, as shown in Figure 3, is established in this paper.

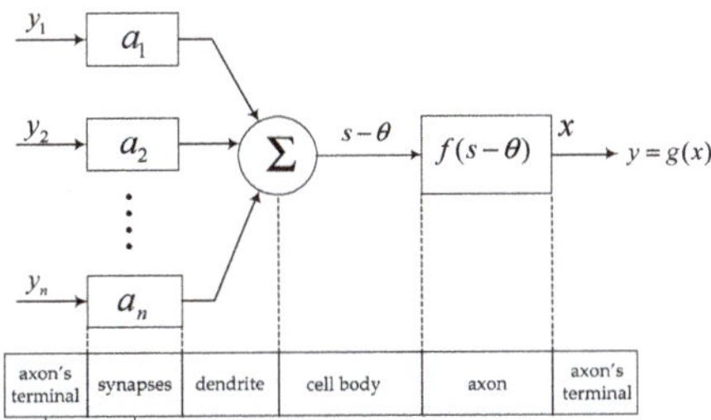

Figure 3. The signal transmission model of the neuron.

If there are n neurons on a neuron's dendrite to form synaptic connections with it, the output signals of these n neurons are $y_1, y_2, \ldots, y_n$, the connection weights of synapses are $a_1, a_2, \ldots, a_n$, the change of membrane potential caused by the input is s and the excitation threshold is θ; the membrane potential is x when reaching the axon's terminal after being transmitted through the whole axon, and the neuron output signal is y. Then, the signal transmission process of the neuron is

$$
\begin{cases}
s = \sum_{i=1}^{n} a_i y_i \\
x = f(s - \theta) \\
y = g(x)
\end{cases}
\tag{10}
$$

where $f(.)$ represents the function of an axon on membrane potential and $g(.)$ represents the nonlinear output function of neurons, which is different for different biological neurons. In this paper, the commonly used threshold function is taken; i.e., $g(x) = \max(0, x)$.

In 1957, W. Rall proposed the equivalent cable model of an axon during signal transmission [22], which regarded the cable as composed of many identical resistance capacitance circuits, considered the current flow in extracellular fluid and the distribution of extracellular potential and finally obtained the distribution of membrane potential in time space $\lambda^2(\partial^2 V/\partial x^2) = \tau(\partial V/\partial t)+V$, where V is the potential of the membrane and λ and τ are the time constant and length constant of the thin film of the cable joint. This model is a second-order partial differential equation, and the solution is relatively complex. In this paper, we only consider the input and output characteristics of the neuron as a signal processing unit instead of the distribution of membrane potential on the whole axon, so we assume that the extracellular potential along the whole axon is always 0—equivalent to grounding—as can be seen in Figure 4. Thus, U is equal to the membrane potential. Given the output change of an input signal after the membrane potential is transmitted through the whole axon, we can calculate $f(.)$ as shown in Formula (10).

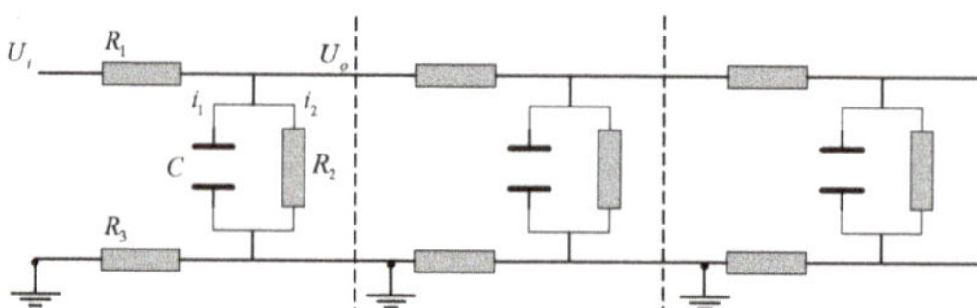

Figure 4. The cable model without considering the extracellular potential.

First, we analyze the basic resistance capacitance circuit in Figure 4 separately and calculate the transfer function between the output U_0 and the input U_i of each section. In Figure 4, i is the current, R is the resistance and C is the capacitance [18]. The following equation can be obtained from Kirchhoff's law and Ohm's law.

$$
i_1 = C\frac{dU_o}{dt}
\tag{11}
$$

$$
U_o = i_2 R_2
\tag{12}
$$

$$
U_i = (i_1 + i_2)(R_1 + R_3) + U_o
\tag{13}
$$

Equations (11) and (12) are substituted into Equation (13):

$$
U_i = (C\frac{dU_o}{dt} + \frac{U_o}{R_2})(R_1 + R_3) + U_o
\tag{14}
$$

The Laplace transform of Equation (14) is

$$
U_i(s) = (sCU_0(s) + \frac{U_o(s)}{R_2})(R_1 + R_3) + U_o(s)
\tag{15}
$$

Finally, we obtian the transfer function between U_0 and U_i:

$$G(s) = \frac{U_o(s)}{U_i(s)} = \frac{1}{C(R_1 + R_3)s + \frac{R_1+R_2+R_3}{R_2}} \tag{16}$$

Assume τ and b as Equation (17):

$$\begin{cases} \tau = C(R_1 + R_3) \\ b = \frac{R_1+R_2+R_3}{R_2} \end{cases} \tag{17}$$

Then, Equation (16) is

$$G(s) = \frac{1}{\tau s + b} \tag{18}$$

As can be seen from Figure 4, the total transfer function of the axon to the membrane potential is $F(S) = G^N(S)$. N is the number of basic links on the whole axon. Thus far, the signal transmission model of the neuron has been determined. Finally, the amplitude–frequency and phase–frequency characteristics of $G(S)$ are analyzed to determine the physical characteristics of the neuron self-growing model:

$$G(j\omega) = \frac{1}{\tau j\omega + b} \tag{19}$$

where ω is the angular frequency.

The amplitude–frequency characteristic of Equation (19) is

$$|G(j\omega)| = \frac{1}{\sqrt{\tau^2\omega^2 + b^2}} \tag{20}$$

The phase–frequency characteristic of Equation (19) is

$$\angle G(j\omega) = -\arctan\left(\frac{\tau\omega}{b}\right) \tag{21}$$

(1) Influence of axon length on amplitude and phase lag

Suppose an axon cable is composed of N basic links as shown in Figure 4; then, the amplitude of the input signal is attenuated to $(1/\sqrt{\tau^2\omega^2 + b^2})^N$ after being transmitted through the axon, and the phase lag is $-N\arctan(\tau\omega/b)$. The longer the axon is, the more basic links in the corresponding cable model there should be; in other words, the larger N is. We can see that $b > 1$ from Equation (17), and so the longer the axon is, the larger N is and the larger the amplitude attenuation and phase lag are. The length of the axon can be obtained by the iteration in Equation (8). If the length of the axon corresponding to the basic link of a resistance capacitance loop is l_1 and the total length of the axon is L [23], then the amplitude of the signal will decay to $(1/\sqrt{\tau^2\omega^2 + b^2})^{L/l_1}$ and the phase lag will be $-N\arctan(\tau\omega/b)$ after passing through the whole axon.

(2) Diameter of the axon

As can be seen from Equations (20) and (21), the amplitude–frequency and phase–frequency characteristics of the neuron signal transmission are determined by the two parameters τ and b. In Equations (16) and (17), R_2 is the electrical leakage resistance between the inside and outside of the cell within a basic link, whose resistance value is much higher than R_1 and R_3, which is equivalent to the myelin sheath of the outer layer of the axon of biological neurons and plays an insulating role. C is the leakage capacitance between the inside and outside of the cell within a basic link. R_3 is the conductive resistance of the extracellular fluid within a basic link. R_2, R_3 and C are the same for every neuron and do not change with the gradual growth of axons. R_1 represents the resistance of the axon core within a basic link, which can be obtained according to the resistance law:

$$R_1 = \rho \frac{l_1}{s} \tag{22}$$

where ρ is the resistivity of the inner core of the axon, s is the cross-sectional area of the axon—$s = \pi^2 d/4$ since the cable model is cylindrical—and d is the diameter of the axon. The axon diameters of different neurons determine the amplitude–frequency and phase–frequency characteristics of the basic link in the cable model.

(3) Connection weight to the postsynaptic neuron

Since $|G(j\omega)| < 1$, the cable model shown in Figure 4 must cause the attenuation of signal amplitude during signal transmission. However, we occasionally want a single neuron to transmit the signal without changing its amplitude when controlling, only changing its phase. For example, in the CPG control network of the joint of a legged robot, each joint is controlled by a motor neuron with joint angular displacement output, and the output signals of the motor neuron in the symmetrical position of the body need to have the same amplitude, because the joint angle of these joints has the same range. If all the intermediate neurons in the network do not change the amplitude of the signal during signal transmission, the output of each joint with the same amplitude can be obtained from the input with the same amplitude. In this paper, a proportional amplification link is artificially added after each basic link to prevent the change of signal transmission amplitude, as shown in Figure 5.

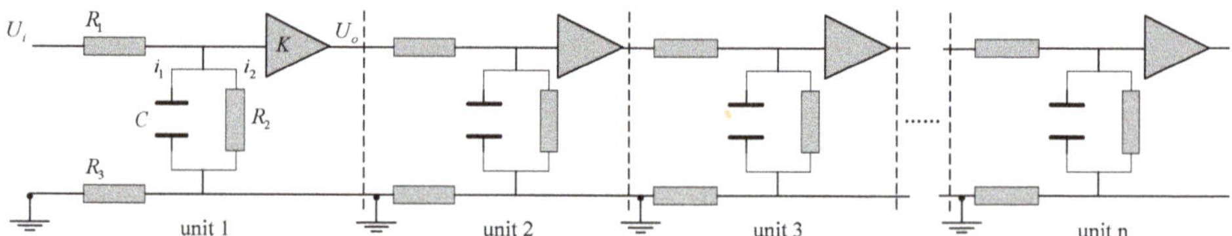

Figure 5. Cable model with a proportional amplification link artificially added.

According to Equation (20), we can obtain K in Figure 5:

$$K = \frac{1}{|G(j\omega)|} = \sqrt{\tau^2\omega^2 + b^2} \tag{23}$$

As can be seen from Equation (23), the amplitude of the output signal from the original single basic link remains constant after passing the proportional amplification link. The output signal of the whole axon is equivalent to multiplying the original output by K^N. According to the definition of connection weight, this is equivalent to the connection weight between the neuron and the postsynaptic neuron:

$$a = K^N = \left(\sqrt{\tau^2\omega^2 + b^2}\right)^N \tag{24}$$

If the input signal of the neuron is an ideal impulse signal or step signal, then we can say that $\omega = 0$. According to Equation (17), Equation (24) is

$$a = b^N = \left(\frac{R_1 + R_2 + R_3}{R_2}\right)^N \tag{25}$$

Because R_2 is much larger than R_1 and R_3, we know that b tends to 1. According to $N = L/l_1$, we can select l_1 reasonably so that the order of magnitude of N will not be too large, and this will mean that the connection weights between all the neurons in the network meet $a \in (1, 2)$.

In Equation (18), the transition time of the first-order system's response to the impulse signal and the step signal—called the delay—is proportional to the time constant τ, which is assumed to be $k\tau$, where k is a constant coefficient related to the allowable error value of the response. If there are N basic links on the whole axon, as shown in Figure 5, the total delay of the output signal relative to the input signal will be $Nk\tau = Lk\tau/l_1$, which is equivalent to the delay of the output of a basic link with a time constant of $L\tau/l_1$ to the input:

$$Tr = L\tau/l_1 \tag{26}$$

Therefore, when the input signal is an impulse signal or a step signal, the self-growing model of a neuron with multiple physical properties can be simplified as follows: the whole neuron is regarded as a basic link, and the length of the axon obtained from the self-growing is used as the total time constant T_r of the neuron to the input signal. A random number from $(1, 2)$ is selected as the connection weight value a between the neuron and the postsynaptic neuron. In addition, the ends of the axons of biological neurons diverge and form synaptic connections with dendrites of many other neurons; thus, multiple axons of the neurons are made to grow directly from the cell body without considering the choice of branching points to simplify the model.

4. Structure of Local CPG Network

(1) Study of neuron model

In 1987, Matsuoka proposed a mathematical neuron monomer model and studied the "oscillator" network output with different numbers of neurons and different connection structures based on it [24]. The mathematical model of the neuron monomer is

$$\begin{cases} T_r\frac{dx}{dt} + x = s - bf \\ y = g(x - \theta), \quad (g(x) = \max\{0, x\}) \\ T_a\frac{df}{dt} + f = y \end{cases} \tag{27}$$

where x represents the membrane potential of the neuron; s stands for neuron input; y is the output of the neuron; f is the quantity reflecting the fatigue effect of neurons; b is a constant coefficient representing the size of fatigue effect; θ presents the membrane potential threshold that activates the neuron, and without losing generality, we can take $\theta = 0$; $g(x)$ is the nonlinear output function; and T_r represents the time constant of the rise of the membrane potential caused by the input signal. T_a represents the time constant of the neuronal fatigue effect. Since the physical characteristics of neurons of the model were not taken into account, there is no concept of the axon, and thus the membrane potential will not change after axon transmission. We continue to regard x as the membrane potential at the end of the axon, and so this model differs from the model established in Equation (10) by only one fatigue characteristic characterized by f and T_a. Set s as the step input signal with n amplitude of 1, $T_r = 1$, $b = 10$, $T_a = 10$, and calculate the change of the output signal of the neuron monomer model over time; the results are shown in Figure 6.

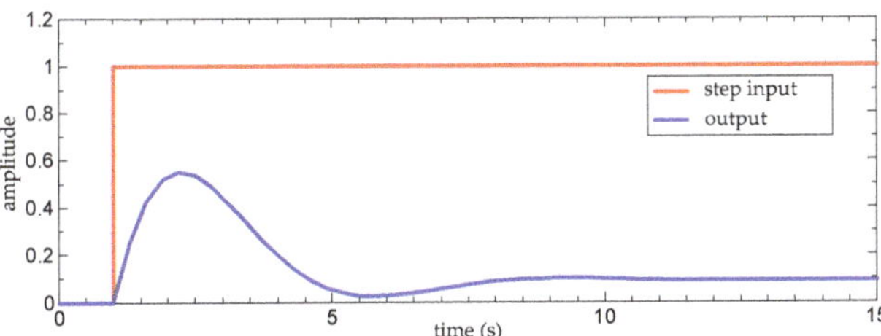

Figure 6. Step response of the neuron model.

As can be seen from Figure 6, first, the neuron is activated and the membrane potential rises; then, the membrane potential declines slowly due to the fatigue effect and remains at a low level. Although the step signal continues, the neuron will not be activated again. We can see that, under appropriate parameters, the mathematical model of the neuron monomer conforms to the signal transmission characteristics of biological neurons.

(2) Double neural oscillator model

Connect two neuron monomer models and make the two inhibit each other as shown in Figure 7. If N1 and N2 are the neurons, the hollow endpoint connection presents the excitatory connection and the solid endpoint connection presents the inhibitory connection; S1 and S2 are the step input signals with an amplitude of 1 with slightly different step times. If S1 and S2 have the same step time, the two neurons will have the same output and will not produce an oscillation because of the symmetrical structure of the network.

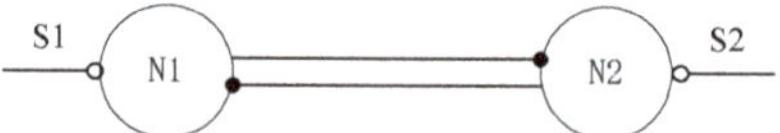

Figure 7. Network of two neurons that inhibit each other.

The mathematical model of each neuron monomer is

$$
\begin{cases}
T_{ri}\frac{dx_i}{dt} + x_i = s_i - b_i f_i - \sum\limits_{j=1}^{n} a_{ji} y_j \\
y_i = \max\{0, x_i\} \\
T_{ai}\frac{df_i}{dt} + f_i = y_i
\end{cases}
\tag{28}
$$

where $\sum a_{ji} y_j$ are the inputs from other neurons. If the input is excitatory, the operator preceding it has a plus sign, with a minus sign for the inhibitory input.

Assume $T_{r1} = T_{r2} = 1$, $b_1 = b_2 = 2.5$, $a_{21} = a_{12} = 1.5$ and $T_{a1} = T_{a2} = 12$ and calculate the variation of the output signal of the oscillator model with time. The results are shown in Figure 8.

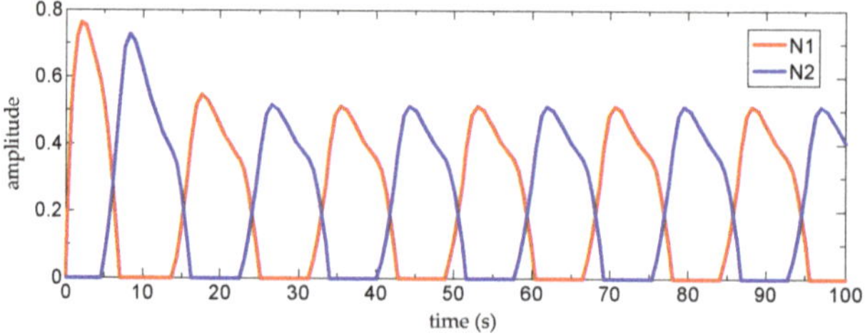

Figure 8. Reciprocal output of two neurons.

In the spinal CPG network of mammals, the interactivity of neurons responsible for the coordination of the left and right sides of the body and the coordination of the extensor and flexor muscles of each joint inhibits the output [25]. It can be seen in Figure 5 that the reciprocal inhibition outputs of the two neurons have the same form as the reciprocal inhibition outputs of the neurons in the biological CPG network, indicating that the neuron monomer model in Equation (26) can reflect the characteristics of the reciprocal inhibition output caused by the reciprocal inhibition structure in the biological CPG network.

(3) Connection structure of local CPG network

As can be seen from the analysis above, as long as the neuron model represented by Equation (27) is composed of inhibitory connections among the neurons and appropriate parameters are taken, the interactive inhibitory output between the neurons will be obtained. Since each neuron monomer in the neuron model represented in Equation (27) has two time constants, *Tr* and *Ta*, and each neuron in the simplified model of self-growing neurons established in Section 3 has only one time constant, *Tr*, we cannot make the growth algorithm of single neuron grow into a local CPG network which has the function of controlling a robot joint. In order to solve this contradiction, we modify the original neuron monomer model, as shown in Equation (29), according to the characteristics of the neuron connections in a biological CPG network.

$$\begin{cases} T_{ri}\frac{dx_i}{dt} + x_i = s_i - \sum_{j=1}^{n} a_{ji}y_j \\ y_i = \max\{0, x_i\} \end{cases} \tag{29}$$

The original neuron monomer model is described with the structure shown in Figure 6. We can see that the excitatory connection from the input neurons to the output neurons in Figure 9 is equivalent to the input signal *S* of the original model. The inhibition from the Renshaw Cell to the output neurons is equivalent to the fatigue characteristic in the original model; thus, the input and output characteristics of signal transmission of the structure shown in Figure 9 are essentially the same as those in the Matsuoka neuron monomer model.

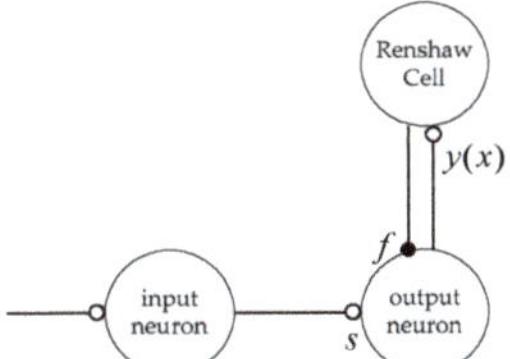

Figure 9. Equivalent model of the original neuron monomer model.

According to Equation (29), the modified neuron monomer model only contains one time constant, *Tr*, and each axon contains a connection weight *a* to the postsynaptic neuron, which is consistent with the simplified model of self-growing neurons established in Section 3. Moreover, the connection structure shown in Figure 6 has the anatomical basis of the biological CPG network. Known as an inhibitory intermediate neuron directly connected to motor neurons, the Renshaw Cell is involved in the coordination of most extensor and flexor muscles as part of the ipsilateral inhibitory network [14,25]. Moreover, the structure in Figure 6 can be seen in many models of spinal motor nerve circuits [26,27].

A local CPG network for the control of a quadruped robot's hip joint can be formed with four identical structures, as shown in Figure 10. In Figure 10, N1, N2, N3 and N4 are the output neurons, respectively corresponding to the four hips of the quadruped robot. N5, N6, N7 and N8 are input neurons, and S1, S2, S3 and S4 are step input signals of the same amplitude. N9, N10, N11 and N12 are four output neurons connected with Renshaw Cells, *Tri* is the time constant of each neuron and $a_{i,j}$ is the connection weight of the synapse between the axon of neuron *i* and the neuron *j*; the connection weights between peripheral neurons are not shown in Figure 10. The network structure is the growth target of the local CPG network self-growing algorithm.

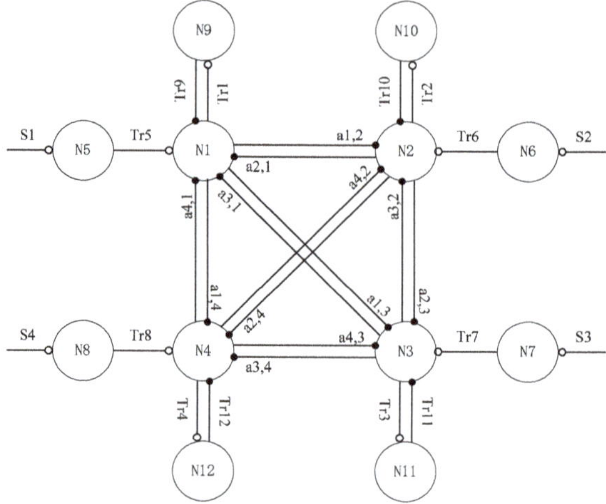

Figure 10. Local Central Pattern Generator (CPG) network connection structure for the hip joint control of a quadruped robot.

(4) Output of local CPG network

One of the basic features of a biological CPG network is that it can obtain a periodic output signal with a certain rhythm from a simple non-rhythmic input signal. In order to verify that the network growth model proposed in this paper can obtain growth results with biological CPG network characteristics, a CPG network growth experiment was carried out and is presented in this section. Assume that the radius r of a single neuron in the growth model is 0.1 units of length and the growth space is a cube with a side length of 5; 12 neurons were randomly distributed in the growth space as shown in Figure 11.

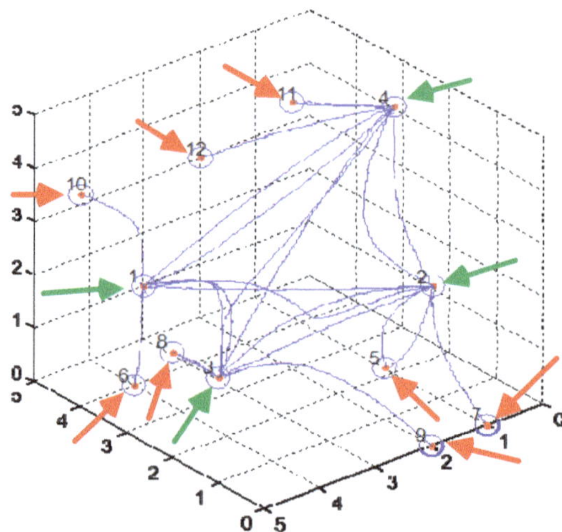

Figure 11. Twelve randomly distributed neurons and the connection structure arising from the network.

In Figure 11, a circle represents a neuron, and the number (1, 2, … , 12) next to it shows the number of neurons. The neurons numbered 1, 2, 3 and 4 are the output neurons, represented by N1, N2, N3 and N4 (as shown by the green arrow in Figure 11); the neurons numbered 5–12 are peripheral neurons, represented by N5–N12. Numbers N5, N6, N7 and N8 are input neurons (as shown by the red arrow in Figure 11). The solid blue lines are the growing connections, as shown in Figure 11.

For simplicity, take $l_1 = r = 0.1$, $\tau = 10$. Putting the length of the blue solid line in Figure 11 into Equation (26), the value of Tr can be obtained as shown in Table 1.

Table 1. The time constant of the axon in the network.

Time Constant	Length of Corresponding Axon	Value
T_{r1}	Length of axon between output neuron N1 and Renshaw Cell N6	2.1801
T_{r2}	Length of axon between output neuron N2 and Renshaw Cell N5	2.5868
T_{r3}	Length of axon between output neuron N3 and Renshaw Cell N8	1.1984
T_{r4}	Length of axon between output neuron N4 and Renshaw Cell N11	1.9691
T_{r5}	Length of axon of Renshaw Cell N5	2.8071
T_{r6}	Length of axon of Renshaw Cell N6	2.2821
T_{r7}	Length of axon of input neuron N7	3.7401
T_{r8}	Length of axon of Renshaw Cell N8	1.2096
T_{r9}	Length of axon of input neuron N9	5.4620
T_{r10}	Length of axon of input neuron N10	4.8037
T_{r11}	Length of axon of Renshaw Cell N11	1.8388
T_{r12}	Length of axon of input neuron N12	4.6752
$T_{r13}-T_{r24}$	Length of interactional axon between N1, N2, N3 and N4	0

Step input signals with amplitude of 1 were input to the four input neurons. Observe the output signals of the network; a representative growth result was taken for discussion. The connection structure of the network growth is shown in Figure 11. In Figure 11, each neuron axon contains two parameters: the time constant Tr and the connection weight a. Since there are 24 axons in the network, there are 48 parameters with 24 time constants and 24 connection weights in the network. Such a large number of parameters in the network is not conducive to the study of the rule between the output and growth results of the network. In order to facilitate the analysis, the axons in the network are divided into two categories: in the first category are the axons that inhibit the connection between the output neurons, with a total of 12; the second category are the axons connected with the peripheral neuron (the input neuron and the Renshaw Cell) between the output neuron, with a total of 12. For the first type of axons, only the connection weights in the growth results are used, and the time constants are all set to 0. For the second type of axon, only the time constant in the growth result is used, and the connection weights are all fixed values. This halves the network's parameter variables. The time constants of the axons of each neuron after a certain growing process are shown in Table 1.

According to Equation (25), the connection weight $a \in (1, 2)$ can be obtained. In order to simplify the calculation, a random number between (1, 2) is taken as the value of a. The connection weights obtained are shown in Table 2, and the connection weights between the output neuron and the peripheral neuron are set to 1 and 2, which meet $a \in [1, 2]$. The output result of the entire network is calculated according to Equation (29), as shown in Figure 12.

Table 2. The connection weight of neuronal axons in the network.

$a_{10,1}$	$a_{7,2}$	$a_{9,3}$	$a_{12,4}$	$a_{1,6}$	$a_{2,5}$	$a_{3,8}$	$a_{4,11}$	$a_{6,1}$	$a_{5,2}$	$a_{8,3}$	$a_{11,4}$
1	1	1	1	1	1	1	1	2	2	2	2

$a_{1,2}$	$a_{1,3}$	$a_{1,4}$	$a_{2,1}$	$a_{2,3}$	$a_{2,4}$	$a_{3,1}$	$a_{3,2}$	$a_{3,4}$	$a_{4,1}$	$a_{4,2}$	$a_{4,3}$
1.24	1.49	1.64	1.90	1.23	1.72	1.78	1.15	1.92	1.77	1.16	1.31

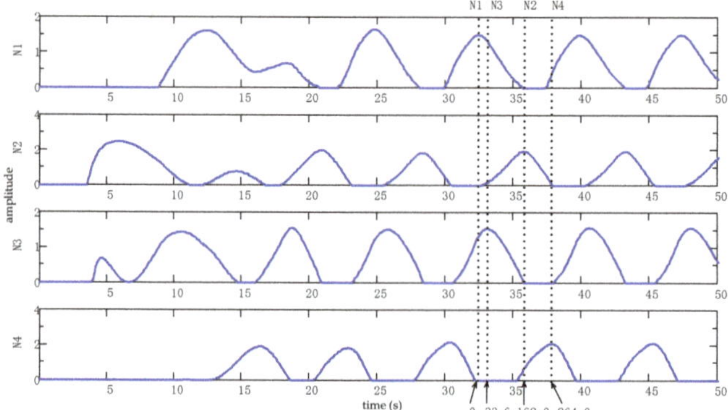

Figure 12. Output signal of the local CPG network.

As can be seen from Figure 12, the output signals of the four output neurons have the same period and phase difference and a similar range of amplitude after a period of stability. Furthermore, it is verified that the self-growing network model established in this paper can obtain a local CPG network which can satisfy the biological CPG network input and output characteristics, and the output of the network can be used for the control of the hip joint of quadruped robot. If the output signal of neuron N1 is used as the basis to calculate the phase difference, it can be determined that the output phase difference between N2 and N1 is 168.0°, the output phase difference between N3 and N1 is 33.6° and the output phase difference between N4 and N1 is 264.0°. The ideal phase differences of the corresponding hip joints of the quadruped robot at the walk gait are 180°, 90° and 270°. We can see that the phase difference between N3 and N1 is large, but the other two phase differences are very close, which shows that the local CPG network obtained by self-growing can output the corresponding control signals for the hip joint to the quadruped robot at a walking gait.

5. Experiment and Analysis

(1) Quadruped robot prototype

The experimental platform of the quadruped robot is shown in Figure 13. There are three joints on each leg: the hip joint α, knee joint β and ankle joint θ. The size of the platform is 1.2 m × 0.5 m × 1.4 m, and the weight is 150 kg. Figure 13a is the simulation prototype of the quadruped robot, and Figure 13b is the experiment of the quadruped robot.

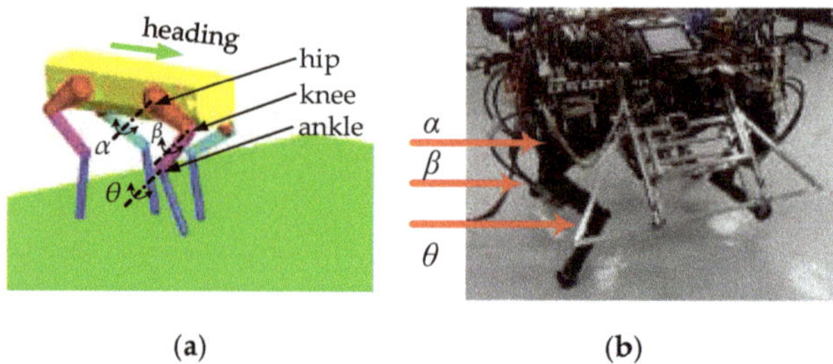

(a)   (b)

Figure 13. Quadruped robot. (**a**) Simulation prototype of quadruped robot. (**b**) Experimental prototype of quadruped robot.

(2) Foot trajectory and joint trajectory planning

Since the ideal angular displacement of the hip joint can be approximated as a cosine function [14], we assume that the control signal of the hip joint is $\alpha = 3\cos(2\pi t/T)$. The trajectory of the foot is a cycloid, as shown in Figure 14b. The height of the robot's hip joint is $h = 100$ cm from the ground, and the step length of each stride is 30 cm. The former leg is taken as an example to calculate the joint angular displacement and that of the knee and ankle joints, as shown in Figure 14a; thus, we obtain the following formula:

$$\left. \begin{array}{l} l_1 \cos\alpha - l_2 \cos(\beta - \alpha) + l_3 \cos[\theta - (\beta - \alpha)] = x \\ l_1 \sin\alpha + l_2 \sin(\beta - \alpha) + l_3 \sin[\theta - (\beta - \alpha)] = h - y \end{array} \right\} \tag{30}$$

where $l_1 = 45$ cm, $l_2 = 60$ cm and $l_3 = 50$ are the lengths of the robot's thigh, middle leg and calf; x and y are the analytic coordinates of the foot trajectory, as shown in Figure 14c, where α and h are known. Thus, Equation (30) is a system of two equations with two unknowns, β and θ, whose solution is unique. The calculated relationship between β, θ and α is shown in Figure 15. α ranges from 30° to 50°, which can be seen from the solid blue line in Figure 15, in order to make sure that the quadruped robot has a large stability margin when walking. The joint trajectory planning process of the hind leg of the robot is the same as that of the foreleg except for the different range of joint angles and the different shape of the curve.

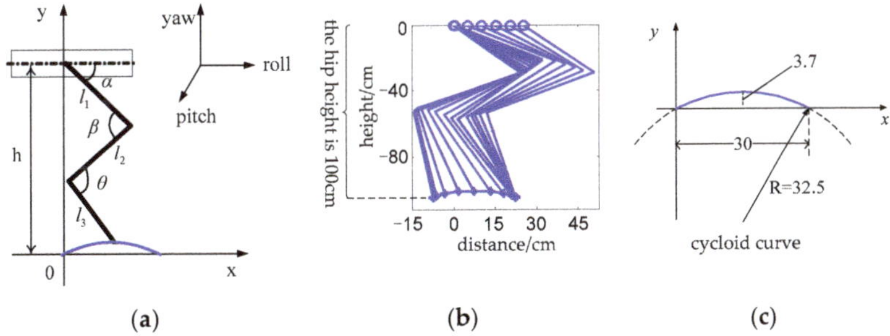

(a)   (b)   (c)

Figure 14. Foot trajectory of robot. (**a**) Structure diagram of one leg. (**b**) Experimental prototype of quadruped robot. (**c**) The analytic coordinates of foot trajectory.

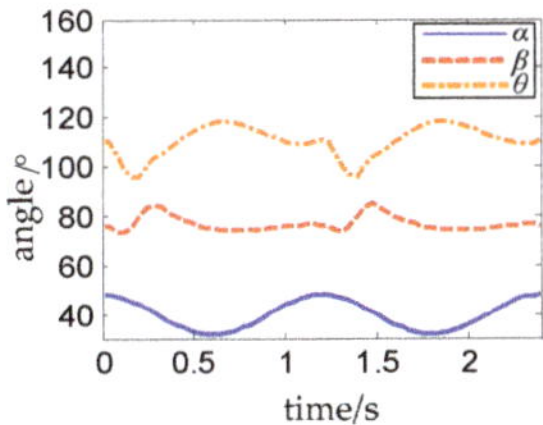

Figure 15. Trajectory planning of the knee and ankle joints.

Thus, the experimental prototype of the quadruped robot used for local CPG network gait control has been completed. It is only necessary to input periodic control signals corresponding to the angular displacements of the four hip joints into the prototype, and then the control signals of the knee and ankle joints of each leg will be obtained according to the corresponding relations in Figure 15; thus, we can drive the robot prototype to walk.

(3) Experiment on simulation prototype and experimental prototype

The random selected parameters $a \in [1, 2]$ of the self-growing network are shown in Table 3. The time constant Tr is shown in Table 1. After calculating the output of the network and cosine to fit the output signal according to Equation (29), the result is shown in Figure 16. We can see that the output of the network is definitely the interaction oscillations between N1, N3, N2, and N4. N1 and N3 are input to the left front and the right hind hip joint of the simulation prototype, while N2 and N4 are input to the right front and the left hind hip joint of simulation prototype; then, the trot gait of the robot is implemented, and the simulation decompositions are shown in Figure 17a. As can be seen from Figure 17a, from $t = 0.2$ s to $t = 1.7$ s, the left front leg and the right hind leg are in the swing phase and swing forward; from $t = 1.9$ s to $t = 3.4$ s, the right front leg and the left hind leg are in the swing phase and swing forward. Each swing time is about 1.6 s.

Table 3. The connection weight of neuronal axons in the network.

$a_{9,1}$	$a_{8,2}$	$a_{12,3}$	$a_{11,4}$	$a_{1,10}$	$a_{2,6}$	$a_{3,7}$	$a_{4,5}$	$a_{10,1}$	$a_{6,2}$	$a_{7,3}$	$a_{5,4}$
1	1	1	1	1	1	1	1	2	2	2	2

$a_{2,1}$	$a_{3,1}$	$a_{4,1}$	$a_{1,2}$	$a_{3,2}$	$a_{4,2}$	$a_{1,3}$	$a_{2,3}$	$a_{4,3}$	$a_{1,4}$	$a_{2,4}$	$a_{3,4}$
1.19	0	1.22	1.25	1.56	0	0	1.41	1.88	1.84	0	1.64

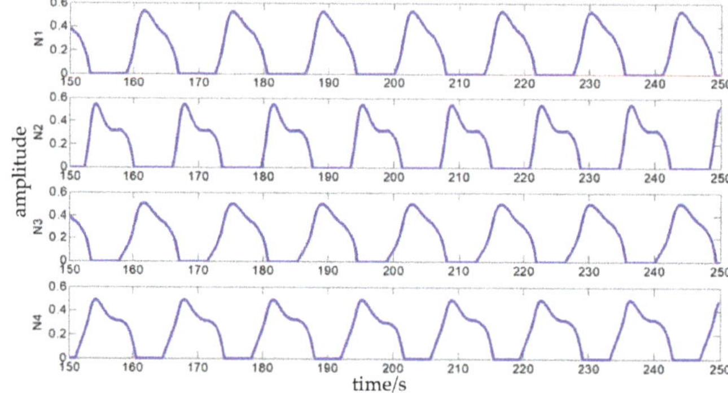

Figure 16. Output of self-growing network in trot gait.

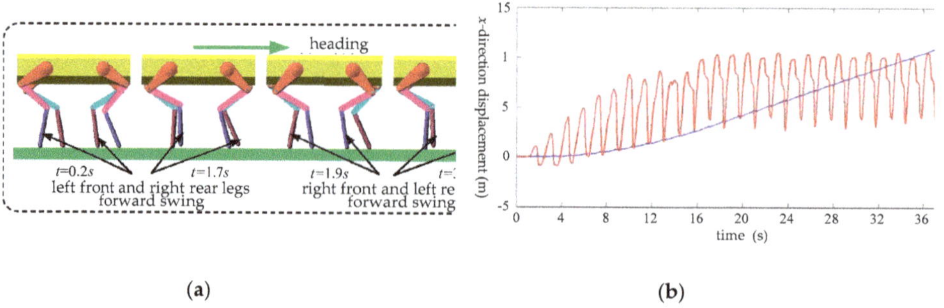

(**a**)   (**b**)

Figure 17. Simulation decomposition diagram of the trot gait of the simulation prototype. (**a**) Simulation diagram of trot gait. (**b**) The y-direction velocity and x-direction displacement curve of a quadruped robot.

The y-direction velocity and x-direction displacement of the simulation prototype are shown in Figure 17b, where the red solid line is the y-direction velocity curve and the blue solid line is the x-direction displacement curve. It can be seen from Figure 17b that the simulation prototype enters a stable walking state gradually, and the y-direction velocity and x-direction displacement reach a stable change when $t = 16$ s.

In Figure 16, N1 and N3 in the output of the self-growing network are input to the left front and the right hind hip joint of the experimental prototype, respectively, while N2 and N4 are input to the right front and the left hind hip joint of the experimental prototype respectively. The trot gait decompositions of the experimental prototype obtained are shown in Figure 18. Combining Figure 17a and Figure 18, we can see that the experimental prototype and the simulation prototype have the same gait cycle (about $T = 1.7$ s) under the control of the same self-growing network, and the self-growing network can achieve stable control of the trot gait of the simulation prototype and the experimental prototype.

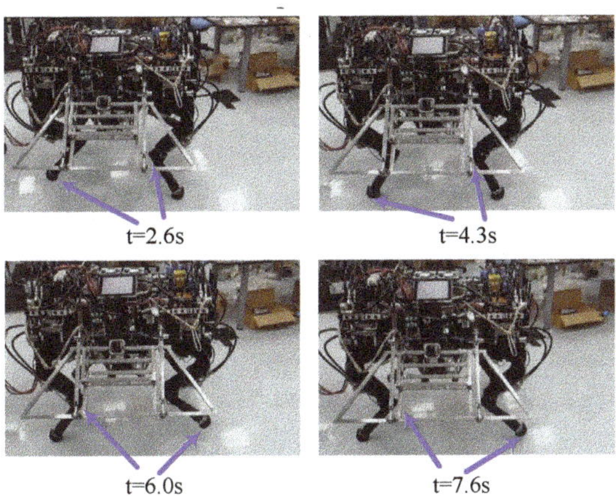

Figure 18. Trot gait diagram of experimental prototype.

The experiment above shows that the quadruped robot control system based on a CPG neural network self-growing algorithm can accomplish the growth of the network, the signal output of the network and the control of a robot's joint, and then realize the rhythmic movement of a quadruped robot. This shows that the CPG neural network self-growing algorithm can be well applied to the rhythm control of a robot. The experiment preliminarily simulated the process from growth to mastery of a certain rhythmic movement of the CPG neural network and verifies the feasibility of the idea of realizing robot rhythmic movement control by combining the microscopic mechanism of the growth and the macro characteristics of the output of the neural network.

6. Conclusions

In this paper, the biological neuron axon could continuously elongate and turn, and the growth of the back part did not change the shape of the previously used grown part. Combined with the law of gravitation, random motion, the Rall cable model and the Matsuoka oscillator model, a local CPG self-growing network with multiple physical characteristics was established. The fusion of a self-growing network and CPG has been realized, the control ability of self-growing network was presented, and the motion control of a quadruped robot was completed. Our conclusions are as follows.

(1) An axon self-growing algorithm based on universal gravitation and random motion was established, the Rall Cable Model for studying the potential distribution of an axon membrane was analyzed and simplified, and multiple physical properties of the synapse (including resistance, capacitance, axon length and diameter, etc.) were combined with the Rall Cable Model to establish a simplified model of single neuron signal transmission, as shown in Equation (19). The multi-physical properties obtained by the neuron self-growth model (such as the synapse length calculated by Equation (9) and the resistance R1 calculated by Equation (22)) were added to Equation (17) to calculate parameters τ and b of Equation (19). On this basis, the entire synapse system of the neuron's self-growth is taken as a whole, as well as the key parameters in the control model: the delay time constant Tr and the synapse, as shown in Equations (25) and (26). The connection weight corresponds to the axon

length and axon diameter in the growth model, which allows the combination of the growth and development process of the neuron and the control function.

(2) By analyzing the Matsuoka oscillator model (shown in Equation (27)) and comparing the difference between the mathematical model of its neuron monomer and the model built in this article, a local CPG network topology is proposed as shown in Figure 10. This topology is regarded as the goal of network self-growth. A neural network growth model based on gravitational field control is established, meaning that the network can grow the target topology from any initial position of the neuron, as shown in Figure 11, to obtain the rhythm output signal corresponding to the angular displacement of the hip joint of the quadruped robot, as shown in Figure 12.

(3) The quadruped robot motion control simulation prototype and experimental prototype are used as shown in Figure 13: the output signal of the local CPG network (as shown in Figure 16) is used as the hip joint control signal of the simulation prototype and the experimental prototype to control the quadruped robot simulation. The prototype and the experimental prototype realized walking with a trot gait, as shown in Figures 17 and 18, which verified the feasibility of the local CPG network self-growth model with multi-physical characteristics proposed in this paper for the gait control of a quadruped robot.

Since the gait control of the quadruped robot by the CPG network in this article is still an open-loop control system, it was not possible to change the gait of the robot through environmental feedback information or to adapt to the complex road conditions and change the rhythm and mode of the network output accordingly. Therefore, in order to give the self-growth-based network model higher practical value when applied to the motion control of a footed robot, the research work that needs to be done in the next stage is as follows:

(1) Continue to improve the self-growth model of neurons, considering more physical characteristics that can be linked to the control model parameters, so that the built model can be constantly close to the complexity of the biological neural network in terms of its microstructure and control function. For example, in this paper, the diameter of the neuron axon is constant. The fundamental reason for this is that only a one-dimensional size of the axon can be obtained by using the trajectory of the particle in the gravitational field as the axon model. The biological mechanism that determines the diameter must be introduced into the self-growth model to obtain a biologically reasonable diameter, so that the control model does not lose its general applicability.

(2) Study how many parameters obtained by network self-growth affect the network output. In the simulation experiment, we found that the time constant only affects the shape, amplitude and period of the output signal and does not affect the mutual inhibition mode, which is mainly affected by the connection weight. Moreover, not every time the network grows can the interactive inhibition output between output neurons be obtained. On occasion, individual or all output neurons will output a constant value signal after reaching stability; that is, the acquisition of the interactive inhibition output is important for the network growth parameters. This is conditional, but we still do not know the necessary condition.

(3) Study the micro-mechanism of interaction with the environment and the adjustment of the upper central system during the development of the CPG network and construct feedforward and feedback pathways for network development, so that the growth and development of the network can change the connection weights between synapses in biological neural networks by learning. Based on the research of (2), weight change is carried out in the direction in which the desired network output can be obtained to realize closed-loop control.

Author Contributions: Conceptualization, M.L. (Ming Liu) and F.Z.; methodology, M.L. (Ming Liu) and F.Z.; software, M.L. (Ming Liu); validation, M.L. (Ming Liu), M.L. (Mantian Li) and F.Z.; formal analysis, M.L. (Ming Liu); investigation, M.L. (Ming Liu); resources, P.W.; data curation, W.G.; writing—original draft preparation, M.L. (Ming Liu), M.L. (Mantian Li) and F.Z.; writing—review and editing, L.S.; visualization, M.L. (Ming Liu); supervision, M.L. (Mantian Li) and L.S.; project administration, M.L. (Mantian Li) and F.Z.; funding acquisition, F.Z. All authors have read and agreed to the published version of the manuscript.

Funding: This work was supported by the Natural Science Foundation of China (Grant No.61773139), The Foundation for Innovative Research Groups of the National Natural Science Foundation of China (Grant No.51521003), Shenzhen Science and Technology Research and Development Fundation (Grant No.JCYJ20190813171009236) and Shenzhen Science and Technology Program (Grant No.KQTD2016112515134654).

Conflicts of Interest: The authors declare no conflict of interest.

References

1. Endo, G.; Morimoto, J.; Matsubara, T.; Nakanishi, J.; Cheng, G. Learning CPG-based Biped Locomotion with a Policy Gradient Method: Application to a Humanoid Robot. *Int. J. Robot. Res.* **2008**, *27*, 213–228. [CrossRef]
2. Maufroy, C.; Kimura, H.; Takase, K. Integration of Posture and Rhythmic Motion Controls in Quadrupedal Dynamic Walking Using Phase Modulations Based on Leg Loading/Unloading. *Auton. Robot.* **2010**, *28*, 331–353. [CrossRef]
3. Ijspeert, A.J. A connectionist central pattern generator for the aquatic and terrestrial gaits of a simulated salamander. *Boil. Cybern.* **2001**, *84*, 331–348. [CrossRef] [PubMed]
4. Von Twickel, A.; Pasemann, F. Reflex-Oscillations in Evolved Single Leg Neuron Controllers for Walking Machines. *Nat. Comput.* **2007**, *6*, 311–337. [CrossRef]
5. Noble, F.K.; Potgieter, J.; Xu, W.L. Modelling and simulations of a central pattern generator controlled, antagonistically actuated limb joint. In Proceedings of the 2011 IEEE International Conference on Systems, Man, and Cybernetics, Anchorage, AK, USA, 9–12 October 2011; pp. 2898–2903.
6. Manoonpong, P.; Pasemann, F.; Wörgötter, F. Sensor-Driven Neural Control for Omnidirectional Locomotion and Versatile Reactive Behaviors of Walking Machines. *Robot. Auton. Syst.* **2008**, *56*, 265–288. [CrossRef]
7. Suter, D.M.; Miller, K.E. The Emerging Role of Forces in Axonal Elongation. *Prog. Neurobiol.* **2011**, *94*, 91–101. [CrossRef]
8. Mortimer, D.; Pujic, Z.; Vaughan, T.; Thompson, A.W.; Feldner, J.; Vetter, I.; Goodhill, G.J. Axon Guidance by Growth-Rate Modulation. *Proc. Natl. Acad. Sci. USA* **2010**, *107*, 5202–5207. [CrossRef]
9. Pearson, Y.E.; Castronovo, E.; Lindsley, T.A.; Drew, D.A. Mathematical Modeling of Axonal Formation Part I: Geometry. *Bull. Math. Boil.* **2011**, *73*, 2837–2864. [CrossRef]
10. Nima, D.; Liu, Y. Modelling Axon Growth Using Driven Diffusion. *APS* **2019**, *2019*, L70–L305.
11. Marinov, T.; Sánchez, H.A.L.; Yuchi, L.; Adewole, D.O.; Cullen, D.K.; Kraft, R.H. A Computational Model of Bidirectional Axonal Growth in Micro-Tissue Engineered Neuronal Networks (micro-TENNs). *Silico Boil.* **2020**, *14*, 15–29. [CrossRef]
12. Kassraian-Fard, P.; Pfeiffer, M.; Bauer, R. A Generative Growth Model for Thalamocortical Axonal Branching in Primary Visual Cortex. *PLoS ONE Comput. Boil.* **2020**, *16*, e1007315. [CrossRef] [PubMed]
13. Gafarov, F. Neural electrical activity and neural network growth. *Neural Netw.* **2018**, *101*, 15–24. [CrossRef] [PubMed]
14. Kiehn, O. Development and Functional Organization of Spinal Locomotor Circuits. *Curr. Opin. Neurobiol.* **2011**, *21*, 100–109. [CrossRef] [PubMed]
15. Zjajo, A.; Hofmann, J.; Christiaanse, G.J.; Van Eijk, M.; Smaragdos, G.; Strydis, C.; De Graaf, A.; Galuzzi, C.; Van Leuken, R. A Real-Time Reconfigurable Multichip Architecture for Large-Scale Biophysically Accurate Neuron Simulation. *IEEE Trans. Biomed. Circuits Syst.* **2018**, *12*, 326–337. [CrossRef] [PubMed]
16. O'Donnell, C. Simplicity, Flexibility, and Interpretability in a Model of Dendritic Protein Distributions. *NEURON.* **2019**, *103*, 950–952. [CrossRef]
17. Koene, R.A.; Tijms, B.M.; Van Hees, P.; Postma, F.; De Ridder, A.; Ramakers, G.J.A.; Van Pelt, J.; Van Ooyen, A. NETMORPH: A Framework for the Stochastic Generation of Large Scale Neuronal Networks With Realistic Neuron Morphologies. *Neuroinformatics* **2009**, *7*, 195–210. [CrossRef]
18. Moser, T. Low-Conductance Intercellular Coupling between Mouse Chromaffin Cells In Situ. *J. Physiol.* **1998**, *506*, 195–205. [CrossRef]
19. Mortimer, D.; Fothergill, T.; Pujic, Z.; Richards, L.J.; Goodhill, G.J. Growth Cone Chemotaxis. *Trends Neurosci.* **2008**, *31*, 90–98. [CrossRef]
20. Borisyuk, R.M.; Cooke, T.; Roberts, A. Stochasticity and Functionality of Neural Systems: Mathematical Modelling Of Axon Growth in the Spinal Cord of Tadpole. *Biosystems* **2008**, *93*, 101–114. [CrossRef]

Appl. Sci. **2020**, *10*, 5497

21. Pearson, Y.E. *Discrete and Continuous Stochastic Models for Neuromorphological Data*; Rensselaer Polytechnic Institute: New York, NY, USA, 2009.

22. Isler, Y. A Software for Simulating Steady-State Properties of Passive Dendrites Based on the Cable Theory. *Comput. Methods Programs Biomed.* **2007**, *88*, 264–272. [CrossRef]

23. Bennett, M.R.; Fernandez, H.; Lavidis, N.A. Development of the Mature Distribution of Synapses on Fibres in the Frog Sartorius Muscle. *J. Neurocytol.* **1985**, *14*, 981–995. [CrossRef] [PubMed]

24. Matsuoka, K. Mechanisms of Frequency and Pattern Control in the Neural Rhythm Generators. *Boil. Cybern.* **1987**, *56*, 345–353. [CrossRef] [PubMed]

25. Kiehn, O. Locomotor Circuits in the Mammalian Spinal Cord. *Annu. Rev. Neurosci.* **2006**, *29*, 279–306. [CrossRef] [PubMed]

26. Van Heijst, J.; Vos, J. Self-Organizing Effects of Spontaneous Neural Activity on the Development of Spinal Locomotor Circuits in Vertebrates. *Boil. Cybern.* **1997**, *77*, 185–195. [CrossRef]

27. Van Heijst, J.J.; Vos, J.E.; Bullock, D. Development in a Biologically Inspired Spinal Neural Network for Movement Control. *Neural Netw.* **1998**, *11*, 1305–1316. [CrossRef]

Article

Quadrupedal Robots' Gaits Identification via Contact Forces Optimization

Gianluca Pepe [1,*], Maicol Laurenza [1], Nicola Pio Belfiore [2] and Antonio Carcaterra [1]

[1] Department of Mechanical and Aerospace Engineering, Sapienza University of Rome, 00184 Rome, Italy; maicol.laurenza@uniroma1.it (M.L.); antonio.carcaterra@uniroma1.it (A.C.)
[2] Department of Engineering, University of Roma Tre, via Della Vasca Navale 79, 00146 Rome, Italy; nicolapio.belfiore@uniroma3.it
* Correspondence: gianluca.pepe@uniroma1.it

Abstract: The purpose of the present paper is the identification of optimal trajectories of quadruped robots through genetic algorithms. The method is based on the identification of the optimal time history of forces and torques exchanged between the ground and the body, without any constraints on leg kinematics. The solutions show how it is possible to obtain similar trajectories to those of a horse's walk but obtaining better performance in terms of energy cost. Finally, a map of the optimal gaits found according to the different speeds is presented, identifying the transition threshold between the walk and the trot as a function of the total energy spent.

Keywords: gait optimization; quadruped robot; genetic algorithm; quadrupedal locomotion; evolutionary programming; optimal contact forces; cost of transport

check for updates

Citation: Pepe, G.; Laurenza, M.; Belfiore, N.P.; Carcaterra, A. Quadrupedal Robots' Gaits Identification via Contact Forces Optimization. *Appl. Sci.* **2021**, *11*, 2102. https://doi.org/10.3390/app11052102

Academic Editor: Alessandro Gasparetto

Received: 31 December 2020
Accepted: 21 February 2021
Published: 27 February 2021

Publisher's Note: MDPI stays neutral with regard to jurisdictional claims in published maps and institutional affiliations.

1. Introduction

Nature has always been a source of inspiration for engineers and scientists who tried to replicate or mimick natural bio-mechanisms. In fact, this can be a smart strategy, since any natural living being is a highly optimized system over millions of years of evolution [1–3]. Magnificent examples of engineering bio-inspired devices date from very ancient Magna Greece times, as the Archita's dove automaton, and a large number of automata within the Erone's tradition. The masterful Leonardo's machines, such as the lion or the knight automaton, or the flying machines, some inspired by the flapping wings of birds, are also astonishing examples.

To date, the development of automatic systems, in particular quadrupedal robots, has found great motivation in the field of human assistance. To assist and cooperate with humans, robots must be able to move easily in workplaces originally designed according to human needs [4]. For example, climbing stairs, avoiding obstacles, or opening doors [5,6]. Among the many technical difficulties to be faced and solved, one of the most important aspects is related to portability and its duration. This scenario justifies the need to optimize the power and energy of locomotion to make these systems more efficient, compact, characterized by limited energy consumption, and using small size and weight actuators.

Interesting questions arise: is it possible to conceive more efficient quadrupeds than those proposed by nature? Is it possible to disclose a different kind of gaits for a quadrupedal mechanism as the product of a strict optimization process? For example, depending on speed, energy consumption, or jerk minimization. Many engineers and robot designers assume the gait of animals is optimal since they would have been able to survive the competition and natural selection [7–10]. For this reason, the quadruped robot tends to be designed inspired by the examples of nature [11,12]. However, biological kinematics of locomotion is not reproducible directly by a legged robot, since biological muscles are extraordinarily efficient components, and animals can produce energy for their actuation by grabbing nutritive elements from the environment, transformed into actuation

255

energy by a digestive process. This permits them to have a relatively light energy storage system. Moreover, artificial sensors are still inferior, in terms of number, miniaturization, and sensitivity against the ones owned by animals [13].

To date, designers tend to create robots that resemble nature as closely as possible, such as imposing specific gaits such as walking, trotting, or galloping and developing mathematical optimization processes that only concern sub-systems, to overcome the technological limitations [14,15]. There are many different approaches for performing the optimization. In [16], for example, the authors use harmonic functions to describe foot trajectory and tune some parameters such as amplitude, phase, and frequency. In [17,18] the authors generate leg trajectories for overcoming small terrain irregularities by optimizing consumed energy. In [19] the authors optimize contact forces to walk on a high-slope terrain, but the gait is imposed a priori.

Evolutionary computation, such as genetic algorithms (GAs), is often a natural choice for the gait optimization of legged robots since it uses optimization methods based on evolutionary theory [20]. Also, the controls of the mentioned cases, tend to mimic or approach the natural movement of the quadrupeds found in nature. Typical controls are central pattern generators [21,22], hybrid zero dynamics [23], or the usual PD [24].

The present work breaks from all the problems relating to technological limitations and lays the foundations of how it is possible to identify, a priori, optimal gaits through parametric optimization algorithms. The authors propose an algorithm that does not depend either on the kinematic chains of the legs or power actuators. The method directly handles the contact forces that the hypothetical legs exchange with the ground, identifying the sequence that guarantees minimum energy consumption and smooth motion during a single stride cycle. The genetic algorithm (GA) is used to minimize the kinetic and potential energy as well as restraining the jerk, identifying the optimal profile of normal and friction forces. The equations of motion are solved through some periodicity constraints of the gait, thus ensuring a stable and continuous motion. The optimal gaits identified are very similar to those found in nature. The authors have optimized the gait of a heavy and bulky horse-like robot, obtaining a walking trot as the optimal gait, and then comparing it with experimental data from real horses. Additionally, a gait transition analysis is performed as a function of locomotion speed and gait cycle length, showing why it is convenient to walk or trot.

The paper is organized as follows: Section 2 describes the dynamic model and ground reaction force shapes; Section 3 defines the optimization parameters and objective functions while introducing stability constraints to assure a periodic motion; Section 4 provides the resulting optimal gait, along with the comparison with horse experiments and identification of gait transition; Section 5 discusses the limitations and future challenges of the method and finally the conclusions.

2. Mathematical Model for the Gait Optimization

The considered quadrupedal model is sketched in Figure 1, where the body, suspended on four legs, transmits forces and moments thanks to the interaction with the ground.

The optimization model proposed in this paper consists of an optimal gait identification, capable of moving the quadrupedal body through the succession of four alternating thrusts generated by legs with some desirable characteristics. This approach is innovative. In fact, the kinematic linkages and actuators of the leg mechanism are not directly involved but left to a second engineering phase of design, not investigated in the present paper.

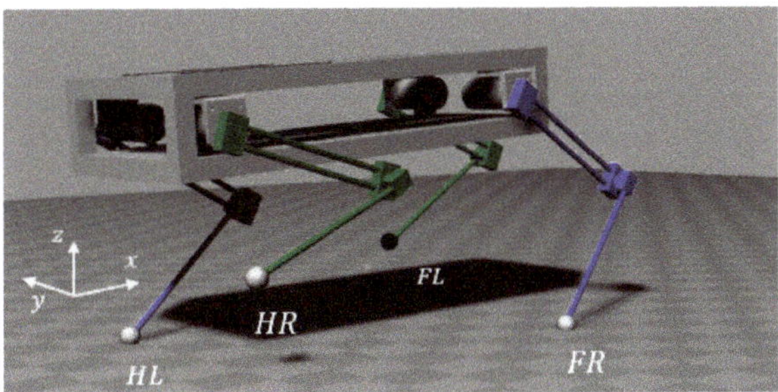

Figure 1. 3D view of the quadruped.

The legs are labelled as *FL* (front left), *FR* (front right), *HL* (hind left), *HR* (hind right). The period during which a single leg maintains its contact with the ground is the *conctact phase* (*CP*), while the one during which the leg is flying in the air moving towards the next contact point, is the *swing phase* (*SP*). In general, these times can be set differently for each leg.

The legs sequencing, sketched in Figure 2, originates a periodic motion of each leg, of period T. The time at which each leg touches the ground is $t_i \in [0, T]$, where $i = FR, FL, HR, HL$. The time duration of the contact phase is T_i, and the time duration of the swing phase is $T - T_i$.

Newton–Euler equations of rigid body dynamics are:

$$m\,\ddot{r} = f_e$$
$$I\dot{\omega} = m_e - \omega \times (I\omega)$$
$$\dot{\theta} = E(\theta)\omega$$

(1)

where the first equation holds in the fixed reference frame, the second in the body reference frame (principal inertia axes), $r = [x, y, z]$ identifies the gravity center position of the quadruped body of total mass m, f_e is the total external force, $I = diag[I_x, I_y, I_z]$ the matrix of inertia, m_e is the total external moment, $\theta = [\psi, \theta, \phi]$, the Tait–Bryan angles (yaw pitch and roll), ω is the angular velocity, and E is the Jacobian matrix.

Assuming small pitch and roll angles and vanishing yaw, the equations can be reduced into the same fixed reference frame:

$$m\ddot{x} = F_x$$
$$m\ddot{y} = F_y$$
$$m\ddot{z} = F_z$$
$$I_x\ddot{\phi} = M_x$$
$$I_y\ddot{\theta} = M_y$$

(2)

where F_x, F_y, F_z, M_x, M_y are the total force and moment transmitted by the four legs *FR*, *FL*, *HR*, and *HL*, respectively.

When the quadruped moves on the ground, the leg-ground contact produces a reaction force. Its vertical component F_z supports the quadrupedal weight, the horizontal component F_x drives the longitudinal motion of the body, and F_y affects the lateral body oscillations.

The vertical force F_{z_i} for each leg acts only in the *CP*; that implies $F_{z_i}(t_i) = 0$ and $F_{z_i}(t_i + T_i) = 0$, that is, its time history, during the *CP*, starts and ends with vanishing values, associated with the incipient contact and incipient detachment of the leg with and from the ground, respectively. The force-time history along the *CP* is identified through three additional points $\left[P_{1_i}; P_{2_i}; P_{3_i}\right]$ (besides the two previously identified) and is fitted by a suitable spline (Figure 3a).

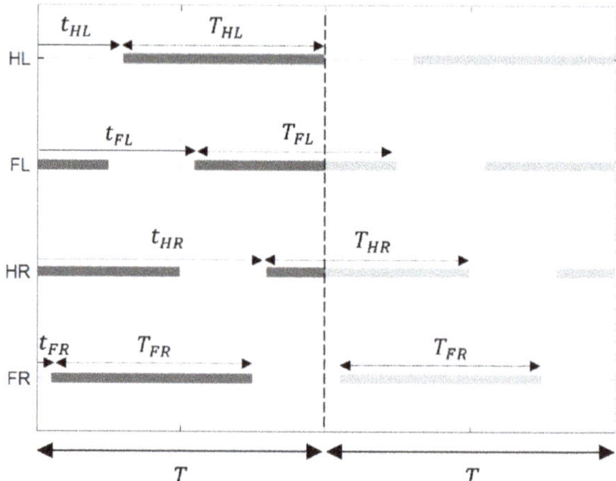

Figure 2. Time diagram of the quadruped's locomotion over the entire period T shows the times t_i at which each leg touches the ground and the time duration T_i of each contact phase. The subscripts $i = FR, FL, HR, HL$ are FL (front left), FR (front right), HL (hind left), and HR (hind right).

The longitudinal and lateral ground reactions for each leg are assumed to belong to the static friction cone. With this assumption, any contact slip condition is absent, and the contact model is purely conservative. Therefore:

$$\sqrt{F_{x_i}^2 + F_{y_i}^2} \leq \mu_s F_{z_i} \tag{3}$$

where μ_s is the static friction coefficient.

For the planar forces F_{x_i} and F_{y_i}, one can introduce suitable constitutive relationships (Figure 3). Several experiments available in the technical literature [25,26] suggest the following gait dynamics for leg locomotion. When the leg touches the ground, in the first phase, a backward longitudinal reaction brakes the body run, followed by a second phase, in which the leg accelerates the body through a forward longitudinal reaction force (Figure 3d). Moreover, the amplitude of the longitudinal reaction force is roughly proportional to the normal vertical force. A similar force trend holds for the lateral force that is responsible for the unavoidable lateral staggering of the quadruped's body. For both longitudinal and lateral forces, one can introduce the suitable factorization form as constitutive relationships:

$$F_{x_i}(t) = \mu_{x_i}(t)F_{z_i}(t)$$
$$F_{y_i}(t) = \mu_{y_i}(t)F_{z_i}(t) \tag{4}$$

that use the proportionality of the planar forces to the vertical one. Moreover, the role of the factors $\mu_{x_i}(t)$, $\mu_{y_i}(t)$ is that of reproducing the time history of the planar force, $F_{x_i}(t)$ or $F_{y_i}(t)$, as reported in Figure 3c,d, modulating the amplitude of the vertical reaction $F_{z_i}(t)$. In particular, $\mu_{x_i}(t)$, $\mu_{y_i}(t)$ are vanishing outside of the CP. Within the CP, in the first part of the contact, $\mu_{x_i}(t)$ takes negative values to reproduce the braking effect, while, in the second part, it becomes positive, to reproduce the acceleration effect.

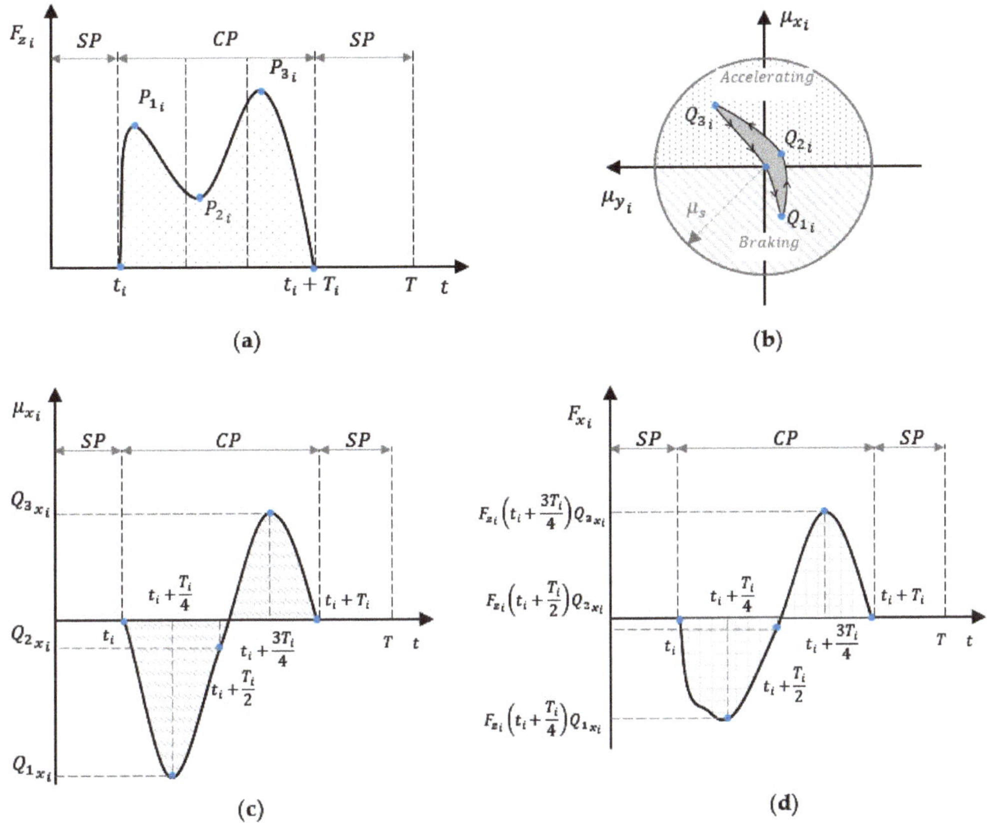

Figure 3. (**a**) Vertical ground force; (**b**) trend of the coefficient of adhesion inside the static friction cone; (**c**) longitudinal adhesion coefficient as a function of time; and (**d**) longitudinal ground force.

Combining Equations (4) and (3), the two modulating factors must satisfy the condition:

$$\sqrt{\mu_{x_i}^2 + \mu_{y_i}^2} \leq \mu_s \tag{5}$$

Looking at Figure 3b, this implies that at any time the point $Q_{j_i}\left[\mu_{x_{j_i}}(t),\ \mu_{y_{j_i}}(t)\right]$ remains confined in the circle of radius μ_s over the plane $\mu_{x_i}(t)$, $\mu_{y_i}(t)$. Moreover, we assume a correlation between the modulating factors during the gait. This implies, concerning Figure 3b, the point $Q_{j_i}\left[\mu_{x_{j_i}}(t),\ \mu_{y_{j_i}}(t)\right]$ describes a closed curve in the plane $\mu_{x_{j_i}}(t)$, $\mu_{y_{j_i}}(t)$ as time is spent, and Q completes the loop in a period T. More precisely, the time histories of $\mu_{x_i}(t)$, $\mu_{y_i}(t)$ are determined by a spline passing through the points $\left[Q_{1_i}; Q_{2_i}; Q_{3_i}\right]$ selected by the designer within the contact cone (see Figure 3b).

The expressions for the moments M_x and M_y follow directly from the leg's reactions. From Figure 4, let's consider, for example, the FR foot at the incipient contact phase and foot position (x_{FR}, y_{FR}, z_{FR}):

$$\begin{aligned} M_y &= F_x \Delta z_{FR} - F_z \Delta x_{FR} \\ M_x &= F_y \Delta z_{FR} + F_z \Delta y_{FR} \end{aligned} \tag{6}$$

where $\Delta x_{FR}(t) = x_{FR} - x$, $\Delta z_{FR}(t) = z_{FR} - z$ and $\Delta y_{FR} = y_{FR} - y$.

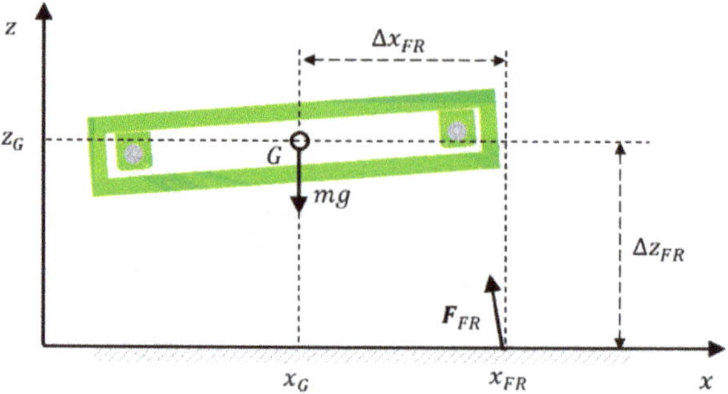

Figure 4. Point of application of the leg force FR and distances Δx_{FR} and Δz_{FR} referred to the center of gravity G.

3. Optimization Model

Usually, in the legged locomotion, the stability of a gait is guaranteed by using a criterion such as zero-moment point [27,28]. Other studies face the stability problem with the Poincare map [29,30] or ground reference points [31]. In this paper, unlike the classical approaches, the gait stability is guaranteed by satisfying periodic limit cycle conditions.

The optimization operates over a single period T using a genetic algorithm (GA) which provides the stride length, frequency, and velocity, and additionally the time history of the contact forces. Therefore, a vector of gait parameters p_{GA} is optimized by using the GA algorithm. Let us explicitly set p_{GA}. The relative phases of the leg motion are t_i, the normalized time duration T_i of the CP is $\beta_i = \frac{T_i}{T}$, the time history of the normal, longitudinal and lateral contact forces are determined by the interpolation of the points P_{j_i} and Q_{j_i} with $j = 1, 2, 3$, respectively. Therefore:

$$p_{GA} = \left[t_i, \beta_i, P_{j_i}, Q_{j_i} \right] \tag{7}$$

The structure of the optimization process involves the optimal vector p_{GA} and, using several constraints derived from the equations of motion, it looks for solutions that minimize a multi-objective cost function related to the dissipated energy and looking for smooth forces trends. Let us examine in detail these constraints.

Note, as a preliminary consideration, that we can impose the continuity kinematic conditions over the stride period:

$$\begin{aligned}
y(0) &= y(T); \quad z(0) = z(T) \\
\dot{x}(0) = \dot{x}(T); \quad \dot{y}(0) &= \dot{y}(T); \quad \dot{z}(0) = \dot{z}(T) \\
\phi(0) &= \phi(T); \quad \theta(0) = \theta(T) \\
\dot{\phi}(0) &= \dot{\phi}(T); \quad \dot{\theta}(0) = \dot{\theta}(T)
\end{aligned} \tag{8}$$

Moreover, the average speeds over the period T are $\bar{\dot{x}} = V$, $\bar{\dot{y}} = 0$, $\bar{\dot{z}} = 0$, $\bar{\dot{\theta}} = 0$, $\bar{\dot{\phi}} = 0$, where V is the average longitudinal speed of the quadruped body.

We start with the vertical equation of motion (an identical procedure applies to the other equations of motion) that produces:

$$m\ddot{z} = \sum_i F_{z_i}(P_{j_i}) - mg \quad \rightarrow \quad \dot{z} = \frac{1}{m} \int_0^t \sum_i F_{z_i}(P_{j_i})\, dt - gt + C \tag{9}$$

where g is the gravity acceleration and C is an integration constant. Therefore:

$$\dot{z}(T) = \frac{1}{m} \int_0^T \sum_i F_{z_i}(P_{j_i})\, dt - gT + C \tag{10}$$

and

$$\dot{z}(0) = C \tag{11}$$

Using the condition $\dot{z}(0) = \dot{z}(T)$, it follows:

$$\frac{1}{T} \int_0^T \sum_i F_{z_i}(P_{j_i})\, dt = mg \tag{12}$$

The force-time history is set by randomly selecting the point P_{j_i} by the GA and using a spline interpolation. Further integration of Equation (10) produces:

$$\begin{aligned}
\dot{z}(t) &= \frac{1}{m} \int_0^t \sum_i F_{z_i}(P_{j_i})\, d\tau - gt + \dot{z}(0) \rightarrow z(t) \\
&= \frac{1}{m} \int_0^t \int_0^\tau \left(\sum_i F_{z_i}(P_{j_i}) - mg \right) dt d\tau + \dot{z}(0)t + z(0)
\end{aligned} \tag{13}$$

and:

$$z(T) = \frac{1}{m} \int_0^T \int_0^\tau \left(\sum_i F_{z_i}(P_{j_i}) - mg \right) dt\, d\tau + \dot{z}(0)T + z(0) \tag{14}$$

Using the periodicity condition $z(0) = z(T)$, it follows:

$$\dot{z}(0) = -\frac{1}{mT} \int_0^T \int_0^\tau \left(\sum_i F_{z_i}(P_{j_i}) - mg \right) dt d\tau \tag{15}$$

Equation (15) automatically satisfies the condition of zero average speed $\bar{\dot{z}} = 0$. In fact:

$$\begin{aligned}
\dot{z}(t) &= \frac{1}{m} \int_0^t \sum_i F_{z_i}(P_{j_i})\, d\tau - gt + \dot{z}(0) \rightarrow \bar{\dot{z}} \\
&= \frac{1}{mT} \int_0^T \int_0^\tau \left(\sum_i F_{z_i}(P_{j_i}) - mg \right) dt\, d\tau + \dot{z}(0) = 0
\end{aligned} \tag{16}$$

The longitudinal dynamics takes into account two inertia effects: one related to the longitudinal motion of the robot body mass m_B, and the second due to the reciprocating motion of the legs of mass m_{p_i}. The total robot mass m is consequently:

$$m = m_B + \sum_i m_{p_i} \tag{17}$$

Let us make the point about the leg motion. Looking at Figure 5, it appears that, during the contact phase, the horizontal speed of the leg is negative, and becomes positive during the swing phase. Therefore, along the gait period, the leg mass behaves as an oscillator that, attached to the primary body mass, generates a back and forth horizontal inertia force.

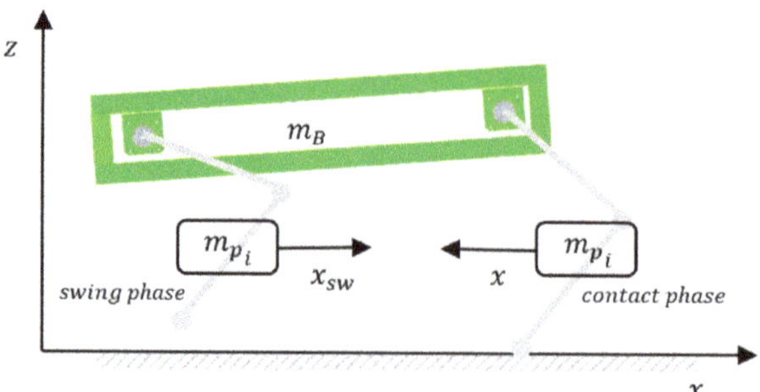

Figure 5. Illustration of the reciprocating longitudinal leg motion during the swing and contact phases, respectively.

The mathematical counterpart of this physical behaviour appears as follows:

$$m_B \ddot{x} - \sum_i m_{p_i} w_i(t) \ddot{x} = \sum_i F_{x_i} - \sum_i m_{p_i}(1 - w_i(t)) \ddot{x}_{sw_i} \tag{18}$$

x_{sw_i} describes the leg mass longitudinal motion during the swing phase. The term $m_{p_i}(1 - w_i(t))\ddot{x}_{sw_i}$ takes into account the inertia force during the swing phase, while, complementarily, $m_{p_i} w_i(t)\ddot{x}$ accounts for the added mass effect during the contact phase, within which the leg motion depends essentially on the body motion. The switch function $w_i(t)$ is defined as:

$$w_i(t) = \begin{cases} 1 & if \;\; the \; i^{th} \; leg \; is \; in \; CP \\ 0 & if \;\; the \; i^{th} \; leg \; is \; in \; SP \end{cases} \tag{19}$$

that activates alternatively the inertia effects of the contact and swing phase, respectively. The swing motion $x_{sw_i}(t)$ is designed through β_i and T by using a spline with initial and final slope equal to the average body velocity V (as an example see Figure 6).

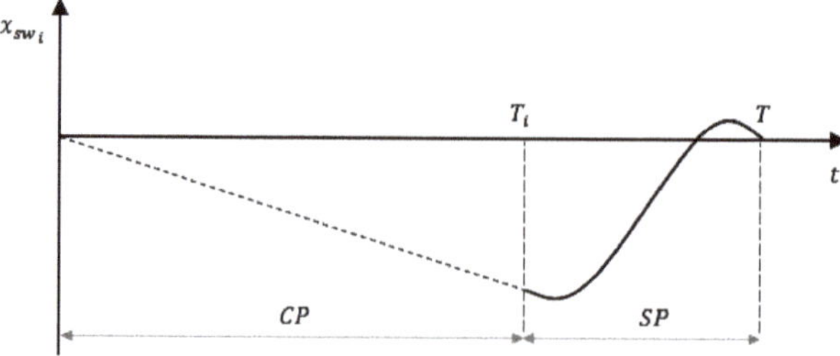

Figure 6. An example of the time history of x_{sw_i} of the leg.

The analogous Equation (12) is determined for the longitudinal motion using the condition $\dot{x}(0) = \dot{x}(T)$:

$$\int_0^T \frac{\sum_i F_{x_i}\left(P_{j_i}, Q_{x_{j_i}}\right) - \sum_i m_{p_i}(1 - w_i(t))\ddot{x}_{sw_i}}{m_B - \sum_i m_{p_i} w_i(t)} dt = 0 \tag{20}$$

Then the average speed $\bar{\dot{x}} = V$ over the period T is:

$$\bar{\dot{x}} = V = \frac{1}{T} \int_0^T \int_0^\tau \frac{\sum_i F_{x_i}\left(P_{j_i}, Q_{x_{j_i}}\right) - \sum_i m_{p_i}(1 - w_i(t))\ddot{x}_{sw_i}}{m_B - \sum_i m_{p_i} w_i(t)} d\tau \, dt + \dot{x}(0) \tag{21}$$

that, together with the periodicity condition, produces:

$$\dot{x}(0) = \dot{x}(T) = V - \frac{1}{T}\int_0^T \int_0^\tau \frac{\sum_i F_{x_i}\left(P_{j_i}, Q_{x_{j_i}}\right) - \sum_i m_{p_i}(1 - w_i(t))\ddot{x}_{sw_i}}{m_B - \sum_i m_{p_i} w_i(t)} d\tau \, dt \tag{22}$$

If the same procedure is applied to the lateral motion, it follows:

$$m\ddot{y} = \sum_i F_{y_i}\left(P_{j_i}, Q_{y_{j_i}}\right) \rightarrow \int_0^T \sum_i F_{y_i}\left(P_{j_i}, Q_{y_{j_i}}\right) dt = 0 \tag{23}$$

derived under the condition $\dot{y}(0) = \dot{y}(T)$. Using the other periodicity condition $y(0) = y(T)$, and $\dot{\bar{y}} = 0$, we obtain:

$$\dot{y}(0) = -\frac{1}{mT}\int_0^T \int_0^\tau \sum_i F_{y_i}\left(P_{j_i}, Q_{y_{j_i}}\right) dt \, d\tau \tag{24}$$

We can apply the same procedure to the remaining cardinal equations. Using the conditions $\dot{\theta}(0) = \dot{\theta}(T)$, $\dot{\phi}(0) = \dot{\phi}(T)$, the following constraints are obtained:

$$\begin{aligned} \int_0^T \sum_i M_{y_i}(x_0) \, dt = 0 \\ \int_0^T \sum_i M_{x_i}(y_0) \, dt = 0 \end{aligned} \tag{25}$$

These can be solved in closed form in terms of the initial conditions x_0 e y_0. The arms of the moments are:

$$\begin{aligned} \Delta x_i(t, x_0) &= x_i(t) - x(t, x_0) \\ \Delta y_i(t, y_0) &= y_i(t) - y(t, y_0) \\ \Delta z_i(t, z_0) &= z_i - z(t, z_0) \end{aligned} \tag{26}$$

Assuming a flat ground, $z_i = 0$, and

$$\begin{aligned} x(t, x_0) &= \int_0^t \int_0^\tau \frac{F_x}{m_B - \sum_i m_{p_i} w_i(t)} \, dt d\tau + \dot{x}_0 t + x_0 \\ y(t, y_0) &= \frac{1}{m}\int_0^t \int_0^\tau F_y \, dt d\tau + \dot{y}_0 t + y_0 \\ z(t, z_0 = h) &= \frac{1}{m}\int_0^t \int_0^\tau (F_z - mg) \, dt d\tau + \dot{z}_0 t + h \end{aligned} \tag{27}$$

Finally, using again the same technique with the conditions $\phi(0) = \phi(T)$, $\theta(0) = \theta(T)$, the final equations yield:

$$\begin{aligned} \dot{\theta}(0) = \dot{\theta}(T) &= -\frac{1}{I_y T}\int_0^T \int_0^\tau M_y(x_0) \, dt d\tau \\ \dot{\phi}(0) = \dot{\phi}(T) &= -\frac{1}{I_x T}\int_0^T \int_0^\tau M_x(y_0) \, dt d\tau \end{aligned} \tag{28}$$

The previous constraints are based on the linear approximation (see Equation (2)) of the equations of motion, and they provide the initial guess of the kinematic variables at the initial and final times, 0 and T. Moreover, the force and the moments, through their spline approximations, satisfy the set of conditions from (7) to (27). With this starting guess, we can solve the following iterative nonlinear optimization problem:

$$
\begin{aligned}
\text{minimize} \quad & \boldsymbol{\omega}(0) - \boldsymbol{\omega}(T) = \mathbf{0_{3x1}} \\
& \boldsymbol{\theta}(0) - \boldsymbol{\theta}(T) = \mathbf{0_{3x1}} \\
\text{subject to} \quad & \dot{\boldsymbol{\omega}} = \boldsymbol{m_e} - \boldsymbol{\omega} \times (\boldsymbol{I\omega}) \\
& \dot{\boldsymbol{\theta}} = \boldsymbol{E}(\boldsymbol{\theta})\boldsymbol{\omega} \\
& |\boldsymbol{\omega}| < \boldsymbol{\omega_{max}} \\
& |\boldsymbol{\theta}| < \boldsymbol{\theta_{max}} \\
\text{initial guess} \quad & \boldsymbol{\theta_0} = [\theta_0, \phi_0, \psi_0]^T = [0,0,0]^T \\
& \boldsymbol{\omega_0} = \left[\dot{\theta}_0, \dot{\phi}_0, 0\right]^T \\
& \boldsymbol{r_0} = [x_0, y_0, z_0]^T
\end{aligned}
\tag{29}
$$

The unknown vector to be determined is $\boldsymbol{p}_{NLP} = \left[\theta_0, \phi_0, \psi_0, \dot{\omega}_{x_0}, \dot{\omega}_{y_0}, \dot{\omega}_{z_0}, x_0, y_0, z_0\right]$. Figure 7 illustrates the flow chart of the GA solution's scheme. The setting of the GA which led to good balancing between learning and overfitting is a size population of 50 individuals, 110 generations, a crossover of 80% and mutation of 1%. Specifically, the genetic algorithm with multi-objective optimization is used for which simultaneous optimization of multiple, often competing, objectives take place [32,33]. The related pseudocode is shown below:

Algorithms 1 *Pseudo code for the gait optimization algorithm.*

1: $t \leftarrow 0$
2: *Initialization parameters* $\left[T, V, m, I_x, I_y, I_z, L, h, m_p\right]$
3: *GA random population initialization of* $\boldsymbol{p}_{GA}$ *vector*
4: **While** *{GA's stop criterion is not met}*
5:    *Vertical ground force definition with continuity kinematic constraints (periodicity condition)*
6:    *Longitudinal and lateral forces with continuity kinematic and static friction cone constraints*
7:    *Torques with continuity kinematic constraints in the linear case (small angles)*
8:    *Initialization guess parameters for Nonlinear Programming*
9:    *Evaluate Objective function for Nonlinear Programming*
10:   **While** *{Nonlinear Programming's stop criterion is not met and the constraints are satisfied}*
11:     *Sequential Quadratic Algorithm*
12:     *Evaluate Objective function*
13:   **Endwhile**
14:   *Evaluate fitness function for each individuals*
15:   *Select of the individuals*
16:   *Recombine of the individuals*
17:   *Mutate of the individuals*
18:   *Generations of new population*
19:   $t \leftarrow t+1$
20: **Endwhile**

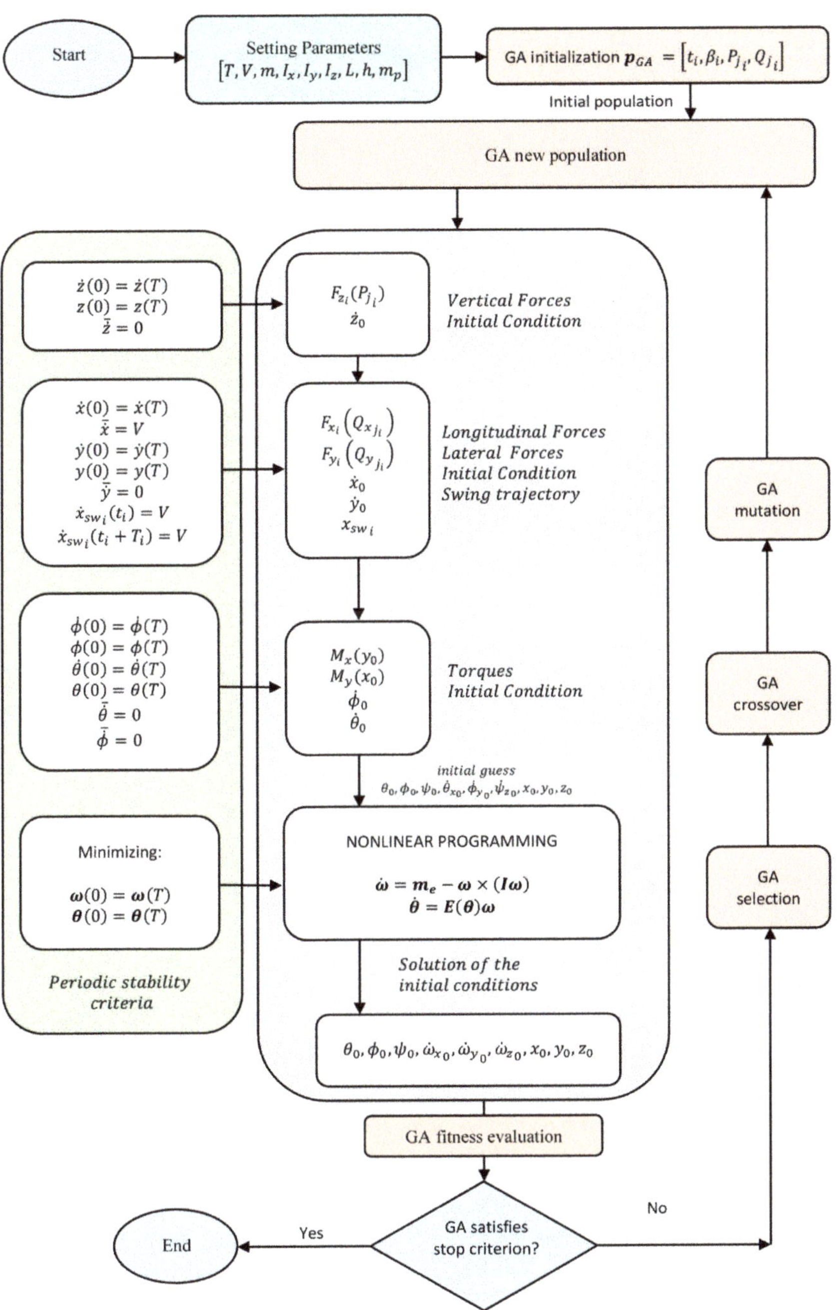

Figure 7. Optimization diagram of the genetic algorithm (GA) algorithm.

Eventually, we introduce the multi-objective function to be minimized for the most efficient gait. The GA's fitness function is defined through the total specific energy E_d, that is, the energy per traveled distance $x(T) - x_0$ over the period, and the ground reaction jerk J_{GRF}.

The specific energy E_d includes the mechanical energy of the body and the swinging leg energy:

$$E_d = \frac{1}{x(T) - x_0} \int_0^T |F_x \, \dot{x}| + |F_y \, \dot{y}| + |F_z \, \dot{z}| + \left|M_x \, \dot{\phi}\right| + \left|M_y \, \dot{\theta}\right| + |mg \, \dot{z}| \\ + \left|\left(\sum_i m_{p_i}(1 - w_i(t))\ddot{x}_{sw_i}\right)\dot{x}_{sw_i}\right| \, dt \tag{30}$$

One can expect power fluctuations along the period with positive and negative values, perfectly balanced since these fluctuations have zero average (the system is conservative). Since E_d integrates the absolute value of the power, it measures the intensity of the power fluctuations. Requiring a small value for E_d means asking for small fluctuations of the power, which implies a benefit in terms of the actuator dimensions and power and thus of energy consumption. On the other hand, the request of a small jerk J_{GRF} assures traditionally smooth contact forces:

$$J_{GRF} = \frac{1}{T} \int_0^T \|\dot{F}\| \, dt \tag{31}$$

where $F = [F_x; F_y; F_z]$.

Multi-objective optimization is used to determine the Pareto frontier, the set of solutions that provides the optimal trade-off between the two criteria.

4. Results

4.1. Optimal Gait Solution

For certain optimization variable sets p_{GA}, the solver may not converge, or the solution may be unrealistic or impractical. A maximum foot stride s_{max} has been imposed to avoid solutions that required unrealistic leg dimensions. The condition was applied to the upper bound of the duty factor: $\beta_{i_{max}} = \frac{s_{max}}{VT}$, where V is the mean target speed of the simulation:

$$0 < \beta_i \le \beta_{i_{max}} \tag{32}$$

The optimization finds an efficient gait (in the sense specified in the previous section) that moves the body at an average speed of 1.35 m/s, with a gait period T of 1 s. The mass properties of the body are selected considering the characteristic parameters of quadrupeds in nature, in particular horses (Table 1, see reference [34]). In addition, some simplification but reasonable assumptions are made, namely: $\beta_{FR} = \beta_{FL} = \beta_F$, $\beta_{HR} = \beta_{HL} = \beta_H$, $F_{z_{FR}} = F_{z_{FL}}$, $F_{z_{HR}} = F_{z_{HL}}$, $F_{x_{FR}} = F_{x_{FL}}$, $F_{x_{HR}} = F_{x_{HL}}$, $F_{y_{FR}} = -F_{y_{FL}}$, and $F_{y_{HR}} = -F_{y_{HL}}$. The obtained solutions are plotted in Figure 8, where a correlation between the energy cost and contact-forces jerk is shown (Pareto frontier). The resulting gaits present a specific leg sequence classified as walk and trot gaits. In nature, walking gaits involve overlapping contact phase such that $\beta > 0.5$. In particular, in a trot gait, the diagonal forelimb and hindlimb move in phase, while in a walking gait the limbs have a lag of 0.25 between the two legs.

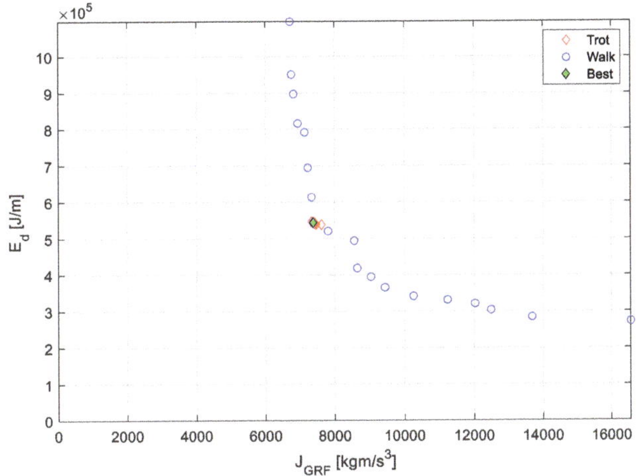

Figure 8. Pareto curve of optimal gaits for $V = 1.35$ m/s and $T = 1$ s.

The best solution according to the Pareto frontier is presented in Figure 9. It can be seen how the body keeps a steady longitudinal speed $\dot{x}$, about the target speed of $V = 1.35$ m/s, maintaining a stable and restrained attitude and assuring the periodicity constraints.

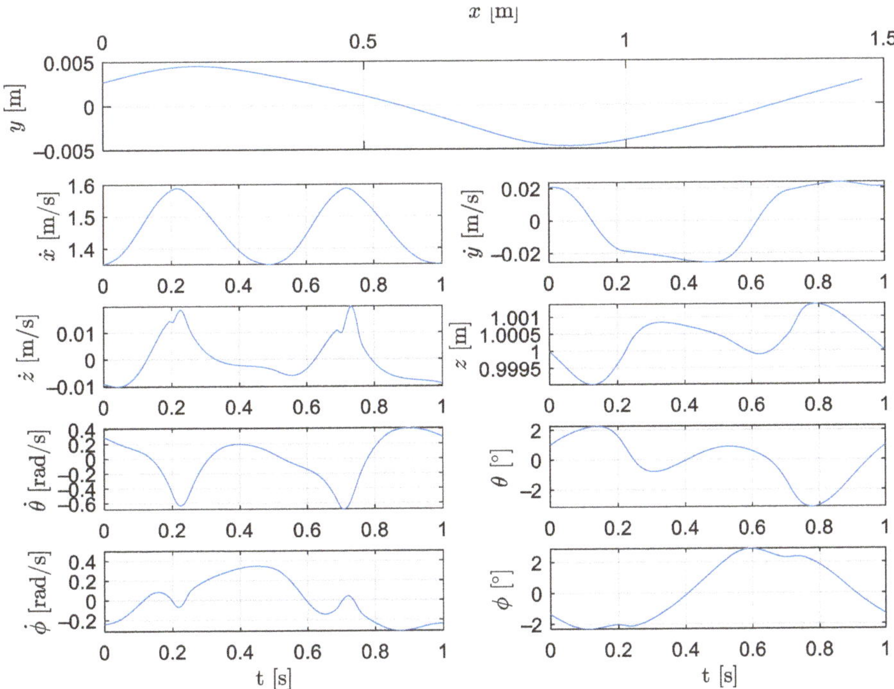

Figure 9. Trajectory and attitude of the selected optimal solution.

This case is classified as a walking trot since it presents a duty cycle $\beta_{FR} = \beta_{FL} = \beta_{HR} = \beta_{HL} = 0.55$, equal for the hind and forelimbs, and $t_i = [0.7; 0.2; 0.2; 0.7]$ (Figure 10). The contact forces are instead shown in Figure 11, and the friction coefficients in Figure 12. It can be observed how the normal forces F_{z_i} have a flat double-hump time history, typical of quadrupeds' locomotion [2,35]. All other significant parameters are shown in Table 1.

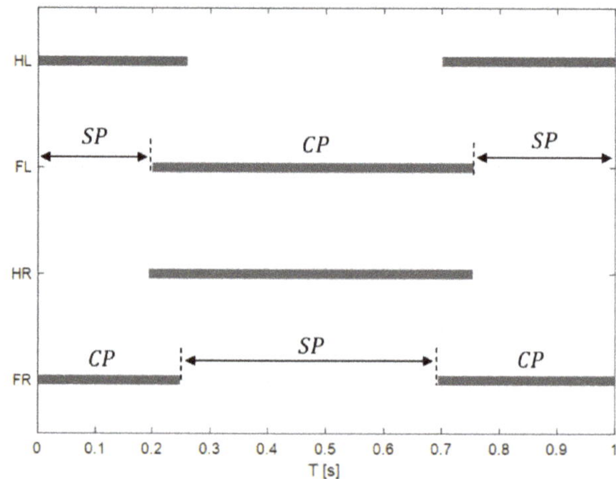

Figure 10. Gait sequence of the optimal walking trot solution: *FL* (front left), *FR* (front right), *HL* (hind left), *HR* (hind right), CP (contact phase), and SP (swing phase).

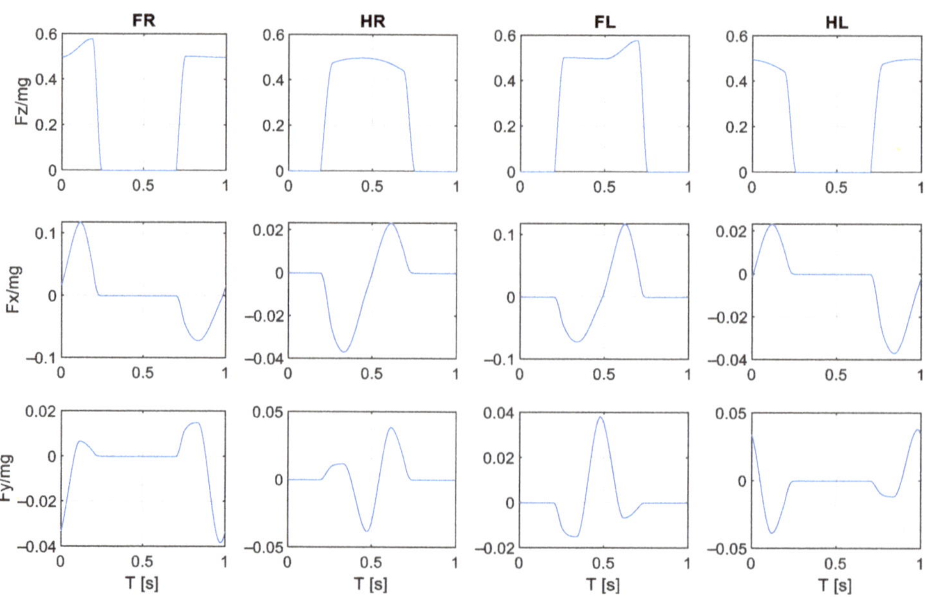

Figure 11. Contact forces normalized over the body weight.

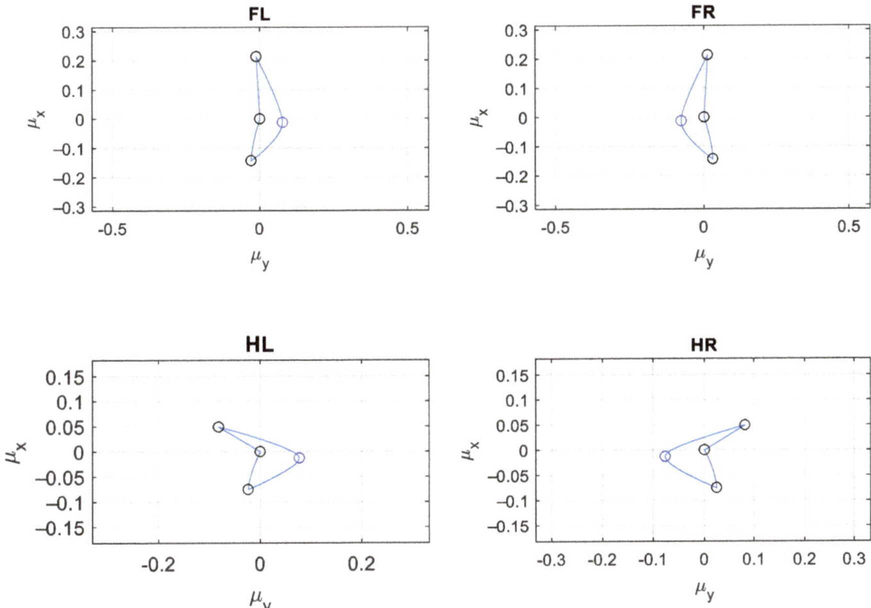

Figure 12. The trend of the coefficient of adhesion for each foot.

Table 1. Simulation setting parameters and results.

Parameter	Value	Unit
m	520	kg
m_p	26	kg
I_x	103	kg m^2
I_y	26.2	kg m^2
I_z	103	kg m^2
L	1.44	m
h	1	m
T	1	s
V	1.35	m/s
$\Delta x_{FR,FL_{max}} ; \Delta x_{FR,FL_{min}}$	1.08; 0.27	m
$\Delta x_{HR,HL_{max}} ; \Delta x_{HR,HL_{min}}$	-0.36; -1.17	m
$s_{FR} = s_{FL} = s_{HR} = s_{HL}$	0.81	m
x_0	0.36	m
y_0	0.0027	m
z_0	1	m
$\beta_{FR} = \beta_{FL} = \beta_{HR} = \beta_{HL}$	0.55	
$t_{FR} = t_{HL}$	0.7	s
$t_{FL} = t_{HR}$	0.2	s

In Figure 13 we can see the trajectory performed by the foot FR during one complete cycle.

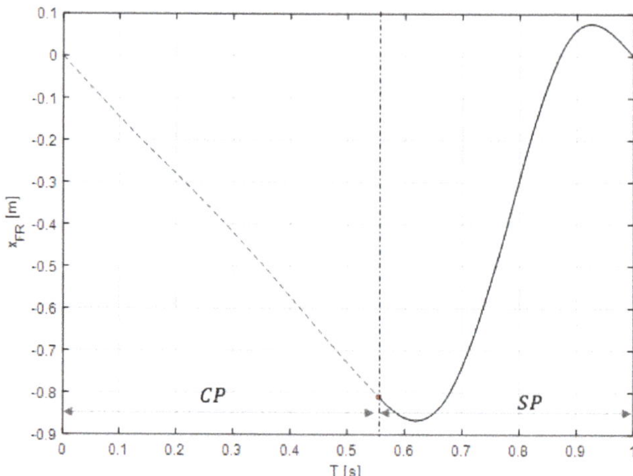

Figure 13. FR foot trajectory during contact and swing phase starting from t_{FR}.

To better assess the variability of the results, a statistical analysis on 10 runs of the optimization algorithm was performed. In all cases, the best gait found was the walking-trot and the ratio between the value of the objective function and its average differs by a maximum of 2.5%, as observed in Figure 14.

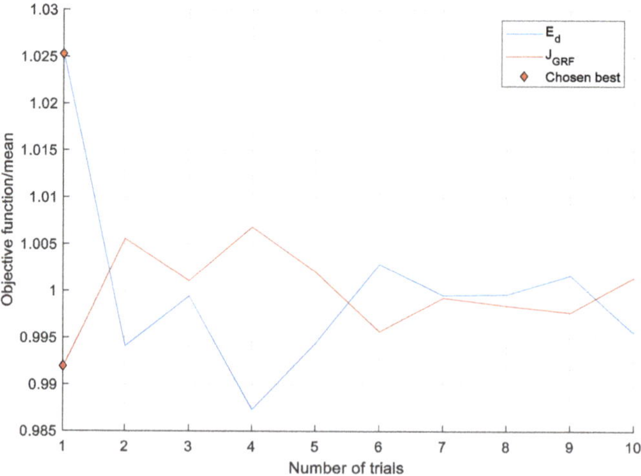

Figure 14. Comparison of fitness functions over the mean value between 10 trials.

Eventually, a comparison between the cumulative mechanical energy of the experimental and numerical optimization data is shown in Figure 15. For the experimental campaign [34], a group of horses with a weight ranging between 417 and 673 kg, was used. For each animal, ten sequences were considered within the speed range from 0.7 to 1.9 m/s. Five markers, positioned along the horses' backbones, identify the vertical and longitudinal motion of their center of mass (CoM_{rigid}). The considered mechanical energy is defined by the expression (a special case of expression (29)):

$$E_{CoM_{rigid}} = \int_0^T \left| F_z \dot{z} \right| + \left| F_x \dot{x} \right| + mg \left| \dot{z} \right| \, dt \tag{33}$$

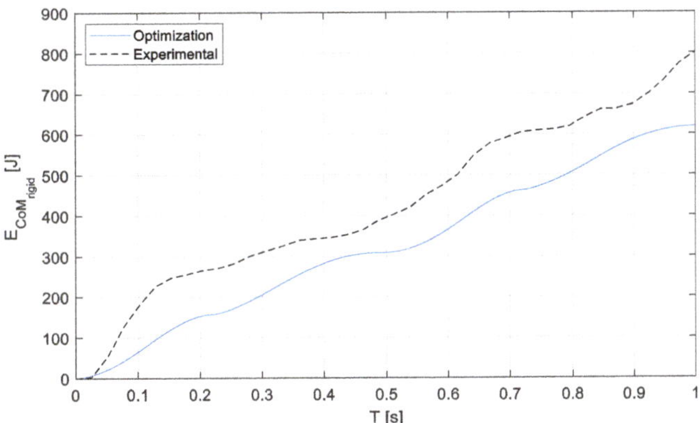

Figure 15. Mechanical energy comparison between experimental data and the optimal solution.

It is clear how the resulting energies are comparable.

4.2. Gaits Analysis for Different Speed and Cycle Duration T

The optimization solutions are tested for different speeds and cycle period T.

Figure 16 depicts the trend of the walking gait representing E_d, following Equation (30), as dependent on the cycle duration, for different speeds between 0.7 and 2 m/s. This map permits finding natural associations between convenient gait period and body velocity, to obtain gaits with a reasonably limited energy cost and smooth forces.

We notice that for $T < 0.5$ s the total energy is very large, because the swing phase has a very limited duration, needing very high energy to relocate the foot at its initial position. On the other hand, at high values for T, again the energy cost increases. Therefore, the map suggests that for any desired speed, there is an intermediate region for the gait period that allows a low energy consumption. Minimum energy expenditures are identified at $T = 1.25$ s for speed 1.35 m/s, at $T = 1$ s for speed 2 m/s, and $T = 1.66$ s for speed 0.7 m/s (blue square, green triangle, and red diamond, respectively).

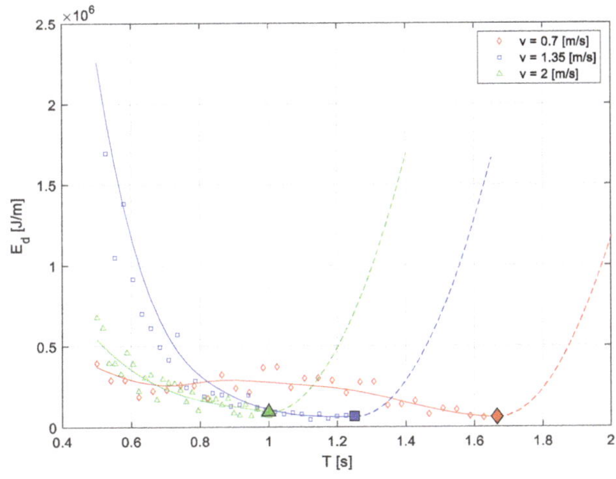

Figure 16. Walk gait energy E_d for different speeds and time cycles T.

In Figure 17, the trend of the trot gait is depicted. We notice the same behavior of low energy cost for the walking gait in the middle region, with an optimal association between periods and speed: $T = 1.25$ s for 1.35 m/s, $T = 1$ s for 2 m/s, and $T = 1.66$ s for 0.7 m/s.

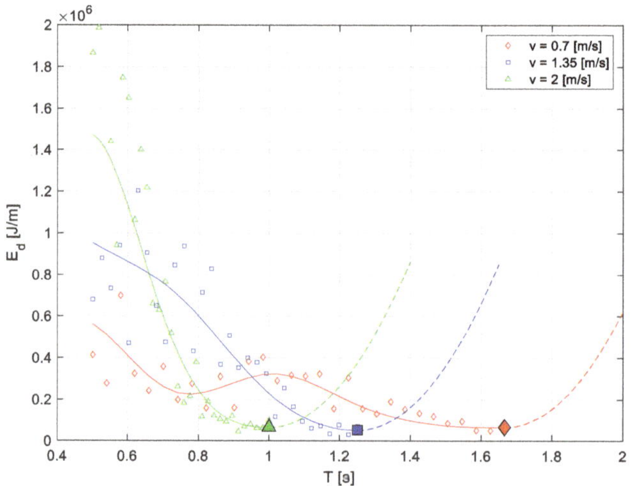

Figure 17. Trot gait energy E_d for different speeds and time cycles T.

Then, taking the minimum points for each gait and comparing them with the energy cost and jerk performances (Figure 18), we see how the transition between walking and trotting takes place at a speed between 0.7 m/s and 1.35 m/s, by using a Pareto criterion. In fact, for the lowest speed, 0.7 m/s, the walk is optimal, in a Pareto sense, since energy cost is lower, and jerk substantially similar. Increasing the speed at 1.35 m/s, a trot performs much better, since both energy and jerk indexes are lower with respect to the walking gait. For higher speeds, the trend of the two curves shows the trot gait is better than walking.

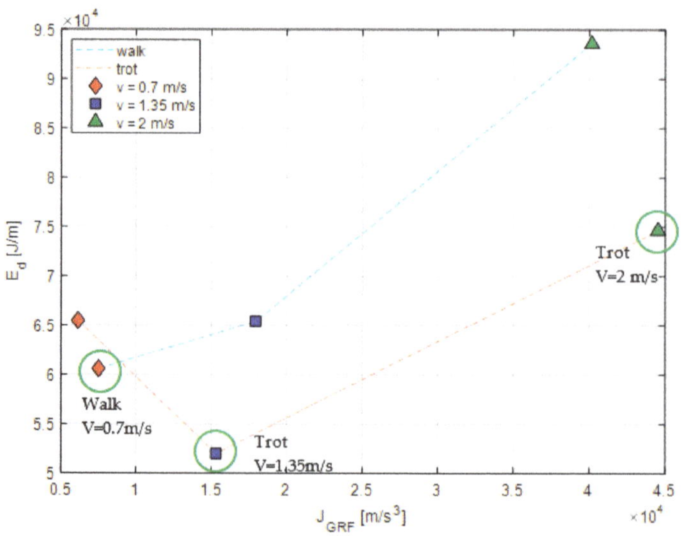

Figure 18. Gait transition from walk to trot based on Pareto's frontier analysis.

5. Discussion

Based on the previous models and results, we can conjecture the optimal gait observed in nature can be derived as an evolutionary effect that combines minimum energy requirements and minimum jerk contact forces, to obtain a gait that is energy-saving and with a moderated mechanical impact on the legs articulations. This point of view has importance also for a quadrupedal robot design that would be aimed at imitating or even enhancing nature's results.

Nevertheless, the previous analysis introduces several approximations with related inaccuracies in the representation of quadrupedal mechanics.

As a first remark, dissipation effects are not taken into considerations. Credible dissipation models need indeed a fine-tuning and require specific and non-trivial experimental tests. This approximation introduces larger errors when increasing the leg's speed since the non-modeled effects of dissipation increase with some power of the velocity. Furthermore, one of the advantages of the previous analysis is the absence of a detailed leg model, that is on the other hand a limitation in the accuracy of the model. Additionally, the absence of a head in the model and its coordinated motion with legs is another approximation when analyzing the quadruped gaits.

Finally, the presence of flexible and deformable elements in the animal mechanical structure is not considered in the model, and these elements absorb elastic energy that plays a role in the global energy balance.

Actuators and their related power are not explicitly included in the animal mechanical model. Besides their obvious importance in the energy balance in the robot design, they guarantee the optimal gait is suitably tracked by the leg motion, with no obvious implications for their performances in terms of supplied maximum force and power, weights, and volumes.

Finally, the contact model is an approximate empirical model, and it will be necessary to check its reliability on the experimental ground.

Future work challenges will be related to overcoming these limitations. More specifically related to quadruped robot design implications, the kinematics and dynamics of leg implementation will be further optimized to be able to pursue the trajectories determined by the optimal gait generator. Non-linear optimal control algorithms recently developed by the authors [36,37] will be used to track kinematic references and forces exchanged on the ground. Such algorithms can be combined with reference trajectory generators [38,39] but can also incorporate the dynamics model and the power actuators by variational feedback control methods [40,41]. Therefore, it is possible to verify and confirm the validity of the results and then extend the work to the optimization of speed changes, changes of gaits, changes of direction, and optimal handling in the presence of obstacles and irregularities of the terrain [42,43].

The present analysis lays the groundwork for the design of new highly efficient quadrupedal robots.

6. Conclusions

This paper presents an investigation related to the optimal gaits for a quadrupedal robot. The interest of this analysis clearly relies on the use of such legged robots in many engineering areas, where the replacement of wheels by legs is highly desirable in an environmental scenario that has an uneven ground with obstacles and in the absence of specific infrastructure that permits a wheeled robot to operate comfortably.

The analysis presents several elements of originality. The optimal gait problem is formulated independently of any specific architecture for the leg mechanisms. The focus is the leg–ground reaction forces that control the quadruped's body motion. A detailed analysis of the contact introduces constitutive relationships for the reaction forces that permit their use in the formulation of the dynamic equation of the body. With the help of the periodicity conditions, that characterize the state variables of the robot involved in its gait and using them an initial guess based on a linearization of the equation of motion, a GA optimization algorithm produces the optimal gaits of the quadruped. A multi-objective function introduces two basic principles of optimization. The first relies on the minimization of the maximum actuators' power and energy consumption. The second relies on the minimization of the contact jerks; that implies a requirement for preventing overload and stressful conditions for the robot mechanisms and its kinematic joints. Their combination, subjected to the Pareto frontier analysis, suggests optimal gaits.

The method is applied to a quadruped the size and mass of which are those of a horse, and the optimal gaits obtained by this optimization method are compared with results found in the technical literature for a real animal. The comparison shows that the different gaits, such as walk and trot, and the convenient transition from the first to the second, depends on the desired speed of the quadruped. The need for a transition comes out as a requirement of energy and jerk optimization; when the speed overcomes a given threshold (located between 0.7 m/s and 1.35 m/s) it is clearly convenient to change the gait. One may speculate this result could explain the origin of the quadruped gaits, as driven by nature, under a very long evolutionary process.

Author Contributions: The optimal gaits of a quadruped robot are part of the Ph.D. Thesis of M.L., under the guidance of G.P. (tutor). The methodology and conceptualization are by G.P. and M.L.; software, and validation, are by M.L. and G.P.; initial suggestions and some conceptualizations about optimal gait problem are by A.C., supervisor of the thesis project, who edited the writing of the manuscript and by N.P.B., former professor at Sapienza, now at University Roma 3. All authors have read and agreed to the published version of the manuscript.

Funding: This research received no external funding.

Institutional Review Board Statement: Not applicable.

Informed Consent Statement: Not applicable.

Conflicts of Interest: The authors declare no conflict of interest.

References

1. Srinivasan, M.; Ruina, A. Computer optimization of a minimal biped model discovers walking and running. *Nature* **2006**, *439*, 72–75. [CrossRef]
2. Dentinger, J.; Street, G. *Animal Locomotion*, 2nd ed.; Biewener, A.A., Patek, S.N., Eds.; Oxford University Press: Oxford, UK, 2018; p. 240. ISBN 978-019874316.
3. Alexander, R.M. Why Mammals Gallop1. *Am. Zool.* **1988**, *28*, 237–245. [CrossRef]
4. Al-dabbagh, A.H.A.; Ronsse, R. A review of terrain detection systems for applications in locomotion assistance. *Robot. Auton. Syst.* **2020**, *133*, 103628. [CrossRef]
5. Shen, T.; Afsar, M.R.; Haque, M.R.; McClain, E.; Meek, S.; Shen, X. A Human-assistive Robotic Platform with Quadrupedal Locomotion*. In Proceedings of the 2019 IEEE 16th International Conference on Rehabilitation Robotics (ICORR), Toronto, ON, Canada, 24–28 June 2019; pp. 305–310.
6. Hyun, D.J.; Park, H.; Ha, T.; Park, S.; Jung, K. Biomechanical design of an agile, electricity-powered lower-limb exoskeleton for weight-bearing assistance. *Robot. Auton. Syst.* **2017**, *95*, 181–195. [CrossRef]
7. Hu, L.; Zhou, C. EDA-Based optimization and learning methods for biped gait generation. *Lect. Notes Control Inf. Sci.* **2007**, *362*, 541–549.
8. Zhai, S.; Jin, B.; Cheng, Y. Mechanical design and gait optimization of hydraulic hexapod robot based on energy conservation. *Appl. Sci.* **2020**, *10*, 3884. [CrossRef]
9. Miandoab, S.N.; Kiani, F.; Uslu, E. Generation of Automatic Six-Legged Walking Behavior Using Genetic Algorithms. In Proceedings of the 2020 International Conference on INnovations in Intelligent SysTems and Applications (INISTA), Novi Sad, Serbia, 24–26 August 2020; pp. 1–5.
10. Kelasidi, E.; Jesmani, M.; Pettersen, K.Y.; Gravdahl, J.T. Locomotion efficiency optimization of biologically inspired snake robots. *Appl. Sci.* **2018**, *8*, 80. [CrossRef]
11. Zeng, X.; Zhang, S.; Zhang, H.; Li, X.; Zhou, H.; Fu, Y. Leg trajectory planning for quadruped robots with high-speed trot gait. *Appl. Sci.* **2019**, *9*, 1508. [CrossRef]
12. Masuri, A.; Medina, O.; Hacohen, S.; Shvalb, N. Gait and Trajectory Optimization by Self-Learning for Quadrupedal Robots with an Active Back Joint. *J. Robot.* **2020**, *2020*, 8051510. [CrossRef]
13. Meng, X.; Wang, S.; Cao, Z.; Zhang, L. A review of quadruped robots and environment perception. In Proceedings of the 2016 35th Chinese Control Conference (CCC), Chengdu, China, 27–29 July 2016; pp. 6350–6356.

14. Yang, K.; Rong, X.; Zhou, L.; Li, Y. Modeling and analysis on energy consumption of hydraulic quadruped robot for Optimal Trot motion control. *Appl. Sci.* **2019**, *9*, 1771. [CrossRef]
15. Kato, T.; Shiromi, K.; Nagata, M.; Nakashima, H.; Matsuo, K. Gait pattern acquisition for four-legged mobile robot by genetic algorithm. In Proceedings of the IECON 2015—41st Annual Conference of the IEEE Industrial Electronics Society, Yokohama, Japan, 9–12 November 2015; pp. 004854–004857.
16. Koco, E.; Glumac, S.; Kovacic, Z. Multiobjective optimization of a quadruped robot gait. In Proceedings of the 22nd Mediterranean Conference on Control and Automation, Palermo, Italy, 16–19 June 2014; pp. 1520–1526.
17. Luneckas, M.; Luneckas, T.; Udris, D.; Plonis, D.; Maskeliunas, R.; Damasevicius, R. Energy-efficient walking over irregular terrain: A case of hexapod robot. *Metrol. Meas. Syst.* **2019**, *26*, 645–660.
18. Luneckas, M.; Luneckas, T.; Udris, D.; Plonis, D.; Maskeliunas, R.; Damasevicius, R. A hybrid tactile sensor-based obstacle overcoming method for hexapod walking robots. *Intell. Serv. Robot.* **2020**. [CrossRef]
19. Focchi, M.; Del Prete, A.; Havoutis, I.; Featherstone, R.; Caldwell, D.; Semini, C. High-slope Terrain Locomotion for Torque-Controlled Quadruped Robots. *Auton. Robot.* **2016**, *41*, 259–272. [CrossRef]
20. Cai, C.; Jiang, H. Performance Comparisons of Evolutionary Algorithms for Walking Gait Optimization. In Proceedings of the 2013 International Conference on Information Science and Cloud Computing Companion, Guangzhou, China, 7–8 December 2013; pp. 129–134.
21. Srisuchinnawong, A.; Shao, D.; Ngamkajornwiwat, P.; Teerakittikul, P.; Dai, Z.; Ji, A.; Manoonpong, P. Neural Control for Gait Generation and Adaptation of a Gecko Robot. In Proceedings of the 2019 19th International Conference on Advanced Robotics (ICAR), Belo Horizonte, Brazil, 2–6 December 2019; pp. 468–473.
22. Rubeo, S.; Szczecinski, N.; Quinn, R. A synthetic nervous system controls a simulated cockroach. *Appl. Sci.* **2017**, *8*, 6. [CrossRef]
23. Ma, W.-L.; Hamed, K.A.; Ames, A.D. First Steps Towards Full Model Based Motion Planning and Control of Quadrupeds: A Hybrid Zero Dynamics Approach. *arXiv* **2019**, arXiv:1909.08124.
24. Ren, D.; Shao, J.; Sun, G.; Shao, X. The complex dynamic locomotive control and experimental research of a quadruped-robot based on the robot trunk. *Appl. Sci.* **2019**, *9*, 3911. [CrossRef]
25. Griffin, T.; Main, R.; Farley, C. Biomechanics of quadrupedal walking: How do four-legged animals achieve inverted pendulum-like movements? *J. Exp. Biol.* **2004**, *207*, 3545–3558. [CrossRef]
26. Hobbs, S.J.; Levine, D.; Richards, J.; Clayton, H.; Tate, J.; Walker, R. Motion analysis and its use in equine practice and research. *Wien. Tierärztliche Mon.* **2010**, *97*, 55–64.
27. Yu, Z.; Zhou, Q.; Chen, X.; Li, Q.; Meng, L.; Zhang, W.; Huang, Q. Disturbance Rejection for Biped Walking Using Zero-Moment Point Variation Based on Body Acceleration. *IEEE Trans. Ind. Inf.* **2019**, *15*, 2265–2276. [CrossRef]
28. Baskoro, A.S.; Priyono, M.G. Design of humanoid robot stable walking using inverse kinematics and zero moment point. In Proceedings of the 2016 International electronics symposium (IES), Denpasar, Indonesia, 29–30 September 2016; pp. 335–339.
29. Gritli, H.; Belghith, S. Walking dynamics of the passive compass-gait model under OGY-based state-feedback control: Analysis of local bifurcations via the hybrid Poincaré map. *Chaos Solitons Fractals* **2017**, *98*, 72–87. [CrossRef]
30. Znegui, W.; Gritli, H.; Belghith, S. Stabilization of the passive walking dynamics of the compass-gait biped robot by developing the analytical expression of the controlled Poincare map. *Nonlinear Dyn.* **2020**, *101*, 1061–1091. [CrossRef]
31. Roan, P.R.; Burmeister, A.; Rahimi, A.; Holz, K.; Hooper, D. Real-world validation of three tipover algorithms for mobile robots. In Proceedings of the 2010 IEEE International Conference on Robotics and Automation, Anchorage, AK, USA, 3–7 May 2010; pp. 4431–4436.
32. Konak, A.; Coit, D.W.; Smith, A.E. Multi-objective optimization using genetic algorithms: A tutorial. *Reliab. Eng. Syst. Saf.* **2006**, *91*, 992–1007. [CrossRef]
33. Tamaki, H.; Kita, H.; Kobayashi, S. Multi-objective optimization by genetic algorithms: A review. In Proceedings of the IEEE International Conference on Evolutionary Computation, Nagoya, Japan, 20–22 May 1996; pp. 517–522.
34. Nauwelaerts, S.; Kaiser, L.; Malinowski, R.; Clayton, H.M. Effects of trunk deformation on trunk center of mass mechanical energy estimates in the moving horse, Equus caballus. *J. Biomech.* **2009**, *42*, 308–311. [CrossRef] [PubMed]
35. Basu, C.; Wilson, A.M.; Hutchinson, J.R. The locomotor kinematics and ground reaction forces of walking giraffes. *J. Exp. Biol.* **2019**, *222*, 308–311. [CrossRef] [PubMed]
36. Pepe, G.; Antonelli, D.; Nesi, L.; Carcaterra, A. Flop: Feedback local optimality control of the inverse pendulum oscillations. In Proceedings of the ISMA 2018—International Conference on Noise and Vibration Engineering and USD 2018—International Conference on Uncertainty in Structural Dynamics, Leuven, Belgium, 17–19 September 2018; pp. 93–106.
37. Pepe, G.; Mezzani, F.; Carcaterra, A.; Cedola, L.; Rispoli, F. Variational control approach to energy extraction from a fluid flow. *Energies* **2020**, *13*, 4913. [CrossRef]
38. Antonelli, D.; Nesi, L.; Pepe, G.; Carcaterra, A. A novel approach in Optimal trajectory identification for Autonomous driving in racetrack. In Proceedings of the 2019 18th European Control Conference (ECC), Naples, Italy, 25–28 June 2019; pp. 3267–3272.
39. Antonelli, D.; Nesi, L.; Pepe, G.; Carcaterra, A. A novel control strategy for autonomous cars. In Proceedings of the 2019 American Control Conference (ACC), Philadelphia, PA, USA, 10–12 July 2019; pp. 711–716.
40. Antonelli, D.; Nesi, L.; Pepe, G.; Carcaterra, A. IMechatronic control of the car response based on VFC. In Proceedings of the ISMA 2018—International Conference on Noise and Vibration Engineering and USD 2018—International Conference on Uncertainty in Structural Dynamics, Leuven, Belgium, 17–19 September 2018; pp. 79–92.

41. Pepe, G.; Roveri, N.; Carcaterra, A. Experimenting Sensors Network for Innovative Optimal Control of Car Suspensions. *Sensors* **2019**, *19*, 3062. [CrossRef]
42. Pepe, G.; Laurenza, M.; Antonelli, D.; Carcaterra, A. A new optimal control of obstacle avoidance for safer autonomous driving. In Proceedings of the 2019 AEIT International Conference of Electrical and Electronic Technologies for Automotive (AEIT AUTOMOTIVE), Torino, Italy, 2–4 July 2019; pp. 1–6.
43. Laurenza, M.; Pepe, G.; Antonelli, D.; Carcaterra, A. Car collision avoidance with velocity obstacle approach: Evaluation of the reliability and performace of the collision avoidance maneuver. In Proceedings of the 5th International Forum on Research and Technologies for Society and Industry: Innovation to Shape the Future, RTSI 2019—Proceedings, Florence, Italy, 9–12 September 2019; pp. 465–470.

Article

Hexapod Robot Gait Switching for Energy Consumption and Cost of Transport Management Using Heuristic Algorithms

Mindaugas Luneckas [1], Tomas Luneckas [1], Jonas Kriaučiūnas [1], Dainius Udris [1], Darius Plonis [2], Robertas Damaševičius [3,*] and Rytis Maskeliūnas [4]

[1] Department of Electrical Engineering, Vilnius Gediminas Technical University, 03227 Vilnius, Lithuania; mindaugas.luneckas@vilniustech.lt (M.L.); tomas.luneckas@vilniustech.lt (T.L.); jonas.kriauciunas@vilniustech.lt (J.K.); dainius.udris@vilniustech.lt (D.U.)

[2] Department of Electronic Systems, Vilnius Gediminas Technical University, 03227 Vilnius, Lithuania; darius.plonis@vilniustech.lt

[3] Faculty of Applied Mathematics, Silesian University of Technology, 44-100 Gliwice, Poland

[4] Department of Multimedia Engineering, Kaunas University of Technology, 51423 Kaunas, Lithuania; rytis.maskeliunas@ktu.lt

* Correspondence: robertas.damasevicius@polsl.pl

Citation: Luneckas, M.; Luneckas, T.; Kriaučiūnas, J.; Udris, D.; Plonis, D.; Damaševičius, R.; Maskeliūnas, R. Hexapod Robot Gait Switching for Energy Consumption and Cost of Transport Management Using Heuristic Algorithms. *Appl. Sci.* **2021**, *11*, 1339. https://doi.org/10.3390/app11031339

Academic Editor: Dario Richiedei
Received: 27 December 2020
Accepted: 29 January 2021
Published: 2 February 2021

Publisher's Note: MDPI stays neutral with regard to jurisdictional claims in published maps and institutional affiliations.

Abstract: Due to the prospect of using walking robots in an impassable environment for tracked or wheeled vehicles, walking locomotion is one of the most remarkable accomplishments in robotic history. Walking robots, however, are still being deeply researched and created. Locomotion over irregular terrain and energy consumption are among the major problems. Walking robots require many actuators to cross different terrains, leading to substantial consumption of energy. A robot must be carefully designed to solve this problem, and movement parameters must be correctly chosen. We present a minimization of the hexapod robot's energy consumption in this paper. Secondly, we investigate the reliance on power consumption in robot movement speed and gaits along with the Cost of Transport (CoT). To perform optimization of the hexapod robot energy consumption, we propose two algorithms. The heuristic algorithm performs gait switching based on the current speed of the robot to ensure minimum energy consumption. The Red Fox Optimization (RFO) algorithm performs a nature-inspired search of robot gait variable space to minimize CoT as a target function. The algorithms are tested to assess the efficiency of the hexapod robot walking through real-life experiments. We show that it is possible to save approximately 7.7–21% by choosing proper gaits at certain speeds. Finally, we demonstrate that our hexapod robot is one of the most energy-efficient hexapods by comparing the CoT values of various walking robots.

Keywords: hexapod robot; path planning; energy consumption; cost of transport; heuristic optimization

1. Introduction

The decrease of energy usage is recognized as one of the principal goals in industrial and manufacturing areas for economic and climate change mitigation reasons [1]. Energy consumption saving in mechatronic and robotic systems has attracted much interest in both industry and academia [2] due to the need to develop energy-efficient and sustainable infrastructure. The existing approaches for increasing energy efficiency include the development of more efficient energy storage technologies for mobile platforms [3], the use of renewable energy resources [4], and also the deployment of innovative computing techniques for controlling digital infrastructure [5].

Mobile robots serve as a great platform for many various tasks that are difficult, dangerous, or even impossible for humans such as space exploration, underground mining, landmine removal, or disaster relief [6,7]. Such tasks require robots to act autonomously in most cases and their work environment is incredibly diverse [8]. The nature of such diversity in environments dictates that mobile robots must adapt to these environments

Appl. Sci. **2021**, *11*, 1339. https://doi.org/10.3390/app11031339 — https://www.mdpi.com/journal/applsci

and not only to one type but a great variety of the terrains. The problem of energy efficiency has been addressed by many different types of robot designs such as flexible (snake-like robots) [9], quadruped robots [10], rovers [11], paddle-driven robots [12], and humanoid robots [13,14].

Terrain diversity is better overcome by legged robots, as they are a lot more adaptable than wheeled or tracked robots [15–17] as dictated by nature (there are no wheeled or tracked animals/insects) [18]. Legged robots can more easily adapt to different types of terrain, their roughness, softness, obstacles, gaps, etc. [19]. However, legged robots still have a lot of challenges to overcome in order to be widely used autonomously in many different applications. These challenges include solving problems for foot-terrain interaction [20–22], gait selection [23,24], walking patterns [25], body stability/balance [17,26], speed [27,28], obstacle avoidance [29,30], motion planning [31,32], trajectory planning [33–35], etc. Robots' autonomy highly depends on their ability to adapt to various types of terrains [36] and their energy efficiency [37,38].

In some studies, vertebrates and invertebrates were observed leading to results stating that oxygen consumption linearly depends on their locomotion speed [39]. A study on modeling the energy consumption of insects [40] and rats [41] suggests that the main reason for changing gaits is to minimize energy consumption. These results were applied to a 2 g mobile robot by [42] and investigated by calculating the cost of transport (CoT). This parameter assesses energy efficiency to transport any type of animal, insect, fish, bird, vehicle or robot from one point to another. CoT allows comparison of different biological or non-biological systems because it is a dimensionless parameter, and thus is widely used in robotics and biology. However, results obtained from experiments with a 2 g mobile robot [42] showed that CoT is significantly larger than in the existing walking robots when an alternating tripod gait is used. In their next work, they presented results of the locomotion and trajectory control technique for microrobots weighing 1.27 g [43]. The Harvard Ambulatory Micro Robot VP series (HAMR-VP) obtained a CoT of 109 which is still a rather large number.

In [44], a method is presented on how to estimate and minimize CoT for legged locomotion. The results of these studies show that it is very important to choose optimal gait to minimize energetic cost. With optimal parameters, CoT is below 1.0 at speeds ranging from 0.1 m/s to 0.35 m/s, but there are no real experiments carried out with either robots or insects. It is difficult to say whether or not results would be the same for walking robots.

Real-time control is proposed in [45] where the authors present a model of a realistic quadrupedal walking machine. The investigation results show that CoT could be reduced by choosing a small duty factor and optimal walking velocity. Also, CoT depends highly on the stroke distance and is less when a stroke is longer. Authors say that the wave gait is only appropriate for low walking speed, and different gaits should be used for higher-speed locomotion. So, to save energy it is very important to select the proper gait.

Additional research, including a duty factor evaluation, was done by [46]. Analysis of the energy usage of a six-legged robot during its locomotion over flat terrain was presented. The results of this investigation showed that the lowest duty factor of the wave gait gives the lowest energy consumption.

CoT depends on the terrain type. Irregular terrain would require more energy expenses than plain terrain. Paper [47] presented a method to optimize the energy use of a hexapod robot on irregular terrain. When the foot trajectories of the legged robot are calculated to minimize energy use during every half a locomotion cycle, it is possible to save about 3.21% of the energy.

The energy efficiency of a hexapod robot wave gait was analyzed in [48]. The author used a simulation model to obtain walking measurements using different modes and varying gait parameters. The results presented in this paper suggest that using a wave gait with different leg sequences is better than using a basic wave gait, although using only wave gait is not efficient, overall, as it is only used for the lower speeds as mentioned before [45].

The analysis of the torque distribution algorithm to decrease the energy cost of the hexapod robot (only for wave gait) is presented in [15]. The proposed scheme allowed reducing energy use from 22% to 39% with the use of fitting robot walking speed and duty factor.

The advantages of gait switching for statically-stable locomotion of a hexapod robot over various terrain types while using a tripod, amble (tetrapod), and wave gaits were explored in [16]. The results showed that on a hard smooth surface at low speeds (up to ~0.1 m/s) the wave gait has higher CoT than the tripod gait. At speeds of ~0.1 m/s, the tripod gait has the lowest CoT, and increasing speed further decreases CoT (at 0.3 m/s CoT = ~15). A similar investigation was done in [49], where locomotion on different surfaces was tested by changing the leg's joint compliance. During each experiment, the robot walked the same distance and the current was measured during its locomotion. Best achieved CoT was 17.2 at a speed of ~0.039 m/s. Wang et al. [50] employed energy usage optimization by introducing equality constraints of dynamic force and moment equilibrium, and inequality constraints of joint torque. Then, they obtained ideal foot forces by solving a quadratic programming (QP) problem.

The novelty and contributions of this paper are as follows:

(1)   A heuristic gait-switching algorithm to minimize the energy consumption of the hexapod robot.
(2)   Adoption of a novel nature-inspired Red Fox Optimization (RFO) for gait parameter selection to minimize the CoT value of the hexapod robot, which has been never done before.
(3)   Experimental study of both algorithms.

The remaining parts of the paper are summarized as follows. Section 2 describes the materials and methods used in our research. Section 3 presents the results of our experiments. Section 4 discusses the results and presents conclusions.

2. Methods and Materials

2.1. Hexapod Robot

For our research, we used an 18 degrees-of-freedom (DoF) six-legged walking robot HexaV4 (Vincross Inc., Beijing, China) (Figure 1) with STM32 microcontroller (STMicroelectronics, Geneva, Switzerland) and Dynamixel AX-12 (ROBOTIS, Seoul, Korea) servo motors. Servo motors are controlled via universal asynchronous receiver-transmitter (UART) by sending them position angles that are calculated from feet coordinates using inverse kinematics. Each servo's motion is controlled via an internal servo controller that is set to default parameters. The robot's body is hexagon shaped to sustain better stability and is made from plastic for lower overall mass (the total weight of the robot is ~1.5 kg). The size of the robot is small: length is 250 mm; width at the front and the back is 100 mm; width at the middle is 190 mm. Since 3D printing is a very useful tool for prototyping the early stages of the robot [51,52], each leg is 3D printed for better durability and divided into three parts: coxa—50 mm; femur—82 mm; tibia—120 mm.

Due to a large variety of different gaits, only four basic hexapod gaits were used for all experiments: tripod gait, wave gait, tetrapod gait, and ripple gait [53]. Used gait time diagrams and leg sequencing is shown in Figure 1b, where the black line represents the leg support phase, and the dashed gap line represents the transfer phase.

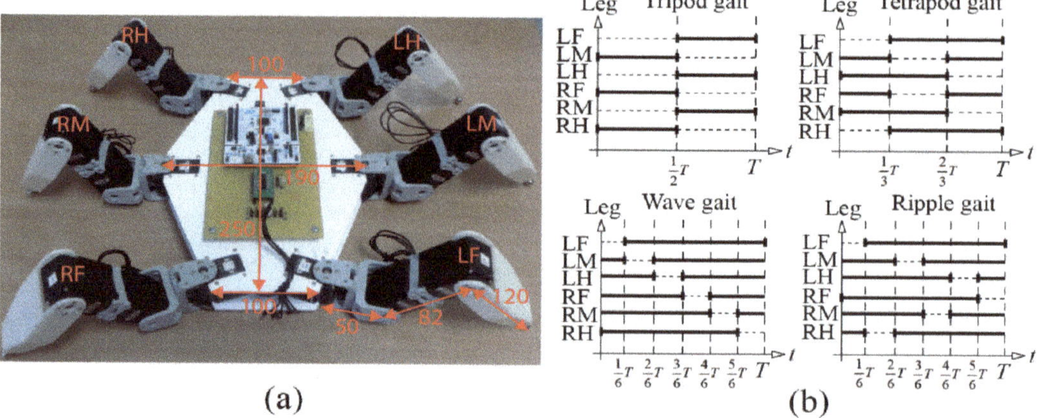

(a) **(b)**

Figure 1. Hexapod robot model HexaV4 (**a**) and gait used in experiments time diagrams (black lines represent leg's support phase, dashed gap—transfer phase) (**b**). Leg abbreviations: LF—left front; LM—left middle; LH—left hind; RF—right front; RM—right middle; RH—right hind; T—gait period. All dimensions are in millimeters.

2.2. Average Energy Consumption Measurement

To accurately evaluate the energy consumption of the hexapod robot, we chose to measure current $I_{avg}\ (t)$ for each speed (even if there is a small difference). First, we measured current values only for each separate robot movement speed, and then calculated the average current value I_{avg} (only for that particular speed). The average value can be calculated by:

$$I_{avg} = \frac{1}{t_2 - t_1} \int_{t_1}^{t_2} I(t)dt \tag{1}$$

As our measurements give us current samples at constant time steps, the average consumed current can be calculated by averaging the measured current samples $I_1 \ldots I_n$:

$$I_{avg} = \frac{\sum_{i=1}^{n} I_i}{n}. \tag{2}$$

Then we can express the average power P_{avg} as:

$$P_{avg} = U \cdot I_{avg}, \tag{3}$$

Here U is the nominal voltage.

2.3. Heuristic Algorithm for Minimum Energy Consumption

To obtain the minimum energy consumption of the hexapod robot, we developed a heuristic algorithm (Figure 2). Since in Section 2.2 we only described experiments where we measured energy consumption dependence on separate robot movement speeds, the algorithm was developed to measure the energy consumption of robots in a realistic movement situation. The algorithm controls how a robot moves using both tripod and wave gaits at certain speed intervals without an additional load. This algorithm was developed for our hexapod robot model, but it could be used as a baseline algorithm for any type of hexapod robot. The only changes that would need to be done to the algorithm are changing the robot's maximum speed and gait changing speed.

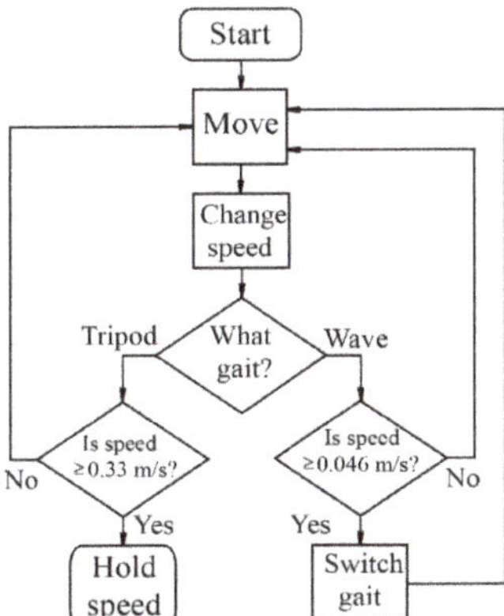

Figure 2. Algorithm for switching between wave and tripod gaits. Wave gait is used when the movement speed is lower than 0.046 m/s; tripod gait is used when the movement speed is higher than 0.046 m/s until the speed reaches a maximum value of 0.33 m/s.

Tripod and wave gaits were selected after conducting a series of tests. During these tests, all four gaits were used with all possible step lengths along with different joint speeds to obtain various robot movement speeds. Each time, the robot's current was measured to estimate energy consumption. The results showed that neither tetrapod nor ripple gaits are energetically efficient. The results also showed that 0.046 m/s speed is the most optimal speed for changing from wave to tripod gait because otherwise, the energy consumption increases. The maximum speed of the robot is 0.33 m/s.

The parameters of the algorithm are set to the minimum: step length is set to 1 cm and servo speeds are set to 12 rpm. These parameters set robot movement speed to 0.00042 m/s. Gait is set to wave gait. After setting initial parameters, the robot starts to move forward. After making 3 full gait cycles, the algorithm increases the robot's movement speed by increasing servo speeds or step length. If movement speed is equal to or greater than 0.046 m/s, the gait type is switched to the tripod gait, otherwise, the robot continues moving forward with the wave gait and further increases speed in the same manner. Gait switching is done using a fixed threshold and is initiated at the end of the gait period to ensure that the previous gait is complete. When tripod gait is selected, the robot moves forward, increasing speed until it reaches a speed of 0.33 m/s, which is maintained until the end of the experiment. Each experiment took 2 min and during the whole time the robot's current was measured using the current shunt sensor mounted on the robot's control system. Over the 2 min, 1500 current values were obtained. This means that the instantaneous current value was measured every 80 ms. Such a sampling rate is adequate to measure current consumption during robot locomotion.

2.4. Calculation of Cost of Transport (CoT)

In mechatronics, it is common to evaluate energy efficiency by verifying CoT [54]. It shows how much power a machine or robot needs to transport itself a certain distance at a certain speed. It is more useful than just evaluating power consumption because the power alone does not provide information about the overall efficiency. In some cases, using more power will allow traveling longer distances at a higher speed, but in other cases, low power consumption might mean slow movement and shorter travel distances. CoT can be calculated by Equation (4):

$$CoT \triangleq \frac{P_{avg}}{mgv} = \frac{U \cdot I_{avg}}{mgv},\tag{4}$$

where m is the weight of the robot/machine with no additional load added [kg], P_{avg} is the average power consumption [W], g—is gravitational force [m/s^2], and v is movement speed [m].

2.5. CoT Optimization Using Red Fox Optimization (RFO) Algorithm

To perform CoT optimization, we adopted the Red Fox Optimization (RFO) algorithm [55]. RFO is a nature-inspired heuristic algorithm that is based on the metaphor of red fox hunting behavior. When crossing the area, the fox takes every opportunity for food and creeps up to the hidden prey until he gets close enough to strike effectively. Exploration of territory in search of food was modeled as a global search when the fox spots the prey in the distance. Before the attack was modeled as a local hunt, the fox traverses through the habitat to get as close as possible to the prey.

The RFO algorithm performs a global search by modeling solution space exploration as the search of food when the fox spots the prey in the distance, whereas local search is modeled as territory traversal aiming to get as close as possible to the prey before attacking it. Formally, the algorithm is described as follows. Let $f \in \mathbb{R}^n$ be the target function of n variables in the solution space, and let $(\overline{x})^{(i)} = \left[(x_0)^{(i)}, (x_1)^{(i)}, \ldots, (x_{n-1})^{(i)} \right]$ mark each point in the space $\langle a, b \rangle^n$, where $a, b \in \mathbb{R}$. Then $(\overline{x})^{(i)}$ is the optimal solution, if the value of function $f\left((\overline{x})^{(i)} \right)$ is a global optimum on $\langle a, b \rangle$.

The search process consists of two stages: variable space exploration and convergence of the solution. The RFO algorithm is summarized in Algorithm 1.

Algorithm 1. *Red Fox Optimization Algorithm*

Inputs: search space $\langle a, b \rangle$, fitness function $f(\cdot)$
Outputs: optimal solution x^{best}
Begin
    Define the number of iterations T, the maximum size of population n,
    Generate randomly the population of n foxes within search space,
    Set iteration count $i := 0$,
    while i $\leq$ T **do**
        For each *fox* **in** *population* **do**
        Sort individuals according to the fitness function,
        Select best individual $\left(x^{best} \right)^i$
        Calculate *reallocation* of individuals
        if *reallocation* is better than the previous position **then**
            Move the *fox*,
        else
            Return the *fox* to previous position,
        end if
        Choose randomly the noticing parameter of the *fox*,
        if *fox* is not noticed **then**
            Calculate reallocation
        else
            Fox stays at his position
        end if
    23: **end for**
    24: Sort *population* according to the fitness function $f(\cdot)$,
    25: Remove worst foxes from the population
    26: Reproduce new foxes
    27 $i := i + 1$
  28: **end while**
  29: return the fittest *fox* x^{best}
End

For our problem of finding the best gait type and gait parameter of the robot, we defined the search space in terms of three independent variables:

$$(\overline{x}) = [(x_1), (x_2), (x_3)], \tag{5}$$

where x_1 is the step length [mm], x_2 is gait resolution, and x_3 is gait type (one of Tripod, Tetrapod, Ripple, or Wave). Optimization aims to find the minimum value of CoT. Therefore, the fitness function is defined by Equation (4).

3. Results

3.1. Power Consumption Measurement Results

The hexapod robot step height and body elevation were set to 1 cm and 10 cm accordingly. These values are the standard configuration for hexapod robot movement and 1 cm is the minimum step height required to execute leg transfer. All experiments were carried out on flat terrain with a soft surface for better contact between the feet and the ground to minimize leg slippage.

The first experiment was done without using any additional load on the robot. The average current dependence on robot movement speed and different gaits is shown in Figure 3a. The speed at which the gait must be switched to sustain the minimal current usage is $v = 0.046$ m/s (Figure 3b). Wave gait is the most energetically efficient for low speeds (0–0.046 m/s) and tripod gait is more efficient for faster movement (0.046–0.33 m/s). The power consumption at low speeds is ~12–13 W (wave gait), and at faster speeds is ~13–28 W (tripod gait). Using proposed gait switching between wave and tripod gaits, ~7.7–21% of energy can be saved.

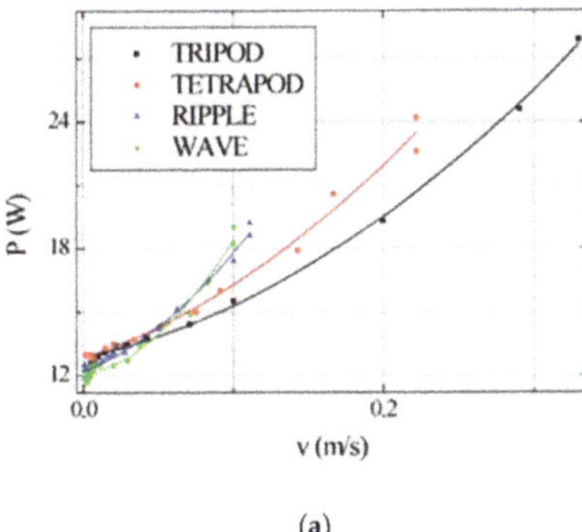

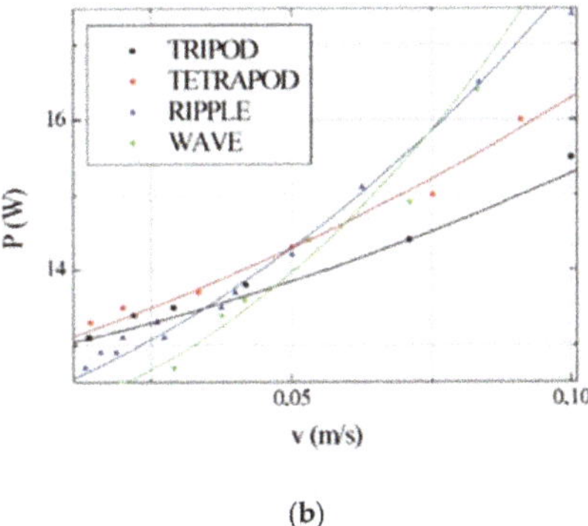

Figure 3. Average power dependence on robot speed and gaits: (**a**) normal view, (**b**) enlarged view. Experiments were carried out without additional load on the robot. Each dot represents a separate experiment with a certain configuration. Lines show polynomial approximation. No gait type switching was used.

To make sure that the recurrence is stable and does not change depending on conditions, we repeated the experiment with 1.16 kg and 2.9 kg loads placed on the top of the robot. Results are shown in Figure 4. All curves for different gaits with 1.16 kg load were arranged in the same way as with no load. The point of gait switching decreased ($v = 0.028$ m/s). Only with a 2.9 kg load does the arrangement of curves start to change. This shows that energy consumption depends on the robot's weight. With increased robot's weight, gait switching speed decreases, i.e., the tripod gait becomes more effective at slower speeds with increased load.

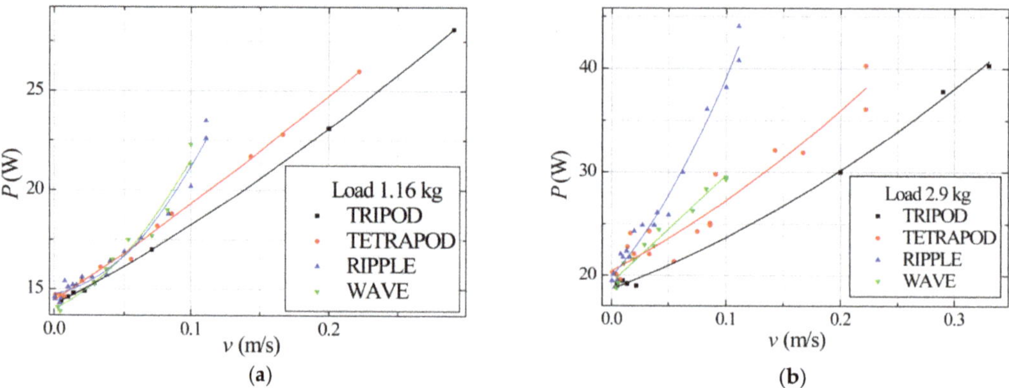

Figure 4. Average power dependence on robot speed and gaits with different robot loads: (**a**) 1.16 kg load, (**b**) 2.9 kg load. Each dot represents a separate experiment with a certain configuration. Lines show polynomial approximation. No gait switching was used.

3.2. Verifying Overall Energy Efficiency

To evaluate our hexapod robot's dependence of CoT on robot movement speed, we applied Equation (5) for each gait, speed, and each obtained power consumption result presented in Figure 5. As we can see from Figure 5a, CoT does not depend on gait type. Even if we take a particular speed, CoT remained the same for each gait. The only difference is that gaits vary in speed intervals. The results in Figure 5b show that the CoT value decreases when the movement speed increases. This means that to obtain lower CoT, it is necessary to switch to the gait with a higher speed interval (e.g., tripod gait). These results along with Equation (4) can be used to optimize the energy consumption according to robot speed.

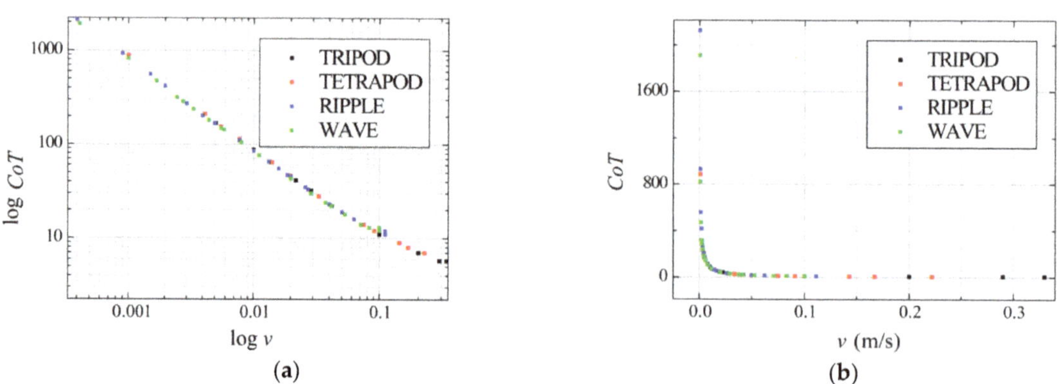

Figure 5. Cost of transport dependence on robot speed: (**a**) logarithmic scale, (**b**) linear scale.

Using only the logarithmic scale, we cannot define the working space of our hexapod robot that is energetically efficient. This is the reason for using the linear scale also (Figure 5b). The robot should be programmed to move only at speeds higher than 0.04 m/s due to CoT having the lowest values in that interval. Between the speed of 0.04 m/s and 0.33 m/s, the value of CoT decreases from ~25 to ~5.75, whereas between the speed of 0.04 m/s down to 0.00042 m/s, the value of CoT rises from ~25 up to ~2000. The tripod gait was the most efficient because it used the least energy, while the robot can reach the highest speed.

To evaluate the efficiency of our hexapod robot, we compared its CoT with different walking robots and micro-robots (Table 1): medium-sized high mobility hexapod robot RHex, Gregor I six-legged robot, a resilient high-speed hexapod robot DASH, tiny quadruped robot MEDIC (Millirobot Enabled Diagnostic of Integrated Circuits), micro hexapod robot HAMR2 from Harvard University, and larger hexapod robot AMOS (Advanced Mobility Sensor Driven-Walking Device). Low CoT value for the HexaV4 robot was obtained due to the good trade-off between the robot's mass, current consumption, and movement speed. In our case, the servos' current was minimal during all experiments (each servo can consume up to ~0.9 A, but the total current consumption of the robot did not exceed 3.5 A).

Table 1. Cost of transport of different walking robots: RHex hexapod robot [56]; Gregor I hexapod [57]; DASH hexapod [58]; MEDIC quadruped robot [59]; HAMR2 hexapod microrobot [42]; AMOS hexapod robot [50]; HexaV4 hexapod robot; Big Dog robot [60].

Name	Rhex	Gregor I	DASH	MEDIC	HAMR2	AMOS	HexaV4	Big Dog
Mass (kg)	7	1	0.0062	0.0055	0.002	5.4	1.5	110
COT	3.7~14	70	14.7	>12,000	128	3.4~11.7	5.75	15

3.3. Maintaining Minimum Energy Consumption

The algorithm shown in Figure 2 was implemented and tested with the hexapod robot several times with and without an additional delay time between speed changes. The speed was increased by the algorithm and the current was measured throughout the tests. The example of the results of the algorithm test is shown in Figure 6. We also varied the delay time to obtain better intervals for calculating the average current. No additional load was used during the experiments. The current time diagram of the full program run without the delay is shown in Figure 7. Due to a large amount of noise, a filter was applied. We used smooth signal processing and the percentile filter method (50%) with 3 points of the window.

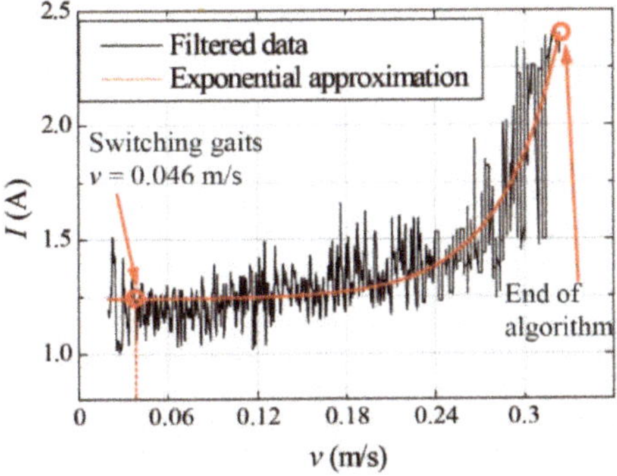

Figure 6. Current-time diagram, achieved by varying robot speed and gaits. Black curve—filtered experimental data; red line—exponential approximation.

By calculating the average current for each interval, we managed to obtain power consumption dependence on the speed of a full program run with a gait switching algorithm. We can see from Figure 7 that power consumption changes from ~12 W to ~24 W. The same results and curve form can be seen in Figure 3, where each result was obtained by a separate experiment. This demonstrates that our proposed method leads to power consumption minimization.

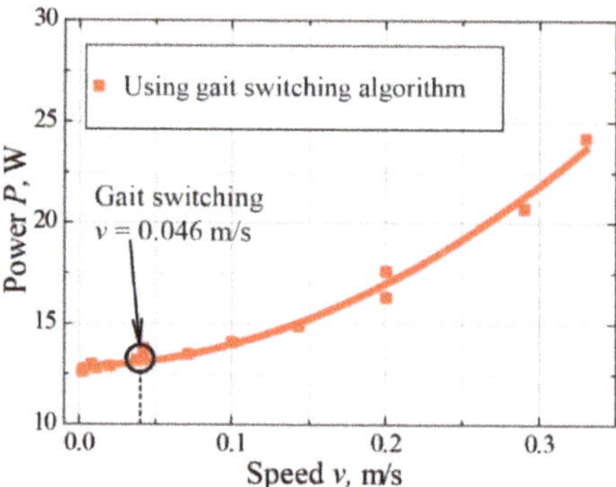

Figure 7. Influence of gait switching on energy consumption. Speed for switching from wave gait to tripod gait is equal to 0.046 m/s.

3.4. Maintaining Minimum Cost of Transport Using Red Fox Optimisation

The algorithm shown in Algorithm 1 was implemented and tested with the hexapod robot several times to calculate the average CoT value. No additional load was used during the experiments. The variable search space and corresponding CoT values are shown in Figure 8. The application of the RFO algorithm allows finding an optimal combination of gait step length and gait resolution for minimal CoT. In our experiment, the optimal solution found corresponds to the step length of 100 mm, resolution of 12 mm, and wave gait type.

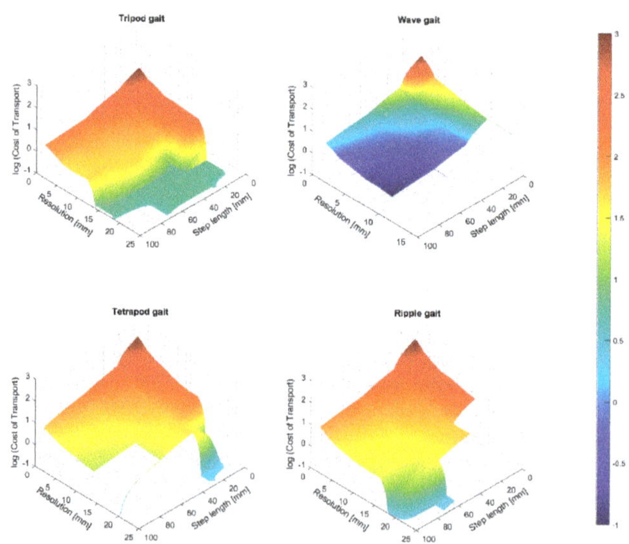

Figure 8. Comparison of Cost of Transport upon gait type, gait step length, and resolution.

4. Discussion and Conclusions

The problem of high energy consumption has always limited the wide-spread use of multi-legged walking robots. The benefits to be achieved by rising task time without upgrading the power supply unit mean that optimization of energy efficiency is an area of significant importance.

In this paper, we analyzed and proposed a method for optimizing the hexapod robot's energy consumption. We performed experiments with different gaits and speeds with no load, 1.16 kg, and 2.9 kg loads. Energetically efficient gaits were selected using power consumption diagrams. Cost of transport dependence on speed was calculated and a hexapod robot working space was established. A robot movement algorithm was also developed for power consumption minimization.

Our results show that neither ripple nor tetrapod gaits are energetically efficient. Only wave gait for slower speeds and tripod gait for fast movement should be used. Maintaining minimum energy consumption requires optimal gait switching. For our hexapod robot, the optimal gait switching speed is 0.046 m/s with no load, and 0.028 m/s with 1.16 kg additional load. By using the gait switching algorithm during movement, we were able to decrease energy consumption by ~7.7–21%.

We adopted Red Fox Optimization (RFO) algorithm to find the best combination of gait step length and gait resolution for minimal CoT. In our experiment, the optimal solution found corresponds to the step length of 100 mm, resolution of 12 mm, and wave gait type.

System resilience in engineering, and especially in robotics, has become an important topic nowadays because even in case of failure, a system is capable of recovering [61]. Walking robots have a wide selection of gaits which makes them resilient to failure; e. g. the ability to move even with one or several damaged legs. In the scope of this research, the hexapod robot can still maintain minimum energy consumption with one or more damaged legs if the correct gait is selected considering the robot's movement speed.

Finally, we observed that the stability of the robot slightly decreased when the movement speed was increased, which is a limitation of the proposed method. However, since the hexapod robot always has at least three legs resting on the surface, the stability criterion [62] is always satisfied.

Author Contributions: M.L. and T.L. contributed to robot development and experiments. J.K. and D.U. contributed to introduction part. D.P., R.D. and R.M. contributed to data analysis and results. All authors have read and agreed to the published version of the manuscript.

Funding: This research received no external funding.

Institutional Review Board Statement: Not applicable.

Informed Consent Statement: Not applicable.

Data Availability Statement: Data is available from the corresponding author upon reasonable request.

Conflicts of Interest: The authors declare no conflict of interest.

References

1. Zaharia, A.; Diaconeasa, M.C.; Brad, L.; Lădaru, G.-R.; Ioanăș, C. Factors Influencing Energy Consumption in the Context of Sustainable Development. *Sustainability* **2019**, *11*, 4147. [CrossRef]
2. Scalera, L.; Palomba, I.; Wehrle, E.; Gasparetto, A.; Vidoni, R. Natural motion for energy saving in robotic and mechatronic systems. *Appl. Sci.* **2019**, *9*, 3516. [CrossRef]
3. Arantes, C.; Sepúlveda, J.; Esteves, J.S.; Costa, H.; Soares, F. Using ultracapacitors as energy-storing devices on a mobile robot platform power system for ultra-fast charging. In Proceedings of the 2014 11th International Conference on Informatics in Control, Automation and Robotics (ICINCO), Vienna, Austria, 1–3 September 2014; pp. 156–164. [CrossRef]
4. Hassan, A.; El Habrouk, M.; Deghedie, S. Renewable Energy for Robots and Robots for Renewable Energy—A Review. *Robotica* **2020**, *38*, 1576–1604. [CrossRef]
5. Okewu, E.; Misra, S.; Maskeliunas, R.; Damasevicius, R.; Fernandez-Sanz, L. Optimizing green computing awareness for environmental sustainability and economic security as a stochastic optimization problem. *Sustainability* **2017**, *9*, 1857. [CrossRef]
6. Kostavelis, I.; Gasteratos, A. Robots in Crisis Management: A Survey. In *Information Systems for Crisis Response and Management in Mediterranean Countries, Proceedings of the 4th International Conference, ISCRAM-Med 2017, Xanthi, Greece, 18–20 October 2017*; Dokas, I., Bellamine-Ben Saoud, N., Dugdale, J., Díaz, P., Eds.; Lecture Notes in Business Information Processing; Springer: Cham, Switzerland, 2017; Volume 301, pp. 43–56. [CrossRef]
7. Jorge, V.A.M.; Granada, R.; Maidana, R.G.; Jurak, D.A.; Heck, G.; Negreiros, A.P.F.; dos Santos, D.H.; Gonçalves, L.M.G.; Amory, A.M. A Survey on Unmanned Surface Vehicles for Disaster Robotics: Main Challenges and Directions. *Sensors* **2019**, *19*, 702. [CrossRef]

8. Liu, P.; Huda, M.N.; Sun, L.; Yu, H. A survey on underactuated robotic systems: Bio-inspiration, trajectory planning and control. *Mechatronics* **2020**, *72*, 102443. [CrossRef]

9. Bing, Z.; Lemke, C.; Cheng, L.; Huang, K.; Knoll, A. Energy-efficient and damage-recovery slithering gait design for a snake-like robot based on reinforcement learning and inverse reinforcement learning. *Neural Netw.* **2020**, *129*, 323–333. [CrossRef]

10. Chai, H.; Rong, X.; Tang, X.; Li, Y.; Zhang, Q.; Li, Y. Gait based planar hopping control of quadruped robot on uneven terrain with energy planning. *J. Jilin Univ.* **2017**, *47*, 557–566. [CrossRef]

11. Zhornyak, L.; Reza Emami, M.; Reza Emami, M. Gait optimization for quadruped rovers. *Robotica* **2020**, *38*, 1263–1287. [CrossRef]

12. Han, J.; Wang, H.; Wang, G.; Ye, X.; Wang, L. Stability and energy consumption of laterally walking gait in crablike robots. *J. Harbin Eng. Univ.* **2017**, *38*, 898–906. [CrossRef]

13. Yang, T.; Zhang, W.; Huang, Q.; Chen, X.; Yu, Z.; Meng, L. A smooth and efficient gait planning for humanoids based on human ZMP. *Robot* **2017**, *39*, 751–758. [CrossRef]

14. Hereid, A.; Hubicki, C.M.; Cousineau, E.A.; Ames, A.D. Dynamic humanoid locomotion: A scalable formulation for HZD gait optimization. *IEEE Trans. Robot.* **2018**, *34*, 370–387. [CrossRef]

15. Agheli, M.; Nestinger, S.S. Foot Force Based Reactive Stability of Multi-Legged Robots to External Perturbations. *J. Intell. Robot. Syst.* **2016**, *81*, 287–300. [CrossRef]

16. Kottege, N.; Parkinson, C.; Moghadam, P.; Elfes, A.; Singh, S.P.N. Energetics-Informed Hexapod Gait Transitions Across Terrains. In Proceedings of the IEEE International Conference on Robotics and Automation (ICRA), Seattle, WA, USA, 26–30 May 2015; pp. 5140–5147. [CrossRef]

17. Jin, B.; Chen, C.; Li, W. Power Consumption Optimization for a Hexapod Walking Robot. *J. Intell. Robot. Syst.* **2013**, *71*, 195–209. [CrossRef]

18. Chai, H.; Rong, X.; Tang, X.; Li, Y. Gait-Based Quadruped Robot Planar Hopping Control with Energy Planning. *Int. J. Adv. Robot. Syst.* **2016**, *13*, 20. [CrossRef]

19. Ma, C.; Yu, F.; Luo, Z. Simulations and Experimental Research on a Novel Soft-Terrain Hexapod Robot. *Int. J. Robot. Autom.* **2015**, *30*, 247–255. [CrossRef]

20. Ding, L.; Gao, H.; Deng, Z.; Song, J.; Liu, Y.; Liu, G.; Iagnemma, K. Foot-terrain interaction mechanics for legged robots: Modeling and experimental validation. *Int. J. Robot. Res.* **2013**, *32*, 1585–1606. [CrossRef]

21. Zhang, T.; Wei, Q.; Ma, H. Position/Force Control for a Single Leg of a Quadruped Robot in an Operation Space. *Int. J. Adv. Robot. Syst.* **2013**, *10*, 137. [CrossRef]

22. Zhu, Y.; Bo, J. Compliance control of a legged robot based on improved adaptive control: Method and experiments. *Int. J. Robot. Autom.* **2016**, *5*, 366–373. [CrossRef]

23. Luneckas, M.; Luneckas, T.; Udris, D.; Plonis, D.; Maskeliunas, R.; Damasevicius, R. Energy-efficient walking over irregular terrain: A case of hexapod robot. *Metrol. Meas. Syst.* **2019**, *26*, 645–660. [CrossRef]

24. Zhang, S.; Xing, Y.; Hu, Y. Composite gait optimization method for a multi-legged robot based on optimal energy consumption. *Chin. Space Sci. Technol.* **2018**, *38*, 32–39. [CrossRef]

25. Zhang, L.; Zhou, C. Optimal three-dimensional biped walking pattern generation based on geodesics. *Int. J. Adv. Robot. Syst.* **2017**, *14*. [CrossRef]

26. Li, M.; Wang, X.; Guo, W.; Wang, P.; Sun, L. System Design of a Cheetah Robot toward Ultra-high Speed. *Int. J. Adv. Robot. Syst.* **2014**, *11*, 5. [CrossRef]

27. Zhang, S.; Liu, M.; Yin, Y.; Rong, X.; Li, Y.; Hua, Z. Static gait planning method for quadruped robot walking on unknown rough terrain. *IEEE Access* **2019**, *7*, 177651–177660. [CrossRef]

28. Sprowitz, A.; Tuleu, A.; Vespignani, M.; Ajallooeian, M.; Badri, E.; Ijspeert, A.J. Towards dynamic trot gait locomotion: Design, control, and experiments with Cheetah-cub, a compliant quadruped robot. *Int. J. Robot. Res.* **2013**, *32*, 932–950. [CrossRef]

29. Xin, Y.; Liang, H.; Mei, T.; Huang, R.; Chen, J.; Zhao, P.; Sun, C.; Wu, Y. A New Dynamic obstacle Collision Avoidance System for Autonomous Vehicles. *Int. J. Robot. Autom.* **2015**, *30*, 278–288. [CrossRef]

30. Luneckas, M.; Luneckas, T.; Udris, D.; Plonis, D.; Maskeliūnas, R.; Damaševičius, R. A hybrid tactile sensor-based obstacle overcoming method for hexapod walking robots. *Intell. Serv. Robot.* **2020**. [CrossRef]

31. Chen, J.; Gao, F.; Huang, C.; Zhao, J. Whole-Body Motion Planning for a Six-Legged Robot Walking on Rugged Terrain. *Appl. Sci.* **2019**, *9*, 5284. [CrossRef]

32. Boscariol, P.; Richiedei, D. Optimization of Motion Planning and Control for Automatic Machines, Robots and Multibody Systems. *Appl. Sci.* **2020**, *10*, 4982. [CrossRef]

33. Gasparetto, A.; Boscariol, P.; Lanzutti, A.; Vidoni, R. Trajectory planning in robotics. *Math. Comput. Sci.* **2012**, *6*, 269–279. [CrossRef]

34. Carabin, G.; Scalera, L. On the trajectory planning for energy efficiency in industrial robotic systems. *Robotics* **2020**, *9*, 89. [CrossRef]

35. Damaševičius, R.; Maskeliūnas, R.; Narvydas, G.; Narbutaitė, R.; Połap, D.; Woźniak, M. Intelligent automation of dental material analysis using robotic arm with jerk optimized trajectory. *J. Ambient Intell. Humaniz. Comput.* **2020**, *11*, 6223–6234. [CrossRef]

36. WeiHai, C.; GuanJiao, R.; JianHua, W.; Dong, L. An adaptive locomotion controller for a hexapod robot: CPG, kinematics and force feedback. *Sci. China Inf. Sci.* **2014**, *57*, 1–18.

37. Xi, W.; Yesilevskiy, Y.; Remy, C.D. Selecting gaits for economical locomotion of legged robots. *Int. J. Robot. Res.* **2015**, *35*, 1140–1154. [CrossRef]
38. Beniak, R.; Tomasz, P. An energy-consumption analysis of a tri-wheel mobile robot. *Int. J. Robot. Autom.* **2016**, *31*. [CrossRef]
39. Nishii, J. Legged insects select the optimal locomotor pattern based on the energetic cost. *J. Biol. Cybern.* **2000**, *83*, 435–442. [CrossRef]
40. Hao, X.; Ma, W.; Liu, C.; Qian, Z.; Ren, L.; Ren, L. Locomotor mechanism of Haplopelma hainanum based on energy conservation analysis. *Biol. Open* **2020**, *9*, bio055301. [CrossRef]
41. Toeda, M.; Aoi, S.; Fujiki, S.; Funato, T.; Tsuchiya, K.; Yanagihara, D. Gait Generation and Its Energy Efficiency Based on Rat Neuromusculoskeletal Model. *Front. Neurosci.* **2020**, *13*, 1337. [CrossRef]
42. Baisch, A.T.; Sreetharan, P.S.; Wood, R.J. Biologically-inspired locomotion of a 2 g hexapod robot. In Proceedings of the 2010 IEEE/RSJ International Conference on Intelligent Robots and Systems, Taipei, Taiwan, 18–22 October 2010; pp. 5360–5365. [CrossRef]
43. Baisch, A.T.; Ozcan, O.; Goldberg, B.; Ithier, D.; Wood, R.J. High speed locomotion for a quadrupedal microrobot. *Int. J. Robot. Res.* **2014**, *33*, 1063–1082. [CrossRef]
44. Nishii, J. An analytical estimation of the energy cost for legged locomotion. *J. Theor. Biol.* **2006**, *238*, 636–645. [CrossRef]
45. Lin, B.S.; Min, S.M. Dynamic modeling, stability, and energy efficiency of a quadrupedal walking machine. *J. Robot. Syst.* **2001**, *18*, 657–670. [CrossRef]
46. Roy, S.S.; Pratihar, D.K. Effects of turning gait parameters on energy consumption and stability of a six-legged walking robot. *Rob. Auton. Syst.* **2012**, *60*, 72–82. [CrossRef]
47. De Santos, P.G.; Garcia, E.; Ponticelli, R.; Armada, M. Minimizing Energy Consumption in Hexapod Robots. *Adv. Robot.* **2012**, *23*, 681–704. [CrossRef]
48. Erden, M.S.; Leblebicioglu, E. Analysis of Wave Gaits for Energy Efficiency. *Auton. Robot.* **2007**, *23*, 213–230.
49. Xiong, X.; Worgotter, F.; Manoonpong, P. Adaptive and Energy Efficient Walking in a Hexapod Robot Under Neuromechanical Control and Sensorimotor Learning. *IEEE Trans. Cybern.* **2016**, *46*, 2521–2534. [CrossRef]
50. Wang, G.; Ding, L.; Gao, H.; Deng, Z.; Liu, Z.; Yu, H. Minimizing the Energy Consumption for a Hexapod Robot Based on Optimal Force Distribution. *IEEE Access* **2020**, *8*, 5393–5406. [CrossRef]
51. Caffola, D.; Ceccarelli, M.; Wang, M.F.; Carbone, G. 3D printing for feasibility check of mechanism design. *Int. J. Mech. Control* **2016**, *17*, 3–12.
52. Devaraja, R.R.; Maskeliūnas, R.; Damaševičius, R. Design and Evaluation of Anthropomorphic Robotic Hand for Object Grasping and Shape Recognition. *Computers* **2021**, *10*, 1. [CrossRef]
53. Haynes, G.C.; Rizzi, A.A. Gaits and gait transitions for legged robots. In Proceedings of the 2006 IEEE International Conference on Robotics and Automation, ICRA 2006, Orlando, FL, USA, 15–19 May 2006; pp. 1117–1122. [CrossRef]
54. Kashiri, N.; Abate, A.; Abram, S.J.; Albu-Schaffer, A.; Clary, P.J.; Daley, M.; Faraji, S.; Furnemont, R.; Garabini, M.; Geyer, H.; et al. An Overview on Principles for Energy Efficient Robot Locomotion. *Front. Robot. AI* **2018**, *5*, 129. [CrossRef]
55. Połap, D.; Woźniak, M. Red fox optimization algorithm. *Expert Syst. Appl.* **2021**, *166*, 114107. [CrossRef]
56. Saranli, U.; Buehler, M.; Koditschek, D.E. RHex: A Simple and Highly Mobile Hexapod Robot. *Int. J. Robot. Res.* **2001**, *20*, 616–631. [CrossRef]
57. Arena, P.; Fortuna, L.; Frasca, M.; Patane, L.; Pavone, M. Realization of a CNN-driven cockroach-inspired robot. In Proceedings of the 2006 IEEE International Symposium on Circuits and Systems, Island of Kos, Greece, 21–24 May 2006; pp. 2649–2652. [CrossRef]
58. Birkmeyer, P.; Peterson, K.; Fearing, R.S. DASH: A dynamic 16g hexapedal robot. In Proceedings of the 2009 IEEE/RSJ International Conference on Intelligent Robots and Systems, St. Louis, MO, USA, 10–15 October 2009; pp. 2683–2689. [CrossRef]
59. Kohut, N.J.; Hoover, A.M.; Ma, K.Y.; Baek, S.S.; Fearing, R.S. MEDIC: A legged millirobot utilizing novel obstacle traversal. In Proceedings of the 2011 IEEE International Conference on Robotics and Automation, Shanghai, China, 9–13 May 2011; pp. 802–808. [CrossRef]
60. Seok, S.; Wang, A.; Chuah, M.Y.; Otten, D.; Lang, J.; Kim, S. Design principles for highly efficient quadrupeds and implementation on the MIT cheetah robot. In Proceedings of the 2013 IEEE International Conference on Robotics and Automation 2013, Karlsruhe, Germany, 6–10 May 2013; pp. 3307–3312.
61. Zhang, T.; Zhang, W.; Gupta, M.M. Resilient Robots: Concept, Review, and Future Directions. *Robotics* **2017**, *6*, 22. [CrossRef]
62. Peng, W.Z.; Mummolo, C.; Kim, J.H. Stability Criteria of Balanced and Steppable Unbalanced States for Full-Body Systems with Implications in Robotic and Human Gait. In Proceedings of the 2020 IEEE International Conference on Robotics and Automation (ICRA), Paris, France, 31 May–31 August 2020; pp. 9789–9795. [CrossRef]

Article

Terrain Estimation for Planetary Exploration Robots [†]

Mauro Dimastrogiovanni [1], Florian Cordes [2] and Giulio Reina [3,*]

[1]   Department of Engineering for Innovation, University of Salento Via per Arnesano, 73100 Lecce, Italy; mauro.dimastrogiovanni@unisalento.it
[2]   DFKI Robotics Innovation Center Bremen Robert-Hooke-Str. 1, 28359 Bremen, Germany; florian.cordes@dfki.de
[3]   Department of Mechanics, Mathematics and Management, Politecnico di Bari, via Orabona 4, 70126 Bari, Italy
*   Correspondence: giulio.reina@poliba.it
†   Parts of this paper were presented at the Third International Conference of IFToMM Naples, Italy, 9–11 September 2020.

Received: 5 August 2020; Accepted: 26 August 2020; Published: 31 August 2020

Abstract: A planetary exploration rover's ability to detect the type of supporting surface is critical to the successful accomplishment of the planned task, especially for long-range and long-duration missions. This paper presents a general approach to endow a robot with the ability to sense the terrain being traversed. It relies on the estimation of motion states and physical variables pertaining to the interaction of the vehicle with the environment. First, a comprehensive proprioceptive feature set is investigated to evaluate the informative content and the ability to gather terrain properties. Then, a terrain classifier is developed grounded on Support Vector Machine (SVM) and that uses an optimal proprioceptive feature set. Following this rationale, episodes of high slippage can be also treated as a particular terrain type and detected via a dedicated classifier. The proposed approach is tested and demonstrated in the field using SherpaTT rover, property of DFKI (German Research Center for Artificial Intelligence), that uses an active suspension system to adapt to terrain unevenness.

Keywords: space robotics; planetary surface exploration; terrain awareness; mechanics of vehicle–terrain interaction; vehicle dynamics

---

1. Introduction

The main challenges that planetary exploration rovers must face refer to: long-range operations in hostile environmental conditions, lack of maintenance, and limited human supervision. It is of primary importance to increase their autonomy level to reduce the reliance on ground control and maximize the mission's scientific return. One of the key technologies for autonomous navigation is the ability to sense and characterize the incoming terrain, avoiding potential hazards. For example, the terrain being traversed may exhibit high deformability and low traction properties due to low packing density and/or limited cohesion. This could result in loss of traction as well as in excessive sinkage that in extreme cases may lead to robot entrapment. For example, in April 2005, the Mars Exploration Rover Opportunity became embedded in a dune of loosely packed drift material and delayed its operations for more than a month. A similar embedding event led to the end of mobility for the Spirit rover in 2010 [1].

Beyond general safety and stability assessment, terrain sensing can be used to improve trajectory tracking by applying methods for slip estimation and compensation (e.g., [2–4]) or traction control (e.g., [5]). Future planetary rovers are expected to be able to extend methods for rough-terrain navigation to infer scientific information of different geological formations [6].

Early research in terrain estimation has relied on forward looking sensors and used limited learning [7]. Monocular and stereo cameras have been the most common sensors used for terrain estimation from a distance [8], followed by lidars and radars [9], which have been often proposed in terrestrial applications [10,11]. The use of exteroceptive sensing leads to the generation of a local digital elevation model (DEM) to obtain and maintain a discrete traversability map. The appearance (e.g., texture, color) of different terrain patches can provide important clues to analyze the surrounding terrain [12].

However, observation of a given terrain from a distance does not provide any information about its impact on the vehicle mobility. It is known that off-road traversability largely depends on the interaction between the robot and the terrain [13]. Dynamic ill-effects including wheel sinkage, slippage and rolling resistance are the result of this complex interplay. For example, ground can be considered drivable based on the geometric elevation map. Yet, the robot can incur serious risks if this terrain offers low traction properties due to high slippage and consequent lack of progression as explained in [14]. An extensive discussion on methods for slippage estimation in planetary rovers can be found in [15], whereas the impact of the irregularity and deformability of the traversed surface on the robot's dynamic response are investigated in [16].

Therefore, recently, methods for terrain estimation have been also presented that use proprioceptive sensing [17,18]. The envisaged idea is that terrain properties can be obtained directly by the rover wheels that serve as tactile sensors. Proprioceptive signals are modulated by the vehicle–terrain interaction and they contain substantial information, which can help to characterize terrain. In addition, learning approaches have been introduced in order to make intelligent autonomous robots adaptive to the site-specific environment [19]. Natural terrains represent a challenging scenario due to variability in surface and lighting conditions, lack of structure, no prior information, and in which learning approaches have proved to be more appropriate than expert rule-based or heuristic strategies. For example, the vertical acceleration was used as the main sensory input to train classifiers based on different learning algorithms, including AdaBoost [20], neural network [21], and Cubature Kalman filtering [22]. Methods that attempt to directly measure some important terrain parameters such as friction angle and cohesion have been proposed using a linear least squares approximation of the classical terramechanics theory [23] or via a Bayesian procedure to deal effectively with the presence of uncertainty [24].

The general learning approach includes data gathering pertaining to the wheel-terrain interaction followed by a mapping stage of proprioceptive data with the corresponding terrain. This functional relationship can help addressing various issues: (a) difficulty in creating a physics-based terrain model due to the large number of variables involved, (b) the mapping from proprioceptive input to a mechanical terrain property is an extremely complicated function, which does not have a known analytical form or a physical model and one possible way to observe it and learn about it is via training examples, (c) a learning approach promotes adaptability of the vehicle's behavior.

This paper presents an approach for terrain identification by gathering important information pertaining to the mechanics of vehicle–environment interaction. The underlying assumption is that characteristic traits of the supporting surface can be extracted using vehicle wheels as "tactile" sensors that generate signals modulated by the physical wheel-soil contact. First, the main features that hold the highest discriminative power for terrain identification are studied. They form a feature space upon which a terrain classifier can be built via support vector machine (SVM). By observing these features during nominal robot operation, different types of surfaces can be discriminated. Following this rationale, conditions of high slippage can also be treated as a particular terrain and detected though a dedicated classifier that represents an additional contribution of this research. The idea for the proposed approach was previously presented in [25], where preliminary results were presented. Here, the method is fully detailed in terms of optimal feature selection and classifier building. It is also generalized to include as well the case of slippage estimation. Finally, extended experimental results are included to quantitatively evaluate the system performance.

Materials and methods used in this research are presented in Section 2, whereas the proprioceptive "traits" and the proposed selection approach for the optimal feature set are described in Section 3. The terrain classifier is described in Section 4 providing experimental results obtained from the rover SherpaTT. Finally, lessons learned, and future developments conclude the paper.

2. Materials and Methods

The system is tested and developed using the rover SherpaTT that is shown in Figure 1. SherpaTT was built by DFKI for long-distance exploration applications [26], negotiation of highly challenging terrains or non-nominal conditions (sinkage in soft soil, getting entangled between rocks or alike), cooperation tasks between heterogeneous rovers in a collaborative sample return mission, search and rescue and/or security missions. SherpaTT is a four-wheeled mobile robot [27] outfitted with an actively articulated suspension system, independent drive and steer wheel motors and a six degrees of freedom (DOF) manipulation arm. The rover is also a hybrid wheeled-leg rover, so it can take advantage of both wheeled and legged locomotion according to the terrain difficulty. SherpaTT has a mass of about 166 kg and a payload capacity of at least 80 kg. Each of the four suspended legs has five DOF that include the rotation of the whole leg about the shoulder or pan axis with respect to the robot body, the two rotations of the inner and outer leg parallelograms, the steer and drive angle of the wheel. SherpaTT's footprint can vary between 1×1m in a folded configuration to 2.4×2.4 m in a maximum square footprint. Various predefined footprint shapes (stances) can be adopted as explained in Figure 2.

Each of the 20 suspension and six arm joints delivers telemetry data at a rate of 100 Hz. The telemetry includes joint position (absolute and incremental), speed, current, PWM duty cycle and two temperatures (housing and motor windings). Additionally, a 6-DOF force-torque sensor (FTS) is placed at the mounting flange of each wheel-drive actuator enabling the direct measurement of the generalized forces that each wheel exchanges with the supporting surface. Active force control for the wheel-ground contact as well as the roll-pitch adaption are two processes of the so-called Ground Adaption Process (GAP) in SherpaTT [28]. Autonomy modules do not need to cope with the limb articulation in rough terrain [29].

Figure 1. SherpaTT in soft sand dunes during Morocco field trials in 2018.

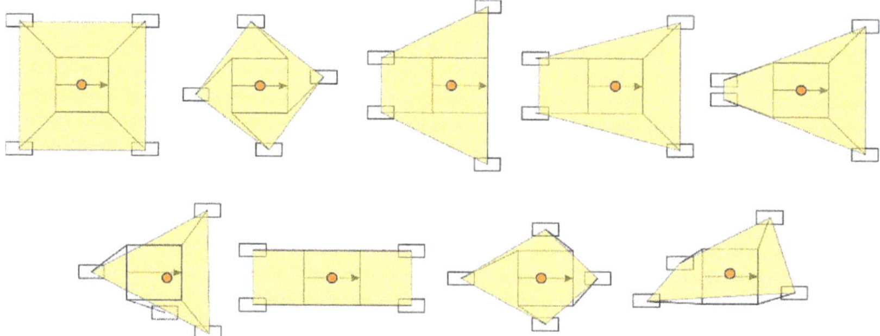

Figure 2. Examples of possible footprint configurations and resulting support polygons.

2.1. Data Sets

Two extensive field trials have been conducted by SherpaTT in the desert of Morocco (2019) and in the desert of Utah (2016). Both sites have been shown to be representative of Mars analogue terrain [30]. During these field trials, different test tracks in natural terrain, GAP-modes and footprints have been tested. Since the focus of those campaigns was different from the aim of this research, only a small portion of the data logs can be effectively used for terrain estimation purposes. Table 1 sums up the datasets available for terrain and slippage classification, clarifying specific terrain, total run time and logged sensors. Specifically, the only available run in the Morocco campaign is performed on sandy dunes of flat/moderate slopes, while Utah data logs are available on flat, moderate sloped and steep sloped rocky terrains. Utah moderate sloped terrain presents an inclination less than 10 deg, while steep sloped terrain had a grade up to 28 degrees. All the available experimental runs are performed following a straight path. Sensor signals are logged together with a synchronized timestamp. Four sensor modalities are available on SherpaTT: the four FTSs, the body's IMU, a Differential Global Positioning System (DGPS) on the rover body and joint telemetry. The FTSs are six axial sensors able to measure the three cartesian components of forces and moments acting on each wheel. The IMU outputs the rover's attitude ad accelerations, while DGPS provides an absolute position of the rover and it is used for ground truthing. Joint position and speed, as well as each joint's electrical current and PWM duty cycle, are available for this study.

Table 1. Datasets gathered by SherpaTT during field trials; for each test run, terrain description, testing time and available logged sensors are indicated.

Run ID	Test Site	Terrain Type	Slope	Run Time [s]	FTS	IMU	DGPS	Encoders
R01	Morocco	Sand	Flat/Moderate	381.5	✓	✗	✗	✓
R02	Utah	Rock	Flat	214.8	✓	✓	✓	✓
R03	Utah	Rock	Flat	214.8	✓	✓	✓	✓
R04	Utah	Rock	Moderate	152.4	✓	✓	✗	✓
R05	Utah	Rock	Steep	453.9	✓	✓	✓	✓
R06	Utah	Rock	Steep	457.7	✓	✓	✓	✓
R07	Utah	Rock	Steep	481.2	✓	✓	✓	✓

Proprioceptive data streams have been associated with a sequence of terrain patches with the same length. In this work, a patch length of 0.3 m is considered. In such a way, for each terrain patch, a set of features has been computed, as extensively explained in the following section.

2.2. Data Pre Processing

Each sensor modality could have a different sample rate. Table 2 specifies sensors logging frequencies. The highest frequency is adopted for the definition of a common absolute timestamp: a frequency of 100 Hz corresponding to one observation each of 0.01 s. Signals with lower sample rates are linearly interpolated to obtain signals at 100 Hz.

Table 2. Sensor sample rates.

Sensor Modality	Sample Rate [Hz]
FTS	100
IMU	100
DGPS	10–20
Encoders	100

In order to have the same time–space association logic for all datasets, and since DGPS is not available on Morocco tests, an odometry-based approach is adopted to relate time-stamp with rover traversed distance. The distance travelled by one of the four rover wheels, d_V, between two successive time-stamps is defined in Equation (1):

$$d_V = \omega R \Delta t \tag{1}$$

where Δt is the sampling interval, ω is the wheel angular speed measured by the encoder and R is the wheel undeformed radius. Sensor data have been subdivided into sub-logs that are associated with each virtual patch. It should be noted that two patches of the same length, in general, may correspond to different actual travel distances due to the extent of wheel slippage. It is also worth mentioning that the use of "distance" windows is used under the underlying assumption of continuous motion of the robot without any stop-and-go manoeuvre.

3. Proprioceptive Sensing

This section presents the set of proprioceptive features used to characterize the properties of a given terrain patch. For each feature, the four statistical moments (mean, variance, skewness and kurtosis) are computed. Data have been manually labelled in terms of two terrain types (sand for Morocco and rocky terrain for Utah) and in terms of three discrete classes of slippage (only for Utah, where slippage can be directly calculated by comparing DGPS with wheel encoders). Data labelling is necessary, as explained later, for the successive stages of feature selection using validity indices and for training the terrain and slippage classifiers.

3.1. Proprioceptive Features

A large amount of sensory data can be gathered from Sherpa and potentially used for classification purposes. However, only a few signals bring significant information related to terrain type. Here, only the most relevant features that we found are discussed. From the FTSs, three forces and three moments are available for each wheel. Among these, we retain the longitudinal force F_X and the wheel drive torque T_D. The wheel drive torque can be also estimated indirectly from the electrical current, C_D, whereas the wheel angular velocity ω can be obtained from wheel encoders. The three body accelerations a_x, a_y and a_z are estimated from the IMU. The mechanical power P_M and the electrical power P_E are computed, respectively, in Equations (2) and (3):

$$P_M = T_D \omega, \tag{2}$$

$$P_E = V I d_{PWM}, \tag{3}$$

where V, I and d_{PWM} are, respectively, the motor voltage, current and PWM duty cycle of the wheel drive. Apart from this set of "primary" quantities, a second set of "derivate" features has been found to be effective for terrain characterization. Derivate features can be obtained by combining primary features according to well-known physics-based relationships. As an example, the friction coefficients μ can be estimated using three different sensor data according to Equations (4)–(6):

$$\mu_1 = F_X/F_Z,\tag{4}$$

$$\mu_2 = T_D/RF_Z,\tag{5}$$

$$\mu_3 = C_D/RF_Z,\tag{6}$$

where F_Z is the wheel vertical force and R is the wheel unloaded radius. Another key feature is the so-called wheel speed deviation, SD, that is the absolute value of the difference between the angular speed of each wheel ω and the average angular speed of the four wheels $\overline{\omega}$ as defined by Equation (7):

$$SD = |\omega - \overline{\omega}|,\tag{7}$$

Note that the SD can be extended as well to turning motion, as explained in [31]. Wheel longitudinal stiffness LS is computed as the ratio between the friction coefficient and the wheel slip s. In the simplified case of a linear relation between friction coefficient and wheel slip, this physical quantity should represent the slope of friction—slippage plot. Three different LS_i values are computed in Equation (8) using the three friction coefficients defined before:

$$LS_i = \frac{\mu_i}{s}, \quad i = 1,\ 2,\ 3,\tag{8}$$

The slippage s is univocally estimated for both accelerating and braking and for each wheel based on Equation (9):

$$s = d\left(1 - \left(\frac{V_X}{\omega R}\right)^d\right), \qquad d = \begin{cases} +1 & \omega R - V_X \geq 0 \\ -1 & \omega R - V_X < 0 \end{cases},\tag{9}$$

where V_X is estimated with the DGPS and d is positive for accelerating and negative for braking. An approximate estimate of wheel sinkage z can be obtained by Equation (10) [32]:

$$z = R\cdot\left(1 - \cos\left(2\cdot\frac{\frac{T_D}{R} - F_X}{F_Z}\right)\right),\tag{10}$$

Table 3 sums up the proprioceptive feature space spanned by the above considerations.

Table 3. Proprioceptive features with relative feature ID number and symbol.

FID	Feature	Symbol
F1	Longitudinal Force	F_X
F2	Drive Torque	T_D
F3	Drive Current	C_D
F4	Mechanical Power	P_M
F5	Electrical Power	P_E
F6	Friction Coefficient 1	μ_1
F7	Friction Coefficient 2	μ_2
F8	Friction Coefficient 3	μ_3
F9	Acceleration X	a_X
F10	Acceleration Z	a_Z
F11	Speed Deviation	SD
F12	Longitudinal Stiffness 1	LS_1
F13	Longitudinal Stiffness 2	LS_2
F14	Longitudinal Stiffness 3	LS_3
F15	Sinkage	z

3.2. Statistical Feature and Validity Indices

For the i-th feature, the main statistical moments are estimated: mean E_i, variance σ_i^2, skewness Sk_i and kurtosis Ku_i of the feature signal for a single terrain patch. These statistics are defined according to Equations (11)–(14):

$$E_i = \frac{1}{N} \sum_{n=1}^{N} x_n \tag{11}$$

$$\sigma_i^2 = \frac{1}{N} \sum_{n=1}^{N} (x_n - E_i)^2 \tag{12}$$

$$Sk_i = \frac{1}{N} \frac{\sum_{n=1}^{N} (x_n - E_i)^3}{\left(\sqrt{\sigma_i^2}\right)^3} \tag{13}$$

$$Ku_i = \frac{1}{N} \frac{\sum_{n=1}^{N} (x_n - E_i)^4}{\left(\sqrt{\sigma_i^2}\right)^4} \tag{14}$$

where x_n is the value assumed by the feature at the time-step n and N is the total number of time-steps for the considered terrain patch.

In addition, each terrain patch is labelled for terrain type, sand for Morocco data or rock for Utah data, and slippage level; high slip is for absolute value of slip greater than 0.5 and all other conditions are low slip.

The statistical features are divided into clusters according to patch labels and a validity index value is associated with each statistical feature. Two validity indices are proposed to obtain a quantitative measure of "feature goodness" for terrain patch classification: the *WB* index [33] and a Pearson Coefficient based index (*PC*) [34]. These indices are employed in the feature selection method described in the following section. The *WB* index is computed as:

$$WB = m \cdot \frac{SSW}{SSB}, \tag{15}$$

The sum of square within cluster *SSW* and between clusters *SSB* can be computed as follows:

$$SSW = \sum_{k=1}^{m} \sum_{i=1}^{n_k} (x_i - \mu_k)^2, \tag{16}$$

$$SSB = \sum_{k=1}^{m} n_k (\mu_k - \mu)^2, \tag{17}$$

where x_i is the generic statistical feature value for the patch i, μ_k is the cluster k centroid value, μ is the overall dataset centroid value, n_k is the number of patches in the cluster k and m is the number of clusters. In this study m is 2 since we have two classes for the terrain classifier. *WB* index assumes lower values for higher distance between different cluster centroids and for lower variance within clusters. A good classifier feature allows us to obtain distant and compact clusters, and thus generally corresponds to a low value of the *WB* index. A *PC* index could be computed through the linear regression of a feature against the m classes, giving a progressive numerical value to each class. A higher value of *PC* potentially corresponds to a better classifier feature.

Figures 3 and 4 show, respectively, the WB^{-1} and *PC* indices values for all the statistical features computed both on terrain and slippage labelled datasets. Some features are missing because the correspondent sensor suite is not available in the classification dataset (please refer to Table 1).

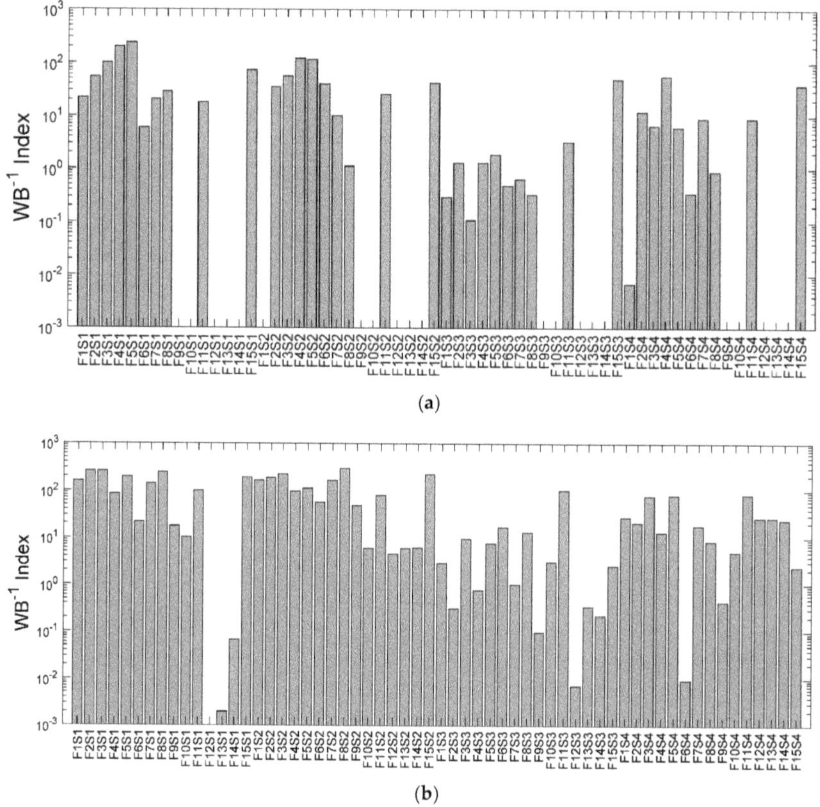

(a)

(b)

Figure 3. WB^{-1} index values for each feature (F1 to F15), for each statistic (S1 to S4, respectively, mean, variance, skewness and kurtosis) and for terrain (**a**) and slippage (**b**) classifier.

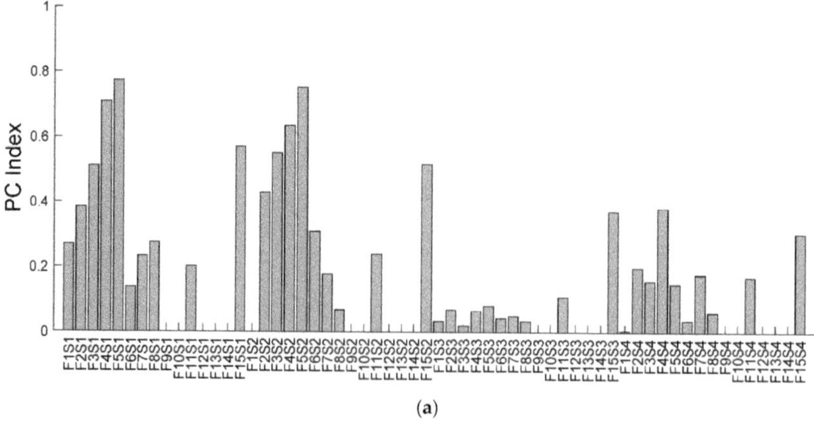

(a)

Figure 4. *Cont.*

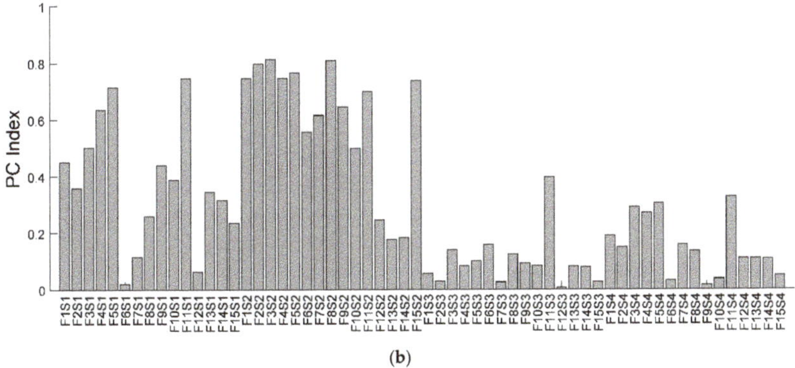

Figure 4. *PC* index values for each feature (F1 to F15), for each statistic (S1 to S4, respectively, mean, variance, skewness and kurtosis) and for terrain (**a**) and slippage (**b**) classifier.

3.3. Feature Selection

In a classification problem, plenty of features can be thought of and used for training. Features can be raw data signals, or a combination of raw signals, signal statistics and so on. However, only a small amount of the available feature set may be actually relevant to address the classification problem.

The selection of the "most significant features" can be tackled in different ways, i.e., as an optimization or a search problem [34]. Those techniques can find a local or global optimum feature set but could be also computationally expensive. The method proposed in this work has a low computational cost, since the number of its iterations equals the total number of features in the initial feature space. For the purpose of this work the initial feature space is made up of 15 features and 4 statistics per feature, for a total number of 60 statistical features.

The feature selection method needs as input the labelled initial feature space, the value of a validity index (VI) associated with each statistical feature and a classifier to be iteratively trained. The VIs employed in this research were the *WB* index and the *PC* index. The classification algorithm is a linear Support Vector Machine (SVM), one of the most adopted techniques for terrain classification problem [18], which guarantees good results with a lower computational effort compared with other techniques such as Convolutional Neural Networks [35]. The feature selection algorithm is described by the following pseudo-code:

01.　　**INPUT**: initial feature set *(Ifs)*, VI vector associated with features *VIv*
02.　　**arrange** *IFs* features in decreasing order of *VIv*
03.　　**initialize** *(F1 score)$_{i-1}$* = 60%, **initialize** reduced feature set *(RFs)* as an empty set
04.　　**for** *i* **from** 1 **to** *(number of features)*
05.　　**add** the *i* feature from *IFs* to the *RFs*
06.　　**train** *n* times the classifier, **compute** average *(F1 score)$_i$*
07.　　**if** *(F1 score)$_i$* − *(F1 score)$_{i-1}$* ≤ *(F1 threshold)*
08.　　**remove** *i* feature from *RFs*
09.　　**end if**
10.　　**update** *(F1 score)$_i$* = *(F1 score)$_{i-1}$*
11.　　**end for**
12.　　**OUTPUT**: *RFs*

At each iteration step, a feature is added to the training set and the algorithm verifies if this new feature leads to a significant increase in the classifier performance. The metric used to quantify the classifier performance is the F1 score. In this work the F1 score percentage improvement threshold for a feature to be accepted has been set to 0.5%. The training phase is repeated three times for each feature ($n = 3$) and an average F1 score of the three trained classifiers is computed. The output of the algorithm is a reduced feature set, including the only features that bring a significant improvement in the classifier performance. The algorithm has been tested both with the PC index and the reciprocal of WB index (WB^{-1}), which allows the first selected features to be potentially the most suitable for the classifier. Figure 5 shows the algorithm results in terms of F1 score for each iteration and for the terrain (above) and slippage (below) classifier. If the performance increase, the new feature is added to the "optimal" or reduced feature set (RFs). It is apparent how the most relevant features are found in the first half of the iterations, where both PC and WB^{-1} have higher values. This figure shows that the chosen validity indices are effective for the selection of relevant features.

Tables 4 and 5 show the F1 score and the number of selected features for the terrain classifier and for different feature selection methods. *PC* and *WB* indicate the feature selection method that uses the singular VI, as described before. In addition, two combined selections have been also tested that use the intersection *Int*(*PC*, *WB*) and the union *Un*(*PC*, *WB*) of the selected feature set obtained from the *PC* and *WB* selection method. Finally, the training method "All" involves the use of the entire feature set.

All the proposed selection methods for the terrain classifier achieve similar F1 score of about 93%. However, the intersection method Int(*PC*,*WB*) is the most suitable both for the F1 score and the size of the reduced feature set. Both single VI methods lead also to similar reduced feature sets. In fact, nine of the features selected by both *WB* and *PC* methods fall into the intersection Int(*PC*,*WB*). Taking all initial features (see "All" method), without any selection phase allows us to reach a good F1 score of 90.4%, but with a higher training set dimensionality and so a higher computational effort.

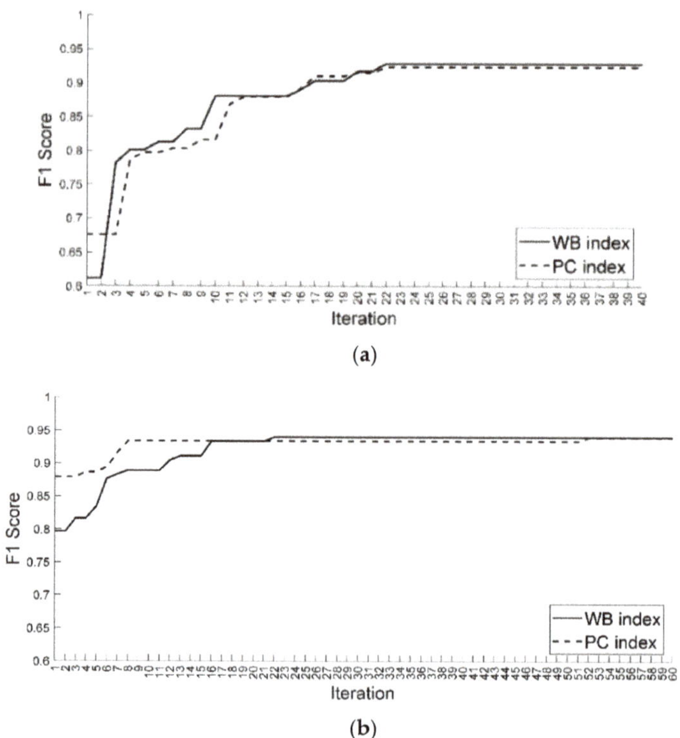

Figure 5. Feature selection performance results for terrain (**a**) and slippage (**b**) classifier.

Table 4. Comparison of different feature selection methods for terrain classification.

	PC	WB	Int(PC,WB)	Un(PC,WB)	All
N° features	11	10	9	12	40
F1 score	0.924	0.929	0.931	0.928	0.904

Table 5. Comparison of different feature selection methods for slippage classification.

	PC	WB	Int(PC,WB)	Un(PC,WB)	All
N° features	6	10	2	14	60
F1 score	0.939	0.940	0.914	0.936	0.765

The same selection procedure has been performed with the slippage classifier. *PC* and *WB* methods select different features. The most relevant features fall into the intersection, which has a slightly lower F1 but a minimal feature set, hence it can be preferred as the selection method. Table 6 collects the reduced feature space for both algorithms.

Table 6. Reduced feature sets for terrain (**a**) and slippage (**b**) classification.

PC	WB	Int(PC,WB)	Un(PC,WB)
F5S1	F5S1	F2S1	F1S1
F4S1	F4S1	F3S1	F2S1
F4S2	F4S2	F4S1	F3S1
F3S2	F3S1	F5S1	F4S1
F3S1	F4S4	F8S1	F5S1
F2S1	F2S1	F11S1	F8S1
F4S4	F8S1	F4S2	F11S1
F8S1	F11S2	F7S2	F3S2
F1S1	F11S1	F4S4	F4S2
F11S1	F7S2		F7S2
F7S2			F11S2
			F4S4
F3S2	F8S2	F1S2	F1S1
F2S2	F3S1	F3S2	F3S1
F4S2	F8S1		F5S1
F1S2	F3S2		F8S1
F11S1	F15S2		F11S1
F1S3	F5S1		F1S2
	F1S2		F2S2
	F1S1		F3S2
	F11S3		F4S2
	F11S2		F8S2
			F11S2
			F15S2
			F1S3
			F11S3

Figure 6 shows the 3D scatter plot of the three statistical features that present the highest value of *WB* and *PC* index, respectively, for the terrain and high-slippage classifier. For visualization purposes and for the terrain classifier only, the corresponding 2D feature distribution is also shown in Figure 7. As seen from these figures, the feature distribution indicates a good separation level between sand and rock and it suggests the implementation of a classification algorithm.

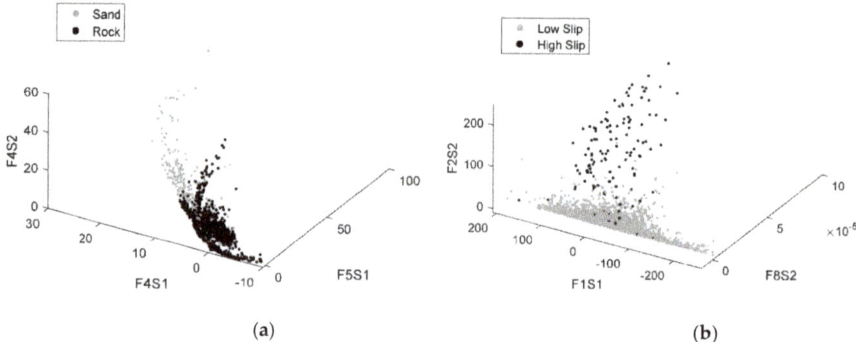

Figure 6. 3D plot of the first three most relevant features selected for terrain (**a**) and slippage (**b**) classification.

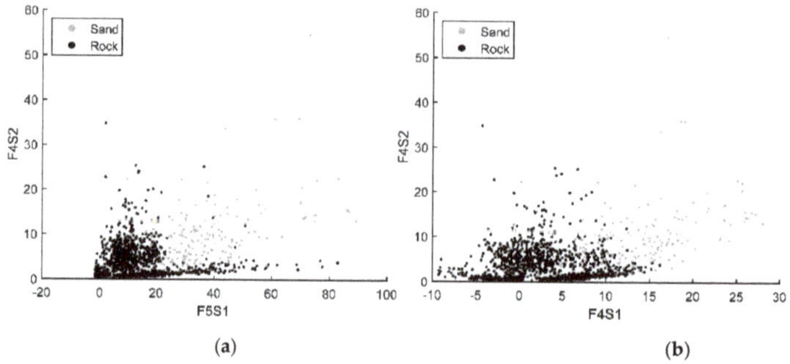

Figure 7. 2D distribution of the most relevant features selected for the terrain classifier: F4S2-F5S1 plane (**a**), F4S2-F4S1 plane (**b**).

4. Classification Task

In this section, the results obtained from the two classifiers for terrain and slippage detection are presented. The classifiers were trained and tested on real data gathered by SherpaTT in the field. The reduced feature sets to be used for training and testing have been obtained with the intersection method described in the previous section (refer to Table 6). For both classifiers, the observations (terrain patches) of the reduced feature dataset have been subdivided into two datasets, according to the patch class. In addition, 75% of the observations of each class dataset were randomly extracted to build the training datasets, while the remaining 25% were then used for testing the trained classifiers [36]. Tables 7 and 8 show the performance metrics of the trained classifiers. The terrain classifier reaches a global accuracy of 96.3% and a global F1 score of 92.2%, while the slippage classifier has an accuracy of 99.0% and an F1 score of 92.7%.

Table 7. Performance metrics of the terrain classifier.

	Precision	Recall	Specificity	F1 Score
Sand	77.6%	97.7%	96.1%	86.5%
Rock	99.7%	96.1%	97.7%	97.9%

Table 8. Performance metrics of the slippage classifier.

	Precision	Recall	Specificity	F1 Score
Low slip	99.7%	99.2%	92.0%	99.5%
High slip	80.8%	92.0%	99.2%	86.0%

The confusion matrix of the training and testing phases are reported in Figures 8 and 9. The training and testing classifier performance are very similar, demonstrating the prediction capability of the two classifiers on new data.

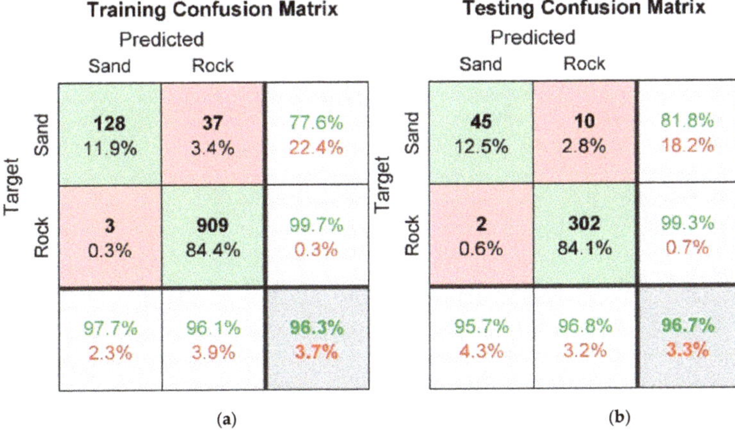

Figure 8. Confusion matrix for training (**a**) and testing (**b**) of the terrain classifier.

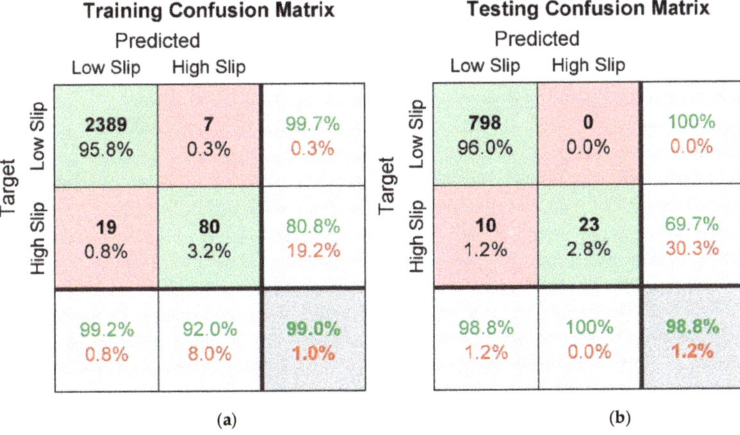

Figure 9. Confusion matrix for training (**a**) and testing (**b**) of the slippage classifier.

5. Discussion and Conclusions

A general approach for terrain and slippage classification has been proposed and validated through field experimental data. A SherpaTT rover testing campaign performed in Morocco and Utah deserts provided proprioceptive datasets on different soil types. A set of features have been defined and extracted from raw experimental data. Two labels for terrain type (Sand–Rock) and slippage level (High Slip–Low Slip) have been associated with each terrain patch, together with a set of statistical features (predictors), obtained from the computation of feature statistical moments on the terrain patch. Different selection methods based on validity index have been tested and proved to be effective for the extraction of relevant features. Different selection methods based on validity index have been tested. The tested validity indices are the *WB* index and a Pearson Coefficient (*PC*)-based index. Both indices are demonstrated to be effective for the extraction of relevant features. Only the features selected by both *WB* and *PC* methods have been kept in the reduced feature set used for classifiers training and testing. Thanks to the selection, the initial statistical feature set of 60 features was reduced to 9 features and to 2 features for slippage classification. The reduced feature set observations, one per terrain patch, were randomly subdivided into a training set and a testing set, including, respectively, 75% and 25% of the observations. The two classifiers were trained on the training set and then tested on the new samples of the testing set, giving as result a global classifier accuracy of 96.7% for terrain classification and 98.8% for slippage classification.

New experimental tests will be performed in the near future on different terrain types and during various rover maneuvers. Those new data will be used to further validate the proposed approach and to improve the classifier generalization. The combination of proprioceptive sensing with exteroceptive perception, as suggested in [18], will be also investigated.

Author Contributions: Conceptualization, M.D., G.R.; Methodology, M.D., G.R.; Data curation and experiments and validation, F.C.; writing—original draft preparation, M.D, G.R.; writing—review and editing, F.C. All authors have read and agreed to the published version of the manuscript.

Funding: This research was funded by the Horizon 2020 European Commission under grant agreement n. 821988 ADE.

Conflicts of Interest: The authors declare no conflict of interest.

References

1. Gonzalez, R.; Iagnemma, K. Slippage estimation and compensation for planetary exploration rovers. State of the art and future challenges. *J. Field Robot.* **2017**, *35*, 564–577. [CrossRef]
2. Filip, J.; Azkarate, M.; Visentin, G. Trajectory Control for Autonomous Planetary Rovers. In Proceedings of the Symposium on Advanced Space Technologies in Automation and Robotics, Leiden, The Netherlands, 20–22 June 2017; pp. 1–6.
3. Reina, G.; Ishigami, G.; Nagatani, K.; Yoshida, K. Odometry Correction Using Visual Slip Angle Estimation for Planetary Exploration Rovers. *Adv. Robot.* **2010**, *24*, 359–385. [CrossRef]
4. Ojeda, L.; Reina, G.; Cruz, D.; Borenstein, J. The FLEXnav precision dead-reckoning system. *Int. J. Veh. Auton. Syst.* **2006**, *4*, 173. [CrossRef]
5. Bussmann, K.; Meyer, L.; Steidle, F.; Wedler, A. Slip Modeling and Estimation for a Planetary Exploration Rover: Experimental Results from Mt. Etna. In Proceedings of the 2018 IEEE/RSJ International Conference on Intelligent Robots and Systems (IROS), Madrid, Spain, 1–5 October 2018; pp. 2449–2456.
6. Sanguino, T.D.J.M. 50 years of rovers for planetary exploration: A retrospective review for future directions. *Robot. Auton. Syst.* **2017**, *94*, 172–185.
7. Goldberg, S.B.; Maimone, M.; Matthies, L. Stereo vision and rover navigation software for planetary exploration. In Proceedings of the IEEE Aerospace Conference; Institute of Electrical and Electronics Engineers (IEEE), Big Sky, MT, USA, 8–15 March 2003.
8. Helmick, D.; Angelova, A.; Matthies, L. Terrain Adaptive Navigation for planetary rovers. *J. Field Robot.* **2009**, *26*, 391–410. [CrossRef]
9. Gingras, D.; Lamarche, T.; Bedwani, J.-L.; Dupuis, É. Rough Terrain Reconstruction for Rover Motion Planning. In Proceedings of the 2010 Canadian Conference on Computer and Robot Vision, Ottawa, ON, Canada, 31 May–2 June 2010; pp. 191–198.
10. Thrun, S.; Montemerlo, M.; Dahlkamp, H.; Stavens, D.; Aron, A.; Diebel, J.; Fong, P.; Gale, J.; Halpenny, M.; Hoffmann, G.; et al. Stanley: The Robot that Won the DARPA Grand Challenge. *J. Field Robot.* **2006**, *23*, 661–692. [CrossRef]
11. Milella, A.; Reina, G.; Underwood, J.; Douillard, B. Combining radar and vision for self-supervised ground segmentation in outdoor environments. In Proceedings of the IEEE International Conference on Intelligent Robots and Systems, San Francisco, CA, USA, 25–30 September 2011; pp. 255–260.
12. Milella, A.; Reina, G.; Underwood, J.; Douillard, B. Visual ground segmentation by radar supervision. *Robot. Auton. Syst.* **2014**, *62*, 696–706. [CrossRef]
13. Bekker, M. The development of a moon rover. *J. Br. Interpl. Soc.* **1985**, *38*, 537.
14. Angelova, A.; Matthies, L.; Helmick, D.; Perona, P. Learning and prediction of slip from visual information. *J. Field Robot.* **2007**, *24*, 205–231. [CrossRef]

15. Gonzalez, R.; Chandler, S.; Apostolopoulos, D. Characterization of machine learning algorithms for slippage estimation in planetary exploration rovers. *J. Terramech.* **2019**, *82*, 23–34. [CrossRef]

16. Reina, G.; Leanza, A.; Messina, A. On the vibration analysis of off-road vehicles: Influence of terrain deformation and irregularity. *J. Vib. Control.* **2018**, *24*, 5418–5436. [CrossRef]

17. Valada, A.; Burgard, W. Deep spatiotemporal models for robust proprioceptive terrain classification. *Int. J. Robot. Res.* **2017**, *36*, 1521–1539. [CrossRef]

18. Reina, G.; Milella, A.; Galati, R. Terrain assessment for precision agriculture using vehicle dynamic modelling. *Biosyst. Eng.* **2017**, *162*, 124–139. [CrossRef]

19. Brooks, C.A.; Iagnemma, K. Self-supervised terrain classification for planetary surface exploration rovers. *J. Field Robot.* **2012**, *29*, 445–468. [CrossRef]

20. Krebs, A.; Pradalier, C.; Siegwart, R. Adaptive rover behavior based on online empirical evaluation: Rover–terrain interaction and near-to-far learning. *J. Field Robot.* **2010**, *27*, 158–180. [CrossRef]

21. Dupont, E.M.; Moore, C.A.; Collins, E.G.; Coyle, E.; Collins, E.G. Frequency response method for terrain classification in autonomous ground vehicles. *Auton. Robot.* **2008**, *24*, 337–347. [CrossRef]

22. Reina, G.; Leanza, A.; Messina, A. Terrain estimation via vehicle vibration measurement and cubature Kalman filtering. *J. Vib. Control.* **2019**, *26*, 885–898. [CrossRef]

23. Iagnemma, K.; Kang, S.; Shibly, H.; Dubowsky, S. Online Terrain Parameter Estimation for Wheeled Mobile Robots with Application to Planetary Rovers. *IEEE Trans. Robot.* **2004**, *20*, 921–927. [CrossRef]

24. Gallina, A.; Krenn, R.; Scharringhausen, M.; Uhl, T.; Schafer, B. Parameter Identification of a Planetary Rover Wheel-Soil Contact Model via a Bayesian Approach. *J. Field Robot.* **2013**, *31*, 161–175. [CrossRef]

25. Dimastrogiovanni, M.; Cordes, F.; Reina, G. Terrain Sensing for Planetary Rovers. In Proceedings of the International Conference of IFToMM ITALY, Naples, Italy, 9–11 September 2020; pp. 269–277.

26. Cordes, F.; Dettmann, A.; Kirchner, F. Locomotion modes for a hybrid wheeled-leg planetary rover. In Proceedings of the 2011 IEEE International Conference on Robotics and Biomimetics, Phuket, Thailand, 7–11 December 2011; pp. 2586–2592.

27. Sonsalla, R.U.; Cordes, F.; Christensen, L.; Roehr, T.M.; Stark, T.; Planthaber, S.; Maurus, M.; Mallwitz, M.; Kirchner, E.A. Field Testing of a Cooperative Multi-Robot Sample Return Mission in Mars Analogue Environment. In Proceedings of the 14th Symposium on Advanced Space Technologies in Robotics and Automation (ASTRA'17), Noordwijk, The Netherlands, 27–28 May 2017.

28. Cordes, F.; Kirchner, F.; Babu, A. Design and field testing of a rover with an actively articulated suspension system in a Mars analog terrain. *J. Field Robot.* **2018**, *35*, 1149–1181. [CrossRef]

29. Cordes, F.; Babu, A.; Kirchner, F. Static force distribution and orientation control for a rover with an actively articulated suspension system. In Proceedings of the 2017 IEEE/RSJ International Conference on Intelligent Robots and Systems (IROS), Vancouver, BC, Canada, 24–28 September 2017; pp. 5219–5224.

30. Clarke, J.; Stoker, C.R. Concretions in exhumed and inverted channels near Hanksville Utah: Implications for Mars. *Int. J. Astrobiol.* **2011**, *10*, 161–175. [CrossRef]

31. Reina, G. Cross-Coupled Control for All-Terrain Rovers. *Sensors* **2013**, *13*, 785–800. [CrossRef] [PubMed]

32. Guo, J.; Guo, T.; Zhong, M.; Gao, H.; Huang, B.; Ding, L.; Li, W.; Deng, Z. In-situ evaluation of terrain mechanical parameters and wheel-terrain interactions using wheel-terrain contact mechanics for wheeled planetary rovers. *Mech. Mach. Ther.* **2020**, *145*, 103696. [CrossRef]

33. Zhao, Q.; Fränti, P. WB-index: A sum-of-squares based index for cluster validity. *Data Knowl. Eng.* **2014**, *92*, 77–89. [CrossRef]

34. Hastie, T.; Tibshirani, R.; Friedman, J. *The Elements of Statistical Learning*; Springer: New York, NY, USA, 2009.

35. Rothrock, B.; Kennedy, R.; Cunningham, C.; Papon, J.; Heverly, M.; Ono, M. SPOC: Deep Learning-based Terrain Classification for Mars Rover Missions. In Proceedings of the AIAA SPACE 2016, Long Beach, CA, USA, 13–16 September 2016.

36. Milella, A.; Marani, R.; Petitti, A.; Reina, G. In-field high throughput grapevine phenotyping with a consumer-grade depth camera. *Comput. Electron. Agric.* **2019**, *156*, 293–306. [CrossRef]

 *applied*
sciences

Article

Dynamic Modeling and Simulation of a Robotic Lander Based on Variable Radius Drums

Matteo Caruso [1,*]**, Lorenzo Scalera** [2]**, Paolo Gallina** [1] **and Stefano Seriani** [1]

[1]   Department of Engineering and Architecture, University of Trieste, via A. Valerio 6/1, 34127 Trieste, Italy; pgallina@units.it (P.G.); sseriani@units.it (S.S.)
[2]   Polytechnic Department of Engineering and Architecture, University of Udine, via delle Scienze 206, 33100 Udine, Italy; lorenzo.scalera@uniud.it
*   Correspondence: matteo.caruso@phd.units.it

Received: 11 November 2020; Accepted: 7 December 2020; Published: 11 December 2020

Abstract: Soft-landing on planetary surfaces is the main challenge in most space exploration missions. In this work, the dynamic modeling and simulation of a three-legged robotic lander based on variable radius drums are presented. In particular, the proposed robotic system consists of a non-reversible mechanism that allows a landing object to constant decelerate in the phase of impact with ground. The mechanism is based on variable radius drums, which are used to shape the elastic response of a spring to produce a specific behavior. A dynamic model of the proposed robotic lander is first presented. Then, its behavior is evaluated through numerical multibody simulations. Results show the feasibility of the proposed design and applicability of the mechanism in landing operations.

Keywords: dynamic modeling; multibody simulation; robotic lander; variable radius drum; impact analysis

1. Introduction

In space exploration, landing on a planet, an asteroid or any other celestial body is an extremely important task and can determine the success or the failure of the entire mission. In the past decades, many missions failed due to the crash of the lander to the ground during this delicate operation. Despite these failed missions, many other missions succeeded and over the years many systems were tested and employed to achieve a soft landing.

The most remarkable mission, known as Apollo 11, from the NASA Agency, had the objective to land a probe transporting humans for the first time on the Moon's surface. Successful landing was obtained first by executing a powered descent maneuver. Then, successful touchdown was achieved by means of a suspension system having a honeycomb shock absorber on each lander-leg strut [1]. During the Vikings mission started in 1975, two twin probes named Viking 1 and Viking 2 were sent to Mars. The landers of the two probes were legged, and a combination of parachute and rocket propulsion was used to achieve soft-landing on the planet surface [2]. The same operations were implemented for the lander used in 1997 for the Cassini–Huygens NASA-ESA-ASI mission, which landed successfully on Titan, one of the moons of Saturn. After the Viking mission, NASA sent further probes to investigate Mars planet. First, it was the turn for NASA's Pathfinder mission, which started in 1996 [3]. Then, in 2004, the NASA's Mars Exploration Rovers Spirit and Opportunity were sent to the red planet. In these three missions, a system based on gas-filled airbags was adopted to achieve soft-landing [4]. Another remarkable mission was the Mars Science Laboratory, which deployed successfully the lander Curiosity on the Mars surface in 2011.

Appl. Sci. **2020**, *10*, 8862; doi:10.3390/app10248862

Successful landing was achieved by adopting a combination of parachute, rockets, and the Sky Crane system. More recently, the Rosetta mission from ESA used a powered descend to land on the comet and a system of anchors to remain attached to it once controlled impact occurred [5].

In parallel to NASA, during the Space Race, the Soviet Space Program, starting with impactors [6], implemented several innovations which led to the first-ever soft-landing on another planet with the lander Luna 9 in February 1966 [7]. In 1970, the Soviet Lunokhod 1 was the first successful rover to land on another planetary body [8]. In the 1990s, the Chinese space agency (China National Space Administration) started a program which enabled the soft-landing of the Chang'e 3 lunar lander and rover Yutu in December 2013 [9]. The lander Chang'e 4 followed in January 2019 [10].

In past years, many authors have studied the problem of soft-landing of landers in different fields. For example, Wang et al. in 2019 studied the use of magnetorheological fluid dampers with semi-active control on damping force in order to adjust to landing conditions [11]. Furthermore, the shock absorber design [12] and the lander stability [13] have been investigated numerically by mean of optimization algorithms. Moreover, legged landers with capabilities of walking on planet surface after the landing have been studied in [14,15].

Two main problems can be addressed to the study of soft landing operations: the modeling of the system dynamics, and the modeling of the interactions between the lander and the soil. Several authors tried to study the soft landing problem by means of simulation software, and chose as a case study a four-legged lander. For example, Xu et al. in 2011 simulated the soft-landing of a lunar lander first using MSC-ADAMS considering the lunar soil as a rigid body, and then MSC-DYTRAN considering the lunar soil as a flexible body [16]. Wan et al. in 2010, chose MSC-DYTRAN to simulate soft-landing of a lunar lander with the finite elements method and a nonlinear transient dynamics approach [17]. Zheng et al. in 2018, simulated the impact of a legged lander using explicit finite element analysis using LS-DYNA and ABAQUS softwares [18]. Liang et al. in 2011, used an explicit nonlinear finite element method implemented in ABAQUS to study the impact of a legged lander and aluminum honeycomb shock absorber with the lunar soil [19]. Other authors in [20–24] studied the impact of the lander with ADAMS considering the soil either as an elastic or a rigid body.

In this paper, we present the dynamic modeling and simulation of a robotic lander, based on variable radius drums (VRDs), which inherits from the concept outlined by Seriani in [25]. The proposed mechanism consists of a non-reversible mechanism that allows a landing object to decelerate in the phase of impact with ground. A VRD is a device characterized by the variation of the spool radius along its profile, as defined by Seriani and Gallina in [26]. In this context, it is used to ensure a constant force and a controlled deceleration of the structure during landing. VRDs were studied for many applications, as, for instance, in gravity-compensation systems [27], to guide a load along an horizontal path, or to improve locomotion of a legged robot [26]. Scalera et al. in 2018, studied a cable-based robotic crane based on VRDs capable of moving a load through a planar working area [28]. Moreover, Fedorov and Birglen in 2018 employed two antagonistic VRDs in a cable-suspended robot to steer the end-effector along a desired pick-and-place trajectory. The same concept was also adopted for the static balancing of a pendulum [29]. In these cases, the rationale for the use of the VRD is to generate geometrical trajectories in a mechanism. Conversely, in this work we employ a synthesized VRD to generate a specific force when connected to a linear spring-loaded mechanism. Other examples of the application of VRDs, also known as non-circular pulleys, include cable actuation in soft robot joints [30], devices for upper leg rehabilitation [31], mechanisms for energy saving [32], and force regulation [33].

When landing on celestial bodies, the ground characteristics are an important influence on the mission [34], and as such many attempts are made to determine them in advance [35,36]. Many works detail the soil characterization, in particular concerning its granulometry [37]. In order to build the simulations on a solid baseline, we used an experimental approach to the determination of the characteristics of the soil; a basaltic sand of volcanic origin with average grain size range of 3 to 5 mm which pertains to the coarse sands as described by NASA [38,39].

The main contributions of this paper can be summarized as follows; the design and the numerical validation of a three-legged robotic lander for space applications. The landing mechanism leverages a passive mechanism for impact energy absorption, which is based on VRDs and cables to transfer the impact energy to a preloaded spring. Moreover, the presented passive mechanism is able to ensure a controlled force and acceleration of the structure, thus preventing damage to the supported payload due to high accelerations. We believe this kind of system can find application in real and concrete scenarios considering its reusability, its high payload/lander mass ratio, and its modularity.

Compared to relevant literature, and in particular the work of Seriani [25] which implements an ideal linear spring–damper contact and ideal Coulomb friction models, this work presents several novelties: it provides an experimental determination of the characteristics of soft basaltic sand soil for implementation in a spring–damper ground contact as well as implementing a nonlinear stiffness model; furthermore, it implements experimentally defined static and dynamic friction coefficients. In addition, the numerical implementation of the theoretical model is more stable and thus avoids the implementation of numerical damping of the legs rotation. Moreover, in this work in the 3D lander three ratchets are introduced, functioning as retention mechanisms that make the system non-reversible. In particular, their role is to limit the return rotation of the legs once they are fully rotated and the payload is completely decelerated; this keeps the lander from bouncing after landing. Additionally, limiting the rotation altogether, it forces the lander to keep its center of mass closer to the ground, thus increasing stability. Moreover the ratchets prevent the discharge of the energy stored in the preloaded springs, thus avoiding high bounces of the lander.

The rest of the paper is organized as follows. In Section 2, the dynamic modeling of the robotic lander based on VRDs is introduced, whereas in Section 3, the setup of the ADAMS numerical simulations is explained. Section 4 reports the simulation results for two test cases: the impact on an horizontal surface and the impact on an inclined surface. In Section 5, the results are discussed and compared with the theoretical model. Finally, the conclusions are given in Section 6.

2. Dynamic Modeling of the Robotic Lander

The lander structure studied in this work is based on the concept outlined in [25]. In addition, a ratchet mechanism is introduced to block the rotation of the legs when the structure of the lander starts to move upwards at the end of the deceleration phase. As shown in Figure 1, the lander is composed of three legs arranged with an angle of $2/3\,\pi$ from the others and having a fixed offset from the center of mass of the lander, and of an hexagonal body on which the payload is fixed. From Figure 1, it can be seen that the lander is symmetrical with respect to a vertical axis passing through its center of mass. Therefore, to define the dynamics of the system, a simplified model of a third of the lander is considered.

2.1. Variable Radius Drum Mechanism

A schematic view of the lander-leg mechanism at the beginning of the impact is illustrated in Figure 2. The main parts composing each lander-leg mechanism are a leg, a fixed radius pulley with radius r_1, attached to the leg, a VRD, another fixed radius pulley with radius r_2, attached to the VRD, and a preloaded spring. Moreover, two cables are used in the mechanism. The extremities of the first cable are attached to the VRD and to the fixed radius pulley (red line in Figure 2). The extremities of the other cable are attached to the preloaded spring, which is connected to the lander chassis, and to the fixed radius pulley connected to the VRD (green line in Figure 2). This specific layout allows the VRD to shape the elastic response of the preloaded spring. More precisely, the preloaded spring generates a reactive torque M acting on the fixed-radius pulley attached to the leg. As the response of a preloaded spring is discontinuous, the reactive torque M can be approximated by the following continuous and differentiable function [25,40,41],

$$M = \frac{2}{\pi} M_0 \arctan\left(f\vartheta\right) + k\vartheta \tag{1}$$

where M_0 is the preloaded torque, f an approximation factor, k the spring stiffness, and ϑ the relative rotation of the leg from its rest position ϑ_r.

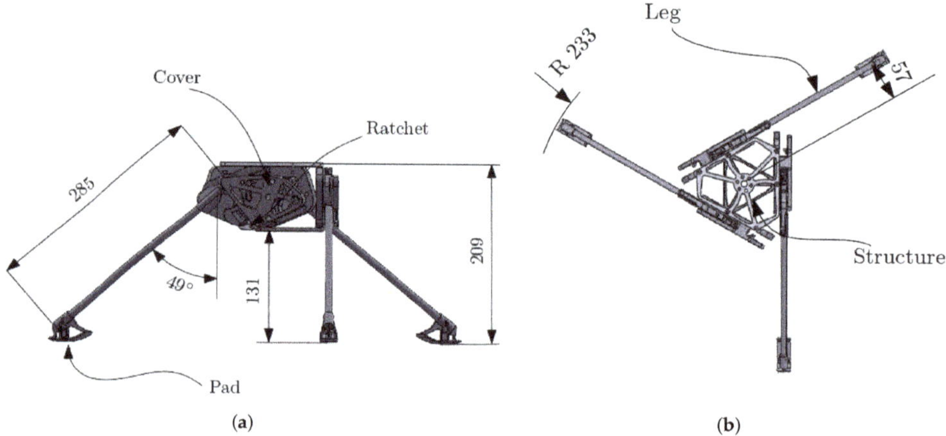

Figure 1. The computer-aided design (CAD) model of the proposed three-legged lander, along with labels for its main parts and main dimensions: (**a**) lateral view and (**b**) top view.

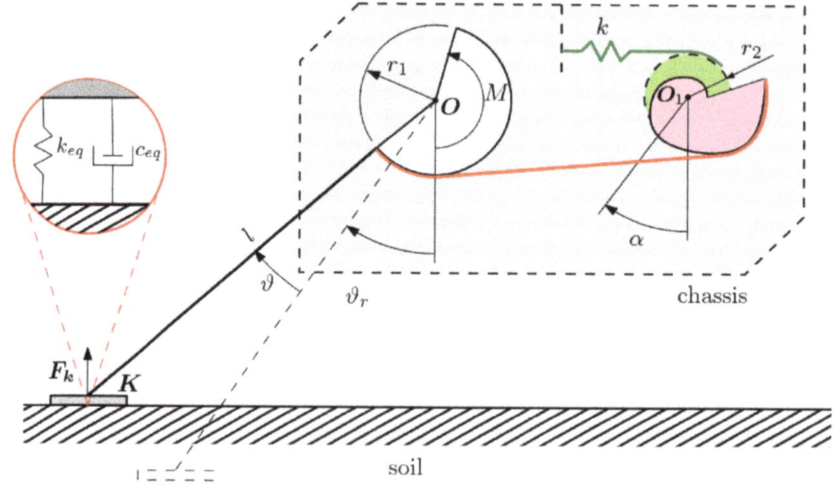

Figure 2. Schematic view of the lander-leg mechanism at the beginning of the impact with soil with detailed view of the model of the normal contact force as a spring–damper system.

In the following, the profile of the VRD is synthesized as explained in [26,28] in order to guarantee a constant force acting on the pads of the lander during the deceleration phase, that is, $F_k(\vartheta) = $ cost. A generic point of the VRD profile P_m, expressed in its local coordinates, is defined by

$$P_m = T(\alpha)\begin{Bmatrix} c_d \\ 0 \end{Bmatrix} + T(\alpha)T(-\gamma)T\left(-\frac{\pi}{2}\right)\begin{Bmatrix} l_t \\ 0 \end{Bmatrix} \tag{2}$$

where α, with reference to Figure 2, is the relative rotation of the VRD; l_t is the distance between the point P_m and the idle pulley center O, expressed by

$$l_t = \frac{c_d \sin\gamma}{1 + \dfrac{\frac{d^2g}{d\alpha^2}}{\sqrt{c_d^2 - \left(\frac{dg}{d\alpha}\right)^2}}} \tag{3}$$

where $T \in \mathbb{R}^{2\times 2}$ is a generic rotation matrix, and γ is defined as

$$\gamma = \arccos\left(\frac{1}{c_d}\frac{dg}{d\alpha}\right) \tag{4}$$

With reference to Figure 2, c_d is defined as the distance between O and O_1, whereas $g(\alpha)$ is the cable wound function. In order to synthesize the VRD for this application, the expression of the cable wound function $g(\alpha)$ and its first- and second-order derivatives with respect to α need to be derived. From Figure 2, the wound cable function and its derivatives assume the following expressions.

$$g(\alpha) = r_1\vartheta, \qquad \frac{dg(\alpha)}{d\alpha} = r_1\frac{d\vartheta}{d\alpha}, \qquad \frac{d^2g(\alpha)}{d\alpha^2} = r_1\frac{d^2\vartheta}{d\alpha^2} \tag{5}$$

By looking at Figure 2, and by applying the principle of virtual works, it is easy to demonstrate that the derivative of ϑ with respect to α is

$$\frac{d\vartheta}{d\alpha} = \frac{||F_s||r_2}{||(K - O) \times F_k||} \tag{6}$$

where F_s is the force exerted by the preloaded spring. Starting from Equation (6) and its second-order derivative, it is possible to synthesize the VRD, given the pad–soil contact force F_k.

2.2. Dynamic Model

The dynamic model of the system can be obtained by considering only a third of the lander thanks to its axial symmetry. A bidimensional model is considered and the displacement of the lander is constrained to be only along the vertical axis. As a starting point for the definition of the dynamic model, the kinetic energy of the system is computed as follows,

$$K = \frac{1}{6}m\dot{x}^2 + \frac{1}{2}I_{GA}\dot{\vartheta}^2 + \frac{1}{2}m_a\left(\dot{x}^2 + \dot{x}\dot{\vartheta}l\cos\vartheta + \frac{l^2}{4}\dot{\vartheta}^2\right) \tag{7}$$

where m is the mass of the lander, x is the vertical displacement of its center of mass, I_{GA} is the moment of inertia of the leg computed in its center of mass, m_a is the mass of the leg, and l is the leg length.

The potential energy of the system $\mathcal{U}$ accounts for the gravitational terms of the leg and the structure and the nonlinear response of the VRD, modeled as a nonlinear torsional spring. It results as

$$\mathcal{U} = -m_a g \cdot (G_A - K) - mg \cdot (G - K) + \frac{2}{\pi} M_0 \left(\vartheta \arctan (f\vartheta) - \frac{1}{2f} \log \left(1 + (f\vartheta)^2 \right) \right) + \frac{1}{2} k\vartheta^2 \quad (8)$$

where g is the gravitational force field. The considered model is a single-degree-of-freedom model. The generalized coordinate ϑ is chosen to describe the entire mechanical system. Then, by defining the Lagrangian function as $\mathcal{L} = \mathcal{K} - \mathcal{U}$, the Lagrangian equation can be written. This results in a second-order ordinary differential equation that represents the equation of motion of the system, as follows,

$$\frac{\partial}{\partial t} \left(\frac{\partial \mathcal{L}}{\partial \dot{\vartheta}} \right) - \frac{\partial \mathcal{L}}{\partial \vartheta} = \sum_{i=1}^{n} Q_{c,i} + \sum_{j=1}^{l} Q_{n.c,j} \quad (9)$$

where $Q_{c,i}$ and $Q_{n.c.,j}$ are the Lagrangian components of the i-th conservative and j-th non-conservative forces, respectively, which are not directly included in the Lagrangian function. In this specific case, those forces contains the components of the contact forces that take place between the lander pad and the ground. The contact force is composed by the normal and frictional components whose modeling is explained in detail below. Note that the generalized force $Q_{c,q}$ for a given force F_c, with point of application x_c, and a generalized coordinate q is expressed as follows,

$$Q_{c,q} = F_c \cdot \frac{\partial x_c}{\partial q} \quad (10)$$

Equation (9) is obtained using Equations (7), (8), and (10), and represents the dynamic model of the lander in the phase of impact with ground. The dynamic model of the lander for the phase between the release of the lander and the touchdown is the one for a free falling object. This model is used to compute the state variables of the lander at the end of the free falling stage, i.e., the beginning of the impact problem. These computed states becomes the boundary initial conditions for the ordinary differential equations resulting from Equation (9). The main effect of the VRD is to generate a nonlinear torque applied to the leg joint in order to keep constant the force acting on the pads of the lander. The nonlinear torque obtained from the VRD synthesis is shown in Figure 3.

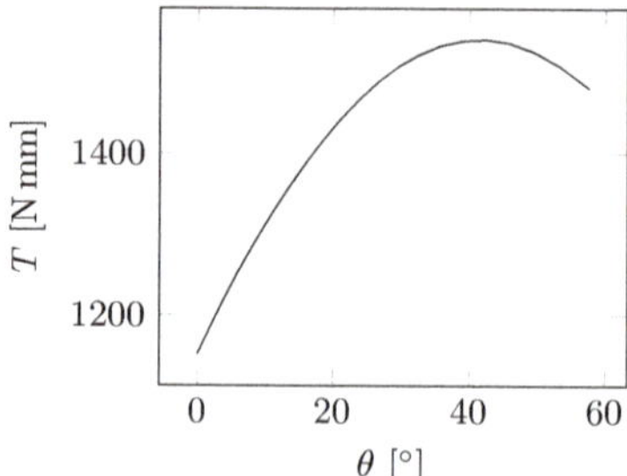

Figure 3. Nonlinear, softening, spring response obtained from VRD synthesis.

2.3. Soil Characterization

In order to completely define the dynamic model of the lander, the properties of the soil need to be analyzed.

In this work, we consider a single type of soil whose characteristics are determined experimentally; while mission-specific conditions of the ground can differ greatly, in this preliminary phase of experimentation we have elected to consider only one type.

In particular, the normal contact force between the pads of the lander and the soil is modeled as a non-linear spring damper system as depicted in Figure 2, and is described by the following [13],

$$\begin{cases} F_k = -k_{eq}x^\delta - c_{eq}\dot{x}, & x \leq 0 \\ F_k = 0, & x > 0 \end{cases} \tag{11}$$

where k_{eq} is the equivalent stiffness of the soil, c_{eq} is the equivalent damping factor of the soil, δ the force exponent responsible of the nonlinear behavior of the spring, and x and $\dot{x}$ are the depth and the speed of penetration of the pad into the soil, respectively. In contacts problems a further force that plays an important role is the frictional force acting between the pad and the soil. In this context, the Coulomb friction model is used, whose expression is reported in

$$\begin{cases} F_f = -\mu_s N, & \dot{y} = 0 \\ F_f = -\mu_k N, & \dot{y} \neq 0 \end{cases} \tag{12}$$

where $\dot{y}$ is the speed acting along a plane parallel to both the surfaces in contact, N is the normal force acting between the parts in contact and having direction coincident with the normal of the previously defined plane, μ_s is the static friction coefficient, and μ_k is the dynamic friction coefficient.

In order to conduct accurate and comparable simulations, the parameters k_{eq}, c_{eq}, μ_s, and μ_k need to be estimated. For what concern the parameters that describe the nonlinear spring damper system, some preliminary tests have been performed to estimate the modulus of subgrade reaction k. This can be considered directly related to the soil equivalent stiffness and is defined as follows,

$$k = \frac{q}{\Delta} \tag{13}$$

where q is the applied pressure and Δ is the penetration depth. More precisely, the test is conducted as follows; a circular plate with dimensions comparable with the lander pad, is positioned on the flat sample soil surface. Subsequently, a 1 kg steel weight is positioned above the circular plate and the deflection at static conditions is measured. The measures, as the deflections are truly little and submillimeter, are taken by mean of image analysis. Comparison between the pictures taken before applying the load and the one taken after the load is applied is done, as can be seen in Figure 4b. The aforementioned test is repeated 13 times, and the measured penetration depth is found to be $\Delta = (0.141 \pm 0.049)$ mm. Therefore, choosing the penetration depth to be the mean value and by using Equation (13), it is possible to estimate the modulus of subgrade reaction k and then by geometric considerations estimate the soil equivalent stiffness k_{eq} [22]. The equivalent stiffness k_{eq} is then estimated to be $k_{eq} \approx 70\,\mathrm{N\,mm^{-1}}$. Finally, the equivalent damping factor c_{eq} instead is estimated as $0.1 \div 1\%$ of the equivalent stiffness factor [22]. Thus, in this work we have assumed $c_{eq} = 0.7\,\mathrm{N\,s\,mm^{-1}}$.

For what concerns the second kind of parameters which describe the friction model, in a first analysis the static friction coefficient can be estimated as $\mu_s \approx \tan\phi$, where ϕ is the internal friction angle of the soil. The angle has been measured experimentally by conducting 15 tests. More precisely, the test is conducted as follows; a small quantity of the sample soil is slowly poured on a flat surface, and a picture of the heap vertical profile is taken. At this point, the internal angle of the soil is determined as can be seen in

Figure 4a and is found to be $\phi = (38.82 \pm 2.02)°$. We elected the angle to be the mean value, consequently $\mu_s \approx 0.804$.

Kinetic friction coefficient μ_k on the other hand is assumed to be a fraction of the measured static friction coefficient. In this case, $\mu_k = 0.7\mu_s = 0.563$.

All the required lander design parameters are summarized in Table 1. Considering that every part of the lander is fabricated in Polylactic Acid (PLA) polymer, the inertial parameters can be computed, and it is possible to solve the ordinary differential equation resulting from Equation (9), and then evaluate the contact force, which is required for the synthesis of the VRD.

Table 1. Parameters used in the design of the lander.

Parameter	Value	Parameter	Value
r	5.00×10^{-2} m	g	$9.81 \, \mathrm{m\,s^{-2}}$
r_1	3.20×10^{-3} m	H	0.450 m
m	0.500 kg	f	1.00×10^4
m_a	3.70×10^{-2} kg	M_0	1.54 N m
l_c	25.3 cm	l_m	27.1 cm
k_s	$50.0 \times 10^3 \, \mathrm{N\,m^{-1}}$	Material	PLA

(a)

(b)

Figure 4. Experimental characterization of the soil: (**a**) measure of the internal angle ϕ; (**b**) measure of the penetration depth. The scale unit is centimeters in both images.

3. Simulated Test Bed

In this section, the modeling of the multi-body system (MBS) representing the lander and the setup for the numerical simulations are presented. Both the entire lander model and the soil are modeled, respectively, as a MBS and a rigid body, and the result of this modeling process can be seen in Figure 5.

The MBS representing the lander model is constructed by connecting, with appropriate joints, each single rigid body which is functional to the correct behavior of the global system. The functional bodies, which globally compose the MBS of the lander, are the lander hexagon structure, its three legs, the three ratchet mechanisms and finally the three covers, respectively, indicated in yellow, gray, red, and green in Figures 5 and 6. Their geometrical properties are summarized in Table 2, whereas the joints between the parts are summarized in Table 3. It is important to report that the non-functional parts of the lander together with the payload are included in the model by adding their mass properties to the lander structure body. Moreover, the joints used are considered as ideal joints, i.e., without considering friction and deformations.

Finally, for what concerns the soil rigid body, it is modeled as a box having dimensions $(10 \times 0.05 \times 1)$ m^3 and positioned below the lander structure center mass.

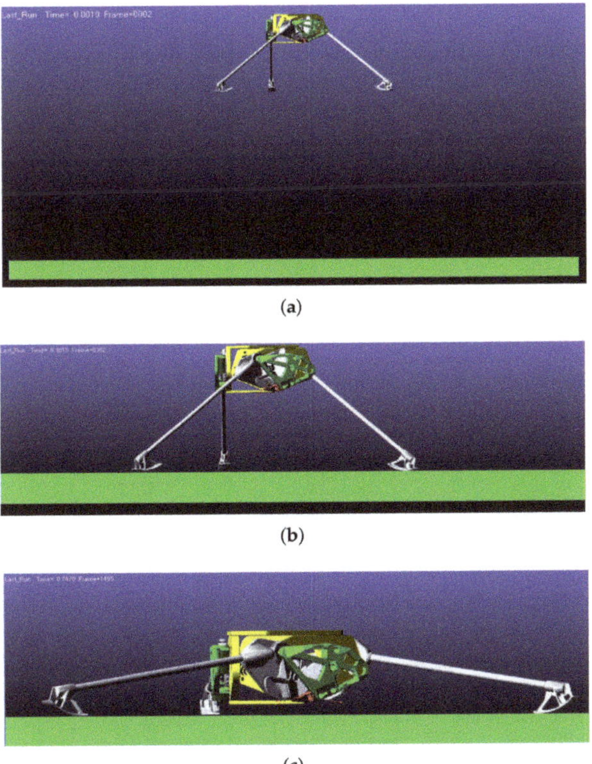

(a)

(b)

(c)

Figure 5. View of the lander MBS and the soil modeled in ADAMS: (**a**) the initial condition of the simulation; (**b,c**) the lander configuration at the beginning and at the end of an impact simulation on a flat surface, respectively.

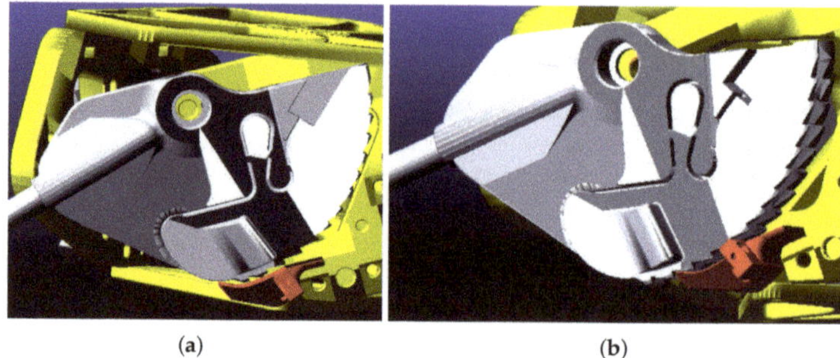

Figure 6. View of the leg joint and the ratchet mechanism used in ADAMS. (**a**) The ratchet and the leg joint at rest position; (**b**) the ratchet blocking the opposite rotations of the leg thanks to the contact force between the ratchet and the leg.

Table 2. Geometrical properties of the main functional rigid bodies composing the lander MBS model. All the inertia components are expressed in kg mm^2.

Bodies	m [kg]	$\mathbf{I}_{xx}$	$\mathbf{I}_{yy}$	$\mathbf{I}_{zz}$
Structure	0.5	430.14	684.78	430.14
Leg	3.7×10^{-2}	374.29	10.67	364.37
Ratchet	5×10^{-4}	9.40×10^{-2}	0.11	3.49×10^{-2}
Cover	1×10^{-3}	17.00	11.49	6.45

Table 3. Joint type used between each pair of rigid bodies.

First Body	Second Body	Joint Type
Structure	Leg	Revolute
Structure	Ratchet	Revolute
Structure	Cover	Fixed
Soil	Ground	Fixed

Once all the bodies and their joints are defined, forces acting on the model should be added. These forces can be split up between external forces and contact forces. The external forces implemented in the simulation environment are a gravitational force field, a linear torsional spring between each ratchet and the structure, and a nonlinear torque applied between each leg and the lander structure. The linear torsional spring between each ratchet and the structure is introduced in order to simulate the elastic deflection of the ratchet from its rest position. In order to provide values that describe the torsional spring, the ratchet is considered as a cantilever, which deflects under the action of the contact force between the ratchet and the leg. Therefore, by using the elastic theory, it is possible to define an equivalent torsional stiffness coefficient $k_{r,eq}$ for the spring, whose expression is $k_{r,eq} = EI/l$. E is the Young elastic modulus of the PLA, I is the moment of inertia of the resistant cross-area of the ratchet, and l is the length of the cantilever. Finally, a stiffness coefficient of $k = 288\,\text{N mm rad}^{-1}$ is set for the springs. The nonlinear torsional spring between each leg and the structure is introduced in order to simulate the resistant torque applied to the leg joint generated by the action of the VRD pulley, as stated in Section 2. For the implementation in ADAMS, instead of using directly Equation (1), a cubic interpolating polynomial function is used in order to approximate the response of the VRD pulley, with coefficients $a_0 = 1147$, $a_1 = 18.6761$, $a_2 = -0.2175$, and $a_3 = -1.0823 \times 10^{-4}$.

The contact forces implemented in the simulation environment are the forces between each leg and the ratchet next to it, the forces between each leg and the cover next to it, and those between each leg and the soil. Every contact is defined as a solid-to-solid contact force which is composed by a normal contact force following the relation shown in Equation (11), and a frictional force following the Coulomb model Equation (12).

The definition of contacts between the legs and the ratchets is fundamental for the correct functioning of the retention mechanism. In particular, when the structure is decelerating, the leg rotates in the admissible direction and this force makes the ratchet to bend, allowing the leg to rotate. When the structure stops its deceleration and starts to move upwards, instead, the tip of the ratchet mechanism connects with the tooth present on the side of the leg joint, allowing the ratchet to limit the opposite rotation of the leg joint. This can be immediately seen from Figure 6a,b. The contacts defined between the legs are required to implement a physical lower end-stop to the revolute joint connecting the leg to the structure. By looking at Figure 6a,b, on the side surface of the leg joint a small part is modeled which goes in contact with the cover and thus limits the joint rotation. Finally, the contacts between the box model representing the soil and each of the three legs are the most important actors for the whole soft-landing simulation and need to be modeled accurately. The parameters that need to be set for the contact are the stiffness coefficient, the nonlinear force exponent, the damping coefficient, the penetration depth, and the static and dynamic friction coefficients. The parameters have been estimated in Section 2.3.

4. Simulation Results

In this section, we present the numerical simulations that are conducted on the model of the three-legged lander prototype. The dynamic behavior of the system is simulated using the multi-body software MSC ADAMS 2019. Two simulated cases are considered in this work:

- **Case (1): impact simulation on flat surface.** In this simulation the behavior is investigated of the lander free-falling from a fixed height and impacting on a flat horizontal surface. In particular, we analyze the contact force distribution, the lander deceleration, the angular displacement of the leg, and the ratchet action.
- **Case (2): topple simulation.** In this simulation, the behavior is investigated of the lander impacting on an inclined surface, after a free-fall from a fixed height. Subsequently, the critical angle is found at which the lander topples.

4.1. Case (1): Impact Simulation on Flat Surface

In this simulation, the lander is let free to fall from a fixed height and impacts with the box representing the soil model. Figure 5a–c shows the initial, impact moment, and final configurations, respectively, of the lander MBS in the simulation over a flat surface. In this case, a trial and error strategy is adopted, in order to investigate the maximum admissible drop height and considering the payload of the lander to be 0.5 kg. Moreover, in this subsection the most representative results that are obtained for the maximum drop height, which is found to be $H = 0.45\,\text{m}$, are reported. A simulation duration of 1 s has been determined, as it is enough to catch the transient phenomenon taking place between the beginning of the simulation and the lander MBS stabilization.

Figure 7a–c reports, respectively, the lander MBS center of mass displacement, speed, and acceleration along the vertical axis, that is the axis of motion of the free falling lander. Figure 8a–c reports, respectively, the rotations of each of the leg structure joints, the torque generated by each of the three VRDs, and the contact forces exerting between each pad and the soil during the soft-landing simulation. Finally, Figure 9a,b shows the rotations of each of the ratchet structure joint and the evolution of the kinetic energy of the lander MBS.

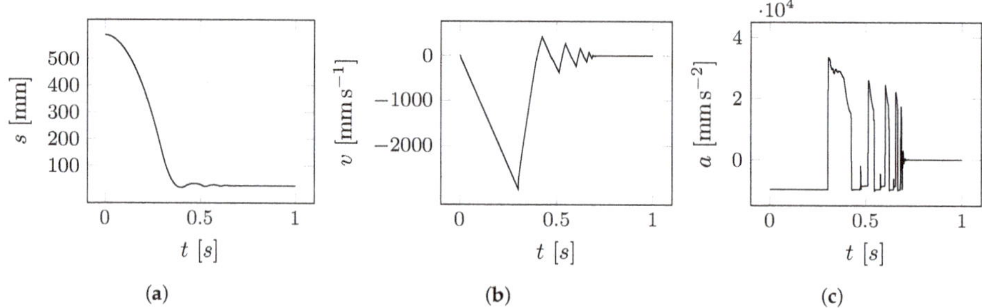

Figure 7. Information about the state of the lander's MBS along the vertical axis: (**a**) the displacement of the center of mass; (**b**) the velocity; (**c**) the acceleration.

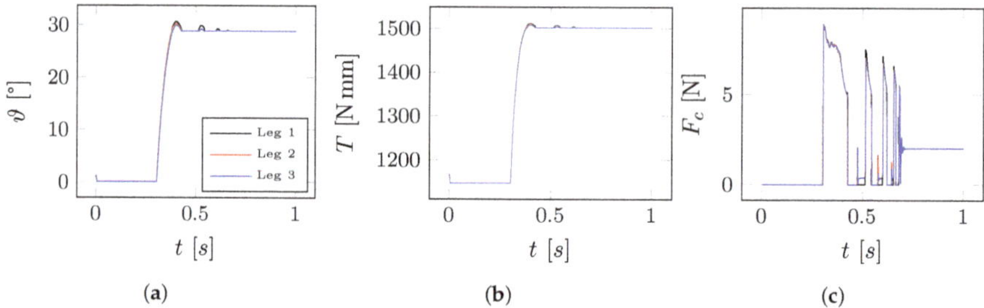

Figure 8. (**a**) The curves representing the angular rotation of each of the three leg structure revolute joint; (**b**) the curves representing the torque response of each of the three VRD's; (**c**) the curves representing the contact force between each of the legs with the soil.

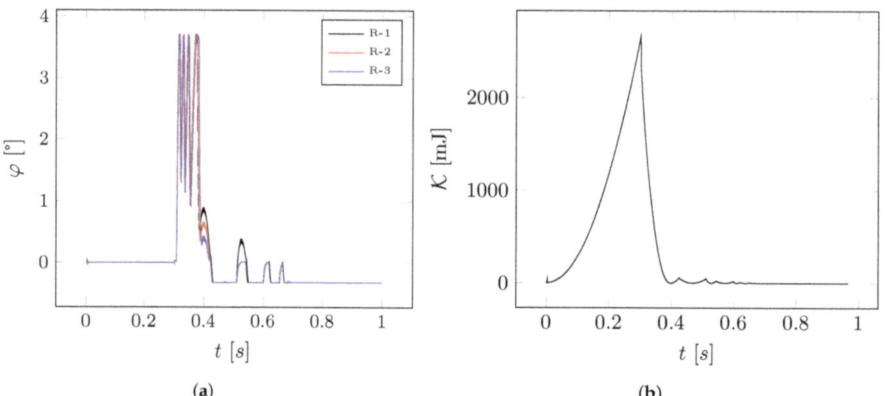

Figure 9. (**a**) The rotation evolution of each of the ratchet-structure joint; (**b**) the kinetic energy evolution over time of the lander MBS

4.2. Case (2): Topple Simulation

In this section, the case of topple simulation is investigated. More precisely, the aim of the numerical tests is to identify at which ground surface angle the lander topple. This angle is named landing critical angle β_{cr}. To achieve this, the same model used from the previous analysis is considered. For every simulation, the model of the ground is rotated by increasing angles around the z axis, and rigidly translated along the y axis in order to keep the drop height constant. The simulations of Case (2) allow to investigate the range of admissible slope angles to achieve safe soft-landing without the lander toppling. It is important to say that all the simulations are conducted using the same maximum drop height of $H = 0.45\,\text{m}$. By pursuing this strategy, it is found that the lander topples at a critical landing angle $\beta_{cr} \approx 51.6°$, as it can be seen in Figure 10. Moreover, some important results for the sample case of an inclined plane simulation with an inclination ratio 1:5 are reported.

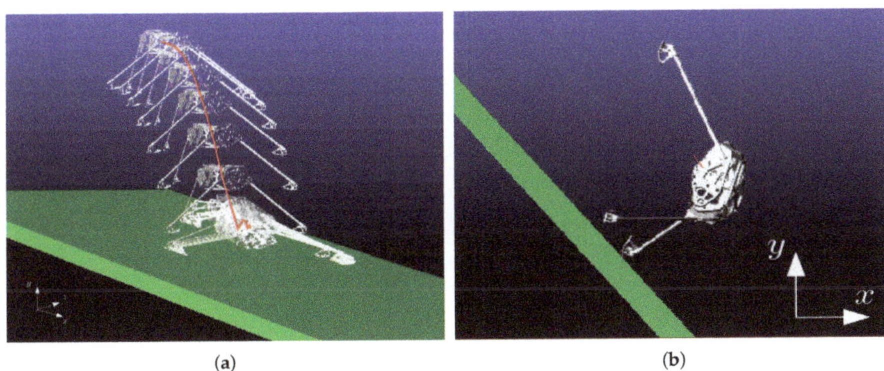

Figure 10. (**a**) A time-series representation of the simulation for the lander MBS soft-landing conducted on an inclined surface; (**b**) the lander MBS toppling at the landing critical angle β_{cr}.

Figure 11a–c reports the lander MBS center of mass displacement, speed, and acceleration along the three axes, respectively. Figure 12a–c reports the rotations of each of the leg structure joints, the torque generated by each of the three VRDs, and the contact forces exerting between each pad and the soil during the soft-landing simulation, respectively. Finally, in Figure 13 the evolution of the kinetic energy of the lander is shown.

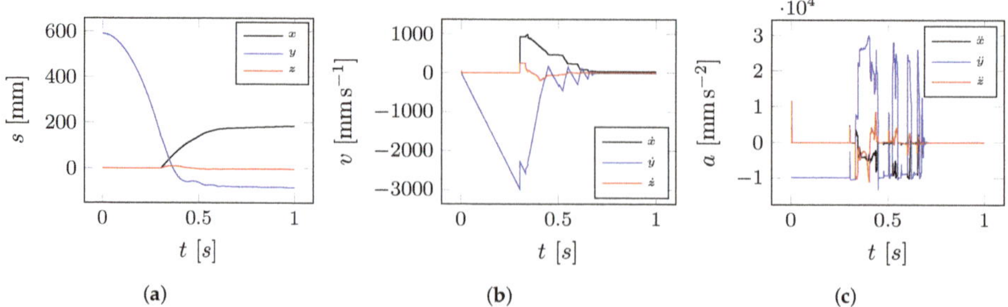

Figure 11. Information about the state of the lander's MBS along the vertical axis: (**a**) the displacement of the center of mass; (**b**) the velocity; (**c**) the acceleration.

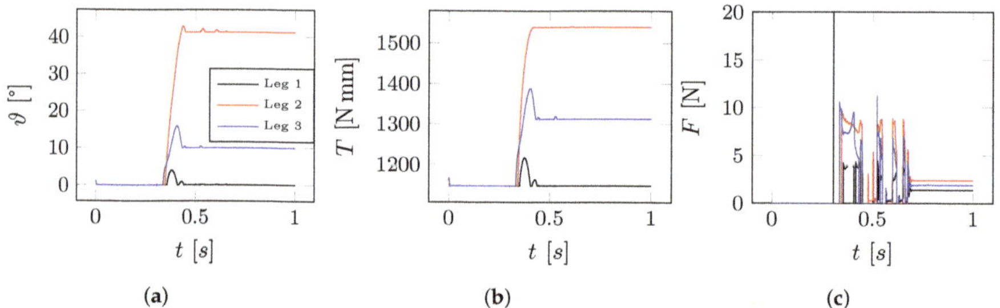

Figure 12. (**a**) The curves representing the angular rotation of each of the three leg structure revolute joint; (**b**) the curves representing the torque response of each of the three VRD's; (**c**) the curves representing the contact force between each of the legs with the soil.

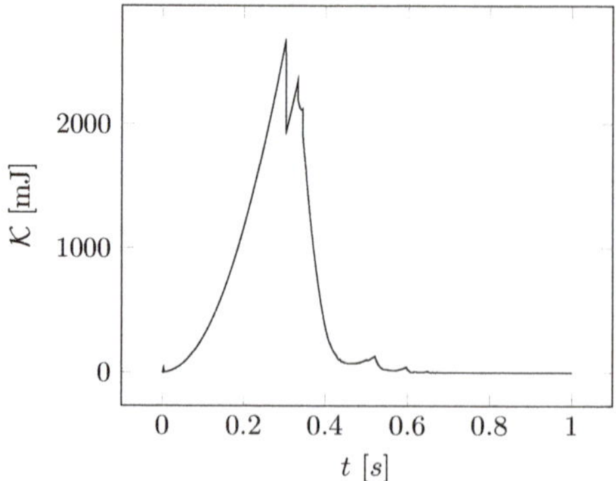

Figure 13. Kinetic energy evolution over time of the lander MBS.

5. Discussion

In this section, the results obtained from the simulations of soft-landing on a flat surface and an inclined surface, reported in Sections 4.1 and 4.2, are discussed. Moreover, some important results obtained with the MATLAB-based numerical integration of the theoretical model developed in Section 2 are reported in order to validate the approach.

5.1. Comparison with the Theoretical Model

Comparisons of the ADAMS simulations with the theoretical model are conducted for two different cases regarding the analytical-based simulation. For the first case, a length of the lander leg equal to l_c is chosen, which corresponds to the conservative value chosen, i.e., the distance from the lander joint axis from the pad closest contact point. For the second case, a length equal to l_m is chosen, which corresponds to the mean distance of the contact point from the joint axis. This is because despite the analytical model, in which the pad is considered a point, the pad in the ADAMS model is a rigid body having its own semicircular geometry. Therefore, while in the analytical model the contact point is always at a fixed distance from the lander joint axis, the geometry of the pad leads to the fact that the contact point keep varying all along the pad profile during the whole contact phase, thus changing the arm lever.

The comparisons are conducted in the scope of the Case (1). Moreover, in order to catch the influence of the ratchets in the global dynamics of the lander, an additional simulation in ADAMS without ratchets is conducted and compared with the others. Figure 14 shows the comparison between the ADAMS simulations and the MATLAB-based analytical simulation for the two leg length value chosen: in the top row the acceleration of the lander is compared, in the middle row the contact force acting between the soil and the lander pad is compared, and in the bottom row the leg-joint rotation is compared. The black curves represent the ADAMS simulations with the ratchet (ADAMS-R), the blue curves represent the ADAMS simulations without ratchets (ADAMS-NR), and finally the red curves represent the simulations based on the analytical model described in Section 2. It is important to remark that the figures refer only to the first bounce of the lander. This is because the lander behavior is different between the simulations after the first bounce. In fact, the analytical model does not implement the ratchet. In ADAMS, on the other hand, this mechanism is present and it affects the behavior of the system after the deceleration phase.

By comparing the curves, similarities in the trends and with the values assumed can be found. More precisely, with the theoretical model the deceleration of the lander during impact assumes values $a \in [3.1, 4.5]g$ for a leg length equal to l_c and values $a \in [2.7, 4.0]g$ for a leg length equal to l_m, whereas with the simulation in ADAMS an acceleration $a \in [3, 3.5]g$ is observed. The same can be observed with the contact force and the leg joint rotation; in particular a contact force $F_c \in [5.4, 10]$N has been observed using the theoretical model and the same is observed in the in ADAMS simulation. Finally, for what concerns the leg joints rotation, it can be seen that in both the theoretical model simulation as well as in the ADAMS simulations, the maximum rotation of the joints reaches a value slightly below 30°. The discontinuities visible in the ADAMS simulation black curves visible from the acceleration and force graphs are due to the action of the ratchets which limits the rotation of the lander legs; in fact, in the curves representing the analytical model (which does not implement the action of the ratchet), these same discontinuities are not present.

Conservative value ($l = l_c$)   Non-Conservative value ($l = l_m$)

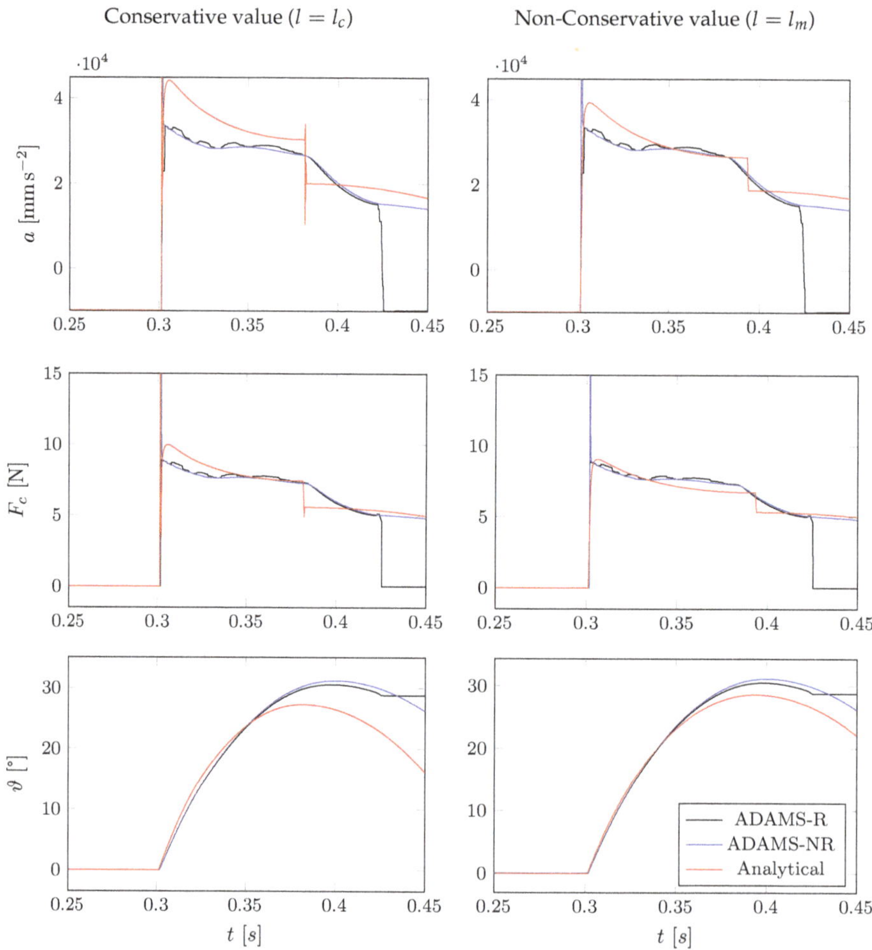

Figure 14. Comparison between ADAMS simulations with and without ratchets and the MATLAB-based analytical simulation for two leg length parameter chosen: in the top row the lander acceleration, in the middle row the contact force between soil and lander pad, and in the bottom layer the evolution of the leg joint rotation.

By comparing the MATLAB-based simulations of the left column with the right column of Figure 14, it can be noticed that the leg length parameter used for the theoretical model, i.e., the distance of the contact point of the pad from the joint axis, has a great influence on the curve's trend of the analytical model. In fact, by using as leg length l_m, the analytical curves better fits the ADAMS ones, especially for the contact force. However, at the beginning of the impact, the analytical curve has a greater peak with respect to the ADAMS ones; this is because at that stage the contact point is further from l_m, and as the impact time goes on, and the legs rotates the contact point comes closer to the mean point until the minimum distance l_c is reached.

By comparing the black curves with the blue curves in Figure 14 it can be clearly seen the influence of the ratchet to the global dynamics of the lander. In particular, from the force and acceleration graphs small humps in the curves can be recognized due to the presence of the ratchet. From the angle graphs, instead, it can be clearly seen the influence of the ratchet: in particular as it exerts a reactive torque which opposes to the leg opening during landing, this leads to the leg opening with a lower angle. It is important to remark that the ratchet has been modeled in order to easily deflect, so in this case the influence of the ratchet is present but it does influence to a lesser extent the global dynamics of the lander. The stiffer the ratchet is, the more it influences the dynamics of the lander, even leading to the limit case of no opening of the legs at all thus imposing enormous vertical accelerations to the structure and the payload.

An other cause which we can address the discrepancies between the MATLAB-based analytical simulation and ADAMS ones is one which is due to the limit that the analytical model is a purely bidimensional model, thus unable to catch 3D related phenomenon. Of particular note is that during the soft-landing the lander structure body slightly rotates around it's vertical axis (allowed given the lander legs configuration), thus part of the lander energy is split into rotational energy and therefore dissipated through lateral friction forces. This just being said can be seen from Figure 15 which shows in the left the lander rotation and in the right its rotational speed during the impact stage. During this stage can be recognized that in 0.15 s the lander rotational energy is almost completely dissipated. Moreover, an other factor which causes discrepancies between the models is that ADAMS implements a nonlinear damper in the contact model; while the analytical model implements a linear damper. This fact causes differences at the initial stage of the contact. In fact the purely linear damper causes greater damping forces in the moment of contact. Namely, the combination of all the factors mentioned above leads to discrepancies between the analytical model and the ADAMS simulations. However, despite the discrepancies between the ADAMS simulations and the analytical model, which after all is a bidimensional idealization of a three dimensional lander, the results of the two models show remarkable adherence, thus validating the approach.

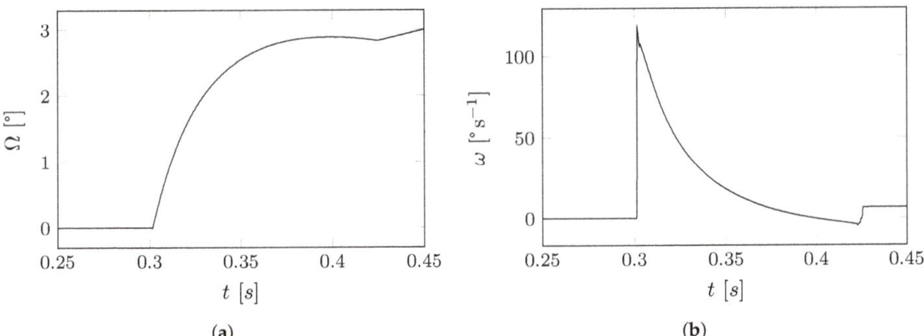

Figure 15. (**a**) The lander structure center mass rotation around its vertical axis and (**b**) its angular speed around the same axis.

5.2. Case (1): Simulation on Horizontal Surface

Figure 7 reports the evolution of the state of the lander MBS during the simulation time. In Figure 7a the loss of height can be seen; Figure 7b shows the evolution and the changes of speed the lander is subjected; finally Figure 7c describes the acceleration of the lander. From these figures it can be seen that at $t \approx 0.3\,\mathrm{s}$, the lander impact the ground with a speed $v \approx 3.0\,\mathrm{m\,s^{-1}}$. Moreover, some sudden changes in speed are observed, which identifies the bounces of the lander after impact. The oscillation speed and amplitude decreases due to the dissipative forces defined in the contact problem. It can be seen that the lander is subjected to peaks of deceleration having amplitude $|a| \in [3, 3.5]g$. It is important to report that the evolution of the acceleration has been slightly filtered in order to cut out the high-frequency peaks of deceleration taking place in specific instants of time due to numerical instabilities. Disregarding these is allowed from a simulation perspective, as their impact on the landing behavior is low and they can easily be filtered out with dampers on the payload attachment points.

When the lander impacts with the soil its legs start to rotate due to the torque generated by the contact force with the ground. As expected, for an impact simulation on an horizontal surface, the legs of the lander rotate uniformly, until the maximum angle $\vartheta \approx 30°$ for each leg. The reactive torques generated have the same trend. These aspects can be observed in Figure 8a,b. From the same figures, the action of the ratchets can be observed. More precisely, when the leg rotation reaches its maximum value, the leg starts to rotate back to the original position. However, the ratchet almost immediately engages and limits this motion, effectively stopping the leg orientation at the current angle; to this angle a fixed torque of $T \approx 1.5\,\mathrm{N\,m}$ is naturally associated. Finally, in Figure 8c the contact forces acting between the soil and the lander leg are shown. In a way similar to the case of acceleration, the contact forces have been slightly filtered in order to remove the high-frequency spikes. In this figure the bounces can be seen and, as expected, the contact force has the same trend as the joint rotation. By looking again at the figure we can say that the first peak of the contact force is the one responsible of dissipating most of the energy of the system, as can be seen in Figure 9b, which represents the kinetic energy of the lander MBS during the whole impact simulation. The other spikes are about the lander stabilizing on the soil surface. However, the amplitude of the contact force is $|F| \in [5.0, 8.7]\mathrm{N}$. Finally, Figure 9a reports the rotation of the ratchet structure revolute joint. The high peaks represent the instants at which the ratchet overcomes the tooth present on the side of the leg, and thus moving to the next tooth. In this case, the ratchet overcomes four teeth. The low peaks instead are due to the leg trying to rotate in the opposite direction, against the engaged ratchet.

5.3. Case (2): Topple Simulation

In the following, the results of an inclined surface sample impact simulation with an inclination of 1:5 are discussed. We disregard the ratchet structure joint rotation. Differently from Case (1), in which the state of the lander was changing only along the vertical axis, in this simulation the state of the lander changes along the other two directions as well. In Figure 11a, the displacement is reported of the lander center of mass along the three directions. In Figure 11b, we can see the speed of the lander along the three directions. Figure 11c shows the acceleration of the center of mass of the lander along the three directions; this has been slightly filtered in order to remove the high-frequency high spikes. While the main component of motion remains along the vertical axis, in this case the others components cannot be neglected. The lander impacts with a vertical velocity $v \approx 3\,\mathrm{m\,s^{-1}}$, and its structure is subjected to peaks of acceleration having modulus $|a| \in [2.1, 3]\,\mathrm{g}$. From the figures the bounces of the lander can be observed before stabilizing.

In Figure 12a, the rotations are reported of each of the three leg structure revolute joints. Contrary to *Case (1)*, the joints rotation have a different behavior between them; the joint relative to the first leg, that is the upstream leg does not rotate at all. On the other hand, the downstream legs exhibit the maximum rotation; that is, $\vartheta \approx 42°$. This reflects also on the reactive torques generated by the lander, as reported in Figure 12b, as the exerted torque is a function of the relative joint rotation.

Moreover, in Figure 12c the behaviors are reported of the contact forces exerting between the soil and each of the the leg pads. From this figure a high peak of force can be observed for the leg 1; this corresponds to the instant when the upstream leg impacts with the soil and acts as a pivot, transforming the translational kinetic energy into rotational kinetic energy. The peaks responsible for the dissipation of most of the kinetic energy can be seen in Figure 13; however, as the force peaks are shifted in time, the dissipation of the kinetic energy has a lower rate. Finally, as we expected, the modulus of these contact forces is not the same. Leg 1 has a lower force, due to the fact that it is the one that is upstream and impact first; in fact, the transformation of its energy from translational to rotational means that less energy goes in the spring. More precisely the force is $F \approx 5\,\text{N}$, whereas for the other two legs which are located downstream the force is higher, reaching $F \approx 10\,\text{N}$.

6. Conclusions

The use of a passive shock absorber, based on the use of VRDs, cables, and preloaded springs, in landers for space applications has been investigated in this paper. The design of a possible lander based on this mechanism has been proposed. In order to prove the effectiveness and the soundness of the analytical model of the dynamics of the lander the behavior of the lander has been investigated numerically, first by implementing the analytical model, and then by mean of the multi-body simulator ADAMS on a physical 3D CAD model of the proposed lander. Seen the similarities between the results obtained using the analytical model and the results obtained in ADAMS we can truly say that the analytical model describes well the behavior of the lander for the case of soft-landing on a flat surface. However, it has been observed that, with respect to ADAMS, the analytical model overestimates the contact force and the acceleration the lander is subject to during impact. This discrepancy might be due to several reasons. The main reasons are the fact that the landing pad shape causes the contact point to change during the leg rotation and that the ADAMS model includes the three ratchets, contrary to the analytical one. Moreover, the contact models are different, i.e., in the analytical model a point to point contact model is used, while in ADAMS it is a solid-to-solid contact model; furthermore, in ADAMS the damper used in the contact is nonlinear while in the analytical model a liner damper is used. Finally, in the simulations it has been observed that this kind of passive landing system guarantees a certain degree of control on the forces and accelerations acting on the lander structure during soft-landing.

Future works foresee the design and fabrication of a real prototype based on the one that has been studied in this paper. Considerations on the ground characteristics will be the basis for the design of the main components of the lander, e.g., the landing pads. Experimental tests will be carried out to further validate the models and to gain insight on the mechanics of the soil. For example, drop tests and deceleration tests will be tailored on those described in this paper and performed using regolith simulants. The results obtained will be compared with the numeric results obtained in this paper.

Moreover, a detailed research about implementing an active control system acting on the preloaded springs will be considered and implemented. Adjusting the preload allows to reshape the VRD response in order to adapt lander to land on different scenarios, or when the ground is at an incline. In this specific situation, a softer action on some legs would allow the lander to avoid toppling or even experience lower deceleration. Other uses of the proposed technology will be explored in the future, especially in fields unrelated to space exploration.

Author Contributions: Conceptualization, M.C. and S.S.; methodology, S.S., L.S., and P.G.; software, M.C. and S.S.; numerical validation, M.C.; experimental investigation, M.C.; resources, S.S. and P.G.; writing—original draft preparation, M.C.; writing—review and editing, S.S., P.G., and L.S.; supervision, funding acquisition S.S. and P.G. All authors have read and agreed to the published version of the manuscript.

Funding: This work has been partially supported by the PRIN project "SEDUCE" n. 2017TWRCNB, and by the internal funding program "Microgrants 2020" of the University of Trieste.

Conflicts of Interest: The authors declare no conflict of interest.

References

1. Rogers, W. *Apollo Experience Report: Lunar Module Landing Gear Subsystem*; NASA: Washington, DC, USA, 1972.
2. O'Neil, W.J.; Rudd, R.P.; Farless, D.L.; Hildebrand, C.E.; Mitchell, R.T.; Rourke, K.H.; Euler, E.A. *Viking Navigation*; Technical Report NASA-CR-162917, JPL-PUB-78-38, NASA, JPL; California Institute of Technology Pasadena: Pasadena, CA, USA, 1979.
3. Behrends, R.; Dillon, L.K.; Fleming, S.D.; Stirewalt, R.E.K. *Mars Pathfinder Airbag Impact Attenuation System*; Technical Report AIAA-95-1552-CP; American Institute of Aeronautics and Astronautics: East Lansing, MI, USA, 2006.
4. Cao, P.; Hou, X.; Xue, P.; Tang, T.; Deng, Z. Research for a modeling method of mars flexible airbag based on discrete element theory. In Proceedings of the 2017 2nd International Conference on Advanced Robotics and Mechatronics (ICARM), Hefei and Tai'an, China, 27–31 August 2017; pp. 351–356. [CrossRef]
5. Accomazzo, A.; Lodiot, S.; Companys, V. Rosetta mission operations for landing. *Acta Astronaut.* **2016**, *125*, 30–40. [CrossRef]
6. Woods, D. Review of the Soviet Lunar exploration programme. *Spaceflight* **1976**, *18*, 273–290.
7. Efanov, V.; Dolgopolov, V. The Moon: From research to exploration (to the 50th anniversary of Luna-9 and Luna-10 Spacecraft). *Sol. Syst. Res.* **2017**, *51*, 573–578. [CrossRef]
8. Malenkov, M. Self-propelled automatic chassis of Lunokhod-1: History of creation in episodes. *Front. Mech. Eng.* **2016**, *11*, 60–86. [CrossRef]
9. Li, S.; Jiang, X.; Tao, T. Guidance summary and assessment of the Chang'e-3 powered descent and landing. *J. Spacecr. Rocket.* **2015**, *53*, 258–277. [CrossRef]
10. Wu, W.; Yu, D.; Wang, C.; Liu, J.; Tang, Y.; Zhang, H.; Zhang, Z. Technological breakthroughs and scientific progress of the Chang'e 4 mission. *Sci. China Inf. Sci.* **2020**, *63*, 1–4. [CrossRef]
11. Wang, C.; Nie, H.; Chen, J.; Lee, H.P. The design and dynamic analysis of a lunar lander with semi-active control. *Acta Astronaut.* **2019**, *157*, 145–156. [CrossRef]
12. Zhou, J.; Jia, S.; Qian, J.; Chen, M.; Chen, J. Improving the buffer energy absorption characteristics of movable lander-numerical and experimental studies. *Materials* **2020**, *13*, 3340. [CrossRef]
13. Liu, Y.; Song, S.; Wang, C. Multi-objective optimization on the shock absorber design for the lunar probe using nondominated sorting genetic algorithm II. *Int. J. Adv. Robot. Syst.* **2017**, *14*, 1729881417720467. [CrossRef]
14. Lin, R.; Guo, W.; Li, M. Novel design of legged mobile landers With decoupled landing and walking functions containing a rhombus joint. *J. Mech. Robot.* **2018**, *10*, 061017. [CrossRef]
15. Lin, R.; Guo, W. Creative design of legged mobile landers with multi-loop chains based on truss-mechanism transformation method. *J. Mech. Robot.* **2020**, *13*, 011013. [CrossRef]
16. Xu, L.; Nie, H.; Wan, J.; Lin, Q.; Chen, J. Analysis of landing impact performance for lunar lander based on flexible body. In Proceedings of the 2011 IEEE International Conference on Computer Science and Automation Engineering, Shanghai, China, 10–12 June 2011; Volume 1, pp. 10–13.
17. Wan, J.; Nie, H.; Chen, J.; Lin, Q. Modeling and simulation of Lunar lander soft-landing using transient dynamics approach. In Proceedings of the 2010 International Conference on Computational and Information Sciences, Chengdu, China, 17–19 December 2010; pp. 741–744.
18. Zheng, G.; Nie, H.; Chen, J.; Chen, C.; Lee, H.P. Dynamic analysis of lunar lander during soft landing using explicit finite element method. *Acta Astronaut.* **2018**, *148*, 69–81. [CrossRef]
19. Liang, D.; Chai, H.; Chen, T. Landing dynamic analysis for landing leg of lunar lander using Abaqus/Explicit. In Proceedings of the 2011 International Conference on Electronic & Mechanical Engineering and Information Technology, Harbin, China, 12–14 August 2011; Volume 8, pp. 4364–4367.
20. Aravind, G.; Vishnu, S.; Amarnath, K.; Hithesh, U.; Harikrishnan, P.; Sreedharan, P.; Udupa, G. Design, analysis and stability testing of Lunar lander for soft-landing. *Mater. Today Proc.* **2020**, *24*, 1235–1243. [CrossRef]
21. Udupa, G.; Sundaram, G.; Poduval, A.P.; Pillai, K.; Kumar, N.; Shaji, N.; Nikhil, P.C.; Ramacharan, P.; Rajan, S.S. Certain investigations on soft lander for Lunar exploration. *Procedia Comput. Sci.* **2018**, *133*, 393–400. [CrossRef]

22. Pham, V.; Zhao, J.; Goo, N.; Lim, J.; Hwang, D.; Park, J. Landing stability simulation of a 1/6 Lunar module with aluminum honeycomb dampers. *Int. J. Aeronaut. Space Sci.* **2013**, *14*, 356–368. [CrossRef]

23. Sun, Y.; Hu, Y.; Liu, R.; Den, Z. Touchdown dynamic modeling and simulation of lunar lander. In Proceedings of the 2010 3rd International Symposium on Systems and Control in Aeronautics and Astronautics, Harbin, China, 8–10 June 2010; pp. 1320–1324.

24. Liu, Y.; Song, S.; Li, M.; Wang, C. Landing stability analysis for Lunar landers using computer simulation experiments. *Int. J. Adv. Robot. Syst.* **2017**, *14*, 1729881417748441. [CrossRef]

25. Seriani, S. A new mechanism for soft landing in robotic space exploration. *Robotics* **2019**, *8*, 103. [CrossRef]

26. Seriani, S.; Gallina, P. Variable radius drum mechanisms. *J. Mech. Robot.* **2015**, *8*, 021016. [CrossRef]

27. Endo, G.; Yamada, H.; Yajima, A.; Ogata, M.; Hirose, S. A passive weight compensation mechanism with a non-circular pulley and a spring. In Proceedings of the 2010 IEEE International Conference on Robotics and Automation, Anchorage, AK, USA, 3–7 May 2010; pp. 3843–3848.

28. Scalera, L.; Gallina, P.; Seriani, S.; Gasparetto, A. Cable-Based Robotic Crane (CBRC): Design and implementation of overhead traveling cranes based on variable radius drums. *IEEE Trans. Robot.* **2018**, *34*, 474–485. [CrossRef]

29. Fedorov, D.; Birglen, L. Differential noncircular pulleys for cable robots and static balancing. *J. Mech. Robot.* **2018**, *10*. 061001. [CrossRef]

30. Suh, J.W.; Kim, K.Y. Harmonious cable actuation mechanism for soft robot joints using a pair of noncircular pulleys. *J. Mech. Robot.* **2018**, *10*, 061002. [CrossRef]

31. Fedorov, D.; Birglen, L. Design of a compliant mechanical device for upper leg rehabilitation. *IEEE Robot. Autom. Lett.* **2019**, *4*, 870–877. [CrossRef]

32. Scalera, L.; Palomba, I.; Wehrle, E.; Gasparetto, A.; Vidoni, R. Natural motion for energy saving in robotic and mechatronic systems. *Appl. Sci.* **2019**, *9*, 3516. [CrossRef]

33. Li, M.; Cheng, W.; Xie, R. Design and experimental validation of two cam-based force regulation mechanisms. *J. Mech. Robot.* **2020**, *12*, 031003. [CrossRef]

34. Grotzinger, J.; Crisp, J.; Vasavada, A.; Anderson, R.; Baker, C.; Barry, R.; Blake, D.; Conrad, P.; Edgett, K.; Ferdowski, B.; et al. Mars Science Laboratory mission and science investigation. *Space Sci. Rev.* **2012**, *170*, 5–56. [CrossRef]

35. Yue, Z.; Di, K.; Liu, Z.; Michael, G.; Jia, M.; Xin, X.; Liu, B.; Peng, M.; Liu, J. Lunar regolith thickness deduced from concentric craters in the CE-5 landing area. *Icarus* **2019**, *329*, 46–54. [CrossRef]

36. Donaldson Hanna, K.; Schrader, D.; Cloutis, E.; Cody, G.; King, A.; McCoy, T.; Applin, D.; Mann, J.; Bowles, N.; Brucato, J.; et al. Spectral characterization of analog samples in anticipation of OSIRIS-REx's arrival at Bennu: A blind test study. *Icarus* **2019**, *319*, 701–723. [CrossRef]

37. Slyuta, E. Physical and mechanical properties of the lunar soil (a review). *Sol. Syst. Res.* **2014**, *48*, 330–353. [CrossRef]

38. Carrier, W.D. Lunar soil grain size distribution. *Moon* **1973**, *6*, 250–263. [CrossRef]

39. Graf, J. *Lunar Soils Grain Size Catalog*; NASA Reference Publication; National Aeronautics and Space Administration, Office of Management, Scientific and Technical Information Program: Washington, DC, USA, 1993.

40. Seriani, S.; Scalera, L.; Gasparetto, A.; Gallina, P. Preloaded structures for space exploration vehicles. In *IFToMM Symposium on Mechanism Design for Robotics*; Springer: Cham, Switzerland, 2018; pp. 129–137.

41. Seriani, S.; Gallina, P.; Scalera, L.; Lughi, V. Development of n-DoF preloaded structures for impact mitigation in cobots. *J. Mech. Robot.* **2018**, *10*, 051009. [CrossRef]

Publisher's Note: MDPI stays neutral with regard to jurisdictional claims in published maps and institutional affiliations.

Article

Design of a Two-DOFs Driving Mechanism for a Motion-Assisted Finger Exoskeleton

Giuseppe Carbone [1,2,*]**, Eike Christian Gerding** [3]**, Burkard Corves** [3]**, Daniele Cafolla** [4]**, Matteo Russo** [5] **and Marco Ceccarelli** [6]

[1] DIMEG, University of Calabria, 87036 Rende, Italy

[2] CESTER, Technical University of Cluj-Napoca, 400114 Cluj-Napoca, Romania

[3] IGMR, RWTH Aachen University, 52062 Aachen, Germany; eike.gerding@rwth-aachen.de (E.C.G.); corves@igmr.rwth-aachen.de (B.C.)

[4] IRCCS Istituto Neurologico Mediterraneo Neuromed, 86077 Pozzilli, Italy; contact@danielecafolla.eu

[5] Faculty of Engineering, University of Nottingham, Nottingham NG7 2RD, UK; matteo.russo@nottingham.ac.uk

[6] LARM2, Laboratory of Robot Mechanics, University of Rome "Tor Vergata", 00133 Rome, Italy; marco.ceccarelli@uniroma2.it

* Correspondence: giuseppe.carbone@unical.it

Received: 12 March 2020; Accepted: 8 April 2020; Published: 10 April 2020

Abstract: This paper presents a novel exoskeleton mechanism for finger motion assistance. The exoskeleton is designed as a serial 2-degrees-of-freedom wearable mechanism that is able to guide human finger motion. The design process starts by analyzing the motion of healthy human fingers by video motion tracking. The experimental data are used to obtain the kinematics of a human finger. Then, a graphic/geometric synthesis procedure is implemented for achieving the dimensional synthesis of the proposed novel 2 degrees of freedom linkage mechanism for the finger exoskeleton. The proposed linkage mechanism can drive the three finger phalanxes by using two independent actuators that are both installed on the back of the hand palm. A prototype is designed based on the proposed design by using additive manufacturing. Results of numerical simulations and experimental tests are reported and discussed to prove the feasibility and the operational effectiveness of the proposed design solution that can assist a wide range of finger motions with proper adaptability to a variety of human fingers.

Keywords: bionic mechanism design; synthesis; exoskeleton; finger motion rehabilitation

---

1. Introduction

Aging of population and stroke incidence are expected to significantly increase in the coming decades and become the second leading cause of disability in Europe as forecast, for example, in [1,2]. Usually, a stroke produces neuro-motory disabilities, including finger impairments. Since the movement of fingers is fundamental in activities of daily life, there is a strong motivation in focusing on finger rehabilitation as a high priority following an injury or a stroke.

Several studies have shown that the rehabilitation after a stroke is faster and more cost effective when using a robotic system as compared to conventional rehabilitation methods, as reported, for example, in [3]. Accordingly, researchers have widely addressed the topic for developing finger exoskeletons and/or similar wearable devices for finger rehabilitation and exercising, as reported, for example, in [4–21].

The "index finger exoskeleton" reported by Agarwal et al. [4] consists of eight linkages that are actuated by two cable drives. The exoskeleton has three DOFs (degrees of freedom) in total, as each linkage between the exoskeleton and finger has one DOF. Each linkage consists of four links and four joints with one DOF each. The exoskeleton allows flexion and extension of all phalanxes of the finger as well as passive abduction and adduction of the first phalanx. This exoskeleton is quickly adjustable and has a low resistance against finger movement when the motors are not activated. The device applies 80 g to the finger, and it has five angle-sensors to monitor link orientations. The index-finger exoskeleton uses a closed-loop torque-control with maximum torque at the first phalanx equal to 250 Nmm. The maximum torque at the second phalanx is 50 Nmm [4].

The exoskeleton reported by Bataller et al. [5] is optimized by an evolutionary synthesis algorithm. The synthesis determines the link length of the mechanism design to fulfill a given coupler curve with high accuracy. The kinematic structure of the device is set before the synthesis. It consists of seven links that are fixed on the back of the hand and on each phalanx, resulting in one DOF in total. A video analysis of the healthy finger motion acquires the coupler curve. The phalanx lengths of a patient's finger are acquired by an X-ray. The proposed exoskeleton is manufactured for each patient individually by 3D-printing and it has one servo motor.

Amadeo is a commercial hand and finger rehabilitation device that has been developed by Tyromotion GmbH [6]. The fingertips can be connected to the caps of the exoskeleton while being adjustable to different finger sizes. Each cap is attached to an automated linear slide. The fingertips can be pushed with a predefined force, and the device can move the finger while they apply no force. In this way, a grasping activity is simulated. The arm and wrist are fixed on the device frame, and the finger caps are connected to a slider. All fingers and the thumb can be treated. The device also allows for several measurements, such as force and range of motion (ROM). In [7], it is reported that 70% of the ROM of a healthy finger is sufficient for rehabilitation motions.

The Script SPO-F exoskeleton [8] is a passive device with no motor but a spring as the actuator. It consists of six links and seven joints with only one DOF. Earlier versions of this exoskeleton have been bulky, complex, and reached a weight of 1.5 kg. The Script SPO-F is actually an exoskeleton for both the wrist and the hand. It is designed for home-usage during rehabilitation treatment. The finger is flexed by a user, and a spring exerts a force on the finger due to the deflection. In contrast to other exoskeletons, it has no predetermined trajectory. The finger can be moved due to a cable connection between the fingertip and the exoskeleton. As the fingertip is fixed, only first and second phalanxes can move. A torque of about 125 Nmm is needed for a 90° flexion of the second phalanx.

The hand exoskeleton version HX [11] is suitable for both the index finger and the thumb. The exoskeleton has five DOFs and is driven by cables. The weight of the index finger module is 118 g, and the total weight lying on the hand is 438 g. The exoskeleton is made of a 3D-printed titanium alloy.

At LARM (Laboratory of Robotics and Mechatronics), a specific research line has been addressing the development of exoskeletons for motion assistance, as reported, for example, in [22–24]. Moreover, [25–27] focus on the fundamentals of the mechanics of grasping as well as the design and validation of anthropomorphic robotic hands. The LARM robotic hands are based on a driving mechanism with linkages that remain within the finger body duringthe finger operation, as reported in [28–30]. The design of such a driving mechanism is the conceptual reference for the exoskeleton solution that is reported in [24] and in preliminary exoskeleton designs, as reported in [31–34].

The main problem with existing exoskeletons is that they are often not wearable by different patients, as in 4,5], are bulky, and the overall equipment is not easily transportable, such as in [6] or is heavy, such as in [11]. Commercial robots, as in [6], are considered too expensive for home rehabilitation use. Further, the Amadeo device is not able to fulfill a complete grasping movement. The solution in [8] has no defined trajectory to move all joints of the finger in a defined way. Accordingly, the authors believe there is still a need for a design procedure that can lead to novel design solutions as based on kinematic analysis and a proper mechanism synthesis referring to the specific task of finger motion assistance.

This paper aims at a systematic design approach towards a novel two-DOFs driving linkage mechanism for a motion assistance finger exoskeleton by presenting a novel design solution for a finger exoskeleton with adaptability to the finger size, as well as cost-oriented design and user-friendly features. The design process is carried out within a specific design procedure. As a first step, the movement of a human finger was characterized by video motion tracking to identify the desired reference finger motions. Then, the relative kinematics of a human finger were obtained based on the acquired data. As a next step, a type synthesis was carried out to identify a mechanism consisting of linkages with two active DOFs as the most convenient solution to assist the motion of a finger along the desired trajectory, as also preliminarily discussed in [33]. This paper also provides FEM analyses that are integrated in the proposed design approach. A graphic/geometric synthesis procedure has been implemented for achieving the dimensional synthesis of the proposed linkage mechanism. Numerical simulations and experimental tests have been carried out and discussed to prove the feasibility and effectiveness of the proposed design solution. The main contribution of this work can be recognized in the design of a proposed novel linkage mechanism that, unlike other existing designs, can drive the three finger phalanxes by using two independent active DOFs that are both driven by rotary servomotors placed on the back of the palm. This configuration allows for a wide range of motion assistance with proper adaptability to a variety of human fingers. The paper content can be outlined as follows: the first section addresses the design requirements for achieving a device for finger exercising/rehabilitation of multiple users; next, the paper deals with the kinematic design of the proposed new device based on a two-DOFs driving linkage mechanism, whose synthesis is described in Section 4; the following section focuses on the mechanical design and construction of a prototype; Section 6 describes an experimental validation with comparisons of numerical and experimental results to assess the feasibility and performance of the proposed device.

2. Design Requirements for a Finger Exoskeleton

To expand the range of suitable patients, a novel proposed exoskeleton should be easy to attach to a finger, adaptable to a wide range of users, and easily portable for home use.

Exoskeletons driven by cables are a common solution in the literature. However, they show a range of drawbacks such as high losses due to friction as well as a high risk of cable failures. Further, cables need to be kept under tension during motion. The cable management system has a negative effect on the portability, as it is often bulky as in [12]. Because of that, servo motors with linkage transmissions are preferred in this work as they can be robust, lightweight, compact, and easy to control. The linkage parts can be easy and cheap to manufacture even with commercial 3D printers. However, the design of such a linkage mechanism requires full understanding of the desired human finger motion assistance.

A human hand is composed of fingers, metacarpus, and carpus. The fingers consist of three phalanxes, except for the thumb, which consists of two phalanxes. The metacarpus is connected to the proximal phalanx. On the fingers, the second link is the medial phalanx. The third link is the distal phalanx. A detailed description of the joints and functionalities of a human hand can be found, for example, in [35,36]. The joints between the intermediate and distal phalanxes are called distal interphalangeal joints (DIP), and the joints between the proximal and intermediate phalanxes are called proximal interphalangeal joints (PIP). The metacarpophalangeal joints (MCP) connect the proximal and metacarpal phalanxes, and the carpometacarpal joints (CMC) connect metacarpal phalanxes and the carpal bones, as shown in the scheme of Figure 1, [36]. In general, flexion reduces the angle between bones or parts of the body, whereas extension increases the angle between the bones of a limb. Abduction is an outward movement of a limb, and adduction is an inward movement of a limb [9]. The MCP joints have two DOFs and they allow flexion and extension as well as abduction and adduction of a finger. The interphalangeal joints PIP and DIP have one DOF each. However, the axis of rotation of these joints is not constant during flexion and extension [10], and the ligaments restrict the movements of the joints. Moreover, DIP and PIP joints cannot be moved independently of each other [35].

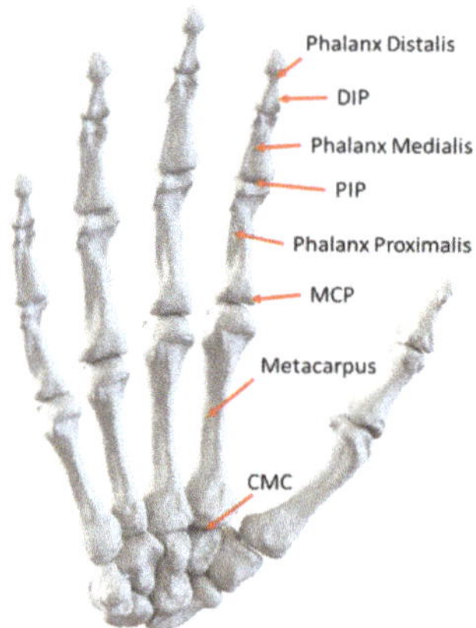

Figure 1. A scheme of bones and joints in a human hand.

A finger exoskeleton can be conveniently designed for motion exercising of the muscles after an injury or a stroke. For this application, the motion of a finger will be assisted by the exoskeleton with a motion trajectory similar to a healthy finger of the same size as proposed in Figure 2. The motion trajectory does not require high accuracy. For example, in [11], a misalignment of a phalanx up to ±8° is deemed acceptable. However, unnatural movements of the finger must be avoided by the mechanism to prevent potential injuries. As an additional design requirement for a finger exoskeleton, its installation space needs to be small enough to be attached onto the finger and not interfere with its motion. For the same reason, the total weight of the system should not exceed 500 g as reported, for example, in [7,13,17]. This value should be considered an upper bound, since a heavy exoskeleton can limit the effectiveness of motion exercising. Indeed, a high weight makes a patient easily tired and not willing to continue the treatment. Thus, it is advisable to limit the number of motors as they are the main source of weight. Motors need to provide a minimum torque of around 200 Nmm as reported previously [3,7].

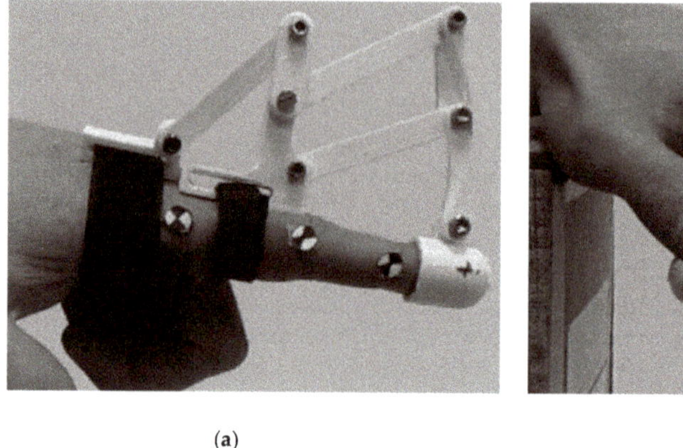
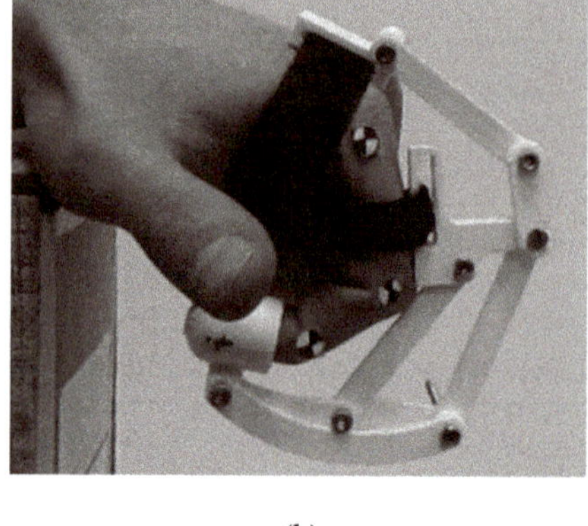

(a) (b)

Figure 2. Desired motion assistance for a finger that moves from fully open (**a**) to fully closed (**b**).

One of the main peculiarities of the proposed novel linkage mechanism design solution is that it can move all three phalanxes of a human finger with two active DOFs. This is mainly achieved by coupling the motion of the last two phalanxes. Accordingly, only two DOFs are needed to replicate whole finger motions. The proposed mechanism is optimized to perform a specific desired motion but it is actually able to perform a wide range of other motion combinations, given its two active degrees of freedom. The proposed design is limited to flexion and extension movements of a finger. However, flexion and extension are identified in literature as the most important for activities of daily life and the first recovery priority in case of finger injury, as also mentioned in [35,36]. Therefore, a motion assistance mechanism at first instance does not need to focus on abduction and adduction movements, and these movements can be safely neglected to achieve a device with a light and simple design.

3. Kinematic Design

The index finger of a healthy subject has been used as a reference to develop and validate the proposed novel exoskeleton. First, grasping motions have been analyzed by using the Computer Vision Toolbox of MATLAB (Mathworks, Natick, MA, USA). Markers have been attached to the joints of the index finger and its fingertip (FT). Distances between the markers have been also measured with a digital caliper for calibration purposes. The camera has been aligned perpendicular to the side of the finger under study, and a fixing frame has been used both for the hand and camera in order to achieve a planar motion of the finger and a fully orthogonal video recording. The acquired videos have been post-processed to collect positions of the joints versus time. Figure 3 shows the experimental setup for the finger motion tracking as well as the definition of the angles and reference frame with the x-axis horizontal oriented from left to right, and the y-axis vertical oriented from top to bottom. The z-axis is defined according to the right-hand rule. The origin of the reference frame is attached to the center of the MCP joint. The angle of the MCP joint is called φ, the angle of the PIP joint is called ε, and the angle of the DIP joint is called τ. The angles have a clockwise positive direction.

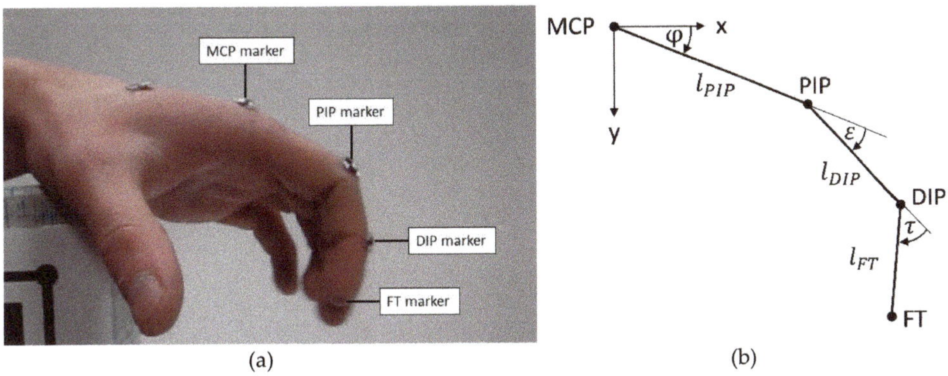

(a) (b)

Figure 3. Experimental setup for finger motion video tracking (**a**) and definition of angles of the finger joints (**b**).

The lengths of the phalanxes of a test subject were identified from set-up measurements as 25 mm for the distal phalanx, 28 mm for the medial phalanx, and 43 mm for the proximal phalanx. Twenty-two frames have been evaluated from the motion tests for a suitable finger motion characterization. Figure 2 shows a detail of a video-captured motion for a healthy human finger movement referring to a typical human finger motion. In particular, the plot in Figure 4 shows trajectories of PIP, DIP, and FT markers during the motion of an index finger that moves from fully open to fully close. These trajectories will be used as the reference motion trajectories for the linkage mechanism of the proposed novel exoskeleton. However, the desired exercising motion usually does not require the full joint feasible rotation range of the motion range of a healthy human finger but only a subset of such a motion.

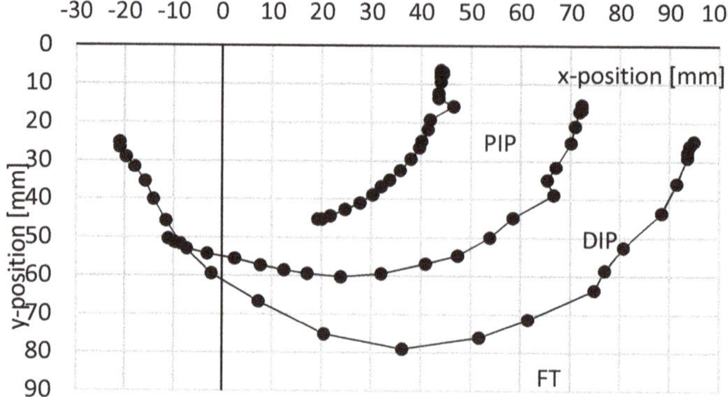

Figure 4. Acquired trajectories of PIP, DIP, and FT markers (as in the set-up of Figure 1 for the finger movement of a healthy human (dots indicate the measured data from the video capture tracking).

Moreover, it is advisable to avoid full finger closing as this would result in an undesired interference with the palm. A specific linkage mechanism has been identified as convenient for mimicking the desired reference finger motion. This proposed mechanism consists of eight links that are arranged as two interconnected four-bar mechanisms. This specific linkage arrangement has been proposed as an Italian patent [34]. The proposed mechanism requires two motors that each drive one of the two four-bar mechanisms. A detailed kinematic scheme is shown in Figure 5. The motors are located at joints D0 and B0. The angle of driving link D0-D is called δ. The angle of the driving link B0-B is called β. Both angles are positive in a clockwise direction. The first four-bar linkage (D0-D-E-MCP) has four links and four revolute joints with one active DOF. The second four-bar linkage (B0-B-A-A0-C) has one active DOF, four links, and four revolute joints.

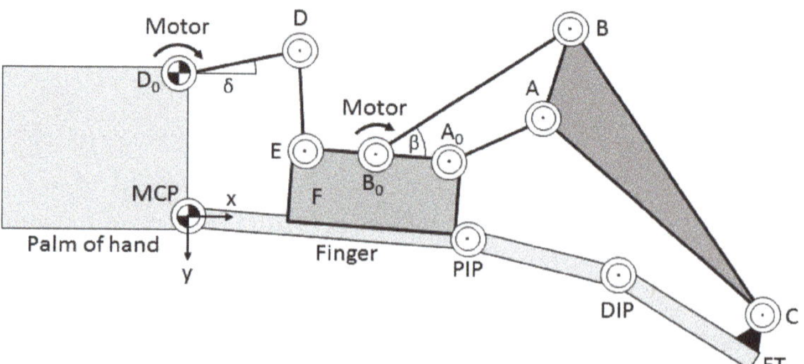

Figure 5. A kinematic scheme of the proposed design with its linkage structure.

The second four-bar linkage drives the point C that is attached to the fingertip (FT). The exoskeleton is also attached to the palm at the fixed frame link D0-MCP and to the first phalanx with the link E-B0-A D, as shown in Figure 5.

For the first linkage, the link lengths need to be chosen by considering the dimensional constraints of the motor and the palm of a hand. Accordingly, the link length D0-MCP is close to the vertical dimension of the selected motor. The link length E-MCP is mostly determined by the geometrical constraints for the attachment of E-B0-A to the first phalanx. The link lengths D0-D and D-E need to be calculated to fulfil the desired motion trajectory of the PIP joint, according to the motion reference in Figure 4. Considering the finger in fully close configuration, one can identify that in this configuration, the longest distance is expected to be D0-E. This distance can be calculated as 63.3 mm from the reference motion data. For achieving this value, the link length $MCP - D_0$ has to be equal to 27.8 mm, the link length D_0-D has to be equal to 32.0 mm, and the link length D-E has to be equal to 58.1 mm. Joints E and A0 on body F have coaxial axes.

To size the link lengths of the second linkage mechanism, a graphic/geometric synthesis dimensional synthesis has been carried out to obtain the desired motion of point C. The procedure is reported as a general formulation, for example, in [35]. The dimensional synthesis starts by calculating the desired relative motion of point C relative to A0 based on the experimental data in Figures 3 and 4 and on the relative position of the joints in the proposed linkage mechanism shown in the scheme of Figure 6. In this scheme, γ is the angle between the horizontal axis and line through A0, α is the angle between the horizontal axis and line through C, and η is the difference between the angles. The relative coordinate system is located in A0 and has its p-axis along the line $MCP - A_0$, as shown in Figure 6.

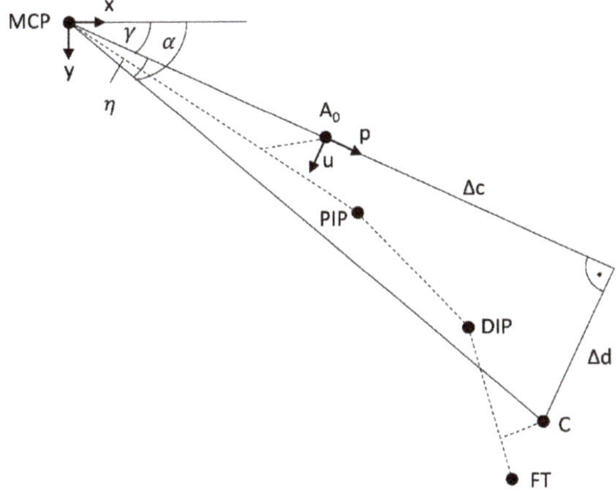

Figure 6. A scheme for determining C coordinates in the second linkage mechanism by considering the known PIP (proximal interphalangeal joint), DIP (distal interphalangeal joint), and FT (fingertip) coordinates.

The u-axis is perpendicular to the p-axis with direction from top to bottom. Δc is the distance between A0 and C along the p-axis, and Δd is the distance between A0 and C along the u-axis.

The coordinates of point PIP are known from the measured trajectory of the finger movement in Figure 4. The displacement of the PIP point and A0 is shown in Figure 7 where φ is the angle between the horizontal axis and line through PIP, while ψ is the angle of MCP from A0 to PIP. The distance between MCP and A0 is l_{A0}, and the distance between MCP and PIP is l_{PIP}. Parameters a and b represent the displacement between A0 and PIP. The coordinates of point A0 can be calculated by using the parameters a and b and the angle γ. Since the angle φ and the length l_{PIP} are known from the calculations of the reference trajectory in Figure 4, ψ can be calculated as

$$\psi = \arctan\left(\frac{b}{l_{PIP} - a}\right) \tag{1}$$

Therefore, referring to Figure 7, the angle γ is given by

$$\gamma = \varphi - \psi \tag{2}$$

The coordinates of A0 can be calculated as

$$x_{A0} = \sqrt{(l_{PIP} - a)^2 + b^2} \cos\gamma \tag{3}$$

$$y_{A0} = \sqrt{(l_{PIP} - a)^2 + b^2} \sin\gamma \tag{4}$$

By squaring and summing Equations (3) and (4), it is possible to obtain

$$l_{A0} = \sqrt{\left(l_{PIP} - a\right)^2 + b^2} \tag{5}$$

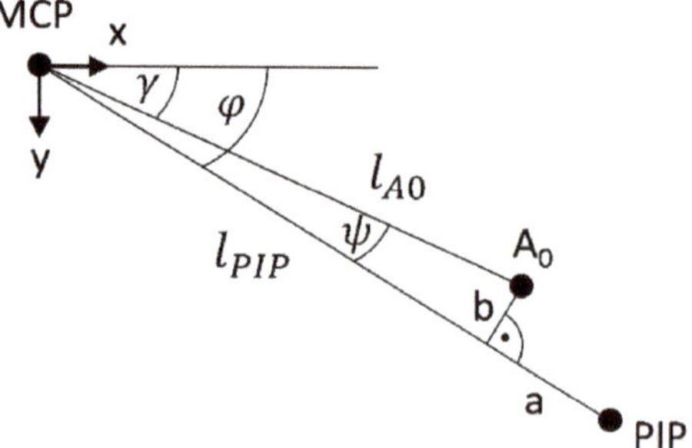

Figure 7. A scheme showing the displacement between A0 and PIP (proximal interphalangeal joint).

A similar procedure is used to calculate the global coordinates of C. Figure 6 shows a sketch for the displacement between FT and C, where λ is the angle between the horizontal axis and the line from DIP to C; θ is the angle between the horizontal axis and the line from DIP to FT. The distance between DIP and FT is l_{FT}, and the distance between DIP and C is l_C. Parameter e is the displacement between FT and C on the line from FT to DIP, and f is the displacement between C and FT perpendicular to that line.

The global x and y coordinates of DIP and FT are known from the captured trajectory in Figure 2. As given in Figure 8, the link orientation, given by angle θ, can be calculated as

$$\theta = \arctan\left(\frac{y_{FT} - y_{DIP}}{x_{FT} - x_{DIP}}\right) \tag{6}$$

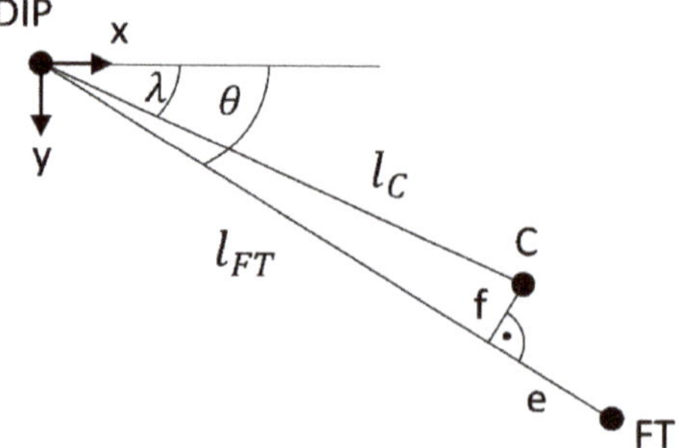

Figure 8. A scheme showing the displacement between FT (fingertip) and C.

Therefore, the angle λ is calculated as

$$\lambda = \theta - \arctan\left(\frac{e}{DIP - FT - f}\right) \tag{7}$$

The global coordinates of C can be calculated as

$$x_C = \sqrt{\left(DIP - FT - f\right)^2 + e^2} \cos\lambda + x_{DIP} \tag{8}$$

$$y_C = \sqrt{\left(\overline{\text{DIP}} - \overline{\text{FT}} - f\right)^2 + e^2} \sin\lambda + y_{\text{DIP}} \tag{9}$$

Consequently, η is calculated as

$$\eta = \alpha - \gamma = \arctan\left(\frac{x_C}{y_C}\right) - \gamma \tag{10}$$

Finally, distances Δc and Δd can be computed as

$$\Delta c = \sqrt{x_c^2 + y_c^2} \cos\eta - l_{A0} \tag{11}$$

$$\Delta d = \sqrt{x_c^2 + y_c^2} \sin\eta \tag{12}$$

The resulting parameters from the above-mentioned calculations for the offset of A0 and C are summarized in Table 1. The parameters are obtained from measurements on the finger of the subject and also referring to an early prototype. The diameter of a reference finger has been measured to calculate the distance from the finger joint to the top of the finger, giving parameters b and f. Parameters a and e were identified from a preliminary prototype, which showed that A0 could be placed directly above the PIP joint, whereas C requires some distance from FT to avoid slipping off the finger during motion. With the calculated positions of joint C, the dimensional synthesis of the second linkage mechanism can be conducted.

Table 1. Design parameters from kinematic calculations in Figures 5 and 6.

Parameter	a	b	e	f
Length [mm]	4	17	15	19

No singular configuration is reachable within the used workspace of this mechanism. This has been verified at the design stage by setting up limits at the feasible transmission angles, and it has also been verified experimentally.

4. Mechanism Synthesis of the Second Linkage

The dimensional synthesis of the second linkage can be outlined as the procedure of determining the remaining lengths of a 4-bar mechanism that guides a point on the coupler curve. One possibility for such a synthesis is the graphical method based on the Burmester theory as described in [37]. This synthesis approach has been selected since it can quickly address the desired features in terms of replicating a desired motion trajectory as given by the reference motion trajectory in Figure 4. The expected accuracy for joints (± 4 degrees) does not justify the use of more complex and time-consuming synthesis methods such as numerical optimization.

The proposed procedure allows an approximation of the desired movement from the measured trajectory in Figure 4 by defining three points on the coupler curve that are reached precisely by the mechanism. The synthesis can be conducted when two lengths of a mechanism and two positions of joints are known. Figure 9a shows the initial problem with all given parameters. The position of joints A0 and B0, lengths $\overline{A_0B_0}$, $\overline{B_0B}$, $\overline{BC}$, and three positions of joint C are given in the initial problem. From the input, the position of the missing joint can be determined. When the positions of all joints are established, the lengths can be determined. Curve k in Figure 9a shows the original movement of joint C, and curve kc in Figure 9b is the movement after the synthesis. With the synthesis in Figure 9b, joint A can be determined, giving the missing lengths $\overline{A_0A}$, $\overline{AC}$, and $\overline{AB}$.

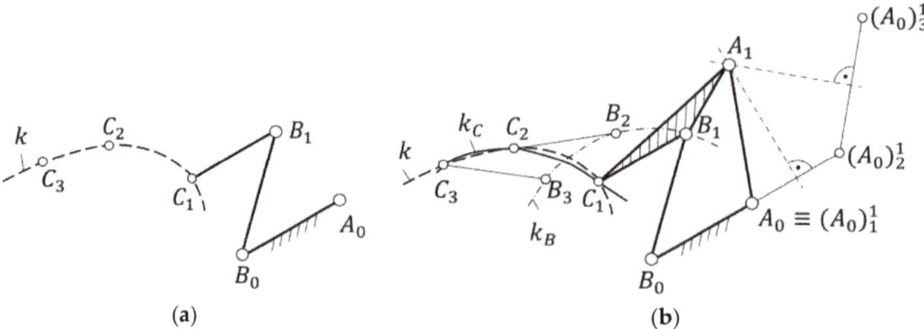

Figure 9. Schemes for a dimensional synthesis: (**a**) the problem to be solved; (**b**) graphical solution as based on the approach that is described in [33].

A pose of a mechanism consists of the positions and the orientations of all links. For example, for pose 1 of the mechanism in Figure 4, joint A is in position A1, joint B is in position B1, and joint C is in Position C1. The graphical procedure for the dimensional synthesis is given in the flow-chart of Figure 10 according to the following steps:

1. The given parameters in Figure 9a are the positions of C1, C2, C3, and B0 as well as the lengths of $\overline{B_0B}$ and $\overline{BC}$;
2. With the information from step 1, joint B is known in pose 1, 2, and 3, as joint B moves on the curve kB. Therefore, the poses of $\overline{B_1C_1}$, $\overline{B_2C_2}$, and $\overline{B_3C_3}$ are known. At this point, joint A is missing to complete the mechanism sizing:
3. Next, one link in one pose is chosen as a reference (in this example, link BC in pose 1);
4. Now, joint A_0 is virtually moved around $\overline{B_1C_1}$ to create A0 that corresponds to position 2 in pose 1 (by moving A0 with respect to the coordinate system of pose 2 into pose 1). The transformation can be done by creating a triangle between C2, A0, and B2 and moving the triangle into the pose of C1 and B1. The resulting position is A_0 from pose 2 in pose 1, written as $(A_0)_2^1$;
5. The same as step 4 is done with A_0 from pose 3 in pose 1, resulting in $(A_0)_3^1$;
6. Finally, A_1 is found by the intersection of the perpendicular bisectors between three positions of A_0. The resulting mechanism guides joint C along curve kC. Curve kC matches curve k in the points C1, C2, and C3 [33];
7. With all known positions of the joints, the link lengths can be determined.
8. Other link lengths/positions can be chosen to improve the matching of curves k and kc.
9. The dimensional synthesis is applied to the second linkage mechanism by using the relative movement of C with respect to A0. The aim is to design a mechanism that matches the desired motion trajectory of point C.

The positions of point C with respect to the u-p coordinate system are calculated by Equations (11) and (12). For the dimensional synthesis, the positions of A0, B0 and the lengths $\overline{A_0A}$ and $\overline{AC}$ are required as well as three reference positions of C within its desired motion trajectory, according to step 1 in Figure 10. A0 and B0 are set as frame joints. A0 has the coordinates (0; 0) with respect to the u-p-coordinate system. B0 is designed with the coordinates (−8; −17.8). The input parameters have been iterated to find a mechanism that fulfills the desired path of C well. After few iterations, sufficient input parameters were found. Positions 1 to 15 of the calculation of point C have been considered for the synthesis, as it gets more difficult to approximate a longer motion path. Furthermore, the identified range of motion is sufficient for rehabilitation purposes.

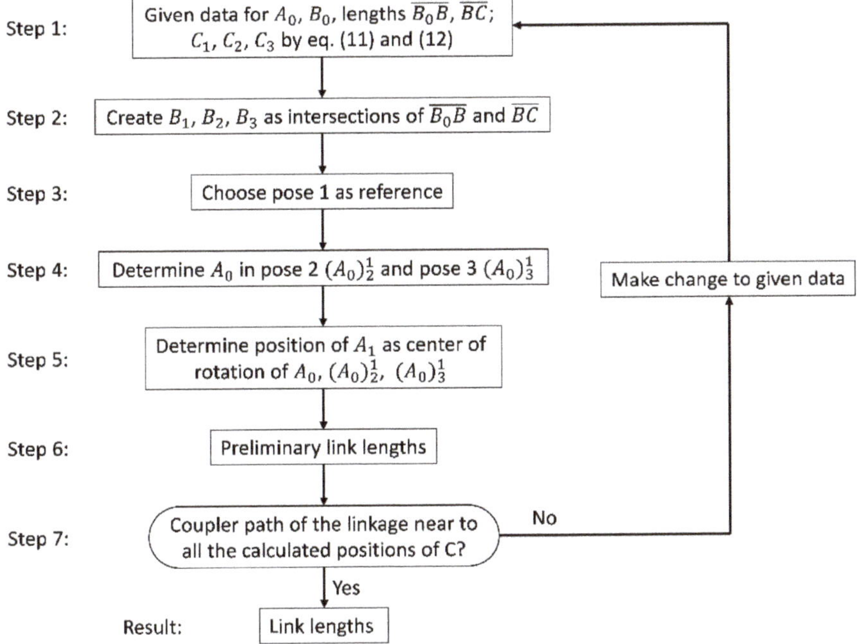

Figure 10. A flowchart for the mechanism synthesis according to the scheme in Figure 9.

The initially chosen link lengths and point coordinates are shown in Figure 11, similar to Figure 4. With this information, positions B2 and B3 can be determined as intersecting points with the known segments of $\overline{B_0B}$ and $\overline{BC}$, as mentioned in step 2. As a result of iterations of the synthesis, $\overline{B_0B}$ has been measured to have a length of 46 mm, and $\overline{BC}$. has been measured to have a length of 53 mm. As input from the positions of C, the first position is chosen as C1, position C2 is in the middle, and position C3 refers to the last used configuration during the finger closing motion. According to step 3, pose 1 is the reference for the synthesis. A0 with respect to position $\overline{B_3C_3}$ is transferred into position $\overline{B_1C_1}$, resulting in $(A_0)_3^1$. $(A_0)_3^1$ is A0 in pose 3 transferred to pose 1. This means that the triangles C3-B3-A0 and C1-B1-$(A_0)_3^1$ are identical. In the same manner, the point $(A_0)_2^1$ can be identified. This corresponds to step 4. The triangles to find the position $(A_0)_3^1$ are given in Figure 12a. The synthesized mechanism and the calculated positions of C are shown in Figure 12b. The previously calculated coordinates for C are marked as crosses, and the computed trajectory of the mechanism is marked as dots. The figure also shows that the trajectory matches well with the desired motion for point C.

The position of A1 is the center point of a circle through the positions A0, $(A_0)_2^1$ and $(A_0)_3^1$, as mentioned in step 5. The resulting position is A1, as pose 1 has been used as a reference. The link lengths can be determined with the obtained position of A1. Since the given data in step 1 have already been identified by a prior iteration, step 7 can be skipped. The complete kinematic design is obtained by combining both linkages. The resulting kinematic design of the whole exoskeleton mechanism is shown in Figure 13. The numerical values are summarized in Table 2, defined according to the scheme in Figure 5. The calculated positions of A0 and C are indicated with plus (+) and cross (x) marks, respectively.

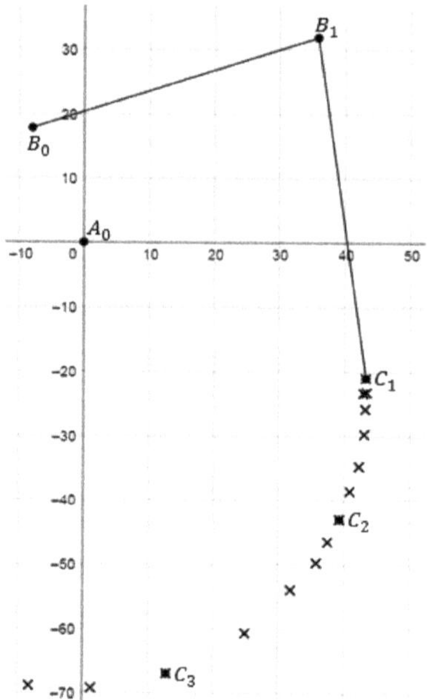

Figure 11. Initial problem of the synthesis for the second linkage with the selected three points C1, C2, C3 along the desired motion trajectory of point C.

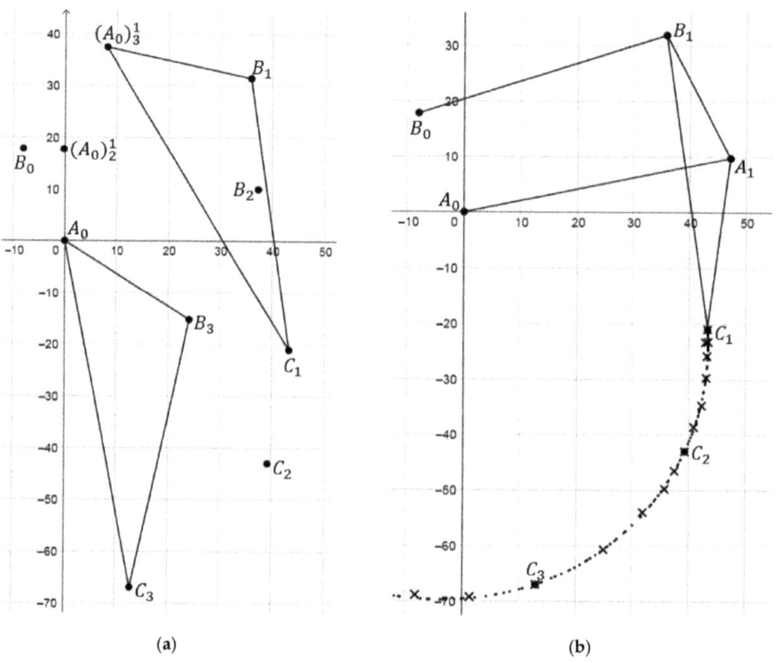

(a)

(b)

Figure 12. A scheme for determining the second linkage: (**a**) Triangles of the relative positions C0; (**b**) synthesized mechanism and resulting trajectory passing through the selected points C1, C2, C3.

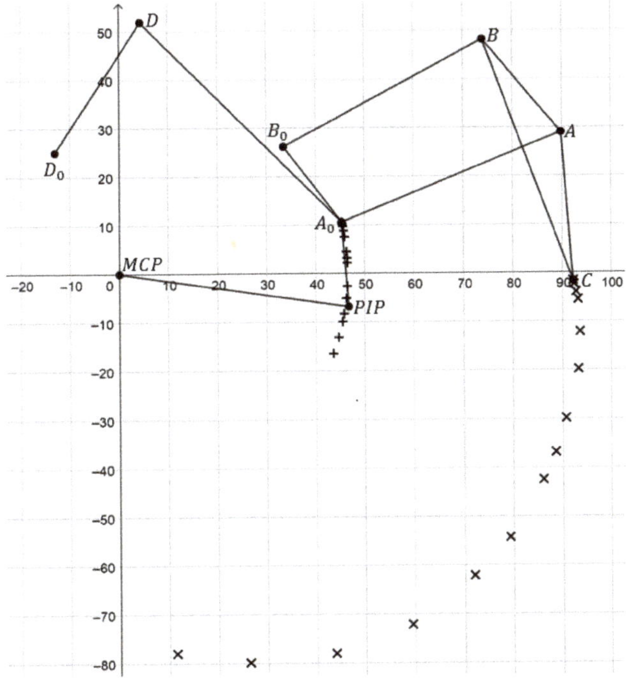

Figure 13. The synthesized kinematic design of the finger exoskeleton in Figure 3 with the trajectories of A0 and C indicated with plus (+) and cross (x) marks.

Table 2. Design parameters and link lengths of the finger exoskeleton for the scheme in Figure 5.

Parameter	Length [mm]	Parameter	Length [mm]	Parameter	Length [mm]
a	4.0	A0-B0	19.5	B-C	53.0
b	17.0	B0-B	46.0	D0-D	32.0
e	15.0	A-B	24.9	D-E	58.1
f	19.0	A-C	30.7	A0-A	48.2
MCP-PIP	43.0	PIP-DIP	28.0	DIP-FT	25.0

This paper reports the proposed graphical procedure for a specific case with nominal biometric measurements. However, the proposed graphical procedure is general in its approach. Accordingly, the same procedure can be performed again for different finger sizes when the phalanx lengths are expected to exceed the adaptability allowed by the proposed design. A chart can be generated with link dimensions for different patient biometrics to adapt the proposed finger exoskeleton to the wearer. Moreover, the proposed synthesis procedure can be also automated by implementing it in a numerical solving algorithm.

5. Mechanical Design and Prototype

Based on the obtained kinematic design, a prototype of the proposed finger exoskeleton has been developed. Since the finger exoskeleton is manufactured by 3D printing, it has been necessary to define the secondary geometric parameters of each linkage, such as link thickness. For this purpose, given the slow speeds, accelerations, and inertias of the application, a specific static analysis has been carried out according to the schemes that are shown in Figure 14. The computation has been carried out by considering as load the maximum motor torque of 216 Nmm, and a maximum force at the connections between finger and exoskeleton equal to about 6 N on the fingertip. This value is calculated by using the principle of virtual powers from the given input torque and also matches previous experiences of similar prototypes.

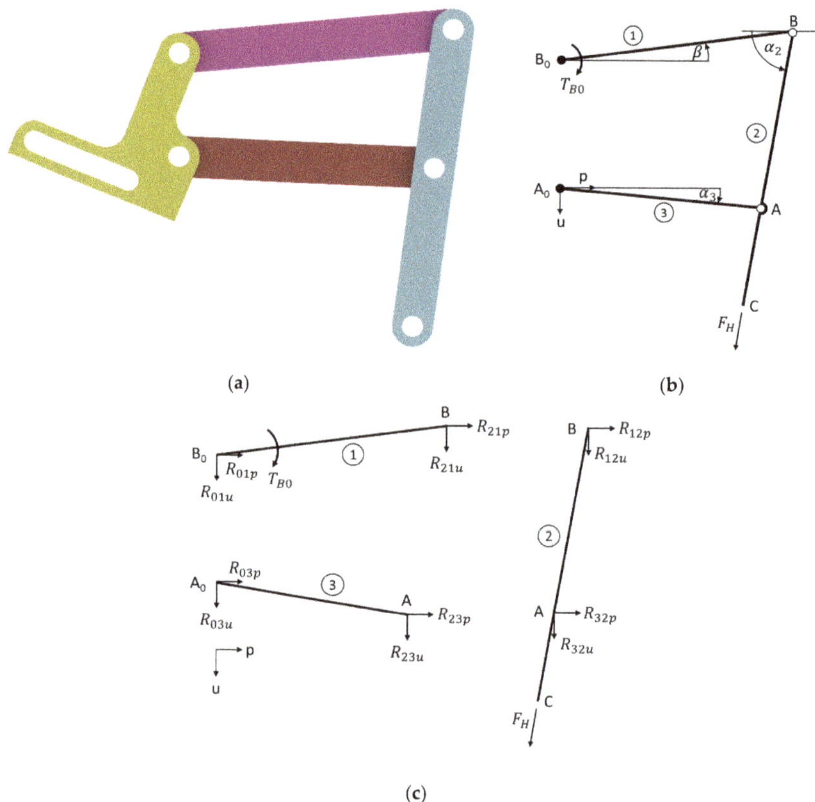

Figure 14. Schemes for force distribution analysis: (**a**) a CAD front view of the second linkage (A0AB0B); (**b**) an overall free body diagram; (**c**) free body diagrams of single elements.

FEM analyses have been carried out iteratively to find a proper linkage cross section and thickness. In particular, final FEM simulations have been carried out by considering the 3D CAD model that is shown in Figure 15. The main link thickness has been set as equal to 2 mm, also based on previous experiences. Similarly, the holes for the joints have been set at a diameter of 4 mm. Therefore, the links need to have a total width of 8 mm and a thickness of 2 mm. Link $\overline{AC}$ is crooked as per Figure 15 in order to avoid collisions with the finger, and link $\overline{BC}$ is crooked to allow for fixation of the joint A. Even though $\overline{ABC}$ behaves kinematically as a single body, it is realized with three different links that can be easily changed to fit the different finger sizes of different users. The link $\overline{BC}$ is manufactured with two beams to increase its stiffness. A CAD design has been elaborated with the above-mentioned design considerations. A functional solution of the mechanism with all its joints is shown in Figure 14. Screws and nuts of M3 size connect the links.

The main merit of the proposed design can be identified in the adaptability to multiple users. This is achieved by slotted holes (Figure 15) in the exoskeleton that allow easy adjustability to users. The slotted holes are used to adapt the finger exoskeleton to the user's biometrics (phalanx lengths) by moving each link to the optimal configuration evaluated through the proposed dimensional synthesis. Cable straps and loop fasteners are used to fix the exoskeleton on the finger, giving some additional adaptability. Accordingly, the proposed exoskeleton is expected to fit users having specific phalanx sizes exceeding ± 10% of the nominal sizes. A design limitation can be identified in the need to replace the links of the device when users are expected to exceed ± 10% of the nominal finger size. This limitation can be partially overcome by preparing sets of replacement links whose sizes are designed to fit with different nominal finger sizes (e.g., for children, male/female adults).

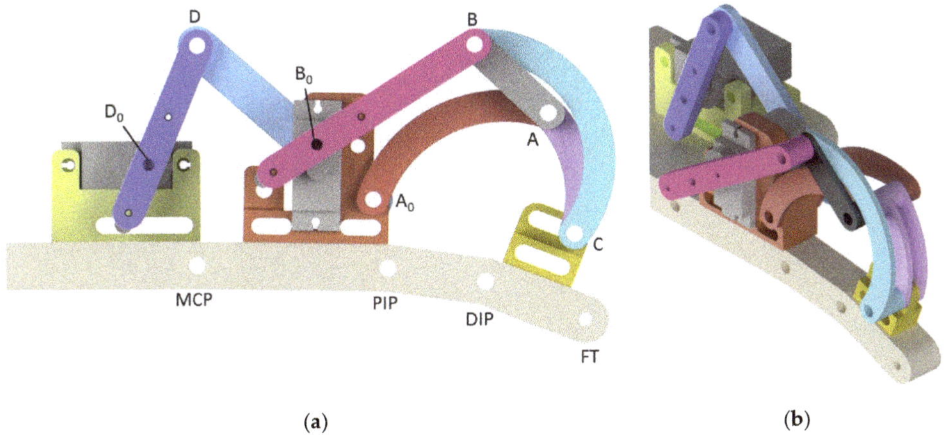

Figure 15. A CAD model of the finger exoskeleton including motors with description of joints (**a**) and angular view (**b**).

Final FEM tests have been performed in SolidWorks 2019 to verify the correctness of the selected cross-sections and minimum required thickness to avoid any failure or plastic deformation, as reported in Figures 16 and 17. A minimum safety factor equal to 1 has been considered in static nodal stress analysis for keeping the overall weight as low as possible. A minimum factor of safety equal 2.8 has been found on shear stress. The chosen material is a commercial poly-lactic acid (PLA) filament that is suitable for additive manufacturing with commercial 3D printers. Its main properties are tensile strength equal to $3 \cdot 10^7$ N/m², elastic modulus equal to $2 \cdot 10^9$ N/m²; Poisson's ration equal to 0.394, mass density equal to 1020 Kg/m³, and shear modulus equal to $3.189 \cdot 10^8$ N/m².

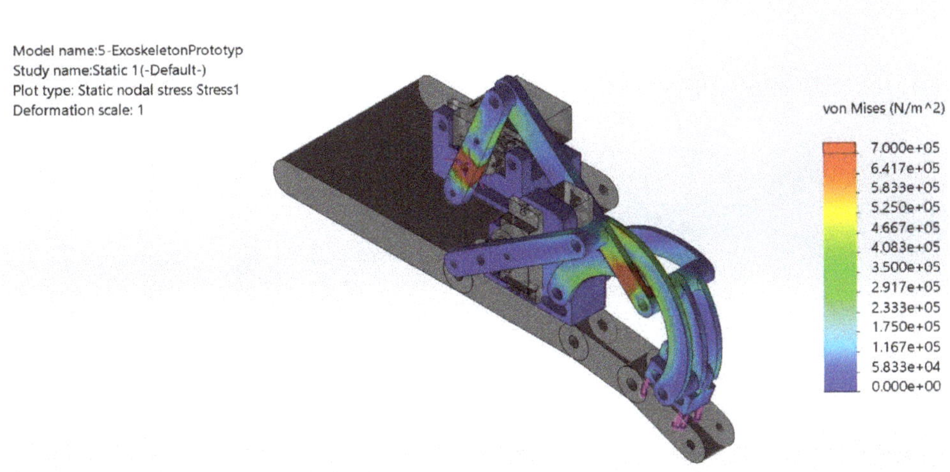

Figure 16. Static nodal FEM analysis of the whole prototype.

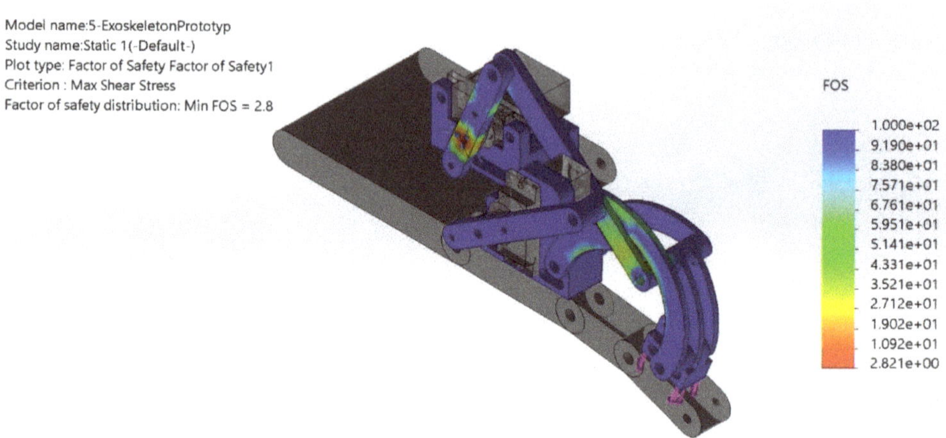

Figure 17. FEM factor of safety calculations based on maximum shear stress.

The size of motors has been chosen to match the results of simulations for the designed mechanism as well as by comparison with similar devices in the literature. Both servo motors have been selected with a nominal torque of 216 Nmm while the desired torque was about 200 Nmm for the first joint and about 150 Nmm for the second joint. Servo motors with a torque of 216 Nmm are integrated into body F and lay on the back of the palm of the human hand. An Arduino microcontroller has been chosen to drive the motors.

The total cost of the system is around 50€, including the servomotors and the microcontroller. The motors are connected to an external power supply, which can be a LiPo battery for easy portability. The exoskeleton has a total weight of 64 g for the parts that are mounted on the finger. This includes the linkage, cable straps, hook and loop fasteners, and servo motors. The whole system has a weight of 175 g, including an Arduino board and all cablings. A full prototype has been built as shown in Figure 18.

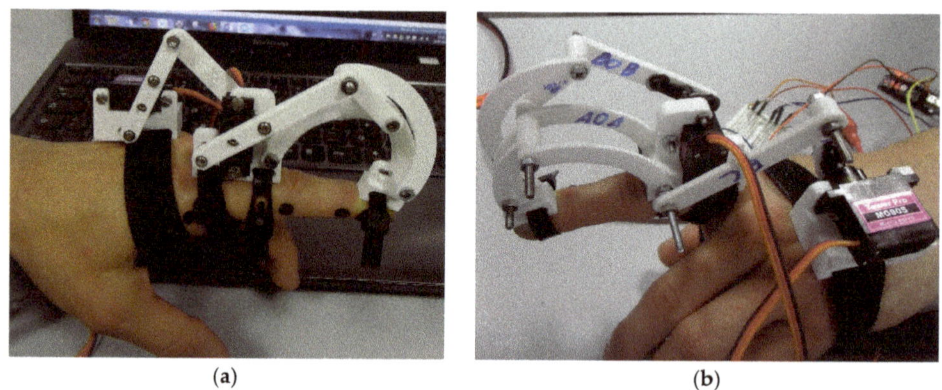

(a)

(b)

Figure 18. The built prototype of the finger exoskeleton, front view (**a**), back view (**b**).

6. Test Results

The built finger exoskeleton prototype has been tested experimentally to prove its feasibility as a finger motion exercising device in terms of it kinematic and operation behaviors. Given the expected slow speed operation, dynamic simulations and tests are not required at this proof-of-concept stage. The finger exoskeleton can be easily worn with Velcro fasteners. The connection between the finger and exoskeleton can be as tight as the subject wishes. Even after long use, it is still comfortable to wear. Also, the calibration procedure is very straightforward. It consists of the following steps: Attaching the exoskeleton to the finger, manually placing the finger in its desired straight configuration; registering this position as the initial configuration; manually placing the finger in its desired fully closed configuration; registering this position as the fully closed configuration. After the above steps, the device is ready to operate within the desired operation range.

The exoskeleton movement has been compared with the reference motion of a healthy human finger, and the angular error has been calculated as also partially reported in [33]. The driving angle of joint D0 is called δ, and the driving angle of joint B0 is called β. The angles of the joints of the finger from the grasping test have been used as an input for a multibody motion simulation of the CAD model on SolidWorks 2019. The simulation acquires the angles of the motor for each position during the finger motion. A comparison of the simulation and the motion of the prototype was used to determine if the exoskeleton prototype moves as planned.

The desired path planning and angular joint coordinates are managed by using an Arduino control board, which is connected to the two servomotors that drive the exoskeleton. The angles of the driving links have been calculated and interpolated for each position that has been experimentally measured during a grasping test, as reported in Figure 4. Then, the obtained motor angles are sent to an Arduino controller, which drives the exoskeleton. The desired motion has been obtained by an offline video post-processing that is carried out with MATLAB, allowing us to measure the angular joint positions in each acquired video frame. This information is converted into a desired set of joint angles versus time. The controller is programmed to move both motors in each desired position with an interpolated step motion. Accordingly, the proposed motion planning consists of passing through a prescribed number of path points with a smooth interpolated motion to reach precision positions. A pause of half a second is set before proceeding with the next step desired precision position to keep the motion safe for the user. The user has the time to easily stop if he/she feels discomfort in any reached configuration. The maximum angular reaches of motors are limited via software limits that are well within the range of the servo motors being equal to 180°. After completing the motion planning, videos of human finger exoskeleton-assisted motions were collected and analyzed. Tests were performed on the same subject as for the initial grasping test in Figures 1 and 3.

Video captures were acquired by tracking both finger joints and exoskeleton joints during a grasping motion. Then, the driving angles and the angles of the finger were determined. The resulting movement of the prototype and the finger is shown in Figure 19. The movement takes approximately 16 s. The proposed tests are performed with a human sitting and his/her forearm fixed on a reference table. The first snapshot Figure 19a refers to the beginning of the movement. Snapshot Figure 19b is taken after 3.2 s, snapshot Figure 19c after 6.5 s, snapshot Figure 19e after 9.7 s, and snapshot Figure 19e after 13.0 s. Snapshot Figure 19f shows the final position of the motion. In this test, the MCP joint moves within a range of 3.1° to 37.6° for a finger flexion. Similarly, the PIP joint goes from 13.0° to 78.7°, and the range of the DIP joint is from 0° to 58.9° for a finger flexion.

The time history of the measured finger joint angles is given in Figure 20. for a finger flexion assisted by the exoskeleton. Namely, Figure 20a–c show the acquired values of angles φ, ε, and τ versus time, respectively. Moreover, (x) markers are reported in Figure 20 to show the simulated angular motions angles of φ, ε, and τ versus time. In the plots of Figure 20, an initial offset can be observed between the measured and simulated values. The maximum velocity and swing frequency of the finger exoskeleton mechanism are defined by means of a reference finger-assisted motion. This motion needs to be very slow to allow safe finger motion assistance, so a fast speed is not desirable. In particular, for the prototype reported in the paper, the average velocity is 5 deg/s, with a swing frequency of 0.07 swings/s. From a practical point of view, this initial offset is mostly due to manufacturing/assembly errors as well as joint clearances. These effects can be seen also as positive in terms of adaptability to a more natural finger trajectory.

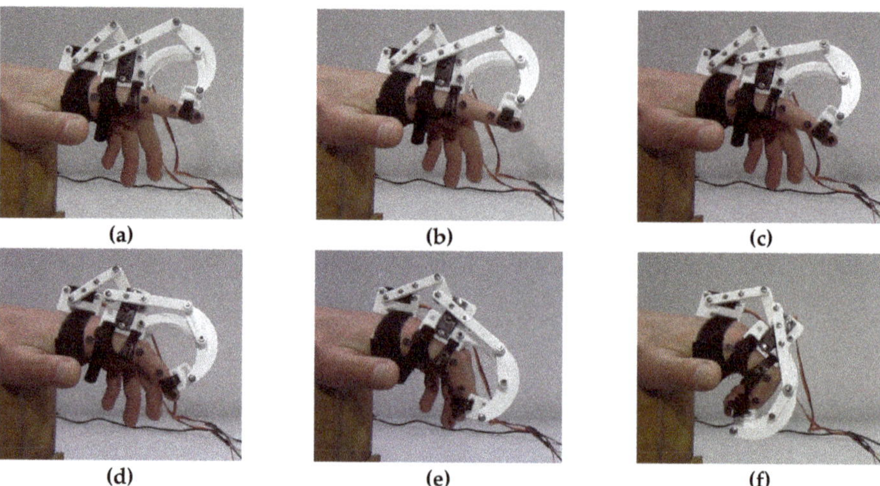

Figure 19. Snapshots of the exoskeleton motion during the test: (**a**) Starting position; (**b**) After 3.2 s; (**c**) After 6.5 s; (**d**) After 9.7 s; (**e**) After 13.0 s; (**f**) Final position.

The absolute error of the angular positions of the finger joints can be calculated by comparing the two curves in Figure 20. To compensate for an initial misalignment, a systematic error can be calculated. Moreover, an angular deviation error can be calculated as the difference between the experimentally measured and simulated values at a given time. For example, the time history of the angle δ is given in Figure 21 including both simulated and experimentally measured values. The experimentally measured angle δ versus time goes from $-47.2°$ to $-12.5°$ in this test. The comparison between the experimentally measured and simulated angle δ versus time shows an initial systematic error of $17.5°$. If this systematic error is compensated, the angular deviation error of angle δ versus time is given in Figure 22. The systematic error is mainly due to the clearance between the exoskeleton and finger, which are connected through velcro straps, whose clearance cannot be eliminated.

The absolute error for the angle δ in Figure 20 is always below ± 4 deg. This absolute error can be considered suitable as it is exactly within the acceptable error given in the design requirements for the proposed motion assistance application. Similarly, the time history of the β angle is given in Figure 21 including both simulated and experimentally measured values. The experimentally measured angle β versus time goes from $-6.5°$ to $64.7°$ in this test. The comparison between the experimentally measured and simulated angle β versus time shows an initial systematic error of $10.8°$. If one removes this systematic error, the angular deviation error of angle β versus time is given in Figure 22. Similarly, a comparison of β from the experimental test and simulation is reported in Figure 23, and the computed deviation error of β is reported in Figure 24. One can note that the absolute error for the angle δ in Figure 22 is below $\pm 4°$ until reaching a critical pose, corresponding to the pose reached after 13 s in the given experiment. Accordingly, operation of the prototype can be considered acceptable within the given design requirements until the closing configuration reaches the critical pose. Further motion needs to be avoided, since it would lead to interference of the human finger with the palm. This limit physically consists of allowing the finger flexion only until the DIP joint is vertically aligned with the MCP joint, referring to the scheme in Figure 5.

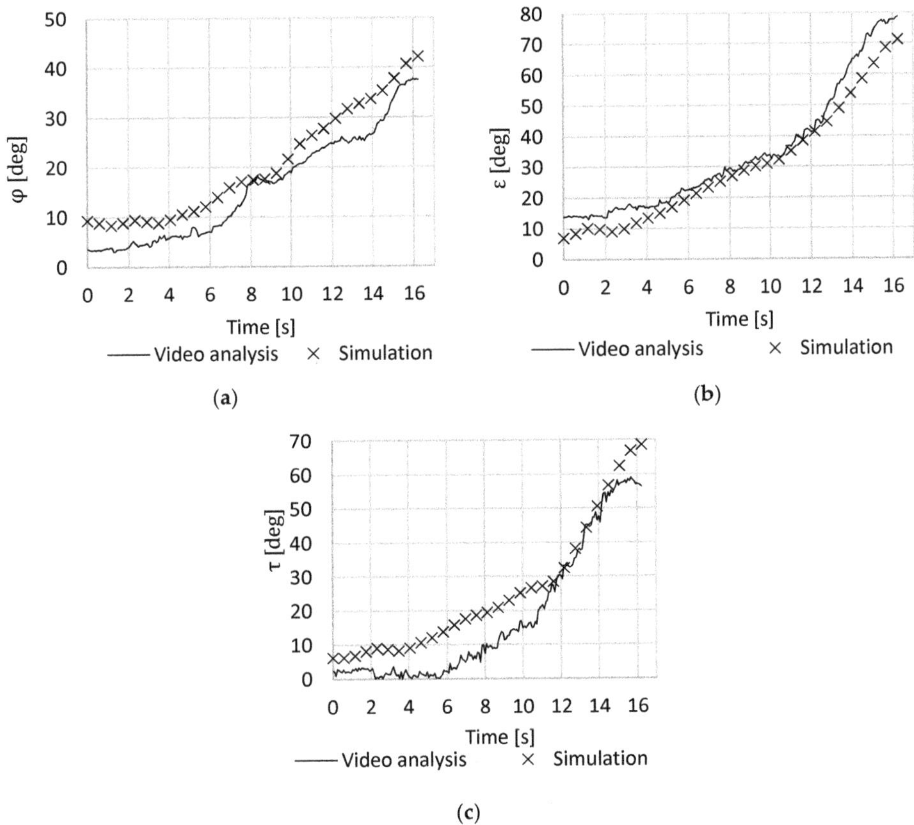

Figure 20. Comparison of finger joint angles in the prototype test (continuous line) and simulation (x marks): (**a**) angle φ versus time; (**b**) angle ε versus time; (**c**) angle τ versus time.

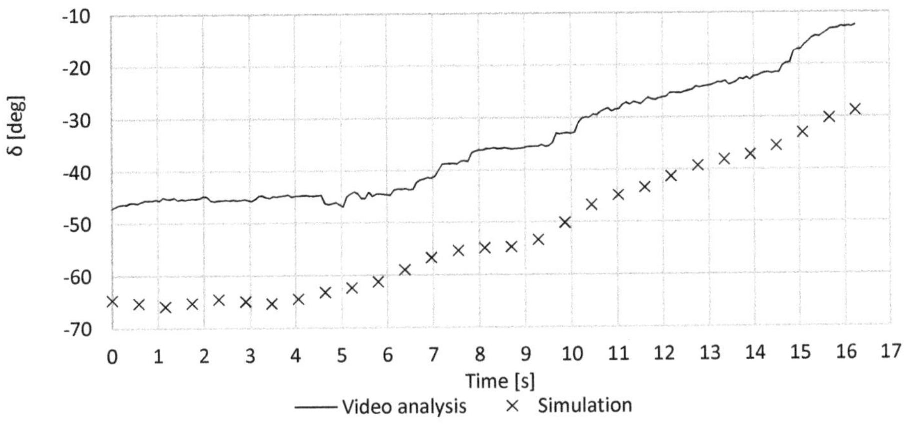

Figure 21. Comparison of δ for the test and simulation versus time.

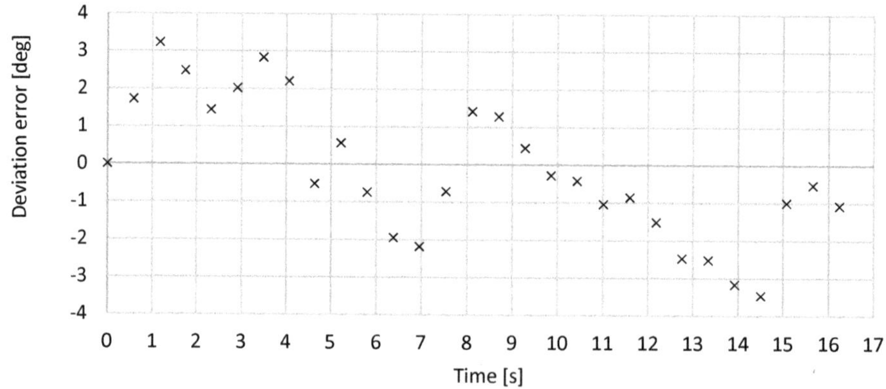

Figure 22. Deviation error of δ versus time.

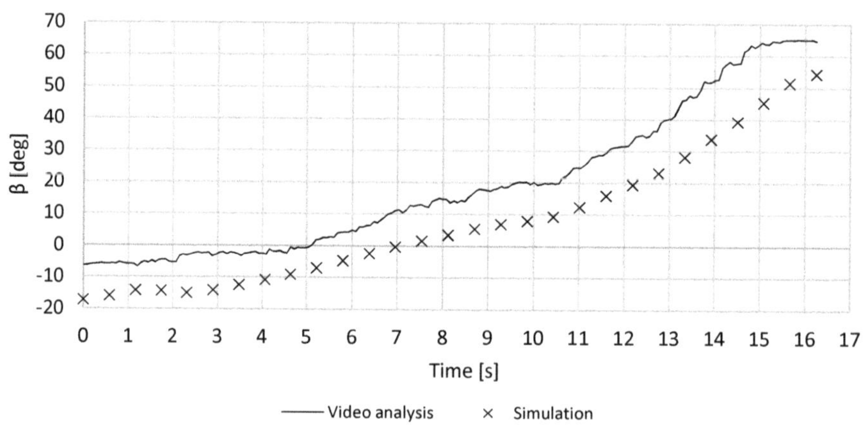

Figure 23. Comparison of β from the experimental test (continuous line) and simulation (x marks) versus time.

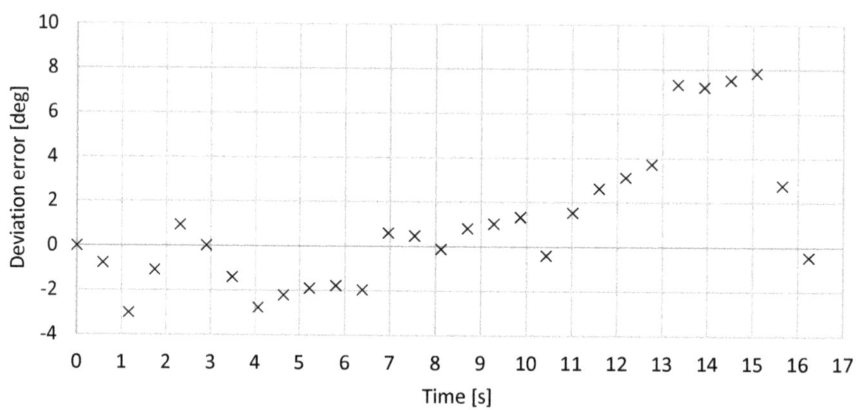

Figure 24. Computed deviation error of β versus time.

Average angular errors can be computed for each joint deviation error, in terms of as a square root average, as reported in Table 3. For all angles, the average error remains within the acceptable range of ± 4 degrees as established in the design requirements. This confirms the proposed finger exoskeleton can provide suitable motion assistance to replicate the desired human finger motion trajectory. The error calculation can be affected by the accuracy of the used method for experimentally measured angles via video capture tracking. This can generate non-negligible errors, in particular, if the tracking camera is not properly aligned or calibrated. For this purpose, the camera has been attached to a fixed frame, and the wrist is fixed, so that the exoskeleton movement is kept fixed in a proper plane. Accordingly, main sources of angular error s can be identified in the used 2D video tracking method as well as in joint clearances and backlashes. However, the clearance and backlash aspects can also provide positive features in terms of flexibility and adaptability of the finger exoskeleton motion to a natural finger motion.

Table 3. Angular errors of the finger joints and driving links.

Angle [deg]	φ	ε	τ	δ	β
	1.8	3.6	5.2	1.8	3.2

7. Conclusions

This paper reports the design of a novel exoskeleton mechanism for finger motion guidance. A human finger motion is analyzed through video motion tracking as a design reference. Then, a novel 2-DOF linkage mechanism is synthesized to mimic the desired finger motion. This linkage mechanism is driven by two independent actuators that can be conveniently placed on the back of the palm. The obtained mechanism with two linkage mechanisms is implemented in a 3D CAD model, and it is rapidly prototyped and assembled into a prototype. The resulting design is achieved with compact and lightweight features, and it can be manufactured in an easy and low-cost manner. The finger exoskeleton has been experimentally validated showing an acceptable error of guiding the desired human finger motion while performing a proper range of motions for assistance and rehabilitation of a large variety of human fingers.

Author Contributions: Conceptualization, G.C. and M.C.; software, E.C.G.; validation, E.C.G., D.C., and M.R.; data curation, E.C.G.; writing—original draft preparation, E.C.G.; writing—review and editing, G.C. and M.R.; supervision, G.C., M.C., and B.C. All authors have read and agreed to the published version of the manuscript.

Funding: This paper presents results from the research activities of the project ID 37_215, MySMIS code 103415 "Innovative approaches regarding the rehabilitation and assistive robotics for healthy ageing" co-financed by the European Regional Development Fund through the Competitiveness Operational Programme 2014–2020, Priority Axis 1, Action 1.1.4, through the financing contract 20/01.09.2016, between the Technical University of Cluj-Napoca and ANCSI as Intermediary Organism in the name and for the Ministry of European Funds.

Acknowledgments: The second author gratefully acknowledges the Erasmus+ program for a period of study he spent at LARM under the supervision of Prof. Marco Ceccarelli and Prof. Giuseppe Carbone.

Conflicts of Interest: The authors declare no conflict of interest.

References

1. United Nations Department of Economic and Social Affairs, Population Division: World Population Prospects: The 2019 Revision, Volume II: Demographic Profiles. 2019. Available online: https://population.un.org/wpp/Publications/Files/WPP2019_PressRelease_EN.pdf (accessed on 9 April 2020).
2. Kaplan, W.; Wirtz, V.; Mantel, A.; Béatrice, P.S.U. Priority medicines for Europe and the world update 2013 report. *Methodology* **2013**, *2*, 99–102.
3. Sale, P.; Lombardi, V.; Franceschini, M. Hand robotics rehabilitation: Feasibility and preliminary results of a robotic treatment in patients with hemiparesis. *Stroke Res. Treat.* **2012**, *2012*, 820931. [CrossRef] [PubMed]
4. Agarwal, P.; Fox, J.; Yun, Y.; O'Malley, M.K.; Deshpande, A.D. An index finger exoskeleton with series elastic actuation for rehabilitation: Design, control and performance characterization. *Int. J. Robot. Res.* **2015**, *34*, 1747–1772. [CrossRef]

5. Bataller, A.; Cabrera, J.A.; Clavijo, M.; Castillo, J.J. Evolutionary synthesis of mechanisms applied to the design of an exoskeleton for finger rehabilitation. *Mech. Mach. Theory* **2016**, *105*, 31–43. [CrossRef]

6. Stein, J.; Bishop, L.; Gillen, G.; Helbok, R. Robot-assisted exercise for hand weakness after stroke: A pilot study. *Am. J. Phys. Med. Rehabil.* **2011**, *90*, 887–894. [CrossRef] [PubMed]

7. Kun, L.I.U.; Hasegawa, Y.; Saotome, K.; Sainkai, Y. Design of a Wearable MRI-Compatible Hand Exoskeleton Robot. In *International Conference on Intelligent Robotics and Applications*; Springer: Cham, Switzerland, 2017; pp. 242–250.

8. Ates, S.; Haarman, C.J.; Stienen, A.H. SCRIPT passive orthosis: Design of interactive hand and wrist exoskeleton for rehabilitation at home after stroke. *Auton. Robot.* **2017**, *41*, 711–723. [CrossRef]

9. Pons, J.L. (Ed.) *Wearable Robots: Biomechatronic Exoskeletons*; John Wiley & Sons: Chichester, UK, 2008.

10. Heo, P.; Gu, G.M.; Lee, S.J.; Rhee, K.; Kim, J. Current hand exoskeleton technologies for rehabilitation and assistive engineering. *Int. J. Precis. Eng. Manuf.* **2012**, *13*, 807–824. [CrossRef]

11. Cempini, M.; Cortese, M.; Vitiello, N. A Powered Finger–Thumb Wearable Hand Exoskeleton with Self-Aligning Joint Axes. *IEEE/ASME Trans. Mechatron.* **2015**, *20*, 705–716. [CrossRef]

12. Li, J.; Zheng, R.; Zhan, Y.; Yao, J. iHandRehab: An Interactive Hand Exoskeleton for Active and Passive Rehabilitation. In Proceedings of the IEEE International Conference on Rehabilitation Robotics, Zurich, Switzerland, 29 June–1 July 2011; pp. 597–602.

13. Aubin, P.M.; Sallum, H.; Walsh, C.; Stirling, L.; Correia, A. A Pediatric Robotic Thumb Exoskeleton for at-Home Rehabilitation: The Isolated Orthosis for Thumb Actuation (IOTA). In Proceedings of the IEEE International Conference on Rehabilitation Robotics, Seattle, WA, USA, 24–26 June 2013; pp. 665–671.

14. Mertz, L. The next generation of exoskeletons: Lighter, cheaper devices are in the works. *IEEE Pulse J.* **2012**, *3*, 56–61. [CrossRef]

15. Jones, C.; Wang, F.; Morrison, R.; Sarkar, N.; Kamper, D. Design and development of the cable actuated finger exoskeleton for hand rehabilitation following stroke. *IEEE/ASME Trans. Mechatron.* **2014**, *19*, 131–140. [CrossRef]

16. Iqbal, J.; Tsagarakis, N.; Caldwell, D. Human hand compatible underactuated exoskeleton robotic system. *Electron. Lett.* **2014**, *50*, 494–496. [CrossRef]

17. Brokaw, E.; Black, I.; Holley, R.; Lum, P. Hand spring operated movement enhancer (handsome): A portable, passive hand exoskeleton for stroke rehabilitation. *IEEE Trans. Neural Syst. Rehabil. Eng.* **2011**, *19*, 391–399. [CrossRef]

18. Polygerinos, P.; Wang, Z.; Galloway, K.C.; Wood, R.J.; Walsh, C.J. Soft robotic glove for combined assistance and at-home rehabilitation. *Robot. Auton. Syst.* **2015**, *73*, 135–143. [CrossRef]

19. Amirabdollahian, F.; Ates, S.; Basteris, A.; Cesario, A.; Buurke, J.; Hermens, H.; Hofs, D.; Johansson, E.; Mountain, G.; Nasr, N.; et al. Design, development and deployment of a hand/wrist exoskeleton for home-based rehabilitation after stroke-script project. *Robotica* **2014**, *32*, 1331–1346. [CrossRef]

20. Nycz, C.J.; Meier, T.B.; Carvalho, P.; Meier, G.; Fischer, G.S. Design Criteria for Hand Exoskeletons: Measurement of Forces Needed to Assist Finger Extension in Traumatic Brain Injury Patients. *IEEE Robot. Autom. Lett.* **2018**, *3*, 3285–3292. [CrossRef]

21. Shen, Z.; Allison, G.; Cui, L. An Integrated Type and Dimensional Synthesis Method to Design One Degree-of-Freedom Planar Linkages with Only Revolute Joints for Exoskeletons. *J. Mech. Des.* **2018**, *140*, 092302. [CrossRef]

22. Copilusi, C.; Ceccarelli, M.; Carbone, G. Design and numerical characterization of a new leg exoskeleton for motion assistance. *Robotica* **2015**, *33*, 1147–1162. [CrossRef]

23. Carbone, G.; Aróstegui Cavero, C.; Ceccarelli, M.; Altuzarra, O. A Study of Feasibility for a Limb Exercising Device. In *Advances in Italian Mechanism Science*; Springer: Cham, Switzerland, 2017; pp. 11–21.

24. Cafolla, D.; Carbone, G. A study of feasibility of a human finger exoskeleton. In *Service Orientation in Holonic and Multi-Agent Manufacturing and Robotics*; Springer: Cham, Switzerland, 2014; pp. 355–364.

25. Ceccarelli, M. *Fundamentals of Mechanics of Robotic Manipulation*; Kluwer/Springer: Dordrecht, The Netherlands, 2004.

26. Carbone, G. (Ed.) *Grasping in Robotics*; Springer: London, UK, 2013.

27. Russo, M.; Ceccarelli, M.; Corves, B.; Hüsing, M.; Lorenz, M.; Cafolla, D.; Carbone, G. Design and Test of a Gripper Prototype for Horticulture Products. *Robot. Comput. Integr. Manuf.* **2017**, *44*, 266–275. [CrossRef]

28. Carbone, G.; Ceccarelli, M. Experimental Tests on Feasible Operation of a Finger Mechanism in the LARM Hand. *Int. J. Mech. Based Des. Struct. Mach.* **2008**, *36*, 1–13. [CrossRef]

29. Carbone, G.; Iannone, S.; Ceccarelli, M. Regulation and Control of LARM Hand III. *Robot. Comput. Integr. Manuf.* **2010**, *26*, 202–211. [CrossRef]

30. Yao, S.; Ceccarelli, M.; Carbone, G.; Zhan, Q.; Lu, Z. Analysis and Optimal Design of an Underactuated Finger Mechanism for LARM Hand. *Front. Mech. Eng.* **2011**, *6*, 332–343. [CrossRef]

31. Carbone, G.; Ceccarelli, M. Design of LARM Hand: Problems and Solutions. In Proceedings of the IEEE International Conference on Automation, Quality and Testing, Robotics AQTR, Cluj-Napoca, Romania, 22–25 May 2008.

32. Gerding, E.-C.; Carbone, G.; Cafolla, D.; Russo, M.; Ceccarelli, M.; Rink, S.; Corves, B. Design of a Finger Exoskeleton for Motion Guidance. In *7th European Conference on Mechanism Science EuCoMeS 2018*; Springer: Cham, Switzerland, 2018; Volume 59, pp. 11–18.

33. Gerding, E.-C.; Carbone, G.; Cafolla, D.; Russo, M.; Ceccarelli, M.; Rink, S.; Corves, B. Design and Testing of a Finger Exoskeleton Prototype. In *Advances in Italian Mechanism Science, Mechanisms and Machine Science*; Springer: Cham, Switzerland, 2018; Volume 68, pp. 342–349.

34. Gerding, E.; Marco, C.; Carbone, G.; Daniele, C.; Matteo, R. Mechanism for Finger Exoskeleton. Italian Patent No. IT. 102018000003847, 6 April 2020.

35. Levangie, P.K.; Norkin, C.C. *Joint Structure and Function: A Comprehensive Analysis*; FA Davis: Philadelphia, PA, USA, 2005.

36. ASSH Homepage. American Society for Surgery of the Hand. Available online: https://www.assh.org/ (accessed on 11 November 2019).

37. Kerle, H.; Corves, B.; Hüsing, M. *Mechanism Design: Fundamentals, Development and Application of Non-Constantly Transmitting Gears*; Springer: Wiesbaden, Germany, 2015. (In German)

applied sciences

Article

On the Optimal Synthesis of a Finger Rehabilitation Slider-Crank-Based Device with a Prescribed Real Trajectory: Motion Specifications and Design Process

Araceli Zapatero-Gutiérrez [1,*], Eduardo Castillo-Castañeda [1] and Med Amine Laribi [2]

[1] Centro de Investigación en Ciencia Aplicada y Tecnología Avanzada, Instituto Politécnico Nacional, Querétaro 76090, Mexico; ecastilloca@ipn.mx

[2] Département Génie Mécanique et Systèmes Complexes, Institut PPRIME, Université de Poitiers, 86073 Poitiers, France; med.amine.laribi@univ-poitiers.fr

* Correspondence: araceli_zapatero@hotmail.com or azapaterog1600@alumno.ipn.mx

Abstract: This article discusses the mechanical redesign of a finger rehabilitation device based on a slider-crank mechanism. The redesign proposal is to obtain a smaller and more portable device that can recreate the motion trajectories of a finger. The real finger motion trajectories were recorded using a motion capture system. The article focused on the optimal synthesis of the rehabilitation device mechanism formulated as a classic trajectory generation problem. The proposed approach was combined with the recorded finger movements and solved using the genetic algorithm (GA) method. Optimization criteria and constraints were successively formulated and solved using a mono-objective function.

Keywords: redesign; slider-crank mechanism; optimization; synthesis problem; rehabilitation devices

Citation: Zapatero-Gutiérrez, A.; Castillo-Castañeda, E.; Laribi, M.A. On the Optimal Synthesis of a Finger Rehabilitation Slider-Crank-Based Device with a Prescribed Real Trajectory: Motion Specifications and Design Process. *Appl. Sci.* **2021**, *11*, 708. https://doi.org/10.3390/app11020708

Received: 6 October 2020
Accepted: 30 December 2020
Published: 13 January 2021

Publisher's Note: MDPI stays neutral with regard to jurisdictional claims in published maps and institutional affiliations.

1. Introduction

In traditional physical and occupational therapy, a therapist often assists a patient with a body movement that the patient cannot complete alone [1]. A rehabilitation program aims to help a patient recover the lost capabilities, resulting in greater patient independence [2]. Rehabilitation programs generally consist of three stages. They begin with passive movements, where the therapist applies an external force to the patient due to the lack of muscular activity [3], then the therapist continues with assistive exercises. The patient does most of the exercises, but needs partial assistance from the therapist to perform them correctly. Finally, in resistive exercises, the therapist provides little assistance to the patient who performs muscle contractions against an external resistance [4,5]. The exercises that contribute to the affected part's mobility are mandatory, since it is crucial to avoid the harmful effects of immobilization [2].

This article focused on the rehabilitation of fingers. A finger is composed of a metacarpal and phalanges, two for the thumb and three for the other fingers. Every finger chain is articulated through the carpometacarpal joint (CM). At the distal direction, each metacarpal is articulated with the proximal phalanx through the metacarpophalangeal joint (MP). The thumb has one joint, the interphalangeal joint (IP), which is between the proximal and distal phalanges. The other four fingers have two joints: the proximal interphalangeal joint (PIP) and the distal interphalangeal joint (DIP). The first is located between the proximal and medial phalanges, and the second is located between the medial and distal phalanges [6], as shown in Figure 1.

Hands are the primary tool of human beings in any work environment, hence, they are exposed to all types of accidents or injuries due to diseases [7]. The use of robotic technology offers therapists with a tool to optimize the therapy process and the patient's recovery. It has been proven to be safe, feasible, and effective for recovering the nervous and muscular systems, at least in patients with stroke [8–13].

353

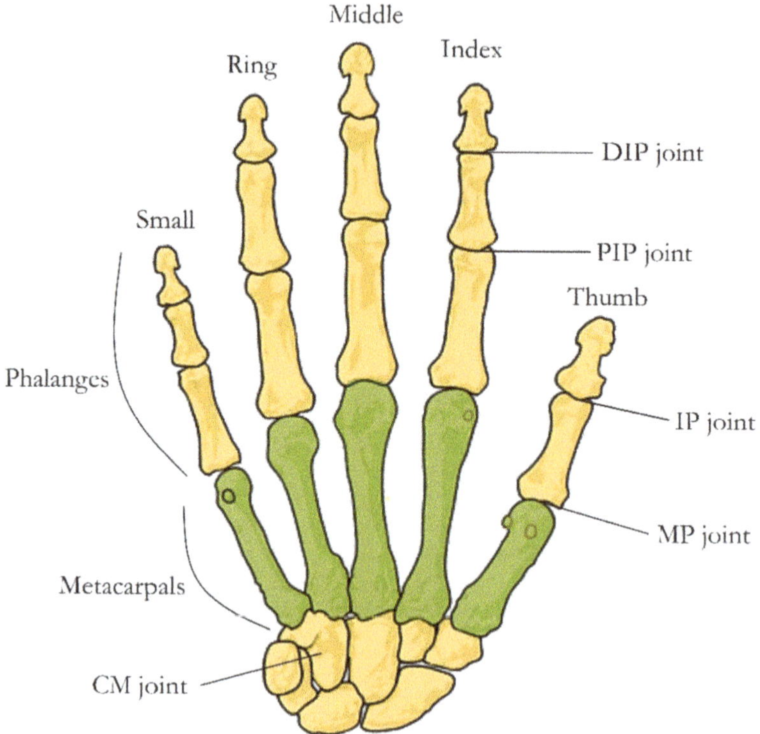

Figure 1. The terminology of bones and joints of the hand.

The development of new technology in robotics for rehabilitation therapy can increase the overall time of therapy, increasing the total patients that are attended to and the quality of therapy with less supervision. There have been many developments around the world in which robots assist patients in performing the desired movements. The articles presented in [14–16] offer a complete classification of assistant robots, separated by the type of limb they assist.

There are many ways of classifying hand rehabilitation robots. One is based on the way motion is transmitted. The transmission allows the actuators to follow the desired hand trajectory. The transmission's election depends on the trajectory that the hand is going to follow. Linkages and cables are common choices [17]. It is also possible to combine different types of transmissions, as in [18]. The screw-nut transmission transforms rotary motion into linear motion. It has a simple design that reduces cost and increases reliability. Another advantage is that the load can be raised by applying minimal effort while providing a precisely controlled linear motion. This main disadvantages of this type of transmission is its low efficiency and the rapid wear of the screw or the nut [19].

Hand rehabilitation robots are also classified based on the type of actuation (e.g., electric motor, pneumatic, hydraulic, and human muscle). The electric motor is the most widely used due to its reliability, availability, and precision [8]. Although pneumatic actuators require minor maintenance, the compressed air needs storage, influencing the device's size and mobility. Hydraulic actuators have good performance because they can generate higher torque than other actuators. However, they require more infrastructure (the oil transmitting pipes and conduits) and space (for the actuation system). The human muscle, as a type of actuation, implies that the impaired hand needs to be activated by functional electrical stimulation to complete the motion [17].

A fundamental aspect of hand rehabilitation robotics is the device hardware design. The design requirements are different from the design of industrial robots. Hardware design involves considerations of the safety, portability, and flexibility of the mechanism to achieve effective rehabilitation for the patient [17].

Finger movement characteristics are essential in device design to specify patient safety. Large devices need hospitals to be equipped with ample rooms to place them. They are also expensive, which reduces the number of units purchased, leading to fewer patients who can use this technology [17].

Exoskeleton-type devices such as those presented in [20,21] require little space because they are mounted on the patient's limb. Usually, the exoskeleton-type devices present more mechanically complex designs because the human joints must coincide with the exoskeleton's joints. Furthermore, that can involve more than one degree of freedom (DOF) in the device. End-effector-type devices such as Amadeo [22], a leading finger rehabilitation device, features a simple and versatile design that easily adjusts to various hand dimensions, however, it needs a large work area to operate. The challenge with the end-effector devices is to make the designs as small as possible to obtain better portability.

From [14], only four devices focus on finger rehabilitation. In [15], seven are end-effector devices for finger rehabilitation, two of which are mentioned in [14]. In [16], they mentioned four systems assisting finger movements, two also mentioned in [14] and one in [15]. The end-effector developments are fewer in comparison with exoskeleton development, resulting in a potential field of research.

Optimization of mechanical and robotic designs is widely used for various applications such as serial manipulators and parallel robots [23–25]. However, it has also gained importance in assistive and rehabilitation devices as well as exoskeleton systems [26–31]. New optimization techniques have been sought to improve the dimensional characteristics and dynamic characteristics of the systems. One of the methods that has gained more attention in the optimization of mechanisms is the genetic algorithm. Due to the quality of the population produced, only the best individuals in each iteration are taken into account.

The authors present the optimization of an end-effector finger rehabilitation device that reproduces the individual trajectory of the fingers, which contributes to the first stage of rehabilitation (assistance for passive movements). Compared to other end-effector devices, the presented rehabilitation mechanism does not imitate the hand's grasping functions.

Ensuring repetitive monitoring of the natural flexion–extension trajectory contributes to proper fit and alignment of the finger joints. The patient recovers the joint mobility simultaneously (MP, PIP, and DIP). Respecting each joint's range of movement can contribute to the gross motor function and fine motor function. From a rehabilitation viewpoint, to improve the range of movement (ROM) at a joint, each joint must be moved through inside its ROM (expressed in degrees) at regular intervals. A continuous passive motion device (like a rehabilitation robot) facilitates the rapid recovery of the neuromuscular system by improving ROM [8,9,32,33].

The goal of this article is the proposal of an optimal redesign for a finger rehabilitation device. The work focused on proposing the optimal dimensions of the existing mechanism's links, allowing for more accurate tracking of a trajectory compared to the existing prototype that uses a screw-nut transmission. Additionally, the original prototype's links are too big concerning the end-effector trajectory's size, which decreases the portability of the prototype.

The rest of this paper is structured as follows. Section 2 contains the finger rehabilitation device's description and kinematic model and presents the characterization and analysis of the real finger trajectories, and finally, the optimization problem. Section 3 presents the optimization results, Section 4 presents the discussion, and, in Section 5, the conclusions are presented.

2. Materials and Methods

Figure 2 shows the redesign process for the previously mentioned finger rehabilitation prototype. The mechanism's redesign focused on finding the optimal dimensions for a new mechanism (with the same architecture as the original) to replicate a real finger's flexion–extension trajectory better.

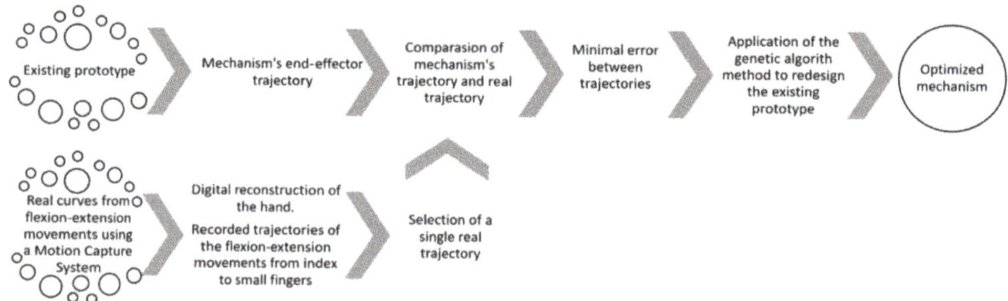

Figure 2. The block diagram for the approach.

The real trajectory analysis was carried out using a motion capture system, with the participation of a group of subjects. The hand's digital reconstruction allowed for the measurement of the finger's flexion–extension movement trajectories. The flexion–extension trajectory of a participant's index finger was selected for comparison with the mechanism's trajectory. This trajectory was chosen due to its large amplitude and equivalence characteristic regarding the other recorded finger motions. The genetic algorithm (GA) method found the dimensions that produced the minimum error between the mentioned trajectories.

2.1. Finger Rehabilitation Device

2.1.1. Existing Prototype

A prototype for assisting flexion–extension movements from the index to small fingers was proposed in 2016 [34]. The prototype consisted of four parallel mounted slider-crank mechanisms (one for each finger), driven by a direct current (DC) motor and a screw-nut transmission, as shown in Figure 3. The end-effector, located in the coupler link, was the contact point between the prototype and the fingertip. The fingertip was attached to the end-effector using an articulated thimble placed at the extremity and was fixed with adhesive fabric. Figure 4 shows how the arm and fingertip are positioned on the mechanism.

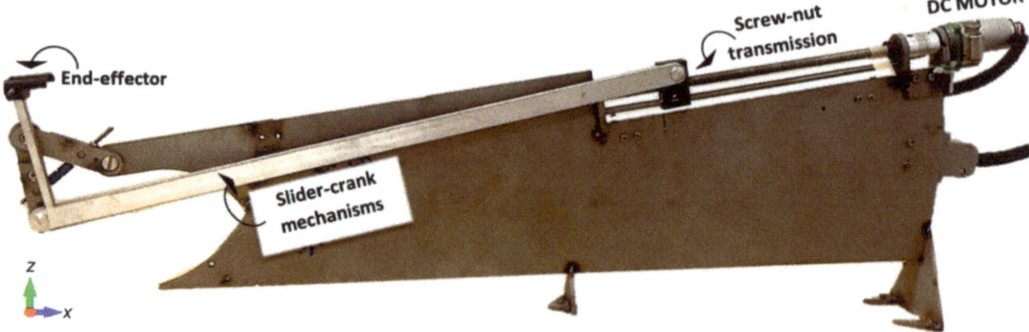

Figure 3. Finger rehabilitation prototype (back view).

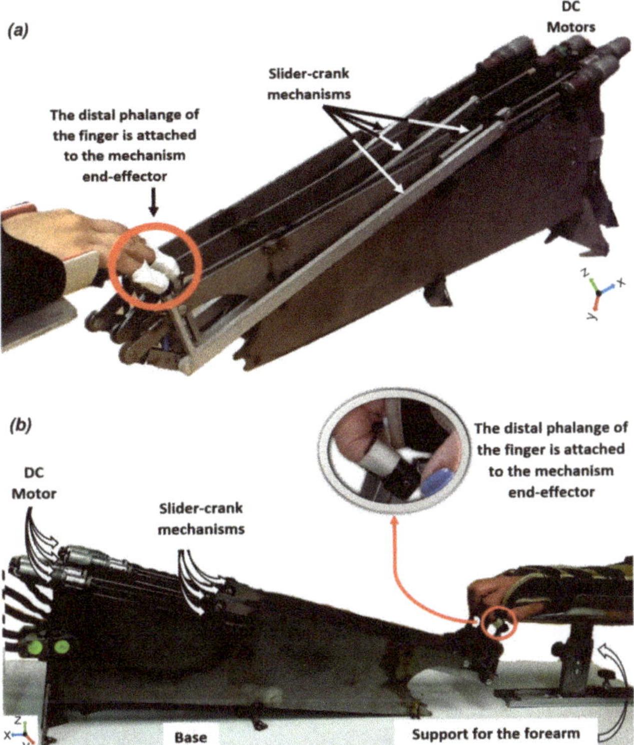

Figure 4. Finger rehabilitation prototype: (**a**) isometric view; (**b**) front view.

The patient's forearm was placed in a support that could be adjusted in the x and z plane. The upper part of the support had an integrated universal wrist–forearm splint. The patient inserted their arm into the splint and adjusted with Velcro straps, which allowed the forearm, wrist, and palm to be entirely fixed on the surface of the support.

The end-effector produces an arc segment, as shown in Figure 5, close to an elliptical trajectory. The end-effector's trajectory aims to closely follow the finger's flexion–extension movement's natural trajectory, which has been studied by several authors [35–37]. This end-effector's trajectory is the defining characteristic of this prototype; other well-known finger rehabilitation devices like Amadeo [22] use linear trajectories [38–40]. This remarkable difference is because the presented rehabilitation mechanism does not imitate the hand's grasping functions.

The elliptical trajectory was compared with the end-effector mechanism's trajectory to propose an optimized size for the prototype. However, this research did not involve mechanical changes or performance improvements to the transmission.

The prototype's original dimensions were obtained using three points of the elliptical trajectory described in the previous section; the full prototype design can be consulted in detail in [34]. The original prototype's elliptical trajectory tried to emulate the natural flexion–extension movement. However, its design did not involve real flexion–extension trajectories. In addition, given its size, it cannot be easily transported. It is essential to ensure that the prototype's end-effector can reproduce real trajectories with minimum error to achieve better rehabilitation results and reducing its dimensions will improve its portability.

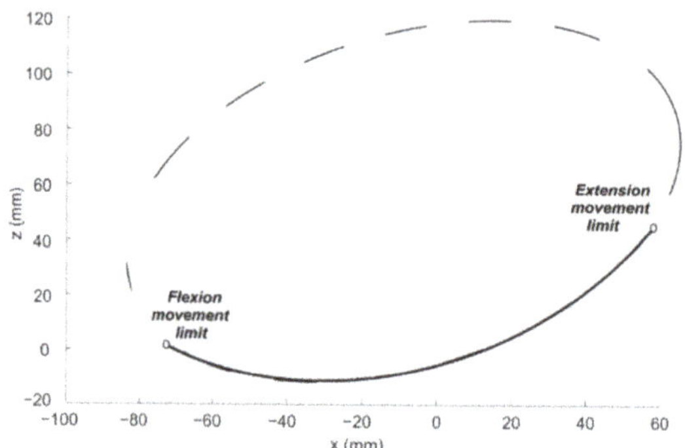

Figure 5. The trajectory of the end-effector prototype.

2.1.2. Architecture and Kinematic Model

Since the four mechanisms are similar, it is sufficient to only analyze one of them. Figure 6 shows the computer-aided design (CAD) of one of the mechanisms and its vector representation. Vector position 1 and 3 correspond to the distance between the points A to B and B to C, respectively. The DC motor provides the angular movement that is converted into a linear movement through the screw-nut transmission. This linear movement is limited by the minimum and maximum position of the flexion and extension points. The kinematics analysis is presented in detail in [41]; for the analytical purposes of the mechanism's redesign, a tag was assigned to every vector, as shown in Table 1.

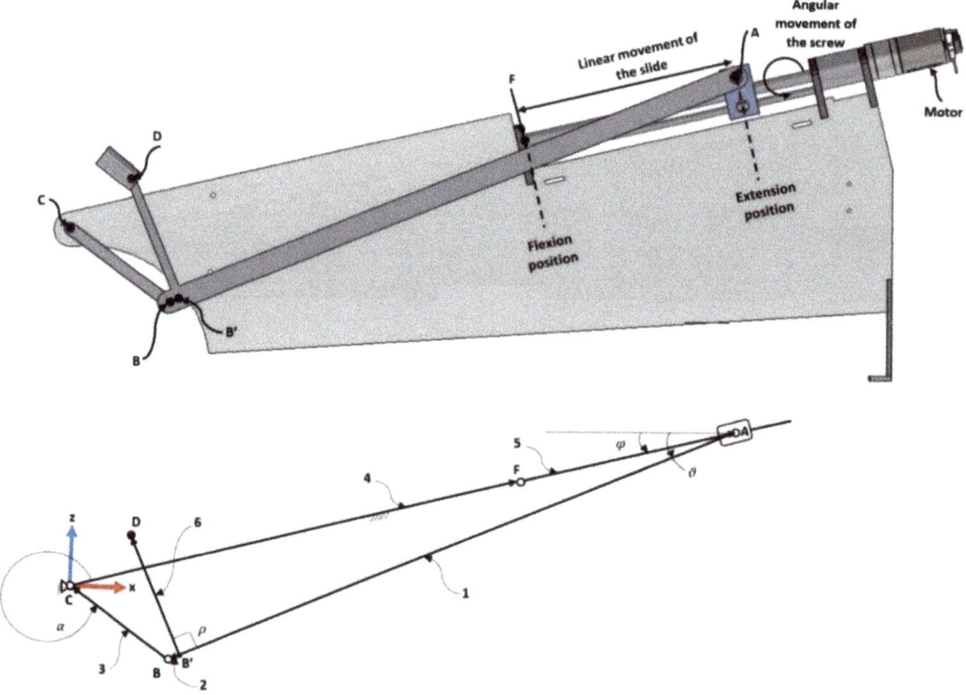

Figure 6. CAD (computer-aided design) and vector representation of the existing prototype.

Vector 4 represents the physical distance between points C and F. Point F represents the minimum position of flexion movement. Point D is the contact point between the patient and the device and describes the elliptical trajectory that was mentioned before. Point D could be considered as a rigid body attached to the AB link and has an angle ρ with respect to AB. Angle φ represents the inclination angle of the mechanism. In the original prototype, $\rho = 90$ degrees and $\varphi = 10$ degrees.

Table 1. Tags for vectors.

Vector	1 From A to B	2 From B′ to B	3 From B to C	4 From C to F	5 From C to A	6 From B′ to D
Tag	a	d	b	f	c	e

Considering Figure 6, Equation (1) defines point D's position coordinates, which is the end-effector's point that describes the elliptical trajectory. Vector 5 (Equation (5)) is defined by a function that includes the length of vector 4 and the distance C_i between the minimum and maximum position values, corresponding to the flexion and extension points. The distance C_i is defined by Equation (6) and can be used for the optimization process. Vector 4 of the original prototype was fixed to 344 mm, while the distance C_i went from 0 to 125 mm. Vectors 2 and 6 are the distances between B to B′ and B to D, respectively.

From Equation (6), C_0 represents the minimum value of C_i and C_m represents the maximal value. C_m is defined by Equation (7), where d_m is the distance the slider needs to generate the desired trajectory. d_m is defined by the sum of the minimal value and the distance between the initial and final points of the trajectory.

$$D = \begin{bmatrix} D_x \\ D_z \end{bmatrix} = \begin{bmatrix} b \cos \alpha + d \cos \vartheta + e \cos \beta \\ b \sin \alpha + d \sin \vartheta + e \sin \beta \end{bmatrix}, \tag{1}$$

where

$$\alpha = 2\pi - \left(arc \cos \left(\frac{a^2 - c^2 - b^2}{-2bc} \right) + \varphi \right), \tag{2}$$

$$\vartheta = \varphi + arc \cos \left(\frac{b^2 - c^2 - a^2}{-2ac} \right), \tag{3}$$

$$\beta = \vartheta + \rho, \tag{4}$$

$$c = f + C_i. \tag{5}$$

The solution for this equation exists if and only if:

$$abs \left(\frac{a^2 - c^2 - b^2}{-2bc} \right) < 1 \text{ and } abs \left(\frac{b^2 - c^2 - a^2}{-2ac} \right) < 1 \quad C_i = \sum_{j=C_0}^{C_m} j, \tag{6}$$

$$C_m = d_m + C_0. \tag{7}$$

2.2. Finger Real Motion

2.2.1. Experimental Setup

To propose an improvement to the actual prototype, we performed several experiments to obtain real curves for the fingers' flexion–extension movements. These real curves were used to evaluate the accuracy of the elliptical trajectory proposed in the original prototype. The finger movements were recorded using the motion capture system Qualisys Track Manager (v2018, Qualisys, Göteborg, Sweden, 2018), which was used to process those records. A set of coordinates for each marker as a function of time was defined to analyze the data. Mokka (v0.6.2, BTK, 2019) and MATLAB software (v2018, MathWorks, Meudon, France, 2018) were used.

The system was calibrated using a reference frame located in the corner on a flat surface. The volunteer placed their forearm, as shown in Figure 7. We recorded the flexion–extension movements from the index to small fingers. A total of six volunteers were involved in the experimentation, both men and women, in the age range of 25 to 40 years old without any finger injury.

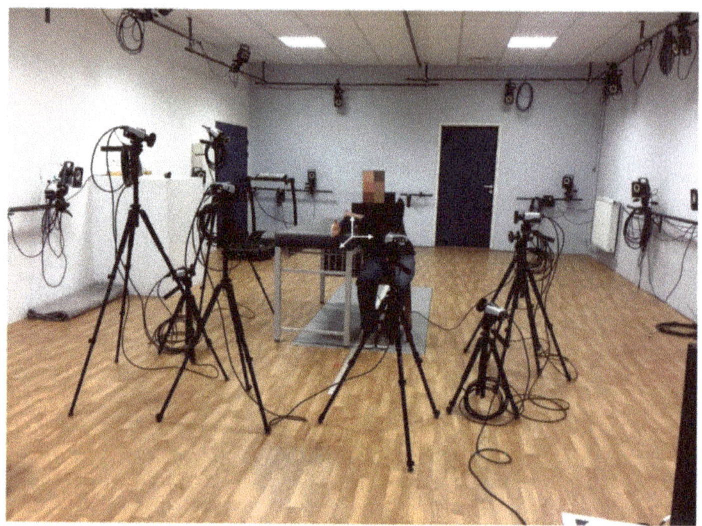

Figure 7. The experimental site with the Qualysis system.

A set of 21 markers was used and installed on the right hand of each volunteer. Table 2 describes each marker's location and the label that was used to reconstruct the hand in digital form. The digital reconstruction of the hand is shown in Figure 8. The digital reconstruction was based on marker locations following the anatomy of the subject.

Table 2. Localization and names of the markers in the hand.

Finger	Marker	Articulation
Thumb	T1	Tip
	T2	IP
	T3	MCP
Index	I1	Tip
	I2	DIJ
	I3	PIJ
	I4	MCP
Middle	M1	Tip
	M2	DIJ
	M3	PIJ
	M4	MCP
Ring	R1	Tip
	R2	DIJ
	R3	PIJ
	R4	MCP
Small	S1	Tip
	S2	DIJ
	S3	PIJ
	S4	MCP
Wrist	W1	Carpal bones
	W2	Carpal bones

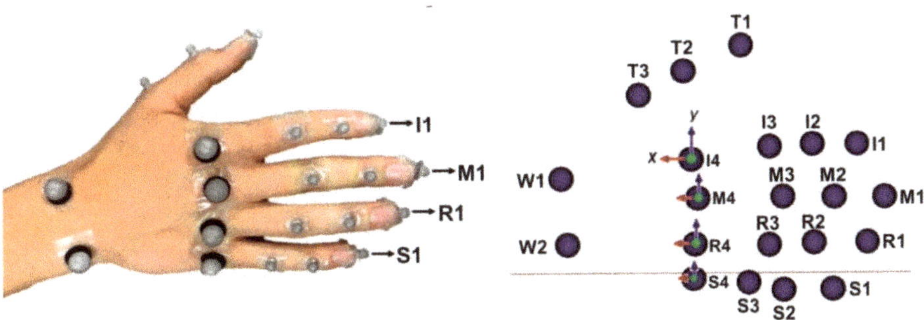

Figure 8. Location of the markers and the digital reconstruction of the hand.

Each volunteer's forearm was placed on a flat surface, with the palm perpendicular to the surface and their fingers in the extended position. The subjects were asked to close and then open their fist, as shown in Figure 9, and to repeat the exercise 10 times. The plane where the flexion–extension movements from the index to small fingers were performed was perpendicular to the plane where the thumb's flexion–extension movements are performed.

(a) (b)

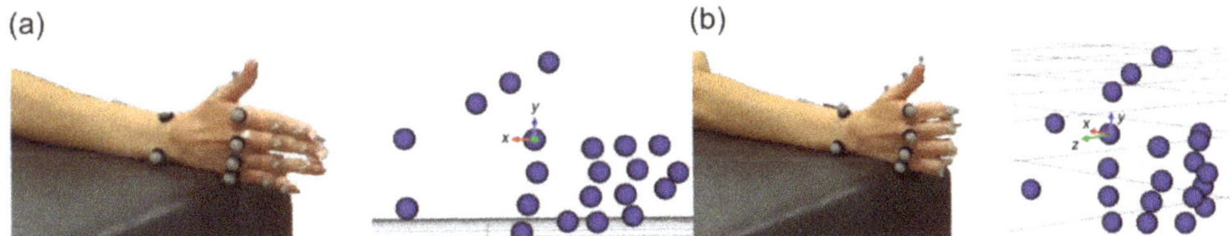

Figure 9. (**a**) Initial position to record. (**b**) Hand in closed fist configuration.

The existing prototype was designed only for the flexion–extension movements from the index to small fingers and considered only the trajectory of the distal phalanx; for this reason, only the marker placed in the distal phalanx for these fingers (I1, M1, R1, and S1) was analyzed.

2.2.2. Results of the Analysis of the Fingers Real Motion

The recorded gesture focused on the flexion–extension movement leading to the fingertip trajectory. This finger movement occurs in the finger plane, which is normal to all joint axes. Consequently, the fingertip trajectory were defined only by the x and z coordinates (as shown in Figure 10). A series of three-dimensional curves, as shown in Figure 11, was obtained using the flexion–extension movement in the xyz space, which corresponded to the recorded motion for a volunteer's index finger.

All subjects were asked to move their finger with a fixed abduction–adduction. Despite this virtual constraint and the low-speed exercise, the obtained fingertip trajectory presented a reduced variation along the y-axes, probed through principal component analysis (PCA). The PCA reduces the dimensionality of a dataset of interrelated variables while retaining the dataset's variation. A new set of variables is generated, a linear combination of the original variables. This new set of variables is called the principal components [42].

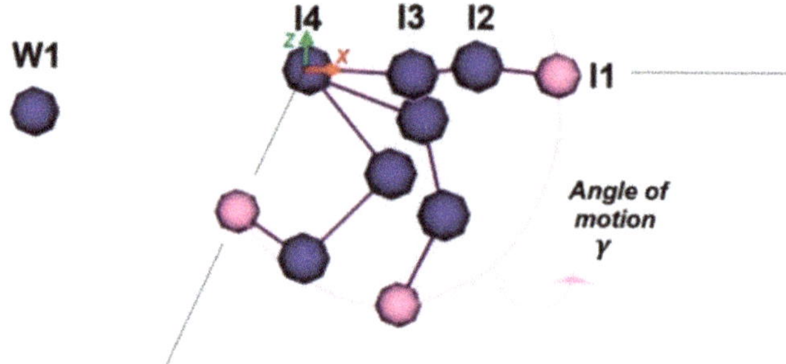

Figure 10. Finger trajectory in the *xz* plane.

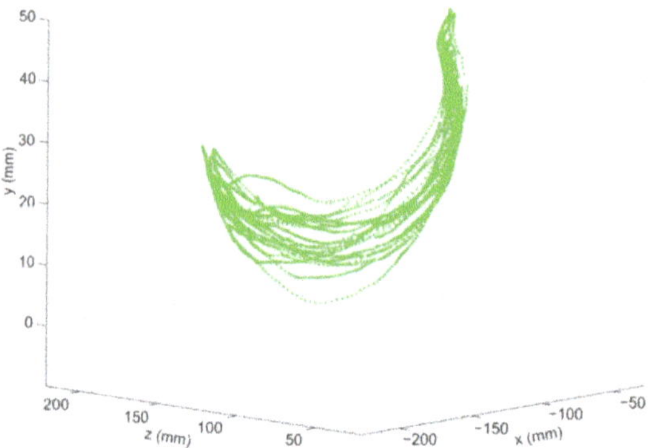

Figure 11. Flexion–extension movement in the space for the index finger of a volunteer.

The PCA indicates which variable is the most valuable for clustering the data. In this article, the original set of variables are the *xyz* coordinates of the *n* point of the flexion–extension movement, which means that three principals components exist. The principal component is obtained by calculating the average values for all the variables, as shown in Equation (8). The original variables are defined as *p*-dimensional vectors that need to be projected onto a *q*-dimensional subspace.

$$\overline{x} = \frac{\sum_{i=1}^{n} x_i}{n} \quad \overline{y} = \frac{\sum_{i=1}^{n} y_i}{n} \quad \overline{z} = \frac{\sum_{i=1}^{n} z_i}{n} \tag{8}$$

The PCA identifies the directions (principal components) along which the variation in the data is maximal. The direction of PC1 represents the most significant variation among the data, PC2 is the second most important direction and is uncorrelated to PC1 and PC3. PC3 represents the less important direction, also uncorrelated with the other principal components. Principal components are normalized eigenvectors of the covariance matrix. These are ordered according to how much of the variation present in the data they contain. The dimensionality of the three-dimensional data can be reduced to two-dimensional data using the first two principal components [43,44].

Table 3 shows the covariance data matrix's eigenvalues and the percentage of each principal component total variance. The eigenvalues measure the amount of variation retained by each principal component [44].

Table 3. Eigenvalues and percentage of the variance of each principal component.

New Set of Variables	Eigenvalues	% of the **Total Variance**	% of the **Cumulative Variance**
PC1	2444.9	82.0531	82.05
PC2	516.5	17.3345	99.39
PC3	18.2	0.6125	100.00

As can be observed in Table 3, the first two components explained 99.39% of all variability, as shown in Figure 12. This proves that the y-axis presents a reduced variation in the flexion–extension movement of the fingers. Due to this, one can conclude that the flexion–extension movement can be assumed in the xz plane. The abduction–adduction movement of the finger was not considered.

(a) (b)

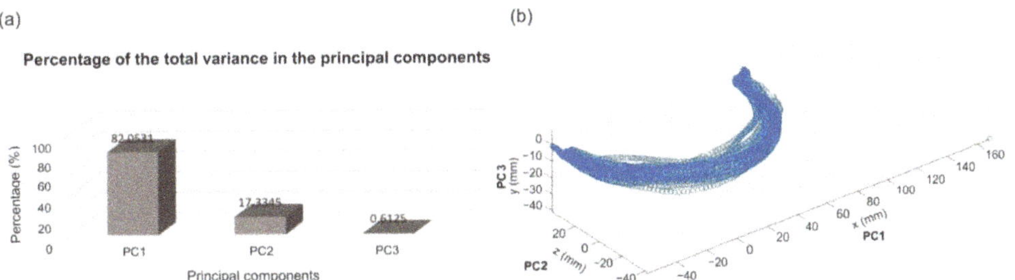

Figure 12. Principal components representation. (**a**) Percentage of the total variance. (**b**) The data represented in the space of the three principal components.

The finger motion, from the index to small, was analyzed. A similarity was observed between all fingertip trajectories, as shown in Figure 13a. An example of graphs obtained for all fingers of one subject is depicted in Figure 13b. The left side of Figure 14 shows all the trajectories recorded in the xyz space. The right side shows one movement in the xz plane to improve the visualization of each finger's trajectories.

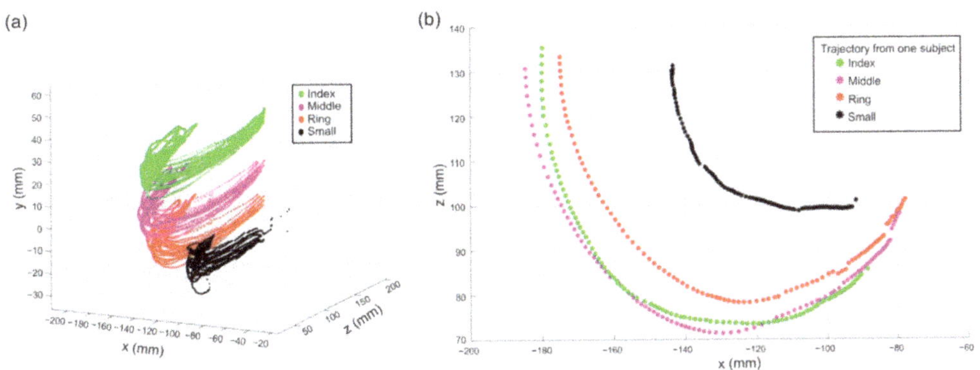

Figure 13. Trajectories for flexion–extension from index to small, (**a**) xyz spatial representation, (**b**) xz planar representation of one participant.

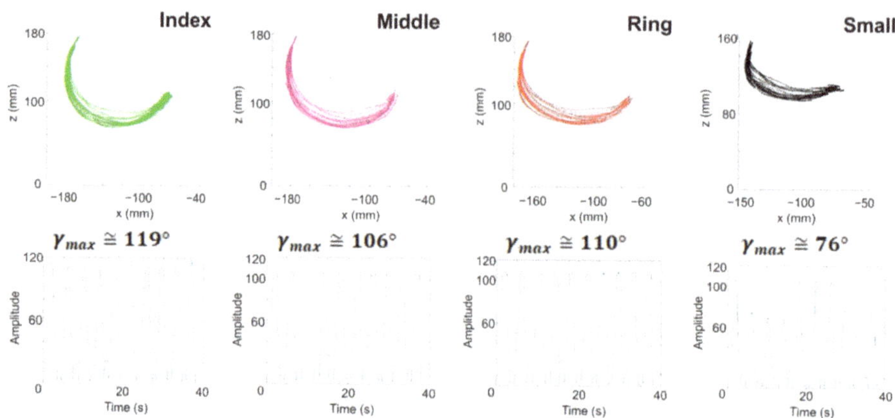

Figure 14. Real trajectories of flexion–extension.

Aside from the trajectories, we computed the range of motion of every finger. We defined the u_{f_i} vector in the finger plane, where $f\epsilon\{I, F, R, S\}$ corresponds to index (I), Middle (M), Ring (R), and Small (S), and the index i corresponds to the frame number from 1 (initial position) to N (last position). The computed vector corresponds to the subtraction between the xz position of the tip marker (I1, M1, R1, S1) and the xz position of the MCP marker (I4, M4, R4, S4), as defined in Equation (9).

$$
\begin{array}{ll}
\text{Finger} & \text{Vector} \\
\text{Index} & u_{I_i} = \text{I1} - \text{I4,} \\
\text{Middle} & u_{M_i} = \text{M1} - \text{M4,} \\
\text{Ring} & u_{R_i} = \text{R1} - \text{R4,} \\
\text{Small} & u_{P_i} = \text{S1} - \text{S4.}
\end{array} \tag{9}
$$

Then, we defined $\hat{u}_{f_i}$ as the normal vector of u_{f_i} as Equation (10) shows. The vector w_f represents the dot product between the initial position and the ith position, in order to compute the angle between these normal vectors, which is expressed in Equation (11).

$$
\hat{u}_f = \frac{u_f}{|u_f|} , \mathbf{w}_f = \left[\sum_{j=1}^{j=2} v_{f1}\cdot v_{fj} \dots \sum_{j=1}^{j=K} v_{f1}\cdot v_{fj} \dots \sum_{j=1}^{j=N} v_{f1}\cdot v_{fj} \right], \tag{10}
$$

$$
\gamma(i) = arc\cos\left(\mathbf{w}_f\right). \tag{11}
$$

Figure 15 shows the trajectories of one subject from the index to small in the xz plane and the maximum and minimum angle of each finger's movement. The maximum angle for the small finger was chosen as $\gamma_{max} \cong 76°$ because several points were outside the trajectory and represented false measurements (Figure 13).

As can be observed from Figure 14, the trajectories for the different fingers were similar in shape but with different maximum angle (γ_{max}) amplitudes. Table 4 shows the values of the angles of movement among the volunteers after several tests.

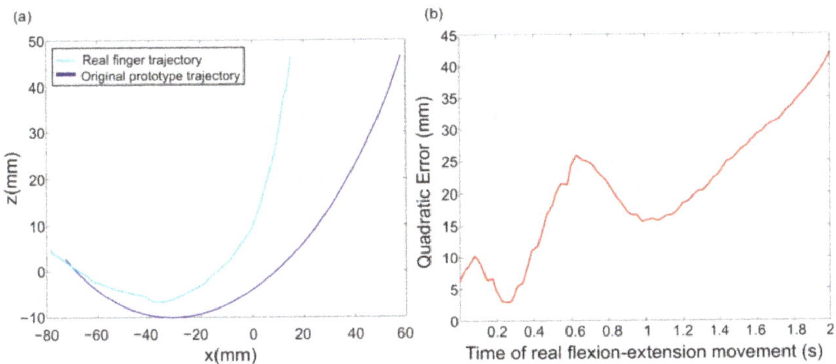

Figure 15. (**a**) Real finger trajectory and the trajectory of the end-effector prototype. (**b**) The quadratic error between trajectories.

Table 4. The angle (°) of movement of every finger among the subjects.

Finger	Subject 1 γ_{max}	Subject 2 γ_{max}	Subject 3 γ_{max}	Subject 4 γ_{max}	Subject 5 γ_{max}	Subject 6 γ_{max}
Index	119°	104°	101°	92°	116°	104°
Middle	106°	117°	120°	92°	122°	130°
Ring	110°	108°	119°	95°	113°	119°
Small	76°	105°	109°	89°	92°	116°

To conclude this section on real finger motion. Figure 15a simultaneously illustrates the original prototype trajectory and index finger trajectory from one subject, which was chosen as the recorded real trajectory. The real finger trajectory did not have a regular elliptical shape, and it seems to have a more closed shape. In contrast, when the end-effector's prototype trajectory maintained an elliptical behavior, an offset between the two trajectories was noted.

It seems feasible that the prototype trajectory follows the real trajectory more closely. Figure 15b shows the quadratic error between the two trajectories. It can be noticed that this error was not negligible and the error was used as an optimization criterion. The end-effector, located at the fingertip, was required to follow a specified curve with a minimum error computed between the obtained and desired trajectories.

2.3. Synthesis Problem

Three types of formulation applied to mechanism synthesis problems can be found in the literature: function generation, trajectory generation, and body guidance [45]. This article focused on trajectory generation based on an optimization process. An optimal mechanism to generate the desired trajectory followed by the fingertip was found. The data from the real finger trajectory, described in the previous section, was used in the computation of the objective function. This allowed us to find the optimal parameter of the solution to reconfigure the existing rehabilitation mechanism. The optimization problem was solved using the genetic algorithm method.

2.3.1. Formulation of the Problem

The optimization problem can be established as the minimization of the error function $E(F)$, which is the sum of the square distance between the *ith* recorded position $(R_{x_{i:N}}, R_{z_{i:N}})$ and the coordinates of point D of the mechanism $(D_{x_{i:N}}, D_{z_{i:N}})$. The error function is expressed by Equation (12). This point corresponds to the end-effector of the slider-crank mechanism. Equation (13) defines the design vector F, which contains the set of variables that should be computed during the optimization procedure.

f_m represents the individual in the genetic algorithm; H_f defines the lower and upper boundaries for each of the design variables; and g_{m1} and g_{m2} are constraints applied to the mechanism.

$$E(F) = \frac{1}{N} \sum_{i=1}^{N} \sqrt{(D_{x_i} - R_{x_i})^2 + (D_{z_i} - R_{z_i})^2}. \tag{12}$$

The optimization synthesis problem was formulated as follows with $m = 10$ parameters for a chosen objective function $E(F)$:

$$\text{Minimize } E(F),$$

Subject to:

$$F = [f_1, \ldots, f_m], \; m = 10, \; f_m \, \epsilon \, H_f = \left[f_m{}^{min}, f_m{}^{max} \right] g_{m1}(F) \leq 1, g_{m2}(F) \leq 1, \tag{13}$$

where $H_f = \left[f_m{}^{min}, f_m{}^{max} \right]$ is the bounding interval for each parameter of vector F. The parameters of vector F are defined in Table 5; most of the parameters are defined in the kinematic section. The variables x_t and y_t represent an offset applied to the real trajectory to relocate it to the same work area of the existing mechanism, allowing for a comparison of both trajectories.

Table 5. Definition of vector F.

f_1	f_2	f_3	f_4	f_5	f_6	f_7	f_8	f_9	f_{10}
a	b	d	e	φ	ρ	f	x_t	y_t	C_i

The kinematic model of the slider-crank mechanism was defined by Equations (1)–(7) in previous sections. Equation (14) defines the inequality constraints to guarantee a feasible solution of the optimization problem and comes from the closed-loop equations. The constraints g_{m1} and g_{m2} are defined in the following formulation:

$$g_{m1} = \left| \frac{a^2 - c^2 - b^2}{-2bc} \right| \leq 1, g_{m2} = \left| \frac{b^2 - c^2 - a^2}{-2ac} \right| \leq 1. \tag{14}$$

2.3.2. Genetic Algorithm Method Implementation and Curves Enhancement

The genetic algorithm (GA) is a probabilistic technique based on the evolutionary theory of natural selection that uses a population of designs rather than a single design at a time. The GA generates a population of individuals at each iteration. These individuals are a combination of the previous population's characteristics (called parents). However, they can also present some mutations in their characteristics. Only the best individuals in the population are used to create a new generation, which allows the approach toward an optimal solution. The GA operators select the next population by computation using a random number generator.

Nevertheless, this algorithm presents a limited accuracy of the final solution. A large number of iterations is needed to obtain a solution [46,47]. Figure 16 shows the scheme of the GA method applied to the problem of mechanism synthesis.

The main goal of the GA is to identify the best value of the design vector F, which contains all the parameters that describe the mechanism. Each design vector is called an individual.

The design vector is evaluated through the fitness function that quantifies the result of the individual's parameters. The best fitness value for a population is the smallest fitness value for any individual in the population [47]. In order to obtain better results in the GA implementation, it is recommended that the input trajectory presents the same step between points. A polynomial fitting method is used and included in the optimal synthesis process.

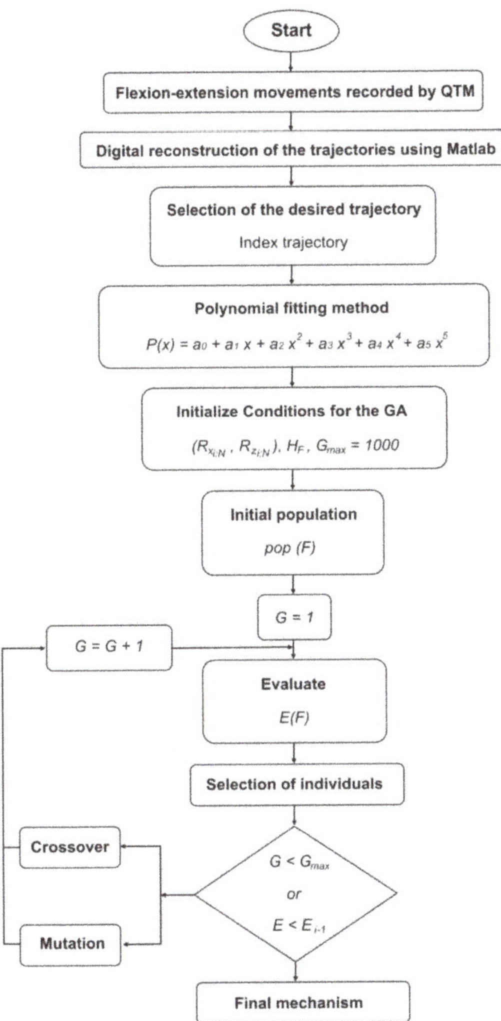

Figure 16. The optimal synthesis process of the rehabilitation device mechanism.

In order to obtain a uniformly distributed trajectory, the desired trajectory was approximated using a polynomial fitting method, as shown in Figure 17 and given by Equation (15), where a_0, a_1, a_2, a_3, a_4, and a_5 are the fitting parameters. An example, based on the index finger's trajectory of subject three, is given in this section. The numerical values of the fitting parameters are defined in Table 6.

$$P(x) = a_0 + a_1 x + a_2 x^2 + a_3 x^3 + a_4 x^4 + a_5 x^5 \tag{15}$$

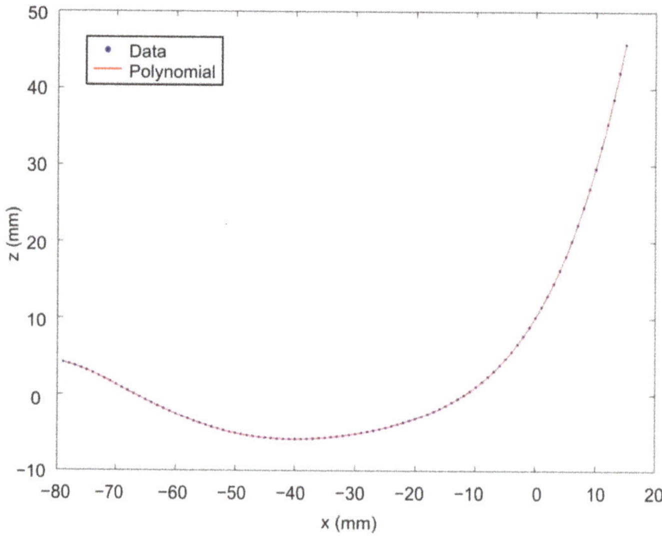

Figure 17. The desired and fitted trajectories.

Table 6. Fitting parameters and errors.

Fitting Parameter	Error
$a_0 = 7.22 \times 10^{-8}$	$5.91 \times 10^{-8} \le a_0 \le 8.53 \times 10^{-8}$
$a_1 = 1.5 \times 10^{-5}$	$1.3 \times 10^{-5} \le a_1 \le 1.6 \times 10^{-5}$
$a_2 = 1.2 \times 10^{-3}$	$0.001085 \le a_2 \le 0.001257$
$a_3 = 5 \times 10^{-2}$	$0.04915 \le a_3 \le 0.05114$
$a_4 = 1.3$	$1.281 \le a_4 \le 1.331$
$a_5 = 10.13$	$9.954 \le a_5 \le 10.31$

A population of 1000 individuals was used, which was manipulated through 1000 generations (G_{max}). The objective function was evaluated 1×10^6 times. The algorithm was allowed to select the design parameter values in the interval stated in Table 7. This interval corresponds to the function H_f. The best fitness value was obtained in the last generation. This solution represents an optimal solution for the mechanism.

Table 7. The lower (mm) and upper (mm) boundaries for each one of the design variables.

H_f	F									
	f_1	f_2	f_3	f_4	f_5	f_6	f_7	f_8	f_9	f_{10}
$f_m{}^{min}$	100	20	5	10	−10	−20	0	−100	−10	−10
$f_m{}^{max}$	1000	600	70	200	120	200	900	4	4	10

3. Improvement of the Existing Mechanism and Results

Figure 18 shows the evolution of the objective function along with the generations. The figure shows the convergence between the best fitness and average fitness values found during the GA iterations. The upper plot (Figure 18a) displays the best score value (black dots) and the mean score (blue dots) versus generation. It showed little progress in lowering the fitness value. The best individual's fitness value remained small throughout the generations. The lower plot (Figure 18b) shows the average distance between individuals at each generation and represents the population's diversity.

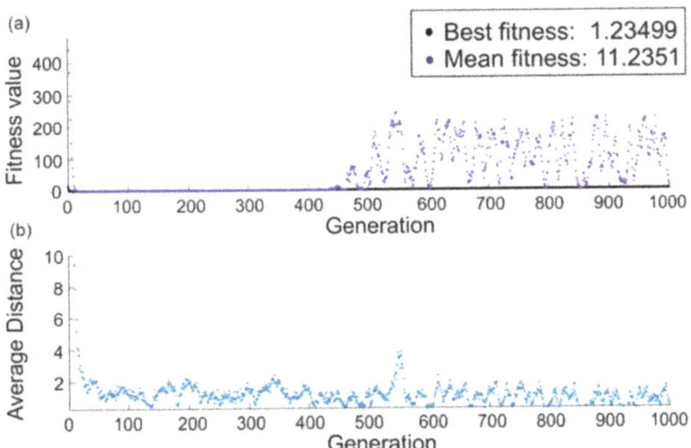

Figure 18. The transition of the fitness function. (**a**) Fitness value plot. (**b**) Average distance plot.

Table 8 shows the comparison between the optimal solution provided for the GA and the prototype's real dimensions. The GA provided a smaller mechanism through the best values of the seven parameters representing the mechanism's dimensions. The optimized mechanism followed the desired trajectory with a minimal error of 1.62 mm. Furthermore, it shows the distance the slider needs to move to reproduce the desired trajectory, which is generally proposed by the designer.

Table 8. Comparison between the existing prototype and the new proposal, units in mm.

F	f_1	f_2	f_3	f_4	f_5	f_6	f_7	f_8	f_9	f_{10}
Prototype	411.19	83.77	7.63	87.17	10	90	344	0	0	125
GA	295.11	54.29	21.11	69.88	−3.56	139.71	246.11	3.99	−6.51	9.99

Figure 19 shows the trajectories among the real data, the polynomial form, and the GA. It can be noticed that even though the error was minimal at the end of the trajectory, the mechanism was unable to reach the last point. This is logical considering the configuration of a slider-crank mechanism, derived from the fact that the slider-crank mechanism cannot close the path of the end-effector even when the distance traveled is increased.

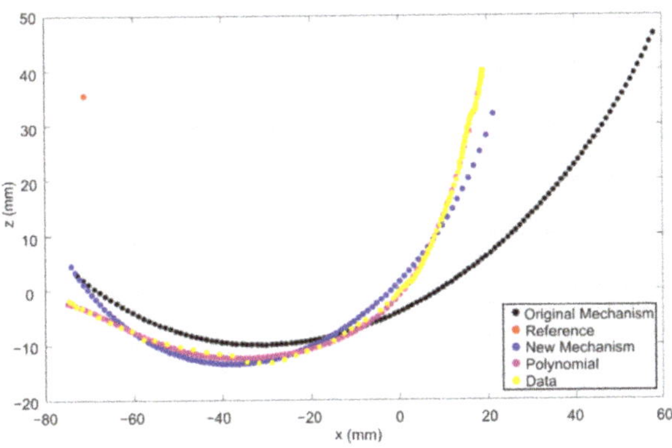

Figure 19. Results of the genetic algorithm (GA).

Figure 20 shows how the error between the trajectories was dramatically reduced when implementing the GA method. Figure 20a shows the error between the real trajectory recorded by the motion capture system and the end-effector's trajectory of the existing prototype (this figure has already been presented at the end of Section 2.2.2). Figure 20b shows the error between the real recorded trajectory and the end-effector's trajectory obtained with the GA method.

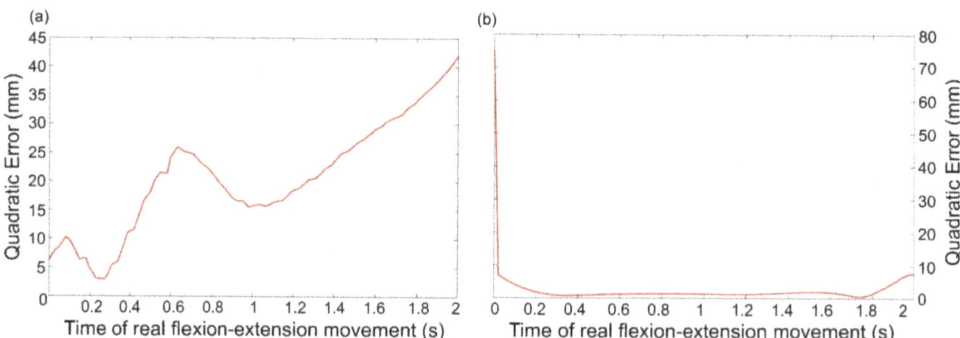

Figure 20. The error between trajectories: (**a**) with the original mechanism, (**b**) with the optimized mechanism.

To validate the quality of the convergence, Figure 21 shows the histogram of the last population. The histogram shows the distribution of the last population in the upper and lower limits of each individual. The frequency represents the number of times that an individual is presented in each limit. The cumulative frequency (expressed in percentage) is the sum of the absolute frequencies and indicates the number of individuals in each interval.

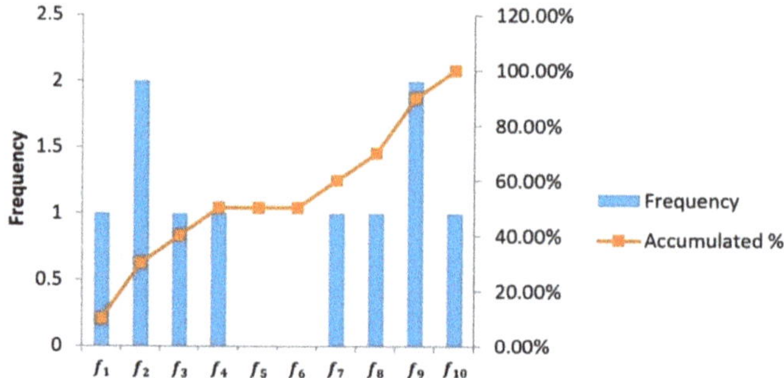

Figure 21. The last population histogram.

Concerning the angle of motion γ, a difference of 2.5 degrees was found between the real trajectory and the end-effector's trajectory of the new mechanism (as is shown in Figure 22).

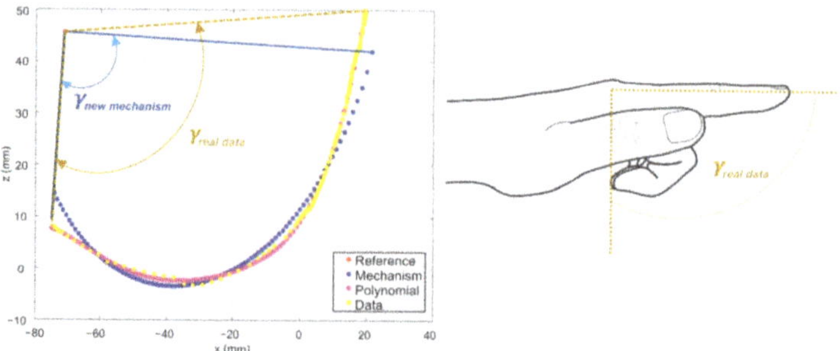

Figure 22. Angles of the amplitude of the trajectories.

Improvement in size over the existing mechanism can be made. Both mechanisms were drawn in extreme positions (corresponding to flexion and extension positions), as shown in Figure 23.

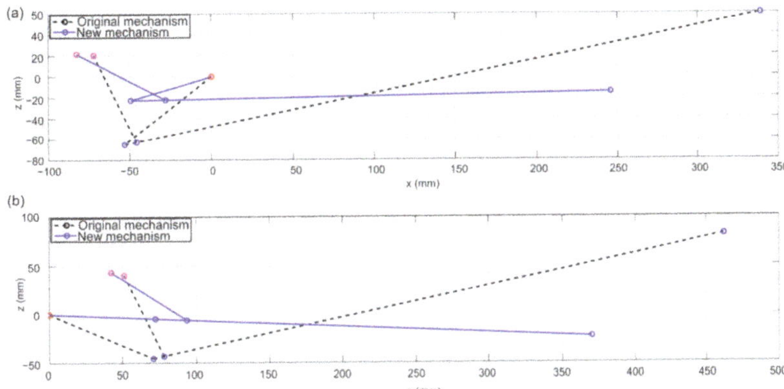

Figure 23. Comparison between the two mechanisms in their extreme positions: (**a**) flexion and (**b**) extension.

In Figure 24, a CAD proposal for the new mechanism and the desired curve is presented. Figure 25 shows the redesign of the four fingers. The distance between mechanisms, measured from the point D of every mechanism, was 28 mm.

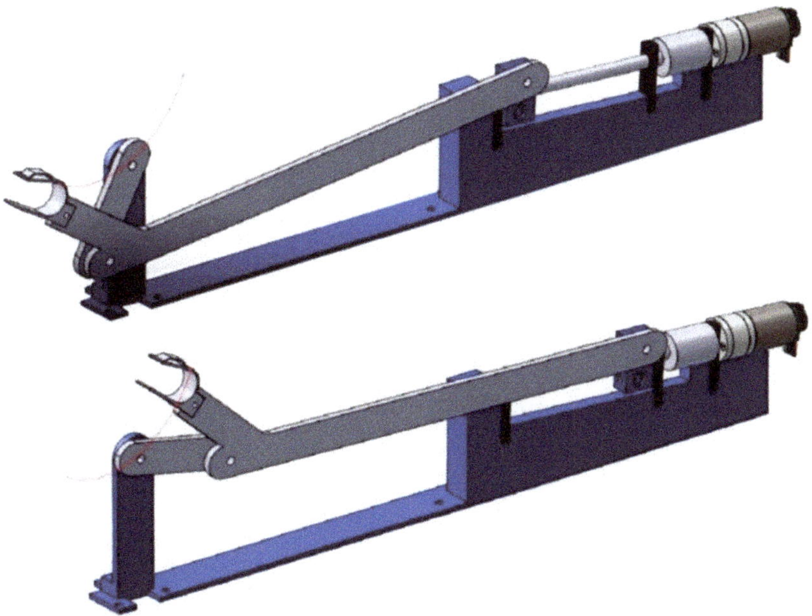

Figure 24. The new mechanism and the trajectory of the end-effector.

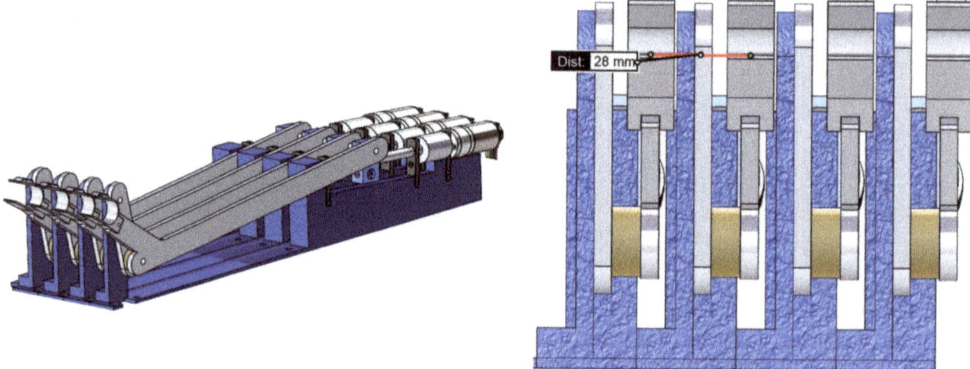

Figure 25. Four mechanisms from the index to small fingers.

4. Discussion

The shape and amplitude of the flexion–extension trajectory had to be considered to design a better mechanism. This is because the mechanism must be able to adjust the range of motion for different flexion–extension amplitudes. Since the optimized mechanism's configuration is an end-effector, it allows for the extrapolation of the result obtained for use with different subjects. The flexion–extension movement is not affected since the trace is at the fingertip.

Using the PCA technique, it was possible to transform the real flexion–extension movement from the xyz space to the xz plane. This simplification of the workspace simplifies the design of the mechanism and does not compromise the trajectory's shape. The use of PCA to derive postural synergies has mainly been used to study the hand's different kinematic configurations during the grasping and manipulation processes [48–51]. This article did not use PCA to reduce kinematic configurations during grasping. It consisted of verifying that the fingers' flexion–extension movement in space could be simplified to a movement in a plane, resulting in a design simplification of the mechanisms that tried to reproduce this trajectory.

The rehabilitation mechanism presented did not seek to emulate the hand's grip functions, but sought to intervene during the first stage of rehabilitation where the physiotherapist repeatedly and independently mobilized the patient's fingers to reduce stiffness and expand the range of motion.

Regarding GA performance, the low measure of population diversity ensured that the best individuals were close to the optimal solution. The number of generations and the population size used also contributed to finding the best solution. The algorithm converges due to the average distance between individuals in terms of fitness value decreasing as generations passed. The capability to converge to an optimal solution means that the objective function was successfully minimized based on the error between the real finger trajectory and the mechanism's trajectory. The error between the original mechanism's trajectories and the real finger increased as the flexion–extension movement was performed. In contrast, this error decreased with the dimensions of the mechanism obtained through the GA.

The amplitude of the trajectory γ of the optimal design had a difference of almost 2.5 degrees, compared with the real trajectory considered for the optimization process. It is assumed that the difference between trajectories is representative of the mechanism's performance. As can be observed in Figure 23, the mechanism trajectory followed the real trajectory very closely. This can be an indicator of mobility recovery in the patient because as the patient has an improvement in the rehabilitation process, the amplitude of the trajectory will increase. It is important to point out that clinical tests are required to evaluate the device's feasibility and potential benefits.

Even though the design results for a slider-crank-based rehabilitation device are promising, this type of mechanism cannot reproduce more closed trajectories. This means that for a certain number of patients, the device cannot meet successful rehabilitation criteria.

One of the issues addressed and successfully handled in this work is that the design vector includes the geometric parameters of the slider-crank mechanism and the distance that the slider needs to displace to generate the desired trajectory. This point is significant because this distance is the input of the mechanism and traditionally considered as known. For example, this distance was fixed in the existing prototype as a consideration of the designer at 125 mm [34].

The mechanism design was limited only to physical therapy because it only helps the patient improve flexion–extension movements. Occupational therapy requires a more complex mechanical design that allows for different finger trajectories to be performed as well as some virtual interaction schemes that simulate daily life activities.

The prototype proposed needs less workspace to operate than Amadeo [22], one of the available commercial finger rehabilitation devices. The optimized prototype is expected to occupy a volume of 0.09 m^3 against the 3 m^3 that Amadeo needs [52]. The optimized prototype pretends to improve every joint's ROM using a near trajectory to the natural flexion–extension trajectory regarding Amadeo's end-effector linear trajectories.

Another available commercial finger rehabilitation device is Digitrainer. The workspace of the DigiTrainer is about 0.017 m^3. However, its configuration uses end-effector rollers that make contact across the entire surface of the finger [53], which can be uncomfortable if the patient has sensitive skin. The optimized prototype presented in this paper avoids sensitive skin problems as the contact is only in the fingertip.

5. Conclusions

This article proposed an improvement of a finger rehabilitation device based on a real trajectory. The slider-crank mechanism is one of the most studied in the theory of mechanisms due to its simplicity, efficiency, and suitability for many applications in which it can be used. The significance of the slider-crank mechanism presented in this article is that the mono-actuator action is applied toward a medical solution. The presented mechanism is a novel finger flexion–extension rehabilitation device.

Its synthesis requires a careful analysis of the natural flexion–extension trajectory of the fingers. The focus of the optimization problem was the mono-objective formulation to minimize the error between the real trajectory and the mechanism's trajectory. This optimal solution allows the mechanism to follow a real trajectory more closely. The optimal dimensions of the finger rehabilitation prototype presented in this article allow for the optimization of the therapist's workspace without compromising the end-effector trajectory's performance.

An experimental setup was performed to obtain real trajectories using a motion capture system and healthy volunteers. The real fingertip trajectory was used as an input to the synthesis problem of the slider-crank mechanism. The presented method is still available for all other subjects and fingers and is also based on the recorded real trajectories.

The kinematic equations were reformulated to be included in the GA method. Using other synthesis methods for this application would mean omitting the design's original purpose: finger rehabilitation. The GA fitness function calculates the distance that the slider must travel to generate a flexion–extension path in the end-effector. This characteristic is a peculiarity of this function because GA proposes the optimal distance between the maximal flexion position and maximal extension position and does not need to be defined by the designer.

Future work will focus on a new topology of mechanisms with more degrees of freedom, which could describe more complex shapes of trajectories. In the next work, the study of transmitted forces on the fingers will also be examined. The idea of studying the behavior of the exchange forces between the patient and the device is to develop a more appropriate control strategy that could address the differences in the fingers' capabilities.

This is an important point because, at this stage, the mechanism can only reproduce the flexion–extension trajectory but cannot deal with situations such as the pain experienced by the patient when trying to complete the movements. A suitable rehabilitation device must consider the relationship between the force provided by the mechanism and finger movement.

Author Contributions: Conceptualization, M.A.L., E.C.C. and A.Z.G.; Methodology, A.Z.G. and M.A.L.; Software, M.A.L. and A.Z.G.; Validation, A.Z.G. and M.A.L.; Formal analysis, A.Z.G., M.A.L., and E.C.C.; Investigation, A.Z.G., M.A.L., and E.C.C.; Resources, M.A.L.; Data curation: A.Z.G., M.A.L., and E.C.C.; Writing-original draft preparation, A.Z.G.; Writing-review and editing: M.A.L., E.C.C., and A.Z.G.; Visualization, M.A.L., E.C.C., and A.Z.G.; Supervision, M.A.L. and E.C.C.; Project administration, M.A.L. and E.C.C.; Funding acquisition, M.A.L. and E.C.C. All authors have read and agreed to the published version of the manuscript.

Funding: This research received no external funding.

Institutional Review Board Statement: Ethical review and approval were waived for this study due to non-invasive experimentation being performed.

Informed Consent Statement: Verbal informed consent was obtained from all subjects involved in the study due to non-invasive experimentation being performed.

Data Availability Statement: Data sharing is not applicable to this article.

Acknowledgments: The authors are grateful to the University of Poitiers, France and the Instituto Politécnico Nacional (IPN), México. As well as the Consejo Nacional de Ciencia y Tecnología (CONACYT), México, for the facilities provided for this research.

Conflicts of Interest: The authors declare no conflict of interest.

References

1. Cooper, R.A.; Dicianno, B.E.; Brewer, B.; LoPresti, E.; Ding, D.; Simpson, R.; Grindle, G.; Wang, H. A Perspective on Intelligent Devices and Environments in Medical Rehabilitation. *Med. Eng. Phys.* **2008**, *30*, 1387–1398. [CrossRef] [PubMed]
2. Freis, N.E. La Rehabilitación En Ortopedia y Traumatología Parte, I. *Rev. Asoc. Argentina Ortop. Traumatol.* **2006**, *71*, 272–277. (In Spanish) [CrossRef]
3. Freis, N.E.; Heinrichs, K. La Rehabilitación En Ortopedia y Traumatología Parte II. *Rev. Asoc. Argentina Ortop. Traumatol.* **2006**, *71*, 362–368. (In Spanish)
4. Dutton, M. Range of Motion. In *Introduction to Physical Therapy and Patient Skills*; McGraw-Hill Education: New York, NY, USA, 2014.
5. Physical Rehabilitation. Available online: https://www.martinpetkov.com/your-opportunity/physical-rehabilitation (accessed on 9 December 2020).
6. Borobia, C. *Valoración Del Daño Corporal. Medicina de Los Seguros. Miembro Superior*; Valoración del Daño Corporal; Masson Elsevier: Barcelona, Spain, 2006. Available online: https://books.google.com.mx/books?id=OClwWvys-ysC (accessed on 19 April 2020).
7. Kim, Y.H.; Choi, J.H.; Chung, Y.K.; Kim, S.W.; Kim, J. Epidemiologic Study of Hand and Upper Extremity Injuries by Power Tools. *Arch. Plast. Surg.* **2019**, *46*, 63–68. [CrossRef]
8. Krebs, H.I.; Edwards, D.; Hogan, N. *Forging Mens et Manus: The MIT Robotic Therapy*; Reinkensmeyer, D.J., Dietz, V., Eds.; Springer International Publishing: Berlin, Germany, 2016; pp. 333–350. [CrossRef]
9. Burgar, C.G.; Lum, P.S.; Shor, P.C.; Machiel Van der Loos, H.F. Development of Robots for Rehabilitation Therapy: The Palo Alto VA/Stanford Experience. *J. Rehabil. Res. Dev.* **2000**, *37*, 663–673.
10. Fasoli, S.E.; Krebs, H.I.; Stein, J.; Frontera, W.R.; Hogan, N. Effects of Robotic Therapy on Motor Impairment and Recovery in Chronic Stroke. *Arch. Phys. Med. Rehabil.* **2003**, *84*, 477–482. [CrossRef]
11. Krebs, H.I.; Volpe, B.T.; Williams, D.; Celestino, J.; Charles, S.K.; Lynch, D.; Hogan, N. Robot-Aided Neurorehabilitation: A Robot for Wrist Rehabilitation. *IEEE Trans. Neural Syst. Rehabil. Eng.* **2007**, *15*, 327–335. [CrossRef]
12. Volpe, B.T.; Ferraro, M.; Krebs, H.I.; Hogan, N. Robotics in the Rehabilitation Treatment of Patients with Stroke. *Curr. Atheroscler. Rep.* **2002**, *4*, 270–276. [CrossRef]
13. Fasoli, S.E.; Adans-Dester, C.P. A Paradigm Shift: Rehabilitation Robotics, Cognitive Skills Training, and Function after Stroke. *Front. Neurol.* **2019**, *10*, 1–10. [CrossRef]
14. Rodríguez-Prunotto, L.; Cano-De La Cuerda, R.; Cuesta-Gómez, A.; Alguacil-Diego, I.M.; Molina-Rueda, F. Terapia Robótica Para La Rehabilitación Del Miembro Superior En Patología Neurológica. *Rehabilitacion* **2014**, *48*, 104–128. [CrossRef]
15. Aggogeri, F.; Mikolajczyk, T.; Kane, J.O. Robotics for Rehabilitation of Hand Movement in Stroke Survivors. *Adv. Mech. Eng.* **2019**, *11*, 1–14. [CrossRef]

16. Maciejasz, P.; Eschweiler, J.; Gerlach-hahn, K.; Jansen-troy, A.; Leonhardt, S. A Survey on Robotic Devices for Upper Limb Rehabilitation. *J. Neuroeng. Rehabil.* **2014**, *11*, 1–29. [CrossRef] [PubMed]
17. Yue, Z.; Zhang, X.; Wang, J. Hand Rehabilitation Robotics on Poststroke Motor Recovery. *Behav. Neurol.* **2017**, *2017*, 3908135. [CrossRef] [PubMed]
18. Wege, A.; Zimmermann, A. Electromyography Sensor Based Control for a Hand Exoskeleton. In Proceedings of the 2007 IEEE International Conference on Robotics and Biomimetics, ROBIO, Sanya, China, 15–18 December 2007; pp. 1470–1475. [CrossRef]
19. Bhandari, V.B. *Design of Machine Elements*; Tata McGraw-Hill Education: New Delhi, Delhi, India, 2007. Available online: https://books.google.com.mx/books?id=d-eNe-VRc1oC (accessed on 31 March 2020).
20. Chiri, A.; Vitiello, N.; Giovacchini, F.; Roccella, S.; Vecchi, F.; Carrozza, M.C. Mechatronic Design and Characterization of the Index Finger Module of a Hand Exoskeleton for Post-Stroke Rehabilitation. *IEEE/ASME Trans. Mechatron.* **2012**, *17*, 884–894. [CrossRef]
21. Pierce, R.M.; Fedalei, E.A.; Kuchenbecker, K.J. A Wearable Device for Controlling a Robot Gripper with Fingertip Contact, Pressure, Vibrotactile, and Grip Force Feedback. *IEEE Haptics Symp. HAPTICS* **2014**, 19–25. [CrossRef]
22. Amadeo®: The Hand Therapy World Champion. Available online: https://tyromotion.com/en/produkte/amadeo/ (accessed on 25 August 2019).
23. Amar, J.; Nagase, K. Design Optimization of Tree-Type Robotic Systems Using Exponential Coordinates and Genetic Algorithms. In Proceedings of the 2020 6th International Conference on Control, Automation and Robotics (ICCAR), Singapore, 20–23 April 2020; pp. 67–73. [CrossRef]
24. Premachandra, H.A.G.C.; Herath, H.M.A.; Suriyage, M.P.; Thathsarana, K.M.; Amarasinghe, Y.W.; Gopura, R.A.R.; Nanayakkara, S.A. Genetic Algorithm Based Pick and Place Sequence Optimization for a Color and Size Sorting Delta Robot. In Proceedings of the 2020 6th International Conference on Control, Automation and Robotics, ICCAR 2020, Singapore, 20–23 April 2020; pp. 209–213.
25. Jamwal, P.K.; Kapsalyamov, A.; Hussain, S.; Ghayesh, M.H. Performance Based Design Optimization of an Intrinsically Compliant 6-Dof Parallel Robot. *Mech. Based Des. Struct. Mach.* **2020**, 1–16. [CrossRef]
26. Zeiaee, A.; Soltani-Zarrin, R.; Langari, R.; Tafreshi, R. Kinematic Design Optimization of an Eight Degree-of-Freedom Upper-Limb Exoskeleton. *Robotica* **2019**, *37*, 2073–2086. [CrossRef]
27. Zhou, L.; Li, Y.; Bai, S. A Human-Centered Design Optimization Approach for Robotic Exoskeletons through Biomechanical Simulation. *Rob. Auton. Syst.* **2017**, *91*, 337–347. [CrossRef]
28. Moosavian, S.A.A.; Nabipour, M.; Absalan, F.; Akbari, V. RoboWalk: Explicit Augmented Human-Robot Dynamics Modeling for Design Optimization. *arXiv* **2019**, arXiv:1907.04114.
29. Hernandez, E.; Valdez, S.I.; Carbone, G.; Ceccarelli, M. Design Optimization of a Cable-Driven Parallel Robot in Upper Arm Training-Rehabilitation Processes. *Mech. Mach. Sci.* **2018**, *54*, 413–423. [CrossRef]
30. Dong, H.; Asadi, E.; Qiu, C.; Dai, J.; Chen, I.M. Geometric Design Optimization of an Under-Actuated Tendon-Driven Robotic Gripper. *Robot. Comput. Integr. Manuf.* **2018**, *50*, 80–89. [CrossRef]
31. Bhupender, B.; Rahul, R. Study and Analysis of Design Optimization and Synthesis of Robotic ARM. *Int. J. Adv. Eng. Manag. Sci.* **2016**, *2*, 239459.
32. Dutton, M. Improving Mobility. In *Dutton's Orthopaedic Examination, Evaluation, and Intervention*; McGraw-Hill Education: New York, NY, USA, 2020.
33. Edgerton, V.R.; Roy, R.R. Robotic Training and Spinal Cord Plasticity. *Brain Res. Bull.* **2009**, *78*, 4–12. [CrossRef] [PubMed]
34. Aguilar-Pereyra, J.F.; Castillo-Castaneda, E. Design of a Reconfigurable Robotic System for Flexoextension Fitted to Hand Fingers Size. *Appl. Bionics Biomech.* **2016**, 1712831. [CrossRef] [PubMed]
35. Yu, Y.; Iwashita, H.; Kawahira, K.; Hayashi, R. Development of Rehabilitation Device for Hemiplegic Fingers by Finger-Expansion Facilitation Exercise with Stretch Reflex. In Proceedings of the 2013 IEEE International Conference on Robotics and Biomimetics, ROBIO 2013, Shenzhen, China, 12–14 December 2013; pp. 1317–1323. [CrossRef]
36. Castillo-Castaneda, E.; Bemardo-Vasquez, A. Personalized Design of a Hand Prosthesis Considering Anthropometry of a Real Hand Extracted from Radiography. In Proceedings of the 2017 IEEE International Conference on Rehabilitation Robotics (ICORR), London, UK, 17–20 July 2017; pp. 1215–1220. [CrossRef]
37. Kamper, D.G. Stereotypical Fingertip Trajectories During Grasp. *J. Neurophysiol.* **2003**, *90*, 3702–3710. [CrossRef]
38. Bishop, L.; Gordon, A.M.; Kim, H. Hand Robotic Therapy in Children with Hemiparesis: A Pilot Study. *Am. J. Phys. Med. Rehabil.* **2017**, *96*, 1–7. [CrossRef]
39. Stein, J.; Bishop, L.; Gillen, G.; Helbok, R. Robot-Assisted Exercise for Hand Weakness After Stroke. *Am. J. Phys. Med. Rehabil.* **2011**, *90*, 887–894. [CrossRef]
40. Treatment Technology. Available online: http://synergicpro.com/en/treatment-en/ (accessed on 9 December 2019).
41. Zapatero-Gutiérrez, A.; Castillo-Castañeda, E. Control Design for a Fingers Rehabilitation Device. In Proceedings of the 2017 IEEE 3rd Colombian Conference on Automatic Control, CCAC 2017, Cartagena, Colombia, 18–20 October 2018; pp. 1–6. [CrossRef]
42. Jolliffe, I.T. *Principal Component Analysis*; Springer Series in Statistics; Springer: New York, NY, USA, 2013. Available online: https://books.google.com.mx/books?id=-ongBwAAQBAJ (accessed on 19 April 2020).
43. Ringnér, M. What Is Principal Component Analysis? *Nat. Biotechnol.* **2008**, *26*, 303–304. [CrossRef]
44. Practical Guide to Principal Component Methods in R. Available online: http://www.sthda.com (accessed on 8 October 2019).

45. Erdman, A.G. Computer-Aided Mechanism Design: Now and the Future. *J. Mech. Des. Trans. ASME* **1995**, *117*, 93–100. [CrossRef]
46. Laribi, M.A.; Mlika, A.; Romdhane, L.; Zeghloul, S. A Combined Genetic Algorithm-Fuzzy Logic Method (GA-FL) in Mechanisms Synthesis. *Mech. Mach. Theory* **2004**, *39*, 717–735. [CrossRef]
47. What Is the Genetic Algorithm? Available online: https://la.mathworks.com/help/gads/what-is-the-genetic-algorithm.html (accessed on 15 April 2019).
48. Ortenzi, D.; Scarcia, U.; Meattini, R.; Palli, G.; Melchiorri, C. Synergy-Based Control of Anthropomorphic Robotic Hands with Contact Force Sensors. *IFAC-PapersOnLine* **2019**, *52*, 340–345. [CrossRef]
49. Ficuciello, F.; Palli, G.; Melchiorri, C.; Siciliano, B. Postural Synergies of the UB Hand IV for Human-like Grasping. *Robot. Auton. Syst.* **2014**, *62*, 515–527. [CrossRef]
50. Palli, G.; Ficuciello, F.; Scarcia, U.; Melchiorri, C.; Siciliano, B. *Experimental Evaluation of Synergy-Based in-Hand Manipulation*; IFAC: Cape Town, South Africa, 2014; Volume 19. [CrossRef]
51. Santello, M.; Flanders, M.; Soechting, J.F. Postural Hand Synergies for Tool Use. *J. Neurosci.* **1998**, *18*, 10105–10115. [CrossRef] [PubMed]
52. Amadeo®in Practice. Available online: https://irp-cdn.multiscreensite.com/91b5b819/files/uploaded/Factsheet_Amadeo_V1_en_screen.pdf (accessed on 13 December 2019).
53. Digitrainer. Available online: https://www.ostracon.gr/wp-content/uploads/2020/12/DigiTrainer_Flyer_english_2019-20_web.pdf (accessed on 13 December 2019).

Article

Safe pHRI via the Variable Stiffness Safety-Oriented Mechanism (V2SOM): Simulation and Experimental Validations [†]

Younsse Ayoubi , Med Amine Laribi *, Marc Arsicault and Saïd Zeghloul

Dept. GMSC, Pprime Institute, CNRS, University of Poitiers, ENSMA, UPR 3346 Poitiers, France;
you.ayoubi@gmail.com (Y.A.); marc.arsicault@univ-poitiers.fr (M.A.); said.zeghloul@univ-poitiers.fr (S.Z.)
* Correspondence: med.amine.laribi@univ-poitiers.fr
† This paper is an extended version of the paper entitled "New Variable Stiffness Safety Oriented Mechanism for Cobots' Rotary Joints", presented at International Conference on Robotics in Alpe-Adria Danube Region, Patras, Greece, 6–8 June 2018.

Received: 14 April 2020; Accepted: 28 May 2020; Published: 30 May 2020

Abstract: Robots are gaining a foothold day-by-day in different areas of people's lives. Collaborative robots (cobots) need to display human-like dynamic performance. Thus, the question of safety during physical human–robot interaction (pHRI) arises. Herein, we propose making serial cobots intrinsically compliant to guarantee safe pHRI via our novel designed device, V2SOM (variable stiffness safety-oriented mechanism). Integrating this new device at each rotary joint of the serial cobot ensures a safe pHRI and reduces the drawbacks of making robots compliant. Thanks to its two continuously linked functional modes—high and low stiffness—V2SOM presents a high inertia decoupling capacity, which is a necessary condition for safe pHRI. The high stiffness mode eases the control without disturbing the safety aspect. Once a human–robot (HR) collision occurs, a spontaneous and smooth shift to low stiffness mode is passively triggered to safely absorb the impact. To highlight V2SOM's effect in safety terms, we consider two complementary safety criteria: impact force (ImpF) criterion and head injury criterion (HIC) for external and internal damage evaluation of blunt shocks, respectively. A pre-established HR collision model is built in Matlab/Simulink (v2018, MathWorks, France) in order to evaluate the latter criterion. This paper presents the first V2SOM prototype, with quasi-static and dynamic experimental evaluations.

Keywords: pHRI; variable stiffness actuator; V2SOM; friendly cobots; safety criteria; human–robot collisions

---

1. Introduction

The currently emerging manufacturing paradigm, known as Industry 4.0, is behind the rethinking of how industrial processes are designed in order to increase their efficiency and flexibility, together with higher levels of automatization [1]. To this end, Industry 4.0 integrates multiple technologies such as the Internet of Things (IoT) [1], artificial intelligence (AI) [1], and cyber-physical systems. Accordingly, robotics researchers are proposing new solutions [2,3] whereby collaborative robots, known as cobots, physically collaborate with well-qualified operators to achieve the goals of this revolution. Hence, the problem of the human subject's safety vs. the robot's high dynamic performances arises. This means that the next generation of widely used cobots should manifest human-friendly attributes. In the literature, two main approaches were proposed to tackle this problem: active impedance control (AIC) and passive compliance (PC). In the former, the impedance [4] is controlled to display safe behavior vis-à-vis the robot's environment, including the humans within it. In the case of a fast human–robot (HR) collision, this approach, because it cannot respond as quickly as within 200 ms [5,6], can allow severe damage to the impacted human [7]. As a result, the PC approach attracted interest due to its instantaneous reaction to any potential impact. This latter, potential impact is defined as a quantification of the maximum impact force a robot can exert in a collision with a stationary object [8]. To emphasize, it is the combination of high mobile inertia and high velocity (i.e., the high kinetic energy) that makes robots dangerous [7,9]. With this in mind, achieving safety without compromising the desired dynamics boils down to reducing the reflected inertia, which is the key feature of the PC approach. Indeed, from a dynamic perspective, integrating passive mechanisms in robot joints decouples a certain colliding inertia from the rest of the robot. This reduces the overall kinetic energy absorbed by the impacted human. In line with this, the series elastic actuator [10] is among the first solutions that has a constant stiffness profile. A passive compliance system composed of purely mechanical elements often provides faster and more reliable responses to dynamic collisions [11]. To enhance the latter's design capacity to react to load variation, a series parallel elastic actuator was introduced by Mathijssen and co-workers [12]. To improve the safety of physical human–robot interaction (pHRI), Zinn et al. presented [13] a distributed macro-mini (DM [2]) actuation system that puts forward low-inertia actuators to interact with the human subject. This allows for both safety and a fast control reaction via the low inertia actuated part. In order to deal with any unsupervised collision while handling variable loads, the concept of the variable stiffness actuator (VSA) emerged. Through the years, several VSAs have been proposed, as discussed in previous studies [3,14]. Their design concepts resulted in different structural paradigms (e.g., serial or antagonistic), different stiffness profiles, and a wide range of power to mass ratios. Herein, a great emphasis is placed on a VSA's stiffness profile vs. the safety aspect. The proposed approach, leading to prototype V2SOM (variable stiffness safety-oriented mechanism) [1], presents the following novelties compared to the literature:

- The stiffness behavior, in the vicinity of zero deflection, is smoothened via a cam follower mechanism.
- The stiffness sharply sinks to maintain, theoretically, as discussed in Section 2, a constant torque threshold in case of a collision.
- The torque threshold, T_{max}, is tunable according to the load variation.

This paper presents the V2SOM design as well as a simulation and experimental validations. V2SOM is primarily conceived to enhance safety in normal working routines, as well as in the case of uncontrolled HR collisions. In Section 2, this aspect is discussed in light of a comparative study between several stiffness profiles. Then, V2SOM's working principle and mechanical structure are illustrated. In Section 3, the impact of V2SOM on human safety is studied in terms of two complementary safety criteria: the head injury criterion (HIC) and impact force criterion (ImpF). The study is carried out via a HR collision model simulation found in previous works [15,16]. The experimental validation of the V2SOM is presented in Section 4. This section includes the quasi-static characterizations as well as the HR collision tests. Section 5 summarizes the outcomes of the present study and gives some future perspectives.

2. Materials and Methods

The VSA's design concept aims to make load-adjustable compliant robots by implementing a variable stiffness mechanism (VSM) in series with the actuation system, as depicted in Figure 1. However, a VSM can simply be described as a tunable spring with a basic nonlinear stiffness profile. With this in mind, we will discuss the properties of the different existing basic stiffness curves and highlight the V2SOM profile that we propose in this work.

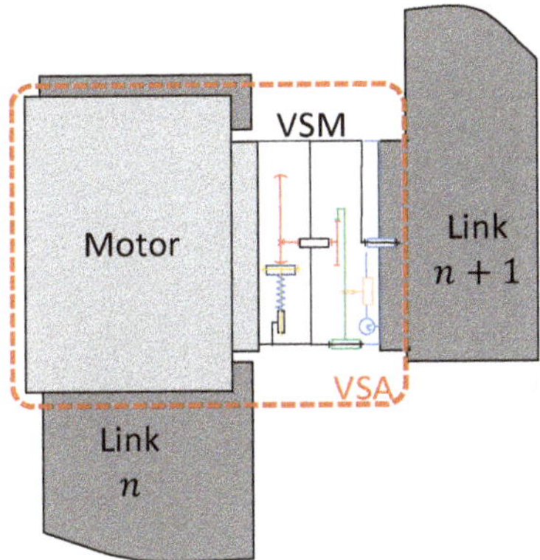

Figure 1. Variable stiffness actuator (VSA) including the variable stiffness safety-oriented mechanism (V2SOM).

2.1. Different VSMs' Basic Stiffness Curves

Previous works on VSAs resulted in design concepts that differ in several aspects, such as mass to volume ratio, elastic energy to mass or volume ratios, working principal, etc. (see [13,17–20] for more details). Herein, we focus on the VSA's stiffness profile. The VSAs mentioned in other studies (see [14,17–20], representative of the current state-of-the-art) can be classified into one of the three categories illustrated in Table 1.

The stiffness profiles of VSAs might be viewed singularly as a tunable basic stiffness profile. These basic stiffness profiles are shown in Figure 2:

- Constant stiffness: this basic stiffness profile is shown in Figure 2a and exemplified in the second column of Table 1 by an actuator with adjustable stiffness (AwAS-II) [19]. At each curve of the AwAS-II characteristic, the stiffness remains practically constant. To adapt this mechanism to a variable load, the torque's slope is tuned via a small motor.
- Biomimetically inspired stiffness, shown in Figure 2b, represents different VSAs, such as the ones developed at the German Aerospace Center (DLR) Institute of Robotics and Mechatronics, the floating spring joint (FSJ) [20], QA-joint (Quasi-Antagonistic Joint) [21], FAS (Flexible Antagonistic Spring) [22] or the one presented by Ayoubi et al. [23]. The third column of Table 1 shows the floating spring joint (FJS) prototype with a stiffness that increases along with the deflection, the same as in biological muscles [23].
- Torque limiter: Park introduced [24,25] the safe joint mechanism (SJM) (see fourth column of Table 1), which is based on a slider-crank mechanism. Its basic stiffness curve corresponds to the one in Figure 2c. This mechanism is supposed to remain stiff under a certain torque level T_1. Upon exceeding it, the stiffness rapidly drops to maintain the torque at T_{max} level, which represents the safety threshold. As the SJM is a slider-crank-based mechanism that is well known for its sensitivity to friction in the vicinity of zero deflection, the safety threshold may be easily exceeded in the case of blunt shocks, as shown by Park and colleagues [26]. For this reason, we introduced the V2SOM basic curve (see Figure 2c,d). In the next section, the full V2SOM characteristic, which allows for coping with load variation, unlike the SJM, is shown. The latter curve presents a finite value stiffness near zero deflection, which continuously drops to display a constant torque threshold T_{max}. The following equation depicts the desired behavior:

$$T_{\gamma}(\gamma) = T_{max}(1 - e^{-s\gamma}),\tag{1}$$

where s is a positive constant and the γ elastic deflection angle is in the range of 0 to $\pi/2$.

Table 1. The three main categories of VSAs according to their stiffness characteristic.

Stiffness Profile	Variable-Constant Stiffness	Variable-Biomimetically Inspired Stiffness	Torque Limiter
Illustrative prototype	Actuator with Adjustable Stiffness (AwAS-II) 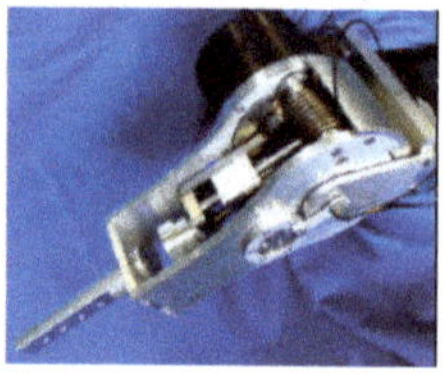	Floating Spring Joint (FSJ) 	Safe Joint Mechanism (SJM) 
Prototype's full characteristic	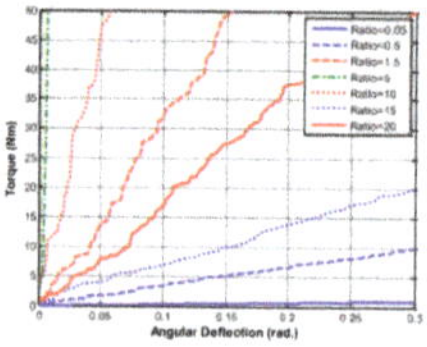	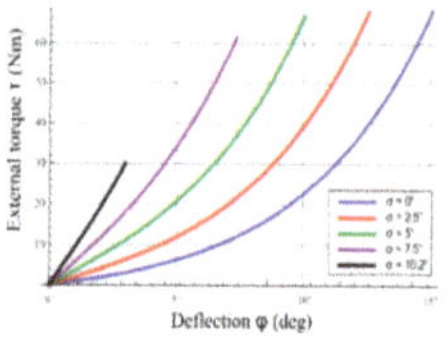	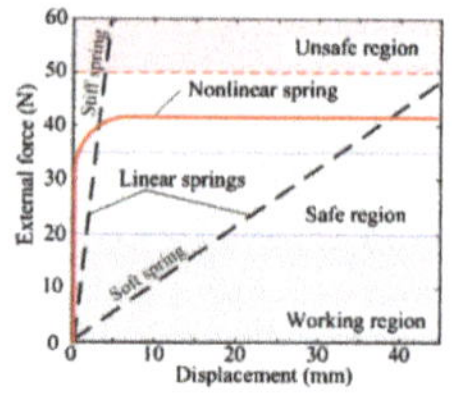

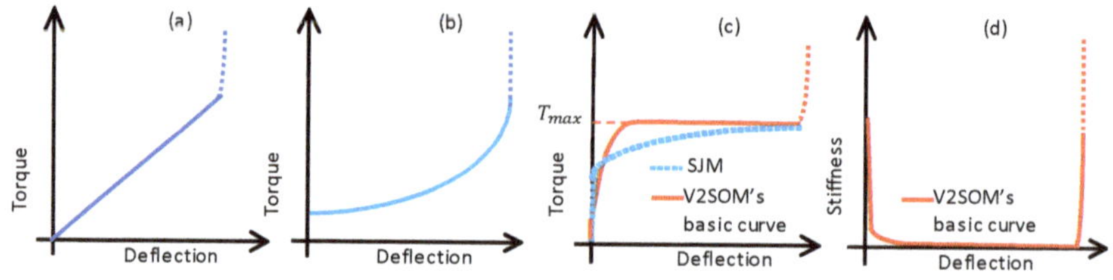

Figure 2. Different basic torque curves and V2SOM stiffness curves: (**a**) Constant stiffness, (**b**) Biomimetically inspired stiffness, (**c**,**d**) V2SOM basic curves.

To compare the three stiffness profiles, we adopted the following factors:

- Factor 1. The maximal stored elastic energy in "normal working conditions," (i.e., not during collision scenarios). This quantifies how much elastic energy is stored in the VSM that can be unleashed as collision kinetic energy, hence increasing the damaging effect of a potential collision.
- Factor 2. Passive torque limitation: in the case of a fast HR collision, it is more convenient to instantaneously contain the exerted torque with the VSM rather than as a control-based reaction [5,6].
- Factor 3. Gravity-induced elastic deflection: this criterion quantifies the VSM's ability to passively limit this elastic deflection, which reduces the controller's compensation action.

Based on the three factors, we ranked the stiffness profiles in Figure 2 accordingly, where the number of '+' reflects the factor's qualitative value. The results are given in Table 2, which shows that the V2SOM basic curve displays better features in normal working conditions (i.e., factors 1 and 3), and in a collision scenario based on factor 2.

Table 2. Results of a comparative analysis of different VSAs' basic torque curves.

Factors	Profile Figure 2a	Profile Figure 2b	V2SOM Basic Profile
Maximal stored elastic energy	+ +	+ + +	+
Passive torque limitation	+ +	+	+ + +
Gravity-induced elastic deflection	+ +	+	+ + +

2.2. Working Principle of V2SOM

V2SOM contains two functional blocks, as depicted in Figure 3a: a nonlinear stiffness generator block (SGB) and a stiffness adjusting block (SAB). To simplify the understanding of the V2SOM kinematic scheme, Figure 3b represents its cross section, which is symmetrical to the rotation axis L_1. The SGB is based on a cam follower mechanism, whereby the cam's rotation γ about the L_5 axis, between $-90°$ and $90°$, induces the translation of its follower according to the slider L_6. Then the follower extends its attached spring. At this level, a deflection angle γ corresponds to a torque value T_γ exerted on the cam. The wide range of this elastic deflection needs to be reduced to a lower range of $-20 \leq \theta \leq 20$, as is widely considered in most VSAs [21,27].

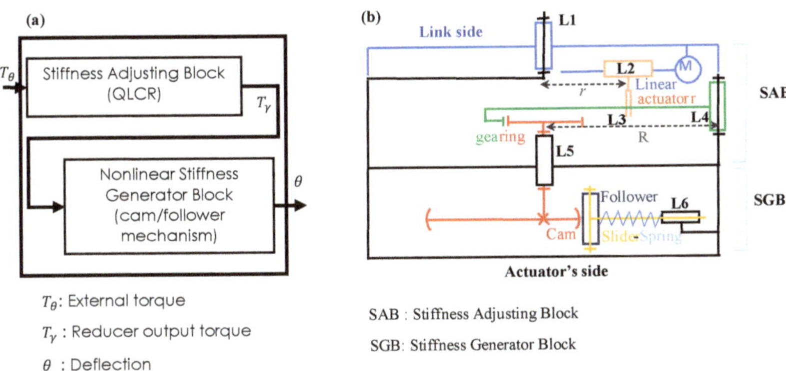

Figure 3. Block representation of the V2SOM, (**a**) Two functional blocks of V2SOM (**b**) V2SOM kinematic scheme.

Figure 4 shows various simplified diagrams necessary to understand the functioning principle of this block. In Figure 4a, the cam follower system is in the resting position, meaning that the springs are relaxed and the cam's applied torque about its rotation axis is $T_\gamma = 0$. Applying a torque $T_\gamma \neq 0$ generates a deflection angle γ and the translation of the sliders supporting the followers, which results in the compression of the springs (Figure 3b).

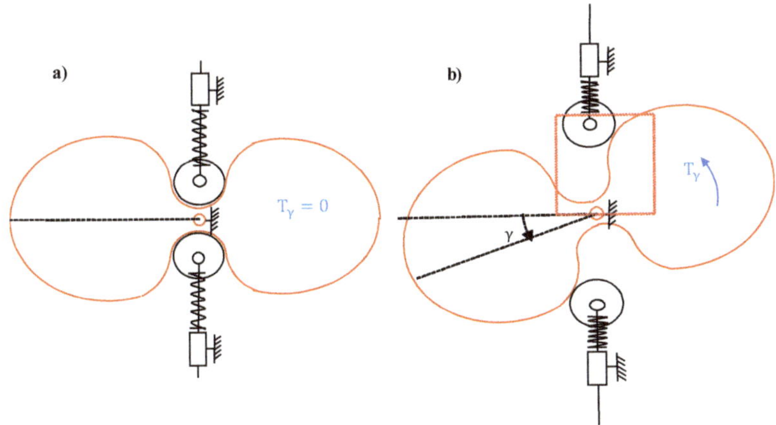

Figure 4. Stiffness generator block: (**a**) at rest $\gamma = 0$ ($T_\gamma = 0$); (**b**) at deflection $\gamma \neq 0$ (T_γ).

The stiffness adjusting block (SAB) acts as a reducer by using a gear ring system, which is considered to be made up of gear mechanisms commonly used to transform the rotary motion into either rotary or linear motion [28]. Furthermore, the SAB serves as a variable reducer thanks to the linear actuator M, which controls the distance r while driving the gears in a lever-like configuration. The reduction ratio of SAB is continuously tunable, allowing V2SOM to cope with the external load T_θ, where the link side makes a deflection angle θ relative to the actuator side.

Figure 5 shows the symmetrical two ring gears (in blue and green in Figure 5a) geared to a central spur gear (in red). These two ring gears are driven by a symmetrical double lever arm system via two rods (in yellow). These rods, which are part of the lever arm L_2, as illustrated in Figure 4, slide freely along the pocket of the ring gear, creating the prismatic joint shown in Figure 5b. Adjusting the position r of the rods along the lever arm via linear actuator M changes the transmission ratio of the lever arm system, and hence the reduction ratio of the SAB. The gears' rotation induces a variation of the rods' position along the pockets, which is characterized by the distance x in Figure 5b.

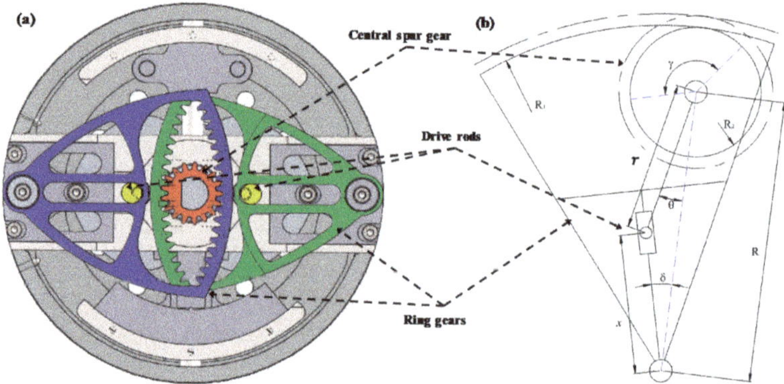

Figure 5. Stiffness adjusting block (SAB) illustration of (**a**) cross section of CAD (computer aided design) model and (**b**) half scheme with a single ring gear.

The V2SOM CAD (computer aided design) model is presented in Figure 6. More details are given in the V2SOM patent [29,30]. Figure 6a shows the two blocks, the stiffness adjustable block and the generator block, with a zoomed-in view. Each block is depicted by additional views of their mechanical parts (Figure 6b–d) with correspondence to the kinematic sketch in Figure 3b. One can identify the linear actuators, the gearing, and cam follower system with its actuation side.

The V2SOM blocks, as shown in Figure 4, are connected rigidly to fulfill separate dedicated tasks:

- The SGB is characterized by the curve of the torque T_γ vs. deflection angle γ. This curve is obtained through the cam profile, the followers, and other design parameters. The basic torque curve leading to the torque characteristic of the V2SOM is depicted in Figure 6a. This basic curve is elaborated with a torque threshold equal to $T_{max} = 2.05$ Nm.
- The SAB is considered a quasi-linear continuous reducer (QLCR) and defined by its ratio expression given in Figure 6b. The ratio is a function of the NL (Nonlinear) factor, the deflection angle γ, and the reducer's tuning parameter r. The NL (Nonlinear) factor is linked to the SAB's internal parameters and can be approximated with a constant when the deflection θ range is between −20 and 20; this process is discussed in the next section.

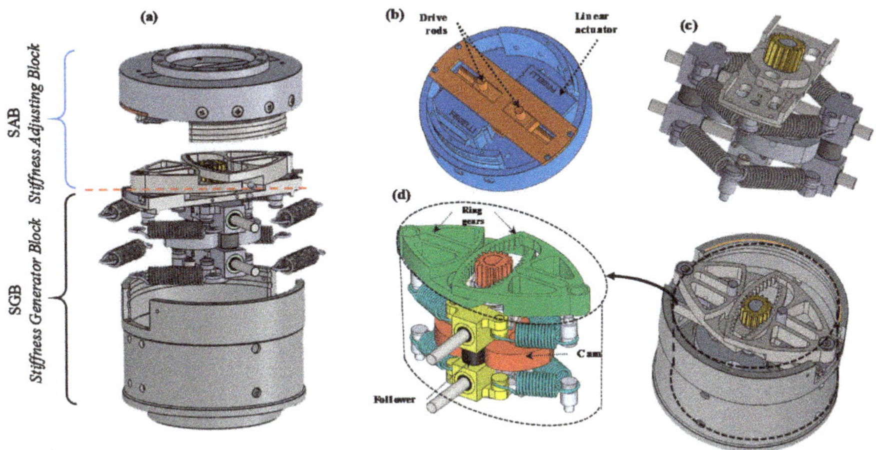

Figure 6. The V2SOM prototype (**a**) Exploded view of V2SOM (**b**) CAD model of SAB (**c**) CAD model of the cam follower mechanism with springs (**d**) CAD model of SGB.

Figure 7c shows the V2SOM characteristics resulting from Figure 7a, with seven increasing reduction ratio settings (seven values of torque tuning). Thanks to the QLCR behavior of the SAB, the curves in Figure 7c follow a formula similar to Equation (1), with the specific tunable constant s and a deflection range.

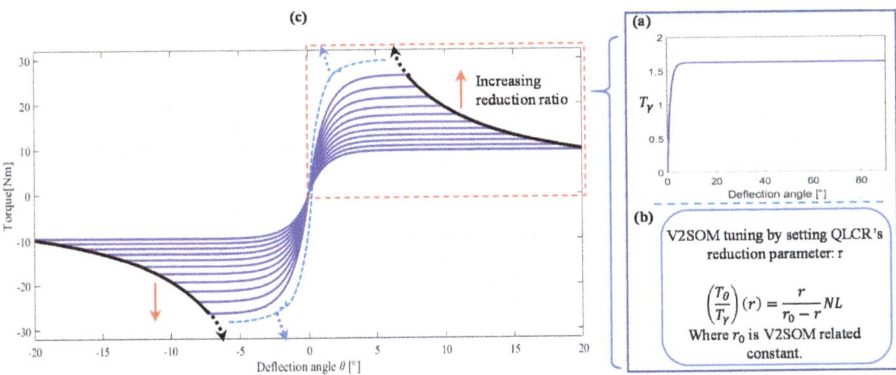

Figure 7. (**a**) Example of V2SOM basic torque curve with (**b**) quasi-linear continuous reducer (QLCR); (**c**) illustration of the V2SOM torque characteristic with seven QLCR settings.

In general, V2SOM has two working modes, between which a transition smoothly takes place in the case of blunt shock, as illustrated in Figure 8. The normal working mode of the V2SOM is the one with linear region, mode (I), which allows the system to avoid a loss of control if the load exceeds the maximum. In case of a collision, mode (II), with the quasi-linear region, is activated. A high stiffness mode (I) is defined within the deflection range $[0, \theta_1]$ and the torque range $[0, T_1]$. The T_1 value defines the torque in normal working conditions. Exceeding this torque value means that the shock absorbing mode is triggered, characterized by a low stiffness that leads to the torque threshold T_{max}. The T_1 torque value, which limits the normal working mode of V2SOM, is an online tunable value. Adjusting this value allows us to cope with possible load variations due to the robot's dynamics.

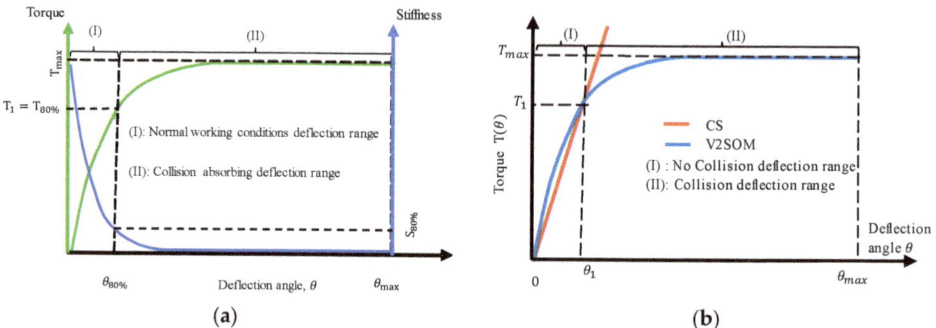

Figure 8. (**a**) V2SOM working modes; (**b**) statically equivalent V2SOM and constant stiffness (CS) profiles.

2.3. Understanding Stiffness and Adjusting the Block's Behavior

The QLCR behavior is exemplified in Figure 9, where two kinds of curves are shown:

- Solid curves $\frac{CT_\theta}{T_\gamma}(\gamma, r)$ represent the QLCR reduction ratio, multiplied by a constant C related to the mechanism's parameters.
- Dotted line curves $IR(\gamma, r)$ represent an ideally equivalent reducer (IR), where the reduction ratio is constant while the deflection angle γ changes. The following equation describes this ideal approximation:

$$IR(\gamma, \ r) = \frac{r}{C(r_0 - r)}. \tag{2}$$

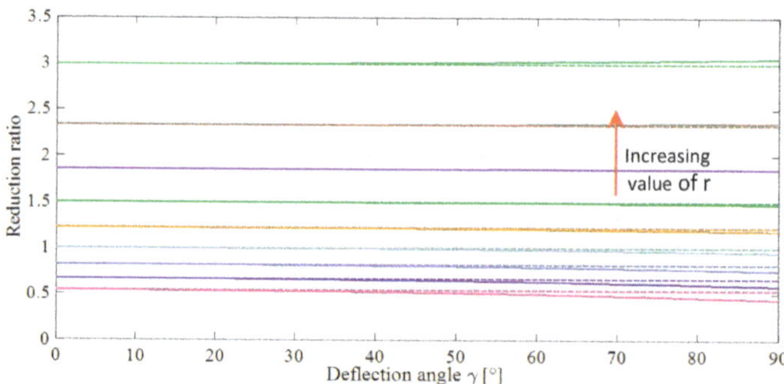

Figure 9. Reduction ratio: QLCR in solid lines: $\frac{CT_\theta}{T_\gamma}(\gamma, r)$ and ideally equivalent reducer (IR) in dotted lines: $IR(\gamma, r)$.

Both curves present the same reduction ratio in the vicinity of zero deflection. This means that QLCR can be considered as an ideal tunable reducer near zero deflection (i.e., not during a collision scenario). Thereupon, QLCR can be tuned using Equation (2). When a collision takes place, the QLCR reduction ratio starts to diverge slightly from the corresponding IR curves. This slight change is not problematic because the function of V2SOM in this phase is to absorb shock energy rather than to precisely set a safety threshold.

At each torque vs. deflection curve of Figure 7 (i.e., for a given reduction ratio *r*), the stiffness passively varies from a high stiffness value near zero deflection to a practically null stiffness for which the torque attains its threshold; for example, see Figure 10b at the setting $\frac{r}{r_0} = 0.475$.

As for the stiffness modulation in the vicinity of zero deflection angle $\left.\frac{dT_\theta}{d\theta}\right|_{\theta=0}$, Figure 10a shows the variation range of the presented prototype. As can be seen, the V2SOM's stiffness near zero deflection allows for a wide modulation range that varies from 670 to 13560 Nm/rad for $\frac{r}{r_0} = 0.4 \to 0.75$. From a theoretical viewpoint, the setting at $r \sim r_0$ gives an infinite stiffness value. However, practically, V2SOM's stiffness is limited by the stiffness of its components. It should be mentioned that the main goal of the V2SOM stiffness profile is to provide a high stiffness in the smaller range (I) to handle a robot's dynamics upon collision (i.e., when it exceeds the tunable threshold, the stiffness drops rapidly and passively to guarantee safe collision absorption).

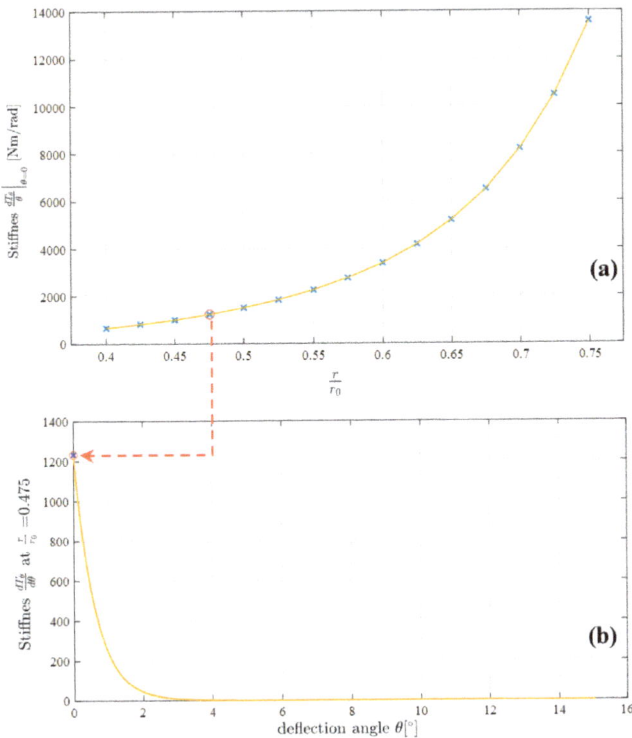

Figure 10. (**a**) Stiffness variation range of the V2SOM; (**b**) stiffness vs. deflection at the setting $\frac{r}{r_0} = 0.475$.

3. Safety Criteria: V2SOM vs. Constant Stiffness

The safety of pHRI is quite problematic in terms of quantification as well as its application to the whole body. However, the abbreviated injury scale (AIS) [31] presents a simple mapping system, from a medical perspective, of different safety criteria, with a unified scale with values ranging from 0 (no injury) to 6 (severe injury or death).

3.1. Safety Criteria

To achieve safe HR collaboration, a level 1 in AIS must be respected. In this work, we consider 0 in AIS as an ergonomic threshold for making human-friendly cobots. ISO/TS-15066 and Newman [32] adopted the most widely considered safety criteria, namely:

- G: The generalized model for brain injury threshold was introduced by Newman (see [33,34]). This index considers both the direction of the impact and the angular accelerations. The G index is valid for 50% of probability of AIS $\geq$ 3, which does not help to evaluate safe and human-friendly HR collisions, where AIS $\leq$ 1.

- NIR (new safety index): This index is quite similar to the HIC formulation. While HIC is generic, NIR is specific to the robot's technical data, which are provided in the manufacturer catalogs [34,35]. In our case, HIC is used to include the cover's effect on safety.
- HIC quantifies high accelerations of the brain (concussion) during blunt shocks even for a short amount of time; for example, HIC_{15} less than 15 ms can cause severe, irreversible health effects [35].
- ImpF (also known as contact force) is quite interesting as it can be applied to the whole body. This value is considered for a specific contact surface with a minimum 2.70 cm^2 area.
- Compression criterion (CC) reflects the damaging effect of human–robot (HR) collisions by means of deformation depth, mainly adopted for the naturally compliant chest and belly regions.

Regarding the data in Table 3, the head region is the most critical part of the human body compared to the trunk region, which is more naturally resilient, as indicated by the CC column. The CC criterion is not relevant to the head region as the skull is quite rigid. In contrast, HIC and ImpF are considered for their complementary aspect of HR shock evaluation. HIC is suitable for internal damage evaluation as it quantifies dangerous brain concussions. ImpF is suitable for external damage evaluation. Note that the HIC is only valid in the case of nonclamped head collision scenarios. On the other hand, a constrained head is a dangerous scenario, as shown by Heinzmann and Zelinsky [8]. Thus, a collaborative workspace should be designed, as note ISO/TS15066 permits, in such a way that free head motion is not compromised; this is the first step to guaranteeing safe pHRI.

Table 3. Safety criteria thresholds for most critical body regions from ISO/TS15066 and Payne [31]. ImpF = impact force; AIS = abbreviated injury scale; CC = compression criterion; HIC = head injury criterion.

Body Region		ImpF (N) for AIS $\leq$ 1	CC (N/mm)	HIC_{15ms} for AIS = 0
Head/Neck	Face	90	75	150
	Neck/sides	190	50	
Trunk	Belly	160	10	

3.2. Human–Robot Collision Model

The human head is the most critical body region when dealing with the safety problems of pHRI. Indeed, some previous works [13,14] have investigated this issue. Furthermore, they proceeded with theoretical modeling of dummy head hardware in crash tests. The resulting model (see Figure 11) has been well-tuned experimentally and validated.

The model in Figure 11 is parameterized according to Wolf et al. [14], with:

- Neck viscoelastic parameters $d_N = 12$ (N·s/m) , $k_N = 3300$ [N/m].
- Head's mass $M_{head} = 5.09$ (Kg) and linear displacement x.
- The contact surface viscoelastic parameters $d_c = 10$ (N·s/m), $k_c = 1500$ (N/m).
- Robot arm contact position l and inertia I_{arm}.
- Rotor inertia I_{rotor}, torque τ_{rotor}, and angular position θ_1.
- VSM's stiffness K and angular deflection $\theta = \theta_1 - \theta_2$.

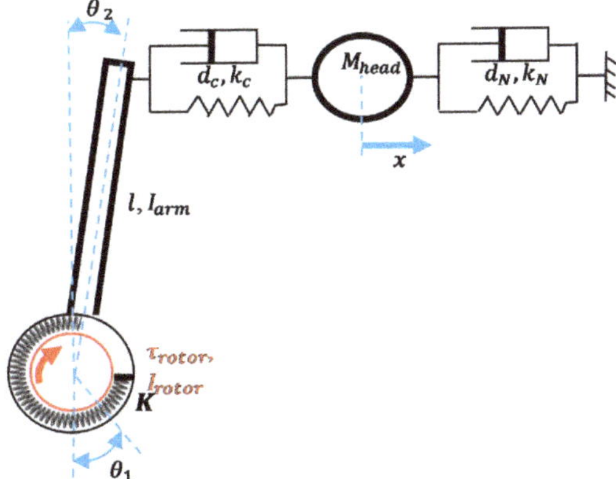

Figure 11. Mechanical model of dummy head hardware collision against a robot arm. Simulation results of human-robot collision.

The collision model allows us to evaluate both ImpF and HIC. The first criterion is directly deduced from the simulation data as the maximum value of the contact force. HIC is evaluated by solving the following optimization problem:

$$HIC_{15} = \max_{t_1,t_2}\left[\left(\frac{1}{(t_2-t_1)}\int_{t_1}^{t_2}\ddot{x}(t)dt\right)^{2.5}(t_2 - t_1)\right],$$

$$\text{Subject to } t_2 - t_1 \leq 15ms,$$

(3)

where $\ddot{x}(t)$ is the head acceleration value at instant t.

In the next sections, to compare between V2SOM and constant stiffness (CS), a VSM is carried out via a simulation of the HR collision model in the Matlab/Simulink platform. Here, the CS value is set so that the two profiles match at a torque value of $0.8\ T_{max}$, as shown in Figure 8. This torque value defines the deflection range of the normal operational mode for the V2SOM, after which the shock-absorbing mode is considered to be triggered.

The following simulations are meant to highlight V2SOM's inertia and torque decoupling capacities in comparison to a statistically equivalent CS-based VSM. These capacities represent the robust passive tackling of a HR collision in the fast and critical phase before the collision detection and reaction take place (e.g., within a range of 15 ms), as quantified via the HIC. These simulations were carried out using parametrization of Table 4 to evaluate both safety criteria (HIC and ImpF).

Table 4. Simulation parameters.

Parameter	Figure 13	Figure 14
(I_{rotor}, I_{arm}) (kgm^2)	$(0.0875{\rightarrow}0.525, 0.14)$	$(0.175, 0.14)$
$(\tau_{rotor}, T_{max}, T_1)$ (Nm)	$(10, 15, 12)$	$(7.5{\rightarrow}30, 15, 12)$
c	37	37
$\dot{\theta}_1\ (\text{rad}\cdot\text{s}^{-1})$	π	π
k_c (N/m)	1500	1500
l (m)	0.6	0.6

3.2.1. Inertia Decoupling

Figure 12 shows that V2SOM presents more than 80% gains on an HIC basis compared to CS. On the other hand, a gain of 10%–40% is noticed for ImpF curves. HIC_{V2SOM} and $ImpF_{V2SOM}$ are steady for a large range of rotor inertia. This property leads us to conclude that V2SOM presents a high inertia decoupling capability compared to a CS-based VSM. Ideally, this characteristic means that the human body, in the case of a HR shock, is subject to only arm-side inertia rather than the heavy resulting arm and rotor inertia.

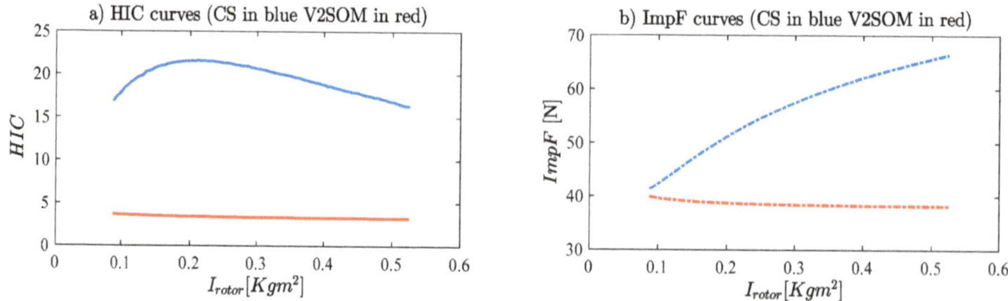

Figure 12. I_{rotor} impact on safety criteria, (**a**) HIC and (**b**) ImpF, for both V2SOM and CS characteristics.

As previously shown by Haddadin et al. [7], lower values of the mobile mass allow for higher velocities to maintain the same safety level. By considering the V2SOM inertia decoupling capability in addition to Haddadin's results, the proposed design allows for better dynamic performance of the cobot without overreaching the safety thresholds.

3.2.2. Torque Decoupling

V2SOM presents a quasi-constant response that can be observed in Figure 13. The large change in motor-applied torque, τ_{rotor}, leads to a significant variation in the HIC and ImpF values of CS-based VSM in comparison with V2SOM, which maintains the same values. The significant variation of 10%–40% in ImpF can be improved with a customized contact surface of the robot arm. However, HIC cannot be reduced and it will be difficult to attenuate the 80% gap between the two responses (Figure 13a). The benefit of V2SOM use, in terms of a reduction in concussions, is visible through HIC mitigation vs. CS use.

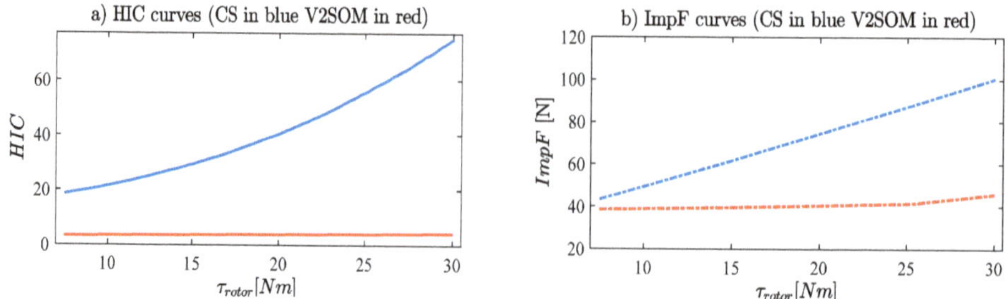

Figure 13. τ_{rotor} impact on safety criteria, (**a**) HIC and (**b**) ImpF, for both V2SOM and CS characteristics.

4. Experimental Validation of V2SOM

In this section a static characterization of the first V2SOM prototype and a preliminary HR collision testing are presented. Figure 14 shows the developed prototype, which is a cylinder 92 mm in diameter and 78 mm in height that weighs about 970 g. A lighter and more compact version is under development, along with a safety-oriented control strategy using V2SOM.

Figure 14. V2SOM first prototype.

4.1. Quasi-Static Characterization of V2SOM

Figure 15 shows the V2SOM characteristics in terms of torque vs. deflection for different settings of the tuning parameter r ranging from 16.6 to 26.6 mm. Quasi-static characterization means the load is applied at a slow rate like a static load [11]. The solid curves represent the theoretical curves, while the experimental curves are represented by dotted lines. The former displays relatively constant torque thresholds, proving that the cam correction brings V2SOM behavior near to its ideal form. The experimental curves deviate slightly from the theoretical ones. This error will be mitigated in upcoming versions by measures such as friction sources analysis and in-depth study of the mechanical parts' deformation. In addition, some parts will be enhanced, such as improving the stiffness profile time change by opting for faster linear actuators. For more details about this prototype, see the table in Appendix A of the mechanical and electrical specifications, which are written according to the recommendations of Grioli et al. [3].

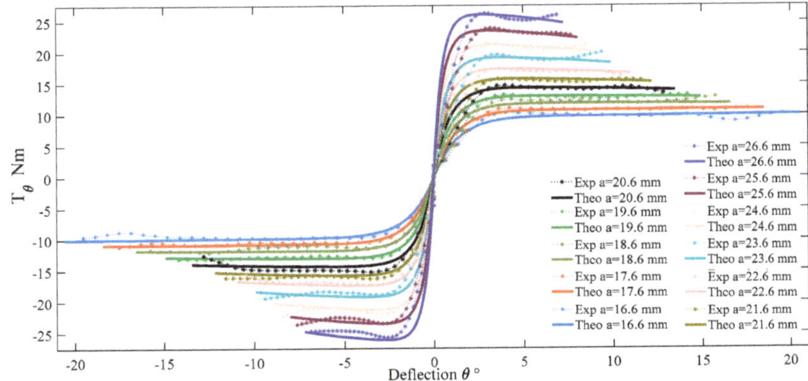

Figure 15. V2SOM characteristics: theoretical (in solid lines) and experimental (in dotted lines) with the 11 different SAB settings.

4.2. Preliminary HR Collision Tests

Currently, a safety-oriented control strategy is in ongoing development, wherein a V2SOM-based cobot functions under preemptive safety conditions in terms of joints' angular velocities. For this purpose, an optimization problem that maximizes the safely reachable angular velocities was formulated by taking into account the actual robot configuration. These constraints result from the simulation and are stored as lookup tables for real-time access.

In this paper, a simple example of tackling a collision with a human subject using V2SOM is demonstrated. The experimental setup is shown in Figure 16.

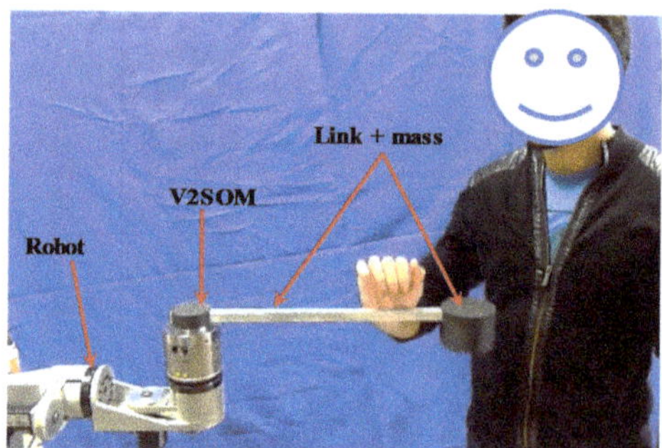

Figure 16. Experimental setup.

In Figure 17, the results of two collision tests are presented. In the first test (Figure 17a,b), the link's inertia is $I_{link} = 0.050$ Kgm^2 with an impact velocity of 90°/s. In the second test, (Figure 17c,d), the link's inertia is $I_{link} = 0.179$ Kgm^2 with an impact velocity of 32°/s. In both cases, V2SOM presents small deflections corresponding to the normal working conditions range (range I; see Figure 8). Upon collision, the safety threshold is exceeded and V2SOM warns of the collision by sending a collision detection message via a CAN (Controller Area Network) bus to the robot's main controller. Hence the robot, through its controller, stops the joint rotation. In the in-between phase (range I and range II in Figure 8), V2SOM passively absorbs the shock's kinetic energy. The elastic end stroke of the V2SOM should not be reached either by a V2SOM maximum storable elastic energy or by a possible slow control reaction. For this reason, both previous features are considered in developing the safety-oriented control strategy.

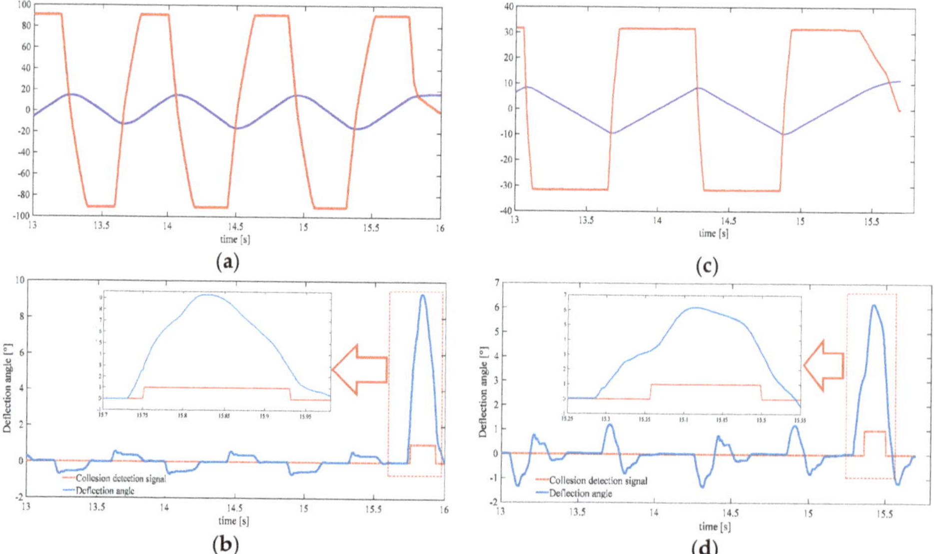

Figure 17. Results of experimental collision tests: (**a**) angular velocity and position of the link in the first test; (**b**) V2SOM deflection angle in the first test; (**c**) angular velocity and position of the link in the second test; (**d**) V2SOM deflection angle in the second test.

The time elapsed between the collision detection (exiting deflection range I) and returning is 180 ms in the first test and 140 ms in the second test.

These illustrative tests are meant first to show the inertia decoupling and torque decoupling capability of the V2SOM and then to shed light on the problem of a collision control strategy that is specifically designed to benefit from the V2SOM properties (e.g., inertia decoupling capacity). This problem is the subject of our upcoming work (for a general review of HR collision control see [4]).

5. Discussion

The simulation allowed us to cover a large range of inertia as well as torque. Based on the obtained results, we can conclude that V2SOM presents a high inertia decoupling capability, which is correlated with the results obtained during the experimentation. Safe physical human–robot interaction is ensured in the case of collision due to the presence of two continuously linked functional modes of V2SOM: high and low stiffness modes. Future work will be focused on the evaluation of the dynamic performance of robots using V2SOM as a way to cope with possible limitations.

6. Conclusions and Future Work

In this paper, we presented V2SOM as the basis of a human-friendly cobot in compliance with safety level 1 on the AIS. V2SOM displays two complementary working modes:

- In normal working conditions, characterized by high stiffness, the proposed mechanism limits the storable elastic energy that increases the absorbed kinetic energy in case of a sudden HR shock. Moreover, this high stiffness limits the elastic deflection to near zero, which eases the robot control.
- In the case of a blunt HR impact, V2SOM stiffness promptly decreases, allowing fast absorption of the collision energy.

From a safety perspective, as shown via the collision simulations, two complementary safety criteria were adopted for internal and external damage evaluation of the human body. Based on these two criteria, HIC and ImpF, important improvements were observed for the V2SOM profile in contrast to an equivalent CS profile. Especially in terms of brain concussion quantification, HIC was reduced by over 70% relative to the results of the CS profile. Moreover, V2SOM presented a high inertia decoupling capacity (i.e., large variation in rotor-side inertia slightly impacts the evaluation of safety on a HIC basis).

The experimental implementation, through the HR collision tests, shows the decoupling capacities of V2SOM. In case of collision, the deflection is detected, and the kinetic energy absorbed, which guarantees safe physical human–robot interaction.

Thus, the present work is focused on an illustrative case study of a one degree-of-freedom robot arm. In future work, different HR collision scenarios will be investigated by considering a generic multi-degrees of freedom (DoFs) robot with V2SOM joints, its dynamics, and the impact location on the robot arm.

7. Patents

WO 2019/043068 A1—7 March 2019: "Mechanical Device with Passive Compliance for Transmitting Rotational Movement," M. Arsicault, Y. Ayoubi, M.A Laribi, S. Zeghloul, and F. Courrèges.

Author Contributions: Conceptualization, Y.A. and M.A.L.; validation, Y.A. and M.A.; designed the experiments and wrote the paper, Y.A., M.A.L., and M.A.; supervised the research work, M.A.L. and S.Z. All authors have read and agreed to the published version of the manuscript.

Funding: This research was funded by ANR (Agence Nationale de la Recherche), grant number ANR-14-CE27-0016, Project: Safety Intelligent Sensor for Cobot (SISCob).

Acknowledgments: This work was supported by the French National Research Agency, convention ANR-14-CE27-0016. This work was sponsored by the French government research program "Investissements d'avenir" through the Robotex Equipment of Excellence (ANR-10-EQPX-44).

Conflicts of Interest: The authors declare no conflicts of interest. The authors declare no conflicts of interest

Appendix A

Table A1. V2SOM mechanical and electrical specifications.

	Mechanical			
1	Lowest Safety Threshold Torque		(Nm)	9.7
2	Safety Threshold Variation Time (from nominal level to completely stiff state)	With/Without Load (prone improvement)	(s)	1.6
3	Maximum Stiffness		(Nm/rad)	∞
4	Minimum Stiffness		(Nm/rad)	~ 0
5	Maximum Elastic Energy		(J)	2.98
6	Maximum Deflection	With Maximum Safety Threshold	(°)	0
		With Minimum Safety Threshold	(°)	20
7	Active Rotation Angle		(°)	$\pm\infty$
8	Angular Resolution		(°)	0.0313
9	Weight		(Kg)	0.970
	Electrical			
10	Nominal Voltage		(V)	12
11	Nominal Current		(A)	0.010
12	Maximum Current		(A)	0.500
	Control			
13	Voltage Supply		(V)	12
14	Nominal Current		(A)	0.105
15	I/O Protocol		CAN (Controller Area Network) (1 Mbit/s)	

References

1. Lu, Y. Industry 4.0: A survey on technologies, applications and open research issues. *J. Ind. Inf. Integr.* **2017**, *6*, 1–10. [CrossRef]
2. Tobe, F. Why Co-Bots Will Be a Huge Innovation and Growth Driver for Robotics Industry. Available online: https://www.aitrends.com/robotics/why-co-bots-will-be-a-huge-innovation-and-growth-driver-for-robotics-industry/ (accessed on 1 January 2016).
3. Grioli, G.; Wolf, S.; Garabini, M.; Catalano, M.G.; Burdet, E.; Caldwell, D.; Carloni, R.; Friedl, W.; Grebenstein, M.; Laffranchi, M.; et al. Variable stiffness actuators: The user's point of view. *Int. J. Robot. Res.* **2015**, *34*, 727–743. [CrossRef]
4. Haddadin, S.; De Luca, A.; Albu-Schaffer, A. Robot Collisions: A Survey on Detection, Isolation, and Identification. *IEEE Trans. Robot.* **2017**, *33*, 1292–1312. [CrossRef]
5. Zinn, M.; Khatib, O.; Roth, B.; Salisbury, J. Playing it safe. *IEEE Robot. Autom. Mag.* **2004**, *11*, 12–21. [CrossRef]
6. Jianbin, H.; Zongwu, X.; Minghe, J.; Zainan, J.; Hong, L. Adaptive Impedance-controlled Manipulator Based on Collision Detection. *Chin. J. Aeronaut.* **2009**, *22*, 105–112. [CrossRef]
7. Haddadin, S.; Albu-Schaffer, A.; Hirzinger, G. The role of the robot mass and velocity in physical human-robot interaction—Part I: Non-constrained blunt impacts. In Proceedings of the IEEE International Conference on Robotics and Automation, Nice, France, 22–26 September 2008.
8. Heinzmann, J.; Zelinsky, A. Quantitative Safety Guarantees for Physical Human-Robot Interaction. *Int. J. Robot. Res.* **2003**, *22*, 479–504. [CrossRef]
9. Haddadin, S.; Albu-Schaffer, A.; Frommberger, M.; Hirzinger, G. The role of the robot mass and velocity in physical human-robot interaction—Part II: Constrained blunt impacts. In Proceedings of the IEEE International Conference on Robotics and Automation, Nice, France, 19 May 2008.

10. Pratt, G.A.; Williamson, M.M. Series elastic actuators. In Proceedings of the IEEE/RSJ International Conference on Intelligent Robots and Systems. Human Robot Interaction and Cooperative Robots, Pittsburgh, PA, USA, 5–9 August 1995.

11. Corral, E.; García, M.; Castejon, C.; Meneses, J.; Gismeros, R. Dynamic Modeling of the Dissipative Contact and Friction Forces of a Passive Biped-Walking Robot. *Appl. Sci.* **2020**, *10*, 2342. [CrossRef]

12. Mathijssen, G.; Cherelle, P.; Lefeber, D.; VanderBorght, B. Concept of a Series-Parallel Elastic Actuator for a Powered Transtibial Prosthesis. *Actuators* **2013**, *2*, 59–73. [CrossRef]

13. Zinn, M.; Roth, B.; Khatib, O.; Salisbury, J.K. A New Actuation Approach for Human Friendly Robot Design. *Int. J. Rob. Res.* **2004**, *23*, 379–398. [CrossRef]

14. Wolf, S.; Grioli, G.; Eiberger, O.; Friedl, W.; Grebenstein, M.; Hoppner, H.; Burdet, E.; Caldwell, D.G.; Carloni, R.; Catalano, M.G.; et al. Variable Stiffness Actuators: Review on Design and Components. *IEEE/ASME Trans. Mechatron.* **2015**, *21*, 2418–2430. [CrossRef]

15. López-Martínez, J.; García-Vallejo, D.; Giménez-Fernández, A.; Torres-Moreno, J.L. A Flexible Multibody Model of a Safety Robot Arm for Experimental Validation and Analysis of Design Parameters. *J. Comput. Nonlinear Dyn.* **2013**, *9*, 011003. [CrossRef]

16. Hyun, D.; Yang, H.S.; Park, J.; Shim, Y. Variable stiffness mechanism for human-friendly robots. *Mech. Mach. Theory* **2010**, *45*, 880–897. [CrossRef]

17. Tagliamonte, N.L.; Sergi, F.; Accoto, D.; Carpino, G.; Guglielmelli, E. Double actuation architectures for rendering variable impedance in compliant robots: A review. *Mechatronics* **2012**, *22*, 1187–1203. [CrossRef]

18. VanderBorght, B.; Albu-Schaeffer, A.; Bicchi, A.; Burdet, E.; Caldwell, D.; Carloni, R.; Catalano, M.G.; Eiberger, O.; Friedl, W.; Ganesh, G.; et al. Variable impedance actuators: A review. *Robot. Auton. Syst.* **2013**, *61*, 1601–1614. [CrossRef]

19. Jafari, A.; Tsagarakis, N.G.; Caldwell, D. AwAS-II: A new Actuator with Adjustable Stiffness based on the novel principle of adaptable pivot point and variable lever ratio. In Proceedings of the IEEE International Conference on Robotics and Automation, Shanghai, China, 9–13 May 2011.

20. Wolf, S.; Eiberger, O.; Hirzinger, G. The DLR FSJ: Energy based design of a variable stiffness joint. In Proceedings of the 2011 IEEE International Conference on Robotics and Automation, Shanghai, China, 9–13 May 2011.

21. Eiberger, O.; Haddadin, S.; Weiß, M.; Albu-Schaffer, A.; Hirzinger, G. On joint design with intrinsic variable compliance: Derivation of the DLR QA-Joint. In Proceedings of the 2010 IEEE International Conference on Robotics and Automation, Anchorage, Alaska, 3 May 2010.

22. Friedl, W.; Chalon, M.; Reinecke, J.; Grebenstein, M. FAS A flexible antagonistic spring element for a high performance over. In Proceedings of the 2011 IEEE/RSJ International Conference on Intelligent Robots and Systems, San Francisco, CA, USA, 25–30 September 2011.

23. Ayoubi, Y.; Laribi, M.A.; Courreges, F.; Zeghloul, S.; Arsicault, M. A complete methodology to design a safety mechanism for prismatic joint implementation. In Proceedings of the 2016 IEEE/RSJ International Conference on Intelligent Robots and Systems (IROS), Daejeon, Korea, 9–14 October 2016.

24. Lan, N.; Crago, P.E. Optimal control of antagonistic muscle stiffness during voluntary movements. *Boil. Cybern.* **1994**, *71*, 123–135. [CrossRef] [PubMed]

25. Park, J.-J.; Song, J.-B. A Nonlinear Stiffness Safe Joint Mechanism Design for Human Robot Interaction. *J. Mech. Des.* **2010**, *132*, 061005. [CrossRef]

26. Park, J.-J.; Lee, Y.-J.; Song, J.-B.; Kim, H.-S. Safe joint mechanism based on nonlinear stiffness for safe human-robot collision. In Proceedings of the 2008 IEEE International Conference on Robotics and Automation, Pasadena, CA, USA, 19–23 May 2008.

27. Petit, F.; Friedl, W.; Höppner, H.; Grebenstein, M. Analysis and Synthesis of the Bidirectional Antagonistic Variable Stiffness Mechanism. *IEEE/ASME Trans. Mechatron.* **2014**, *20*, 684–695. [CrossRef]

28. Meneses, J.; Garcia-Prada, J.C.; Castejon, C.; Rubio, H.; Corral, E. The kinematics of the rotary into helical gear transmission. *Mech. Mach. Theory* **2017**, *108*, 110–122. [CrossRef]

29. Ayoubi, Y.; Laribi, M.A.; Arsicault, M.; Zeghloul, S.; Courreges, F. Mechanical Device with Variable Compliance for Rotary Motion Transmission. WO 2019/043068 A1, 7 March 2019.

30. Ayoubi, Y.; Laribi, M.A.; Zeghloul, S.; Arsicault, M. Design of V2SOM: The safety mechanism for cobot's rotary joints. In *IFToMM Symposium on Mechanism Design for Robotics*; Springer: Cham, Switzerland, 2018; pp. 147–157.

31. Payne, D.A.R.; Patel, S. Levels Of Consciousness In Relation To Head Injury Criteria. 2001. Available online: http://www.eurailsafe.net/subsites/operas/HTML/appendix/Table13.htm (accessed on 1 January 2017).
32. HIC Tolerance Levels Correlated To Brain Injury. Available online: http://www.eurailsafe.net/subsites/operas/HTML/appendix/Table14.htm (accessed on 3 June 2015).
33. Newman, J.A. A generalized acceleration model for brain injury threshold (GAMBIT). In Proceedings of the International IRCOBI Conference, Zurich, Switzerland, 2–4 September 1986.
34. Alén-Cordero, C.; Carbone, G.; Ceccarelli, M.; Echávarri, J.; Muñoz, J.L. Experimental tests in human–robot collision evaluation and characterization of a new safety index for robot operation. *Mech. Mach. Theory* **2014**, *80*, 184–199. [CrossRef]
35. Gao, D.; Wampler, C.W. Assessing the Danger of Robot Impact. *IEEE Robot. Autom. Mag.* **2009**, *16*, 71–74. [CrossRef]

MDPI
St. Alban-Anlage 66
4052 Basel
Switzerland
Tel. +41 61 683 77 34
Fax +41 61 302 89 18
www.mdpi.com

Applied Sciences Editorial Office
E-mail: applsci@mdpi.com
www.mdpi.com/journal/applsci

www.ingramcontent.com/pod-product-compliance
Lightning Source LLC
LaVergne TN
LVHW071611170726
843515LV00009B/2372